AF895793

Studies in Computational Intelligence

Volume 1062

Series Editor

Janusz Kacprzyk, Polish Academy of Sciences, Warsaw, Poland

The series "Studies in Computational Intelligence" (SCI) publishes new developments and advances in the various areas of computational intelligence—quickly and with a high quality. The intent is to cover the theory, applications, and design methods of computational intelligence, as embedded in the fields of engineering, computer science, physics and life sciences, as well as the methodologies behind them. The series contains monographs, lecture notes and edited volumes in computational intelligence spanning the areas of neural networks, connectionist systems, genetic algorithms, evolutionary computation, artificial intelligence, cellular automata, self-organizing systems, soft computing, fuzzy systems, and hybrid intelligent systems. Of particular value to both the contributors and the readership are the short publication timeframe and the world-wide distribution, which enable both wide and rapid dissemination of research output.

This series also publishes Open Access books. A recent example is the book Swan, Nivel, Kant, Hedges, Atkinson, Steunebrink: The Road to General Intelligence https://link.springer.com/book/10.1007/978-3-031-08020-3

Indexed by SCOPUS, DBLP, WTI Frankfurt eG, zbMATH, SCImago.

All books published in the series are submitted for consideration in Web of Science.

George A. Anastassiou

Banach Space Valued Neural Network

Ordinary and Fractional Approximation and Interpolation

Springer

George A. Anastassiou
Department of Mathematical Sciences
University of Memphis
Memphis, TN, USA

ISSN 1860-949X ISSN 1860-9503 (electronic)
Studies in Computational Intelligence
ISBN 978-3-031-16399-9 ISBN 978-3-031-16400-2 (eBook)
https://doi.org/10.1007/978-3-031-16400-2

© The Editor(s) (if applicable) and The Author(s), under exclusive license to Springer Nature Switzerland AG 2023
This work is subject to copyright. All rights are solely and exclusively licensed by the Publisher, whether the whole or part of the material is concerned, specifically the rights of translation, reprinting, reuse of illustrations, recitation, broadcasting, reproduction on microfilms or in any other physical way, and transmission or information storage and retrieval, electronic adaptation, computer software, or by similar or dissimilar methodology now known or hereafter developed.
The use of general descriptive names, registered names, trademarks, service marks, etc. in this publication does not imply, even in the absence of a specific statement, that such names are exempt from the relevant protective laws and regulations and therefore free for general use.
The publisher, the authors, and the editors are safe to assume that the advice and information in this book are believed to be true and accurate at the date of publication. Neither the publisher nor the authors or the editors give a warranty, expressed or implied, with respect to the material contained herein or for any errors or omissions that may have been made. The publisher remains neutral with regard to jurisdictional claims in published maps and institutional affiliations.

This Springer imprint is published by the registered company Springer Nature Switzerland AG
The registered company address is: Gewerbestrasse 11, 6330 Cham, Switzerland

Dedicated to the memory of all of my teachers and mentors

Preface

This monograph is the continuation of author's earlier (i) 2011 monograph, "Intelligent Systems: Approximation by Artificial Neural Networks", Springer, Intelligent Systems Reference Library, Volume 19, and (ii) 2016 monograph, "Intelligent Systems II: Complete Approximation by Neural Network Operators", Studies in Computational Intelligence, Volume 608.

The innovation here is that the neural networks are Banach space valued, induced by a great variety of activation functions deriving from the arctangent, algebraic, Gudermannian and generalized symmetric sigmoid functions.

Thus, in this monograph, all presented is original work by the author given at a very general level to cover a maximum number of different kinds of neural networks: giving ordinary, fractional, fuzzy and stochastic approximations. We exhibit univariate, fractional and multivariate approximations. Iterated-sequential approximations are also studied. The new and important feature here is that the functions under approximation are Banach space valued.

As a result, this monograph is the natural and expected evolution of recent author's research work put in a book form for the first time. The presented approaches are original, and the chapters are self-contained and can be read independently. This monograph is suitable to be used in related graduate classes and research projects.

The motivation to write this monograph came from the following: various issues related to the modeling and analysis of ordinary and fractional order systems have gained an increased popularity, as witnessed by many books and volumes in Springer's program: http://www.springer.com/gp/search?query=fractional&submit=Prze%C5%9Blij and the purpose of our book is to capture at a very general level a deeper formal analysis on some issues that are relevant to many areas, for instance: decision-making, complex processes, systems modeling and control and related areas.

The above are deeply embedded in the fields of mathematics, engineering, computer science, physics, economics, social and life sciences.

Next is listed the author's related recent monographs in fractional analysis and applications:

- George Anastassiou, "Intelligent Computations: Abstract Fractional Calculus, Inequalities, Approximations", Springer, Studies in Computational Intelligence 734, Heidelberg, New York, 2018.
- George Anastassiou, "Nonlinearity: Ordinary and Fractional Approximations by Sublinear and Max-product Operators", Springer, Studies in Systems, Decision and Control 147, Heidelberg, New York, 2018.
- George Anastassiou, "Ordinary and Fractional approximation by non-additive integrals: Choquet, Shilkret and Sugeno integral approximators", Springer, Studies in Systems, Decision and Control 190, Heidelberg, New York, 2019.
- George Anastassiou, "Intelligent Analysis: Fractional Inequalities and Approximations Expanded", Springer, Studies in Computational Intelligence 886, Heidelberg, New York, 2020.
- George Anastassiou, "Generalized Fractional Calculus: New Advancements and Applications", Springer, Studies in Systems, Decision and Control 305, Heidelberg, New York, 2021.
- George Anastassiou, "Constructive Fractional Analysis with Applications", Springer, Studies in Systems, Decision and Control 362, Heidelberg, New York, 2021.
- George Anastassiou, "Unification of Fractional Calculi with Applications", Springer, Studies in Systems, Decision and Control 398, Heidelberg, New York, 2022.
- George Anastassiou, "Abstract Fractional Monotone Approximation, Theory and Applications, Studies in Systems, Decision and Control 411, Heidelberg, New York, 2022.

The complete list of presented topics here follows:

- Algebraic function-based Banach space valued ordinary and fractional neural network approximations,
- Gudermannian function activated Banach space valued ordinary and fractional neural network approximations,
- Generalized symmetrical sigmoid function activated Banach space valued ordinary and fractional neural network approximation,
- Abstract multivariate algebraic function activated neural network approximations,
- General multivariate arctangent function activated neural network approximations,
- Abstract multivariate Gudermannian function activated neural network approximations,
- Generalized symmetrical sigmoid function activated neural network multivariate approximation,
- Degree of Approximation by Kantorovich-Choquet quasi-interpolation neural network operators revisited,
- Degree of Approximation by Kantorovich-Shilkret quasi-interpolation neural network operators revisited,
- Vector Voronouskaya type asymptotic expansions for sigmoid functions induced quasi-interpolation neural network operators revisited,

- Fuzzy Fractional more sigmoid function activated neural network approximations revisited,
- Multivariate Fuzzy Approximation by Neural Network Operators activated by several sigmoid functions revisited,
- Multivariate Fuzzy-Random and stochastic arctangent, algebraic, Gudermannian and generalized symmetric activation functions induced Neural Network Approximations.

The book's results are expected to find applications in the many areas of Applied Mathematics, Computer Science and Engineering, especially in artificial intelligence and machine learning. Other possible applications can be in applied sciences like Statistics, Economics, etc. All in all, what is presented here is a valuable tool for a large range of applications. Therefore, this monograph is suitable for researchers, graduate students, practitioners and seminars of the above disciplines, also to be in all Science and Engineering libraries.

The preparation of the book took place during 2021–22 at the University of Memphis, at stay home of author during the COVID-19 and Omicron outbreak keeping him alive, happy and in control!

The author likes to thank Professor Alina Lupas of the University of Oradea, Romania, for checking and reading the manuscript.

Memphis, TN, USA
July 2022

George A. Anastassiou

Contents

1 Algebraic Function Induced Banach Space Valued Ordinary and Fractional Neural Network Approximations 1
1.1 Introduction 1
1.2 Basics 2
1.3 Main Results 11
References 41

2 Gudermannian Function Induced Banach Space Valued Ordinary and Fractional Neural Network Approximations 43
2.1 Introduction 43
2.2 Basics 44
2.3 Main Results 54
References 83

3 Generalized Symmetrical Sigmoid Function Induced Banach Space Valued Ordinary and Fractional Neural Network Approximations 85
3.1 Introduction 85
3.2 Auxiliary Results 86
3.3 Main Results 98
References 107

4 Abstract Multivariate Algebraic Function Induced Neural Network Approximations 109
4.1 Introduction 109
4.2 Basic 110
4.3 Multivariate General Neural Network Approximations 118
References 141

5 General Multivariate Arctangent Function Induced Neural Network Approximations ... 143
5.1 Introduction ... 143
5.2 Auxiliary Notions ... 144
5.3 Multivariate General Neural Network Approximations ... 152
References ... 176

6 Abstract Multivariate Gudermannian Function Induced Neural Network Approximations ... 177
6.1 Introduction ... 177
6.2 Background ... 178
6.3 Multivariate General Neural Network Approximations ... 186
References ... 202

7 Generalized Symmetrical Sigmoid Function Induced Neural Network Multivariate Approximation ... 205
7.1 Introduction ... 205
7.2 Auxiliary Results (See Also [14]) ... 206
7.3 Multivariate General Neural Network Approximations ... 214
References ... 231

8 Quantitative Approximation by Kantorovich-Choquet Quasi-Interpolation Neural Network Operators Revisited ... 233
8.1 Introduction ... 233
8.2 Background ... 234
8.2.1 About the Arctangent Activation Function ... 236
8.2.2 About the Algebraic Activation Function ... 238
8.2.3 About the Gudermannian Activation Function ... 241
8.2.4 About the Generalized Symmetrical Activation Function ... 243
8.3 Main Results ... 246
References ... 252

9 Quantitative Approximation by Kantorovich-Shilkret Quasi-interpolation Neural Network Operators Revisited ... 255
9.1 Introduction ... 255
9.2 Background ... 256
9.2.1 About the Arctangent Activation Function ... 258
9.2.2 About the Algebraic Activation Function ... 260
9.2.3 About the Gudermannian Activation Function ... 262
9.2.4 About the Generalized Symmetrical Activation Function ... 265
9.3 Main Results ... 268
References ... 272

10 Voronouskaya Univariate and Multivariate Asymptotic Expansions for Sigmoid Functions Induced Quasi-interpolation Neural Network Operators Revisited 275
10.1 Background 275
10.1.1 About the Arctangent Activation Function 275
10.1.2 About the Algebraic Activation Function 279
10.1.3 About the Gudermannian Activation Function 282
10.1.4 About the Generalized Symmetrical Activation Function 285
10.2 Main Results 293
References 303

11 Univariate Fuzzy Fractional Various Sigmoid Function Activated Neural Network Approximations Revisited 305
11.1 Introduction 305
11.2 Fuzzy Fractional Mathematical Analysis Basics 306
11.3 Real Neural Network Approximation 318
11.3.1 About the Arctangent Activation Function Neural Networks 318
11.3.2 About the Algebraic Activation Function Neural Networks 323
11.3.3 About the Gudermannian Activation Function Neural Networks 328
11.3.4 About the Generalized Symmetrical Activation Function Neural Networks 333
11.4 Main Results: Approximation by Fuzzy Quasi-interpolation Neural Network Operators with Respect to Arctangent, Algebraic, Gudermannian and Generalized Symmetrical Activation Functions 339
References 347

12 Multivariate Fuzzy Approximation by Neural Network Operators Induced by Several Sigmoid Functions Revisited 351
12.1 Introduction 351
12.2 Fuzzy Real Analysis Background 352
12.3 About Neural Networks Background 359
12.3.1 About the Arctangent Activation Function 359
12.3.2 About the Algebraic Activation Function 362
12.3.3 About the Gudermannian Activation Function 366
12.3.4 About the Generalized Symmetrical Activation Function 369
12.4 Main Results: Fuzzy Multivariate Neural Network Approximation Based on the Arctangent, Algebraic, Gudermannian and Generalized Symmetrical Activation Functions 377
References 384

13 Multivariate Fuzzy-Random and Stochastic Various Activation Functions Activated Neural Network Approximations 387
13.1 Fuzzy-Random Functions and Stochastic Processes Background 387
13.2 About Neural Networks Background 391
13.2.1 About the Arctangent Activation Function 391
13.2.2 About the Algebraic Activation Function 395
13.2.3 About the Gudermannian Activation Function 398
13.2.4 About the Generalized Symmetrical Activation Function 401
13.3 Main Results 408
References 419

Conclusion 423

Chapter 1
Algebraic Function Induced Banach Space Valued Ordinary and Fractional Neural Network Approximations

Here we research the univariate quantitative approximation, ordinary and fractional, of Banach space valued continuous functions on a compact interval or all the real line by quasi-interpolation Banach space valued neural network operators. These approximations are derived by establishing Jackson type inequalities involving the modulus of continuity of the engaged function or its Banach space valued high order derivative or fractional derivatives. Our operators are defined by using a density function generated by an algebraic sigmoid function. The approximations are pointwise and of the uniform norm. The related Banach space valued feed-forward neural networks are with one hidden layer. It relies as [14].

1.1 Introduction

The author in [1, 2], see Chaps. 2–5, was the first to establish neural network approximation to continuous functions with rated by very specifically defined neural network operators of Cardaliagnet-Euvrard and "Squashing" types, by employing the modulus of continuity of the engaged function or its high order derivative, and producing very tight Jackson type inequalities. He treats there both the univariate and multivariate cases. The defining these operators "bell-shaped" and "squashing" functions are assumed to be of compact suport. Also in [2] he gives the Nth order asymptotic expansion for the error of weak approximation of these two operators to a special natural class of smooth functions, see Chaps. 4–5 there.

The author inspired by [15], continued his studies on neural networks approximation by introducing and using the proper quasi-interpolation operators of sigmoidal and hyperbolic tangent type which resulted into [3–7], by treating both the univariate and multivariate cases. He did also the corresponding fractional cases [8, 9, 13].

© The Author(s), under exclusive license to Springer Nature Switzerland AG 2023
G. A. Anastassiou, *Banach Space Valued Neural Network*,
Studies in Computational Intelligence 1062,
https://doi.org/10.1007/978-3-031-16400-2_1

The author here performs algebraic sigmoidal based neural network approximations to continuous functions over compact intervals of the real line or over the whole $\mathbb{R}$ with valued to an arbitrary Banach space $(X, \|\cdot\|)$. Finally he treats completely the related X-valued fractional approximation. All convergences here are with rates expressed via the modulus of continuity of the involved function or its X-valued high order derivative, or X-valued fractional derivatives and given by very tight Jackson type inequalities. Iterated fractional approximation is also included.

Our compact intervals are not necessarily symmetric to the origin. Some of our upper bounds to error quantity are very flexible and general. In preparation to prove our results we establish important properties of the basic density function defining our operators which is induced by algebraic sigmoidal function.

Feed-forward X-valued neural networks (FNNs) with one hidden layer, the only type of networks we deal with in this chapter, are mathematically expressed as

$$N_n(x) = \sum_{=0}^{n} c_j \sigma\left(\langle a_j \cdot x\rangle + b_j\right), \quad x \in \mathbb{R}^s, \, s \in \mathbb{N},$$

where for $0 \leq j \leq n$, $b_j \in \mathbb{R}$ are the thresholds, $a_j \in \mathbb{R}^s$ are the connection weights, $c_j \in X$ are the coefficients, $\langle a_j \cdot x\rangle$ is the inner product of a_j and x, and σ is the activation function of the network. About neural networks in general read [16, 18, 20]. See also [9] for a complete study of real valued approximation by neural network operators.

1.2 Basics

We consider the generator algebraic function

$$\varphi(x) = \frac{x}{\sqrt[2m]{1 + x^{2m}}}, \quad m \in \mathbb{N}, x \in \mathbb{R}, \tag{1.1}$$

which is a sigmoidal type of function and is a strictly increasing function.

We see that $\varphi(-x) = -\varphi(x)$ with $\varphi(0) = 0$. We get that

$$\varphi'(x) = \frac{1}{\left(1 + x^{2m}\right)^{\frac{2m+1}{2m}}} > 0, \, \forall x \in \mathbb{R}, \tag{1.2}$$

proving φ as strictly increasing over $\mathbb{R}$, $\varphi'(x) = \varphi'(-x)$. We easily find that $\lim_{x\to+\infty} \varphi(x) = 1$, $\varphi(+\infty) = 1$, and $\lim_{x\to-\infty} \varphi(x) = -1$, $\varphi(-\infty) = -1$.

We consider the activation function

$$\Phi(x) = \frac{1}{4}\left[\varphi(x+1) - \varphi(x-1)\right]. \tag{1.3}$$

Clearly it is $\Phi(x) = \Phi(-x)$, $\forall x \in \mathbb{R}$, so that Φ is an even function and symmetric with respect to the y-axis.

Since $x + 1 > x - 1$, we have $\varphi(x+1) > \varphi(x-1)$ and $\Phi(x) > 0$, $\forall x \in \mathbb{R}$.

Also it is

$$\Phi(0) = \frac{1}{2\sqrt[2m]{2}}. \tag{1.4}$$

We observe that

$$\Phi'(x) = \frac{1}{4}\left(\varphi'(x+1) - \varphi'(x-1)\right) =$$

$$\frac{1}{4}\left(\frac{1}{\left(1+(x+1)^{2m}\right)^{\frac{2m+1}{2m}}} - \frac{1}{\left(1+(x-1)^{2m}\right)^{\frac{2m+1}{2m}}}\right), \ \forall x \in \mathbb{R}. \tag{1.5}$$

Let now $x > 0$, then $x > -x$ and $(x+1)^2 > (x-1)^2 \geq 0$, implying $(x+1)^{2m} > (x-1)^{2m} \geq 0$, $m \in \mathbb{N}$, and $1 + (x+1)^{2m} > 1 + (x-1)^{2m} > 0$. Consequently it holds

$$\frac{1}{\left(1+(x-1)^{2m}\right)^{\frac{2m+1}{2m}}} > \frac{1}{\left(1+(x+1)^{2m}\right)^{\frac{2m+1}{2m}}}, \tag{1.6}$$

proving $\Phi'(x) < 0$ for $x > 0$.

That is Φ is strictly decreasing over $(0, +\infty)$.

Clearly, Φ is strictly increasing over $(-\infty, 0)$ and $\Phi'(0) = 0$.

Furthermore we obtain that

$$\lim_{x \to +\infty} \Phi(x) = \frac{1}{4}\left[\varphi(+\infty) - \varphi(+\infty)\right] = 0, \tag{1.7}$$

and

$$\lim_{x \to -\infty} \Phi(x) = \frac{1}{4}\left[\varphi(-\infty) - \varphi(-\infty)\right] = 0. \tag{1.8}$$

That is the x-axis is the horizontal asymptote of Φ.

Conclusion, Φ is a bell symmetric function with maximum

$$\Phi(0) = \frac{1}{2\sqrt[2m]{2}}, \ m \in \mathbb{N}. \tag{1.9}$$

We need

Theorem 1.1 *We have that*

$$\sum_{i=-\infty}^{\infty} \Phi(x - i) = 1, \ \forall x \in \mathbb{R}. \tag{1.10}$$

Proof We observe that

$$\sum_{i=-\infty}^{\infty} (\varphi(x-i) - \varphi(x-1-i)) =$$

$$\sum_{i=0}^{\infty} (\varphi(x-i) - \varphi(x-1-i)) + \sum_{i=-\infty}^{-1} (\varphi(x-i) - \varphi(x-1-i)).$$

Furthermore ($\lambda \in \mathbb{Z}^+$)

$$\sum_{i=0}^{\infty} (\varphi(x-i) - \varphi(x-1-i)) = \tag{1.11}$$

$$\lim_{\lambda\to\infty} \sum_{i=0}^{\lambda} (\varphi(x-i) - \varphi(x-1-i)) \text{ (telescoping sum)}$$

$$= \lim_{\lambda\to\infty} (\varphi(x) - \varphi(x-(\lambda+1))) = 1 + \varphi(x).$$

Similarly, it holds

$$\sum_{i=-\infty}^{-1} (\varphi(x-i) - \varphi(x-1-i)) = \lim_{\lambda\to\infty} \sum_{i=-\lambda}^{-1} (\varphi(x-i) - \varphi(x-1-i))$$
$$= \lim_{\lambda\to\infty} (\varphi(x+\lambda) - \varphi(x)) = 1 - \varphi(x). \tag{1.12}$$

Therefore we derive

$$\sum_{i=-\infty}^{\infty} (\varphi(x-i) - \varphi(x-1-i)) = 2, \ \forall x \in \mathbb{R}, \tag{1.13}$$

and

$$\sum_{i=-\infty}^{\infty} (\varphi(x+1-i) - \varphi(x-i)) = 2, \ \forall x \in \mathbb{R}. \tag{1.14}$$

Adding (1.13) and (1.14) we find

$$\sum_{i=-\infty}^{\infty} (\varphi(x+1-i) - \varphi(x-1-i)) = 4, \ \forall x \in \mathbb{R}. \tag{1.15}$$

Clearly, then

$$\Phi(x-i)=\frac{1}{4}[\varphi(x+1-i)-\varphi(x-1-i)],$$

proving (1.10). ■

We make

Remark 1.2 Because Φ is even it holds

$$\sum_{i=-\infty}^{\infty}\Phi(i-x)=1,\ \forall x\in\mathbb{R}.$$

Hence

$$\sum_{i=-\infty}^{\infty}\Phi(i+x)=1,\ \forall x\in\mathbb{R},$$

and

$$\sum_{i=-\infty}^{\infty}\Phi(x+i)=1,\ \forall x\in\mathbb{R}. \tag{1.16}$$

Theorem 1.3 *It holds*

$$\int_{-\infty}^{\infty}\Phi(x)\,dx=1. \tag{1.17}$$

Proof We observe that

$$\int_{-\infty}^{\infty}\Phi(x)\,dx=\sum_{j=-\infty}^{\infty}\int_{j}^{j+1}\Phi(x)\,dx=\sum_{j=-\infty}^{\infty}\int_{0}^{1}\Phi(x+j)\,dx= \tag{1.18}$$

$$\int_0^1\left(\sum_{j=-\infty}^{\infty}\Phi(x+j)\,dx\right)=\int_0^1 1dx=1.$$

So $\Phi(x)$ is a density function on $\mathbb{R}$. ■

We need

Theorem 1.4 *Let* $0<\alpha<1$, *and* $n\in\mathbb{N}$ *with* $n^{1-\alpha}>2$. *It holds*

$$\sum_{\begin{cases} k=-\infty \\ :|nx-k|\geq n^{1-\alpha}\end{cases}}^{\infty}\Phi(nx-k)<\frac{1}{4m\left(n^{1-\alpha}-2\right)^{2m}},\quad m\in\mathbb{N}. \tag{1.19}$$

Proof We have that

$$\Phi(x) = \frac{1}{4}\left[\varphi(x+1) - \varphi(x-1)\right], \ \forall x \in \mathbb{R}.$$

Let $x \geq 1$. That is $0 \leq x-1 < x+1$. Applying the mean value theorem we get

$$\Phi(x) = \frac{1}{4} \cdot 2 \cdot \varphi'(\xi) = \frac{1}{2\left(1+\xi^{2m}\right)^{\frac{2m+1}{2m}}} > 0, \tag{1.20}$$

where $0 \leq x-1 < \xi < x+1$.

Then,

$$(x-1)^2 < \xi^2 < (x+1)^2$$

$$(x-1)^{2m} < \xi^{2m} < (x+1)^{2m}$$

$$1+(x-1)^{2m} < 1+\xi^{2m} < 1+(x+1)^{2m}$$

$$\left(1+(x-1)^{2m}\right)^{\frac{2m+1}{2m}} < \left(1+\xi^{2m}\right)^{\frac{2m+1}{2m}} < \left(1+(x+1)^{2m}\right)^{\frac{2m+1}{2m}}$$

$$\frac{1}{2\left(1+\xi^{2m}\right)^{\frac{2m+1}{2m}}} < \frac{1}{2\left(1+(x-1)^{2m}\right)^{\frac{2m+1}{2m}}}. \tag{1.21}$$

Hence

$$\Phi(x) < \frac{1}{2\left(1+(x-1)^{2m}\right)^{\frac{2m+1}{2m}}}, \ \forall x \geq 1. \tag{1.22}$$

Thus, we have

$$\sum_{\substack{k=-\infty \\ :|nx-k| \geq n^{1-\alpha}}}^{\infty} \Phi(nx-k) = \sum_{\substack{k=-\infty \\ :|nx-k| \geq n^{1-\alpha}}}^{\infty} \Phi(|nx-k|) <$$

$$\frac{1}{2} \sum_{\substack{k=-\infty \\ :|nx-k| \geq n^{1-\alpha}}}^{\infty} \frac{1}{\left(1+(|nx-k|-1)^{2m}\right)^{\frac{2m+1}{2m}}} \leq \tag{1.23}$$

$$\frac{1}{2}\int_{(n^{1-\alpha}-1)}^{\infty} \frac{1}{\left(1+(x-1)^{2m}\right)^{\frac{2m+1}{2m}}}dx = \frac{1}{2}\int_{n^{1-\alpha}-2}^{\infty} \frac{1}{\left(1+z^{2m}\right)^{\frac{2m+1}{2m}}}dz =: (*).$$

We see that

$$1 + z^{2m} > z^{2m}$$

$$\left(1 + z^{2m}\right)^{\frac{2m+1}{2m}} > z^{2m+1}$$

$$\frac{1}{z^{2m+1}} > \frac{1}{\left(1 + z^{2m}\right)^{\frac{2m+1}{2m}}}. \tag{1.24}$$

Therefore it holds

$$(*) < \frac{1}{2} \int_{n^{1-\alpha}-2}^{\infty} \frac{1}{z^{2m+1}} dz = \frac{1}{2} \int_{n^{1-\alpha}-2}^{\infty} z^{-(2m+1)} dz =$$

$$\frac{1}{2} \left(\frac{z^{-(2m+1)+1}}{-(2m+1)+1} \right) \Bigg|_{n^{1-\alpha}-2}^{\infty} = \frac{1}{2} \left(-\frac{z^{-2m}}{2m} \right) \Bigg|_{n^{1-\alpha}-2}^{\infty} = \tag{1.25}$$

$$\frac{z^{-2m}}{4m} \Bigg|_{\infty}^{n^{1-\alpha}-2} = \frac{\left(n^{1-\alpha} - 2\right)^{-2m}}{4m} - \frac{(\infty)^{-2m}}{4m} = \frac{\left(n^{1-\alpha} - 2\right)^{-2m}}{4m},$$

proving (1.19). ■

Denote by $\lfloor \cdot \rfloor$ the integral part of the number and by $\lceil \cdot \rceil$ the ceiling of the number.

Theorem 1.5 *Let* $[a, b] \subset \mathbb{R}$ *and* $n \in \mathbb{N}$ *so that* $\lceil na \rceil \leq \lfloor nb \rfloor$. *It holds*

$$\frac{1}{\sum_{k=\lceil na \rceil}^{\lfloor nb \rfloor} \Phi(nx - k)} < 2\left(\sqrt[2m]{1 + 4^m}\right), \tag{1.26}$$

$\forall\, x \in [a, b]$, $m \in \mathbb{N}$.

Proof Let $x \in [a, b]$. We see that

$$1 = \sum_{k=-\infty}^{\infty} \Phi(nx - k) > \sum_{k=\lceil na \rceil}^{\lfloor nb \rfloor} \Phi(nx - k) =$$

$$\sum_{k=\lceil na \rceil}^{\lfloor nb \rfloor} \Phi(|nx - k|) > \Phi(|nx - k_0|), \tag{1.27}$$

$\forall\, k_0 \in [\lceil na \rceil, \lfloor nb \rfloor] \cap \mathbb{Z}$.

We can choose $k_0 \in [\lceil na \rceil, \lfloor nb \rfloor] \cap \mathbb{Z}$ such that $|nx - k_0| < 1$.

Therefore we get that

$$\Phi\left(|nx - k_0|\right) > \Phi(1) = \frac{1}{4}\left(\frac{2}{\sqrt[2m]{1+2^{2m}}}\right) = \frac{1}{2\sqrt[2m]{1+2^{2m}}}, \tag{1.28}$$

and

$$\sum_{k=\lceil na \rceil}^{\lfloor nb \rfloor} \Phi\left(|nx - k|\right) > \frac{1}{2\sqrt[2m]{1+2^{2m}}}. \tag{1.29}$$

That is

$$\frac{1}{\sum_{k=\lceil na \rceil}^{\lfloor nb \rfloor} \Phi\left(|nx - k|\right)} < 2\sqrt[2m]{1+4^m}, \tag{1.30}$$

proving the claim. ■

We make

Remark 1.6 We also notice that

$$1 - \sum_{k=\lceil na \rceil}^{\lfloor nb \rfloor} \Phi(nb - k) = \sum_{k=-\infty}^{\lceil na \rceil - 1} \Phi(nb - k) + \sum_{k=\lfloor nb \rfloor + 1}^{\infty} \Phi(nb - k)$$

$$> \Phi\left(nb - \lfloor nb \rfloor - 1\right) \tag{1.31}$$

(call $\varepsilon := nb - \lfloor nb \rfloor$, $0 \leq \varepsilon < 1$)

$$= \Phi(\varepsilon - 1) = \Phi(1 - \varepsilon) \geq \Phi(1) > 0.$$

Therefore

$$\lim_{n \to \to \infty}\left(1 - \sum_{k=\lceil na \rceil}^{\lfloor nb \rfloor} \Phi(nb - k)\right) > 0. \tag{1.32}$$

Similarly, it holds

$$1 - \sum_{k=\lceil na \rceil}^{\lfloor nb \rfloor} \Phi(na - k) = \sum_{k=-\infty}^{\lceil na \rceil - 1} \Phi(na - k) + \sum_{k=\lfloor nb \rfloor + 1}^{\infty} \Phi(na - k)$$

$$> \Phi\left(na - \lceil na \rceil + 1\right) \tag{1.33}$$

(call $\eta := \lceil na \rceil - na$, $0 \leq \eta < 1$)

$$= \Phi(1 - \eta) \geq \Phi(1) > 0.$$

Therefore again

$$\lim_{n\to\infty}\left(1-\sum_{k=\lceil na\rceil}^{\lfloor nb\rfloor}\Phi(na-k)\right)>0. \tag{1.34}$$

Here we find that

$$\lim_{n\to\infty}\sum_{k=\lceil na\rceil}^{\lfloor nb\rfloor}\Phi(nx-k)\neq 1,\ \text{for at least some } x\in[a,b]. \tag{1.35}$$

Note 1.7 *Let $[a,b]\subset\mathbb{R}$. For large enough n we always obtain $\lceil na\rceil\leq\lfloor nb\rfloor$. Also $a\leq\frac{k}{n}\leq b$, iff $\lceil na\rceil\leq k\leq\lfloor nb\rfloor$.*

In general it holds (by $\sum_{i=-\infty}^{\infty}\Phi(x-i)=1,\ \forall\, x\in\mathbb{R}$) that

$$\sum_{k=\lceil na\rceil}^{\lfloor nb\rfloor}\Phi(nx-k)\leq 1. \tag{1.36}$$

Let $(X,\|\cdot\|)$ be a Banach space.

Definition 1.8 Let $f\in C([a,b],X)$ and $n\in\mathbb{N}:\lceil na\rceil\leq\lfloor nb\rfloor$. We introduce and define the X-valued linear neural network operators

$$A_n(f,x):=\frac{\sum_{k=\lceil na\rceil}^{\lfloor nb\rfloor}f\left(\frac{k}{n}\right)\Phi(nx-k)}{\sum_{k=\lceil na\rceil}^{\lfloor nb\rfloor}\Phi(nx-k)},\ x\in[a,b]. \tag{1.37}$$

Clearly here $A_n(f,x)\in C([a,b],X)$.

For convenience we use the same A_n for real valued functions when needed. We study here the pointwise and uniform convergence of $A_n(f,x)$ to $f(x)$ with rates.

For convenience, also we call

$$A_n^*(f,x):=\sum_{k=\lceil na\rceil}^{\lfloor nb\rfloor}f\left(\frac{k}{n}\right)\Phi(nx-k), \tag{1.38}$$

(similarly, A_n^* can be defined for real valued functions) that is

$$A_n(f,x):=\frac{A_n^*(f,x)}{\sum_{k=\lceil na\rceil}^{\lfloor nb\rfloor}\Phi(nx-k)}. \tag{1.39}$$

So that

$$A_n(f,x) - f(x) = \frac{A_n^*(f,x)}{\sum_{k=\lceil na \rceil}^{\lfloor nb \rfloor} \Phi(nx-k)} - f(x) =$$

$$\frac{A_n^*(f,x) - f(x)\left(\sum_{k=\lceil na \rceil}^{\lfloor nb \rfloor} \Phi(nx-k)\right)}{\sum_{k=\lceil na \rceil}^{\lfloor nb \rfloor} \Phi(nx-k)}. \tag{1.40}$$

Consequently, we derive that

$$\|A_n(f,x) - f(x)\| \leq 2\left(\sqrt[2m]{1+4^m}\right) \left\| A_n^*(f,x) - f(x)\left(\sum_{k=\lceil na \rceil}^{\lfloor nb \rfloor} \Phi(nx-k)\right) \right\| =$$

$$2\left(\sqrt[2m]{1+4^m}\right) \left\| \sum_{k=\lceil na \rceil}^{\lfloor nb \rfloor} \left(f\left(\frac{k}{n}\right) - f(x)\right) \Phi(nx-k) \right\|. \tag{1.41}$$

We will estimate the right and hand side of (1.41).

For that we need, for $f \in C([a,b], X)$ the first modulus of continuity

$$\omega_1(f,\delta) := \sup_{\substack{x, y \in [a,b] \\ |x-y| \leq \delta}} \|f(x) - f(y)\|, \ \delta > 0.$$

Similarly, it is defined ω_1 for $f \in C_{uB}(\mathbb{R}, X)$ (uniformly continuous and bounded functions from $\mathbb{R}$ into X), for $f \in C_B(\mathbb{R}, X)$ (continuous and bounded X-valued), and for $f \in C_u(\mathbb{R}, X)$ (uniformly continuous).

The fact $f \in C([a,b], X)$ or $f \in C_u(\mathbb{R}, X)$, is equivalent to $\lim_{\delta \to 0} \omega_1(f,\delta) = 0$, see [11].

We make

Definition 1.9 When $f \in C_{uB}(\mathbb{R}, X)$, or $f \in C_B(\mathbb{R}, X)$, we define

$$\overline{A_n}(f,x) := \sum_{k=-\infty}^{\infty} f\left(\frac{k}{n}\right) \Phi(nx-k), \tag{1.42}$$

$n \in \mathbb{N}$, $x \in \mathbb{R}$,

the X-valued quasi-interpolation neural network operator.

We make

Remark 1.10 We have that

$$\left\| f\left(\frac{k}{n}\right)\right\| \leq \|f\|_{\infty,\mathbb{R}} < +\infty, \tag{1.43}$$

and

$$\left\| f\left(\frac{k}{n}\right)\right\| \Phi(nx-k) \leq \|f\|_{\infty,\mathbb{R}}\, \Phi(nx-k) \tag{1.44}$$

and

$$\sum_{k=-\lambda}^{\lambda} \left\| f\left(\frac{k}{n}\right)\right\| \Phi(nx-k) \leq \|f\|_{\infty,\mathbb{R}} \left(\sum_{k=-\lambda}^{\lambda} \Phi(nx-k)\right), \tag{1.45}$$

and finally

$$\sum_{k=-\infty}^{\infty} \left\| f\left(\frac{k}{n}\right)\right\| \Phi(nx-k) \leq \|f\|_{\infty,\mathbb{R}}, \tag{1.46}$$

a convergent in $\mathbb{R}$ series.

So, the series $\sum_{k=-\infty}^{\infty} \left\| f\left(\frac{k}{n}\right)\right\| \Phi(nx-k)$ is absolutely convergent in X, hence it is convergent in X and $\overline{A_n}(f,x) \in X$. We denote by $\|f\|_{\infty} := \sup_{x\in[a,b]} \|f(x)\|$, for $f \in C([a,b], X)$, similarly it is defined for $f \in C_B(\mathbb{R}, X)$.

1.3 Main Results

We present a set of X-valued neural network approximations to a function given with rates.

Theorem 1.11 *Let* $f \in C([a,b], X)$, $0 < \alpha < 1$, $n \in \mathbb{N} : n^{1-\alpha} > 2$, $x \in [a,b]$, $m \in \mathbb{N}$. *Then*

(i)

$$\|A_n(f,x) - f(x)\| \leq \left(\sqrt[2m]{1+4^m}\right)\left[2\omega_1\left(f, \frac{1}{n^\alpha}\right) + \frac{\|f\|_\infty}{m\left(n^{1-\alpha}-2\right)^{2m}}\right] =: \lambda_1, \tag{1.47}$$

and

(ii)

$$\|A_n(f) - f\|_\infty \leq \lambda_1. \tag{1.48}$$

We get that $\lim_{n\to\infty} A_n(f) = f$, *pointwise and uniformly.*

Proof We see that

$$\left\| \sum_{k=\lceil na\rceil}^{\lfloor nb\rfloor} \left(f\left(\frac{k}{n}\right) - f(x)\right) \Phi(nx-k) \right\| \leq$$

$$\sum_{k=\lceil na\rceil}^{\lfloor nb\rfloor} \left\| f\left(\frac{k}{n}\right) - f(x)\right\| \Phi(nx-k) =$$

$$\sum_{\begin{cases} k=\lceil na\rceil \\ : \left|\frac{k}{n}-x\right| \leq \frac{1}{n^{\alpha}} \end{cases}}^{\lfloor nb\rfloor} \left\| f\left(\frac{k}{n}\right) - f(x)\right\| \Phi(nx-k) +$$

$$\sum_{\begin{cases} k=\lceil na\rceil \\ : \left|\frac{k}{n}-x\right| > \frac{1}{n^{\alpha}} \end{cases}}^{\lfloor nb\rfloor} \left\| f\left(\frac{k}{n}\right) - f(x)\right\| \Phi(nx-k) \leq \tag{1.49}$$

$$\sum_{\begin{cases} k=\lceil na\rceil \\ : \left|\frac{k}{n}-x\right| \leq \frac{1}{n^{\alpha}} \end{cases}}^{\lfloor nb\rfloor} \omega_1\left(f, \left|\frac{k}{n}-x\right|\right) \Phi(nx-k) +$$

$$2\|f\|_{\infty} \sum_{\begin{cases} k=-\infty \\ : |k-nx| > n^{1-\alpha} \end{cases}}^{\infty} \Phi(nx-k) \leq$$

$$\omega_1\left(f, \frac{1}{n^{\alpha}}\right) \sum_{\begin{cases} k=-\infty \\ : \left|\frac{k}{n}-x\right| \leq \frac{1}{n^{\alpha}} \end{cases}}^{\infty} \Phi(nx-k) +$$

$$2\|f\|_{\infty} \sum_{\begin{cases} k=-\infty \\ : |k-nx| > n^{1-\alpha} \end{cases}}^{\infty} \Phi(nx-k) \underset{\text{(by Theorem 1.4)}}{\leq}$$

$$\omega_1\left(f, \frac{1}{n^{\alpha}}\right) + \frac{\|f\|_{\infty}}{2m\left(n^{1-\alpha}-2\right)^{2m}}.$$

That is

$$\left\| \sum_{k=\lceil na \rceil}^{\lfloor nb \rfloor} \left(f\left(\frac{k}{n}\right) - f(x) \right) \Phi(nx - k) \right\| \leq$$

$$\omega_1\left(f, \frac{1}{n^\alpha}\right) + \frac{\|f\|_\infty}{2m\left(n^{1-\alpha} - 2\right)^{2m}}. \tag{1.50}$$

Using (1.41) we derive (1.47). ■

It follows

Theorem 1.12 *Let $f \in C_B(\mathbb{R}, X)$, $0 < \alpha < 1$, $n \in \mathbb{N} : n^{1-\alpha} > 2$, $x \in \mathbb{R}$, $m \in \mathbb{N}$. Then*

(i)

$$\left\| \overline{A_n}(f, x) - f(x) \right\| \leq \omega_1\left(f, \frac{1}{n^\alpha}\right) + \frac{\|f\|_\infty}{2m\left(n^{1-\alpha} - 2\right)^{2m}} =: \lambda_2, \tag{1.51}$$

and

(ii)

$$\left\| \overline{A_n}(f) - f \right\|_\infty \leq \lambda_2. \tag{1.52}$$

For $f \in C_{uB}(\mathbb{R}, X)$ we get $\lim_{n\to\infty} \overline{A_n}(f) = f$, pointwise and uniformly.

Proof We observe that

$$\left\| \overline{A_n}(f, x) - f(x) \right\| = \left\| \sum_{k=-\infty}^{\infty} f\left(\frac{k}{n}\right) \Phi(nx - k) - f(x) \sum_{k=-\infty}^{\infty} \Phi(nx - k) \right\| =$$

$$\left\| \sum_{k=-\infty}^{\infty} \left(f\left(\frac{k}{n}\right) - f(x) \right) \Phi(nx - k) \right\| \leq$$

$$\sum_{k=-\infty}^{\infty} \left\| f\left(\frac{k}{n}\right) - f(x) \right\| \Phi(nx - k) = \tag{1.53}$$

$$\sum_{\substack{k = -\infty \\ : \left|\frac{k}{n} - x\right| \leq \frac{1}{n^\alpha}}}^{\infty} \left\| f\left(\frac{k}{n}\right) - f(x) \right\| \Phi(nx - k) +$$

$$\sum_{\left\{\begin{array}{l} k=-\infty \\ : \left|\frac{k}{n}-x\right| > \frac{1}{n^{\alpha}} \end{array}\right.}^{\infty} \left\| f\left(\frac{k}{n}\right) - f(x) \right\| \Phi(nx-k) \leq$$

$$\sum_{\left\{\begin{array}{l} k=-\infty \\ : \left|\frac{k}{n}-x\right| \leq \frac{1}{n^{\alpha}} \end{array}\right.}^{\infty} \omega_1\left(f, \left|\frac{k}{n}-x\right|\right) \Phi(nx-k) +$$

$$2\|f\|_{\infty} \sum_{\left\{\begin{array}{l} k=-\infty \\ : \left|\frac{k}{n}-x\right| > \frac{1}{n^{\alpha}} \end{array}\right.}^{\infty} \Phi(nx-k) \leq$$

$$\omega_1\left(f, \frac{1}{n^{\alpha}}\right) \sum_{\left\{\begin{array}{l} k=-\infty \\ : \left|\frac{k}{n}-x\right| \leq \frac{1}{n^{\alpha}} \end{array}\right.}^{\infty} \Phi(nx-k) + \frac{2\|f\|_{\infty}}{4m\left(n^{1-\alpha}-2\right)^{2m}} \leq \tag{1.54}$$

$$\omega_1\left(f, \frac{1}{n^{\alpha}}\right) + \frac{\|f\|_{\infty}}{2m\left(n^{1-\alpha}-2\right)^{2m}},$$

proving the claim. ■

We need the X-valued Taylor's formula in an appropiate form:

Theorem 1.13 ([10, 12]) *Let $N \in \mathbb{N}$, and $f \in C^N([a,b], X)$, where $[a,b] \subset \mathbb{R}$ and X is a Banach space. Let any $x, y \in [a,b]$. Then*

$$f(x) = \sum_{i=0}^{N} \frac{(x-y)^i}{i!} f^{(i)}(y) + \frac{1}{(N-1)!} \int_y^x (x-t)^{N-1} \left(f^{(N)}(t) - f^{(N)}(y)\right) dt. \tag{1.55}$$

The derivatives $f^{(i)}$, $i \in \mathbb{N}$, are defined like the numerical ones, see [21], p. 83. The integral $\int_y^x$ in (1.55) is of Bochner type, see [19].

By [12, 17] we have that: if $f \in C([a,b], X)$, then $f \in L_{\infty}([a,b], X)$ and $f \in L_1([a,b], X)$.

In the next we discuss high order neural network X-valued approximation by using the smoothness of f.

Theorem 1.14 *Let $f \in C^N([a,b], X)$, $n, N, m \in \mathbb{N}$, $0 < \alpha < 1$, $x \in [a,b]$ and $n^{1-\alpha} > 2$. Then*

(i)

$$\|A_n(f,x) - f(x)\| \leq \left(\sqrt[2m]{1+4^m}\right)\left\{\sum_{j=1}^{N}\frac{\|f^{(j)}(x)\|}{j!}\left[\frac{2}{n^{\alpha j}} + \frac{(b-a)^j}{2m\left(n^{1-\alpha}-2\right)^{2m}}\right] + \right. \tag{1.56}$$

$$\left.\left[\omega_1\left(f^{(N)}, \frac{1}{n^\alpha}\right)\frac{2}{n^{\alpha N}N!} + \frac{\|f^{(N)}\|_\infty (b-a)^N}{N!m\left(n^{1-\alpha}-2\right)^{2m}}\right]\right\},$$

(ii) assume further $f^{(j)}(x_0) = 0$, $j = 1, ..., N$, *for some* $x_0 \in [a,b]$, *it holds*

$$\|A_n(f,x_0) - f(x_0)\| \leq \left(\sqrt[2m]{1+4^m}\right)\cdot \tag{1.57}$$

$$\left[\omega_1\left(f^{(N)}, \frac{1}{n^\alpha}\right)\frac{2}{n^{\alpha N}N!} + \frac{\|f^{(N)}\|_\infty (b-a)^N}{N!m\left(n^{1-\alpha}-2\right)^{2m}}\right],$$

and

(iii)

$$\|A_n(f) - f\|_\infty \leq \left(\sqrt[2m]{1+4^m}\right)\left\{\sum_{j=1}^{N}\frac{\|f^{(j)}\|_\infty}{j!}\left[\frac{2}{n^{\alpha j}} + \frac{(b-a)^j}{2m\left(n^{1-\alpha}-2\right)^{2m}}\right] + \right.$$

$$\left.\left[\omega_1\left(f^{(N)}, \frac{1}{n^\alpha}\right)\frac{2}{n^{\alpha N}N!} + \frac{\|f^{(N)}\|_\infty (b-a)^N}{N!m\left(n^{1-\alpha}-2\right)^{2m}}\right]\right\}. \tag{1.58}$$

We derive that $\lim_{n\to\infty} A_n(f) = f$, *pointwise and uniformly.*

Proof Next we apply the X-valued Taylor's formula with Bochner integral remainder (1.55). We have (here $\frac{k}{n}, x \in [a,b]$)

$$f\left(\frac{k}{n}\right) = \sum_{j=0}^{N}\frac{f^{(j)}(x)}{j!}\left(\frac{k}{n}-x\right)^j + \int_x^{\frac{k}{n}}\left(f^{(N)}(t) - f^{(N)}(x)\right)\frac{\left(\frac{k}{n}-t\right)^{N-1}}{(N-1)!}dt. \tag{1.59}$$

Then

$$f\left(\frac{k}{n}\right)\Phi(nx-k) = \sum_{j=0}^{N}\frac{f^{(j)}(x)}{j!}\Phi(nx-k)\left(\frac{k}{n}-x\right)^j + \tag{1.60}$$

$$\Phi(nx-k)\int_x^{\frac{k}{n}}\left(f^{(N)}(t) - f^{(N)}(x)\right)\frac{\left(\frac{k}{n}-t\right)^{N-1}}{(N-1)!}dt.$$

Hence

$$\sum_{k=\lceil na\rceil}^{\lfloor nb\rfloor} f\left(\frac{k}{n}\right)\Phi(nx-k) - f(x)\sum_{k=\lceil na\rceil}^{\lfloor nb\rfloor}\Phi(nx-k) = \tag{1.61}$$

$$\sum_{j=1}^{N}\frac{f^{(j)}(x)}{j!}\sum_{k=\lceil na\rceil}^{\lfloor nb\rfloor}\Phi(nx-k)\left(\frac{k}{n}-x\right)^{j} +$$

$$\sum_{k=\lceil na\rceil}^{\lfloor nb\rfloor}\Phi(nx-k)\int_{x}^{\frac{k}{n}}\left(f^{(N)}(t)-f^{(N)}(x)\right)\frac{\left(\frac{k}{n}-t\right)^{N-1}}{(N-1)!}dt.$$

Thus

$$A_n^*(f,x) - f(x)\left(\sum_{k=\lceil na\rceil}^{\lfloor nb\rfloor}\Phi(nx-k)\right) =$$

$$\sum_{j=1}^{N}\frac{f^{(j)}(x)}{j!}A_n^*\left((\cdot-x)^{j}\right) + \Lambda_n(x), \tag{1.62}$$

where

$$\Lambda_n(x) := \sum_{k=\lceil na\rceil}^{\lfloor nb\rfloor}\Phi(nx-k)\int_{x}^{\frac{k}{n}}\left(f^{(N)}(t)-f^{(N)}(x)\right)\frac{\left(\frac{k}{n}-t\right)^{N-1}}{(N-1)!}dt. \tag{1.63}$$

We assume that $b-a>\frac{1}{n^{\alpha}}$, which is always the case for large enough $n\in\mathbb{N}$, that is when $n>\left\lceil (b-a)^{-\frac{1}{\alpha}}\right\rceil$.

Thus $\left|\frac{k}{n}-x\right|\leq\frac{1}{n^{\alpha}}$ or $\left|\frac{k}{n}-x\right|>\frac{1}{n^{\alpha}}$.

Let

$$\gamma := \int_{x}^{\frac{k}{n}}\left(f^{(N)}(t)-f^{(N)}(x)\right)\frac{\left(\frac{k}{n}-t\right)^{N-1}}{(N-1)!}dt, \tag{1.64}$$

in the case of $\left|\frac{k}{n}-x\right|\leq\frac{1}{n^{\alpha}}$, we find that

$$\|\gamma\| \leq \omega_1\left(f^{(N)},\frac{1}{n^{\alpha}}\right)\frac{1}{n^{\alpha N}N!} \tag{1.65}$$

for $x\leq\frac{k}{n}$ or $x\geq\frac{k}{n}$.

We prove it next.

(i) Indeed, for the case of $x \leq \frac{k}{n}$, we have

$$\|\gamma\| = \left\| \int_x^{\frac{k}{n}} \left(f^{(N)}(t) - f^{(N)}(x) \right) \frac{\left(\frac{k}{n} - t\right)^{N-1}}{(N-1)!} dt \right\| \leq$$

$$\int_x^{\frac{k}{n}} \left\| f^{(N)}(t) - f^{(N)}(x) \right\| \frac{\left(\frac{k}{n} - t\right)^{N-1}}{(N-1)!} dt \leq$$

$$\int_x^{\frac{k}{n}} \omega_1 \left(f^{(N)}, |t - x| \right) \frac{\left(\frac{k}{n} - t\right)^{N-1}}{(N-1)!} dt \leq \omega_1 \left(f^{(N)}, \frac{1}{n^{\alpha}} \right) \int_x^{\frac{k}{n}} \frac{\left(\frac{k}{n} - t\right)^{N-1}}{(N-1)!} dt = \tag{1.66}$$

$$\omega_1 \left(f^{(N)}, \frac{1}{n^{\alpha}} \right) \frac{\left(\frac{k}{n} - x\right)^N}{N!} \leq \omega_1 \left(f^{(N)}, \frac{1}{n^{\alpha}} \right) \frac{1}{n^{\alpha N} N!}.$$

(ii) for the case of $x > \frac{k}{n}$, we have

$$\|\gamma\| = \left\| \int_x^{\frac{k}{n}} \left(f^{(N)}(t) - f^{(N)}(x) \right) \frac{\left(\frac{k}{n} - t\right)^{N-1}}{(N-1)!} dt \right\| =$$

$$\left\| \int_{\frac{k}{n}}^{x} \left(f^{(N)}(t) - f^{(N)}(x) \right) \frac{\left(t - \frac{k}{n}\right)^{N-1}}{(N-1)!} dt \right\| \leq$$

$$\int_{\frac{k}{n}}^{x} \left\| f^{(N)}(t) - f^{(N)}(x) \right\| \frac{\left(t - \frac{k}{n}\right)^{N-1}}{(N-1)!} dt \leq \tag{1.67}$$

$$\int_{\frac{k}{n}}^{x} \omega_1 \left(f^{(N)}, |t - x| \right) \frac{\left(t - \frac{k}{n}\right)^{N-1}}{(N-1)!} dt \leq \omega_1 \left(f^{(N)}, \frac{1}{n^{\alpha}} \right) \int_{\frac{k}{n}}^{x} \frac{\left(t - \frac{k}{n}\right)^{N-1}}{(N-1)!} dt =$$

$$\omega_1 \left(f^{(N)}, \frac{1}{n^{\alpha}} \right) \frac{\left(x - \frac{k}{n}\right)^N}{N!} \leq \omega_1 \left(f^{(N)}, \frac{1}{n^{\alpha}} \right) \frac{1}{n^{\alpha N} N!}.$$

We have proved (1.65).

We treat again γ, see (1.64), but differently:

Notice also for $x \leq \frac{k}{n}$ that

$$\left\| \int_x^{\frac{k}{n}} \left(f^{(N)}(t) - f^{(N)}(x) \right) \frac{\left(\frac{k}{n} - t \right)^{N-1}}{(N-1)!} dt \right\| \leq$$

$$\int_x^{\frac{k}{n}} \left\| f^{(N)}(t) - f^{(N)}(x) \right\| \frac{\left(\frac{k}{n} - t \right)^{N-1}}{(N-1)!} dt \leq \tag{1.68}$$

$$2 \left\| f^{(N)} \right\|_\infty \int_x^{\frac{k}{n}} \frac{\left(\frac{k}{n} - t \right)^{N-1}}{(N-1)!} dt = 2 \left\| f^{(N)} \right\|_\infty \frac{\left(\frac{k}{n} - x \right)^N}{N!}$$

$$\leq 2 \left\| f^{(N)} \right\|_\infty \frac{(b-a)^N}{N!}.$$

Next assume $\frac{k}{n} \leq x$, then

$$\left\| \int_x^{\frac{k}{n}} \left(f^{(N)}(t) - f^{(N)}(x) \right) \frac{\left(\frac{k}{n} - t \right)^{N-1}}{(N-1)!} dt \right\| =$$

$$\left\| \int_{\frac{k}{n}}^x \left(f^{(N)}(t) - f^{(N)}(x) \right) \frac{\left(t - \frac{k}{n} \right)^{N-1}}{(N-1)!} dt \right\| \leq$$

$$\int_{\frac{k}{n}}^x \left\| f^{(N)}(t) - f^{(N)}(x) \right\| \frac{\left(t - \frac{k}{n} \right)^{N-1}}{(N-1)!} dt \leq \tag{1.69}$$

$$2 \left\| f^{(N)} \right\|_\infty \int_{\frac{k}{n}}^x \frac{\left(t - \frac{k}{n} \right)^{N-1}}{(N-1)!} dt = 2 \left\| f^{(N)} \right\|_\infty \frac{\left(x - \frac{k}{n} \right)^N}{N!}$$

$$\leq 2 \left\| f^{(N)} \right\|_\infty \frac{(b-a)^N}{N!}.$$

Thus

$$\|\gamma\| \leq 2 \left\| f^{(N)} \right\|_\infty \frac{(b-a)^N}{N!}. \tag{1.70}$$

in the two cases.

Therefore

$$\Lambda_n(x) = \sum_{\substack{k = \lceil na \rceil \\ \left| \frac{k}{n} - x \right| \leq \frac{1}{n^\alpha}}}^{\lfloor nb \rfloor} \Phi(nx - k)\gamma + \sum_{\substack{k = \lceil na \rceil \\ \left| \frac{k}{n} - x \right| > \frac{1}{n^\alpha}}}^{\lfloor nb \rfloor} \Phi(nx - k)\gamma. \tag{1.71}$$

Hence

$$\|\Lambda_n(x)\| \leq \sum_{\begin{cases} k=\lceil na\rceil \\ \left|\frac{k}{n}-x\right| \leq \frac{1}{n^\alpha} \end{cases}}^{\lfloor nb\rfloor} \Phi(nx-k)\left(\omega_1\left(f^{(N)},\frac{1}{n^\alpha}\right)\frac{1}{N!n^{\alpha N}}\right)+ \tag{1.72}$$

$$\left(\sum_{\begin{cases} k=\lceil na\rceil \\ \left|\frac{k}{n}-x\right| > \frac{1}{n^\alpha} \end{cases}}^{\lfloor nb\rfloor} \Phi(nx-k)\right) 2\left\|f^{(N)}\right\|_\infty \frac{(b-a)^N}{N!} \overset{(1.19)}{\leq}$$

$$\omega_1\left(f^{(N)},\frac{1}{n^\alpha}\right)\frac{1}{N!n^{\alpha N}}+\frac{1}{4m\left(n^{1-\alpha}-2\right)^{2m}}2\left\|f^{(N)}\right\|_\infty \frac{(b-a)^N}{N!}=$$

$$\omega_1\left(f^{(N)},\frac{1}{n^\alpha}\right)\frac{1}{N!n^{\alpha N}}+\frac{\left\|f^{(N)}\right\|_\infty (b-a)^N}{N!2m\left(n^{1-\alpha}-2\right)^{2m}}.$$

That is

$$\|\Lambda_n(x)\| \leq \frac{\omega_1\left(f^{(N)},\frac{1}{n^\alpha}\right)}{N!n^{\alpha N}}+\frac{\left\|f^{(N)}\right\|_\infty (b-a)^N}{N!2m\left(n^{1-\alpha}-2\right)^{2m}}, \tag{1.73}$$

$\forall\, x \in [a,b]$.

We further see that

$$A_n^*\left((\cdot-x)^j\right)=\sum_{k=\lceil na\rceil}^{\lfloor nb\rfloor}\Phi(nx-k)\left(\frac{k}{n}-x\right)^j, \tag{1.74}$$

where A_n^* is defined similarly for real valued functions.

Therefore

$$\left|A_n^*\left((\cdot-x)^j\right)\right| \leq \sum_{k=\lceil na\rceil}^{\lfloor nb\rfloor}\Phi(nx-k)\left|\frac{k}{n}-x\right|^j=$$

$$\sum_{\begin{cases} k=\lceil na\rceil \\ \left|\frac{k}{n}-x\right| \leq \frac{1}{n^\alpha} \end{cases}}^{\lfloor nb\rfloor} \Phi(nx-k)\left|\frac{k}{n}-x\right|^j + \sum_{\begin{cases} k=\lceil na\rceil \\ \left|\frac{k}{n}-x\right| > \frac{1}{n^\alpha} \end{cases}}^{\lfloor nb\rfloor} \Phi(nx-k)\left|\frac{k}{n}-x\right|^j \leq$$

$$\frac{1}{n^{\alpha j}}+(b-a)^j\frac{1}{4m\left(n^{1-\alpha}-2\right)^{2m}}. \tag{1.75}$$

That is

$$\left|A_n^*\left((\cdot-x)^j\right)\right| \leq \frac{1}{n^{\alpha j}}+(b-a)^j \frac{1}{4m\left(n^{1-\alpha}-2\right)^{2m}}, \tag{1.76}$$

for $j=1, \ldots, N$.

Putting things together we have proved

$$\left\| A_n^*(f, x)-f(x)\left(\sum_{k=\lceil na\rceil}^{\lfloor nb\rfloor} \Phi(nx-k)\right)\right\| \leq \sum_{j=1}^N \frac{\left\|f^{(j)}(x)\right\|}{j!} \tag{1.77}$$

$$\left[\frac{1}{n^{\alpha j}}+\frac{(b-a)^j}{4m\left(n^{1-\alpha}-2\right)^{2m}}\right]+\left[\omega_1\left(f^{(N)}, \frac{1}{n^\alpha}\right) \frac{1}{n^{\alpha N} N!}+\frac{\left\|f^{(N)}\right\|_{\infty}(b-a)^N}{N!2m\left(n^{1-\alpha}-2\right)^{2m}}\right],$$

that is establishing the theorem. ■

All integrals from now on are of Bochner type [19].

We need

Definition 1.15 ([12]) Let $[a, b] \subset \mathbb{R}$, X be a Banach space, $\alpha>0$; $m=\lceil\alpha\rceil \in \mathbb{N}$, ($\lceil\cdot\rceil$ is the ceiling of the number), $f:[a, b] \rightarrow X$. We assume that $f^{(m)} \in L_1([a, b], X)$. We call the Caputo-Bochner left fractional derivative of order α:

$$\left(D_{*a}^\alpha f\right)(x):=\frac{1}{\Gamma(m-\alpha)} \int_a^x(x-t)^{m-\alpha-1} f^{(m)}(t)\, dt, \ \forall x \in[a, b]. \tag{1.78}$$

If $\alpha \in \mathbb{N}$, we set $D_{*a}^\alpha f:=f^{(m)}$ the ordinary X-valued derivative (defined similar to numerical one, see [21], p. 83), and also set $D_{*a}^0 f:=f$.

By [12], $\left(D_{*a}^\alpha f\right)(x)$ exists almost everywhere in $x \in[a, b]$ and $D_{*a}^\alpha f \in L_1([a, b], X)$.

If $\left\|f^{(m)}\right\|_{L_{\infty}([a, b], X)}<\infty$, then by [12], $D_{*a}^\alpha f \in C([a, b], X)$, hence $\left\|D_{*a}^\alpha f\right\| \in C([a, b])$.

We mention

Lemma 1.16 ([11]) *Let* $\alpha>0$, $\alpha \notin \mathbb{N}$, $m=\lceil\alpha\rceil$, $f \in C^{m-1}([a, b], X)$ *and* $f^{(m)} \in L_{\infty}([a, b], X)$. *Then* $D_{*a}^\alpha f(a)=0$.

We mention

Definition 1.17 ([10]) Let $[a, b] \subset \mathbb{R}$, X be a Banach space, $\alpha>0$, $m:=\lceil\alpha\rceil$. We assume that $f^{(m)} \in L_1([a, b], X)$, where $f:[a, b] \rightarrow X$. We call the Caputo-Bochner right fractional derivative of order α:

$$\left(D_{b-}^\alpha f\right)(x):=\frac{(-1)^m}{\Gamma(m-\alpha)} \int_x^b(z-x)^{m-\alpha-1} f^{(m)}(z)\, dz, \ \forall x \in[a, b]. \tag{1.79}$$

We observe that $\left(D_{b-}^{m}f\right)(x) = (-1)^{m} f^{(m)}(x)$, for $m \in \mathbb{N}$, and $\left(D_{b-}^{0}f\right)(x) = f(x)$.

By [10], $\left(D_{b-}^{\alpha}f\right)(x)$ exists almost everywhere on $[a,b]$ and $\left(D_{b-}^{\alpha}f\right) \in L_1([a,b], X)$.

If $\left\| f^{(m)} \right\|_{L_\infty([a,b],X)} < \infty$, and $\alpha \notin \mathbb{N}$, by [10], $D_{b-}^{\alpha}f \in C([a,b], X)$, hence $\left\| D_{b-}^{\alpha}f \right\| \in C([a,b])$.

We need

Lemma 1.18 ([11]) *Let $f \in C^{m-1}([a,b], X)$, $f^{(m)} \in L_\infty([a,b], X)$, $m = \lceil \alpha \rceil$, $\alpha > 0$, $\alpha \notin \mathbb{N}$. Then $D_{b-}^{\alpha}f(b) = 0$.*

We mention the left fractional Taylor formula

Theorem 1.19 ([12]) *Let $m \in \mathbb{N}$ and $f \in C^{m}([a,b], X)$, where $[a,b] \subset \mathbb{R}$ and X is a Banach space, and let $\alpha > 0 : m = \lceil \alpha \rceil$. Then*

$$f(x) = \sum_{i=0}^{m-1} \frac{(x-a)^i}{i!} f^{(i)}(a) + \frac{1}{\Gamma(\alpha)} \int_a^x (x-z)^{\alpha-1} \left(D_{*a}^{\alpha}f\right)(z)\, dz, \tag{1.80}$$

$\forall\, x \in [a,b]$.

We also mention the right fractional Taylor formula.

Theorem 1.20 ([10]) *Let $[a,b] \subset \mathbb{R}$, X be a Banach space, $\alpha > 0$, $m = \lceil \alpha \rceil$, $f \in C^{m}([a,b], X)$. Then*

$$f(x) = \sum_{i=0}^{m-1} \frac{(x-b)^i}{i!} f^{(i)}(b) + \frac{1}{\Gamma(\alpha)} \int_x^b (z-x)^{\alpha-1} \left(D_{b-}^{\alpha}f\right)(z)\, dz, \tag{1.81}$$

$\forall\, x \in [a,b]$.

Convention 1.21 *We assume that*

$$D_{*x_0}^{\alpha} f(x) = 0, \text{ for } x < x_0, \tag{1.82}$$

and

$$D_{x_0-}^{\alpha} f(x) = 0, \text{ for } x > x_0, \tag{1.83}$$

for all $x, x_0 \in [a,b]$.

We mention

Proposition 1.22 ([11]) *Let $f \in C^{n}([a,b], X)$, $n = \lceil \nu \rceil$, $\nu > 0$. Then $D_{*a}^{\nu} f(x)$ is continuous in $x \in [a,b]$.*

Proposition 1.23 ([11]) *Let $f \in C^m([a,b], X)$, $m = \lceil \alpha \rceil$, $\alpha > 0$. Then $D_{b-}^{\nu} f(x)$ is continuous in $x \in [a,b]$.*

We also mention

Proposition 1.24 ([11]) *Let $f \in C^{m-1}([a,b], X)$, $f^{(m)} \in L_\infty([a,b], X)$, $m = \lceil \alpha \rceil$, $\alpha > 0$ and*

$$D_{*x_0}^{\alpha} f(x) = \frac{1}{\Gamma(m-\alpha)} \int_{x_0}^{x} (x-t)^{m-\alpha-1} f^{(m)}(t)\, dt, \tag{1.84}$$

for all $x, x_0 \in [a,b] : x \geq x_0$.

Then $D_{*x_0}^{\alpha} f(x)$ is continuous in x_0.

Proposition 1.25 ([11]) *Let $f \in C^{m-1}([a,b], X)$, $f^{(m)} \in L_\infty([a,b], X)$, $m = \lceil \alpha \rceil$, $\alpha > 0$ and*

$$D_{x_0-}^{\alpha} f(x) = \frac{(-1)^m}{\Gamma(m-\alpha)} \int_{x}^{x_0} (\zeta - x)^{m-\alpha-1} f^{(m)}(\zeta)\, d\zeta, \tag{1.85}$$

for all $x, x_0 \in [a,b] : x_0 \geq x$.

Then $D_{x_0-}^{\alpha} f(x)$ is continuous in x_0.

Corollary 1.26 ([11]) *Let $f \in C^m([a,b], X)$, $m = \lceil \alpha \rceil$, $\alpha > 0$, $x, x_0 \in [a,b]$. Then $D_{*x_0}^{a} f(x)$, $D_{x_0-}^{a} f(x)$ are jointly continuous functions in (x, x_0) from $[a,b]^2$ into X, X is a Banach space.*

We need

Theorem 1.27 ([11]) *Let $f : [a,b]^2 \to X$ be jointly continuous, X is a Banach space. Consider*

$$G(x) = \omega_1(f(\cdot, x), \delta, [x,b]), \tag{1.86}$$

$\delta > 0$, $x \in [a,b]$.

Then G is continuous on $[a,b]$.

Theorem 1.28 ([11]) *Let $f : [a,b]^2 \to X$ be jointly continuous, X is a Banach space. Then*

$$H(x) = \omega_1(f(\cdot, x), \delta, [a,x]), \tag{1.87}$$

$x \in [a,b]$, is continuous in $x \in [a,b]$, $\delta > 0$.

We make

Remark 1.29 ([11]) Let $f \in C^{n-1}([a,b])$, $f^{(n)} \in L_\infty([a,b])$, $n = \lceil \nu \rceil$, $\nu > 0$, $\nu \notin \mathbb{N}$. Then

$$\left\| D_{*a}^{\nu} f(x) \right\| \leq \frac{\left\| f^{(n)} \right\|_{L_\infty([a,b],X)}}{\Gamma(n-\nu+1)} (x-a)^{n-\nu}, \ \forall x \in [a,b]. \tag{1.88}$$

Thus we observe ($\delta > 0$)

$$\omega_1\left(D^{\nu}_{*a}f,\delta\right) = \sup_{\substack{x,y\in[a,b]\\|x-y|\leq\delta}} \left\|D^{\nu}_{*a}f(x) - D^{\nu}_{*a}f(y)\right\| \leq \tag{1.89}$$

$$\sup_{\substack{x,y\in[a,b]\\|x-y|\leq\delta}} \left(\frac{\left\|f^{(n)}\right\|_{L_\infty([a,b],X)}}{\Gamma(n-\nu+1)}(x-a)^{n-\nu} + \frac{\left\|f^{(n)}\right\|_{L_\infty([a,b],X)}}{\Gamma(n-\nu+1)}(y-a)^{n-\nu}\right)$$

$$\leq \frac{2\left\|f^{(n)}\right\|_{L_\infty([a,b],X)}}{\Gamma(n-\nu+1)}(b-a)^{n-\nu}.$$

Consequently

$$\omega_1\left(D^{\nu}_{*a}f,\delta\right) \leq \frac{2\left\|f^{(n)}\right\|_{L_\infty([a,b],X)}}{\Gamma(n-\nu+1)}(b-a)^{n-\nu}. \tag{1.90}$$

Similarly, let $f \in C^{m-1}([a,b])$, $f^{(m)} \in L_\infty([a,b])$, $m = \lceil\alpha\rceil$, $\alpha > 0$, $\alpha \notin \mathbb{N}$, then

$$\omega_1\left(D^{\alpha}_{b-}f,\delta\right) \leq \frac{2\left\|f^{(m)}\right\|_{L_\infty([a,b],X)}}{\Gamma(m-\alpha+1)}(b-a)^{m-\alpha}. \tag{1.91}$$

So for $f \in C^{m-1}([a,b])$, $f^{(m)} \in L_\infty([a,b])$, $m = \lceil\alpha\rceil$, $\alpha > 0$, $\alpha \notin \mathbb{N}$, we find

$$\sup_{x_0\in[a,b]} \omega_1\left(D^{\alpha}_{*x_0}f,\delta\right)_{[x_0,b]} \leq \frac{2\left\|f^{(m)}\right\|_{L_\infty([a,b],X)}}{\Gamma(m-\alpha+1)}(b-a)^{m-\alpha}, \tag{1.92}$$

and

$$\sup_{x_0\in[a,b]} \omega_1\left(D^{\alpha}_{x_0-}f,\delta\right)_{[a,x_0]} \leq \frac{2\left\|f^{(m)}\right\|_{L_\infty([a,b],X)}}{\Gamma(m-\alpha+1)}(b-a)^{m-\alpha}. \tag{1.93}$$

By [12] we get that $D^{\alpha}_{*x_0}f \in C([x_0,b],X)$, and by [10] we obtain that $D^{\alpha}_{x_0-}f \in C([a,x_0],X)$.

We present the following X-valued fractional approximation result by neural networks.

Theorem 1.30 *Let* $\alpha > 0$, $N = \lceil\alpha\rceil$, $\alpha \notin \mathbb{N}$, $f \in C^N([a,b],X)$, $0 < \beta < 1$, $m \in \mathbb{N}$, $x \in [a,b]$, $n \in \mathbb{N}: n^{1-\beta} > 2$. *Then*

(i)

$$\left\| A_n(f,x) - \sum_{j=1}^{N-1} \frac{f^{(j)}(x)}{j!} A_n\left((\cdot - x)^j\right)(x) - f(x) \right\| \leq$$

$$\frac{2\left(\sqrt[2m]{1+4^m}\right)}{\Gamma(\alpha+1)} \left\{ \frac{\left(\omega_1\left(D_{x-}^{\alpha} f, \frac{1}{n^{\beta}}\right)_{[a,x]} + \omega_1\left(D_{*x}^{\alpha} f, \frac{1}{n^{\beta}}\right)_{[x,b]}\right)}{n^{\alpha\beta}} + \right.$$

$$\left. \frac{1}{4m\left(n^{1-\beta}-2\right)^{2m}} \left(\left\| D_{x-}^{\alpha} f \right\|_{\infty,[a,x]} (x-a)^{\alpha} + \left\| D_{*x}^{\alpha} f \right\|_{\infty,[x,b]} (b-x)^{\alpha} \right) \right\}, \tag{1.94}$$

(ii) if $f^{(j)}(x) = 0$, *for* $j = 1, \ldots, N-1$, *we have*

$$\|A_n(f,x) - f(x)\| \leq \frac{2\left(\sqrt[2m]{1+4^m}\right)}{\Gamma(\alpha+1)}$$

$$\left\{ \frac{\left(\omega_1\left(D_{x-}^{\alpha} f, \frac{1}{n^{\beta}}\right)_{[a,x]} + \omega_1\left(D_{*x}^{\alpha} f, \frac{1}{n^{\beta}}\right)_{[x,b]}\right)}{n^{\alpha\beta}} + \right.$$

$$\left. \frac{1}{4m\left(n^{1-\beta}-2\right)^{2m}} \left(\left\| D_{x-}^{\alpha} f \right\|_{\infty,[a,x]} (x-a)^{\alpha} + \left\| D_{*x}^{\alpha} f \right\|_{\infty,[x,b]} (b-x)^{\alpha} \right) \right\}, \tag{1.95}$$

(iii)

$$\|A_n(f,x) - f(x)\| \leq 2\left(\sqrt[2m]{1+4^m}\right) \cdot$$

$$\left\{ \sum_{j=1}^{N-1} \frac{\left\|f^{(j)}(x)\right\|}{j!} \left\{ \frac{1}{n^{\beta j}} + \frac{(b-a)^j}{4m\left(n^{1-\beta}-2\right)^{2m}} \right\} + \right.$$

$$\frac{1}{\Gamma(\alpha+1)} \left\{ \frac{\left(\omega_1\left(D_{x-}^{\alpha} f, \frac{1}{n^{\beta}}\right)_{[a,x]} + \omega_1\left(D_{*x}^{\alpha} f, \frac{1}{n^{\beta}}\right)_{[x,b]}\right)}{n^{\alpha\beta}} + \right.$$

$$\left.\left. \frac{1}{4m\left(n^{1-\beta}-2\right)^{2m}} \left(\left\| D_{x-}^{\alpha} f \right\|_{\infty,[a,x]} (x-a)^{\alpha} + \left\| D_{*x}^{\alpha} f \right\|_{\infty,[x,b]} (b-x)^{\alpha} \right) \right\}\right\}, \tag{1.96}$$

$\forall\, x \in [a,b]$,

and

(iv)

$$\|A_n f - f\|_\infty \le 2\left(\sqrt[2m]{1+4^m}\right)\cdot$$

$$\left\{\sum_{j=1}^{N-1}\frac{\|f^{(j)}\|_\infty}{j!}\left\{\frac{1}{n^{\beta j}}+\frac{(b-a)^j}{4m\left(n^{1-\beta}-2\right)^{2m}}\right\}+\right.$$

$$\frac{1}{\Gamma(\alpha+1)}\left\{\frac{\left(\sup_{x\in[a,b]}\omega_1\left(D_{x-}^{\alpha}f,\frac{1}{n^\beta}\right)_{[a,x]}+\sup_{x\in[a,b]}\omega_1\left(D_{*x}^{\alpha}f,\frac{1}{n^\beta}\right)_{[x,b]}\right)}{n^{\alpha\beta}}+\right.$$

$$\left.\left.\frac{(b-a)^\alpha}{4m\left(n^{1-\beta}-2\right)^{2m}}\left(\sup_{x\in[a,b]}\left\|D_{x-}^{\alpha}f\right\|_{\infty,[a,x]}+\sup_{x\in[a,b]}\left\|D_{*x}^{\alpha}f\right\|_{\infty,[x,b]}\right)\right\}\right\}. \quad (1.97)$$

Above, when $N=1$ the sum $\sum_{j=1}^{N-1}\cdot=0$.

As we see here we obtain X-valued fractionally type pointwise and uniform convergence with rates of $A_n \to I$ the unit operator, as $n\to\infty$.

Proof Let $x\in[a,b]$. We have that $D_{x-}^{\alpha}f(x)=D_{*x}^{\alpha}f(x)=0$.

From Theorem 1.19, we get by the left Caputo fractional Taylor formula that

$$f\left(\frac{k}{n}\right)=\sum_{j=0}^{N-1}\frac{f^{(j)}(x)}{j!}\left(\frac{k}{n}-x\right)^j+ \quad (1.98)$$

$$\frac{1}{\Gamma(\alpha)}\int_x^{\frac{k}{n}}\left(\frac{k}{n}-J\right)^{\alpha-1}\left(D_{*x}^{\alpha}f(J)-D_{*x}^{\alpha}f(x)\right)dJ,$$

for all $x\le\frac{k}{n}\le b$.

Also from Theorem 1.20, using the right Caputo fractional Taylor formula we get

$$f\left(\frac{k}{n}\right)=\sum_{j=0}^{N-1}\frac{f^{(j)}(x)}{j!}\left(\frac{k}{n}-x\right)^j+ \quad (1.99)$$

$$\frac{1}{\Gamma(\alpha)}\int_{\frac{k}{n}}^{x}\left(J-\frac{k}{n}\right)^{\alpha-1}\left(D_{x-}^{\alpha}f(J)-D_{x-}^{\alpha}f(x)\right)dJ,$$

for all $a\le\frac{k}{n}\le x$.

Hence we have

$$f\left(\frac{k}{n}\right)\Phi(nx-k)=\sum_{j=0}^{N-1}\frac{f^{(j)}(x)}{j!}\Phi(nx-k)\left(\frac{k}{n}-x\right)^{j}+ \tag{1.100}$$

$$\frac{\Phi(nx-k)}{\Gamma(\alpha)}\int_{x}^{\frac{k}{n}}\left(\frac{k}{n}-J\right)^{\alpha-1}\left(D_{*x}^{\alpha}f(J)-D_{*x}^{\alpha}f(x)\right)dJ,$$

for all $x\leq\frac{k}{n}\leq b$, iff $\lceil nx\rceil\leq k\leq\lfloor nb\rfloor$, and

$$f\left(\frac{k}{n}\right)\Phi(nx-k)=\sum_{j=0}^{N-1}\frac{f^{(j)}(x)}{j!}\Phi(nx-k)\left(\frac{k}{n}-x\right)^{j}+ \tag{1.101}$$

$$\frac{\Phi(nx-k)}{\Gamma(\alpha)}\int_{\frac{k}{n}}^{x}\left(J-\frac{k}{n}\right)^{\alpha-1}\left(D_{x-}^{\alpha}f(J)-D_{x-}^{\alpha}f(x)\right)dJ,$$

for all $a\leq\frac{k}{n}\leq x$, iff $\lceil na\rceil\leq k\leq\lfloor nx\rfloor$.

Therefore it holds

$$\sum_{k=\lfloor nx\rfloor+1}^{\lfloor nb\rfloor}f\left(\frac{k}{n}\right)\Phi(nx-k)=\sum_{j=0}^{N-1}\frac{f^{(j)}(x)}{j!}\sum_{k=\lfloor nx\rfloor+1}^{\lfloor nb\rfloor}\Phi(nx-k)\left(\frac{k}{n}-x\right)^{j}+ \tag{1.102}$$

$$\frac{1}{\Gamma(\alpha)}\sum_{k=\lfloor nx\rfloor+1}^{\lfloor nb\rfloor}\Phi(nx-k)\int_{x}^{\frac{k}{n}}\left(\frac{k}{n}-J\right)^{\alpha-1}\left(D_{*x}^{\alpha}f(J)-D_{*x}^{\alpha}f(x)\right)dJ,$$

and

$$\sum_{k=\lceil na\rceil}^{\lfloor nx\rfloor}f\left(\frac{k}{n}\right)\Phi(nx-k)=\sum_{j=0}^{N-1}\frac{f^{(j)}(x)}{j!}\sum_{k=\lceil na\rceil}^{\lfloor nx\rfloor}\Phi(nx-k)\left(\frac{k}{n}-x\right)^{j}+ \tag{1.103}$$

$$\frac{1}{\Gamma(\alpha)}\sum_{k=\lceil na\rceil}^{\lfloor nx\rfloor}\Phi(nx-k)\int_{\frac{k}{n}}^{x}\left(J-\frac{k}{n}\right)^{\alpha-1}\left(D_{x-}^{\alpha}f(J)-D_{x-}^{\alpha}f(x)\right)dJ.$$

Adding the last two equalities (1.102) and (1.103) obtain

$$A_n^*(f,x) = \sum_{k=\lceil na \rceil}^{\lfloor nb \rfloor} f\left(\frac{k}{n}\right) \Phi(nx-k) = \tag{1.104}$$

$$\sum_{j=0}^{N-1} \frac{f^{(j)}(x)}{j!} \sum_{k=\lceil na \rceil}^{\lfloor nb \rfloor} \Phi(nx-k) \left(\frac{k}{n} - x\right)^j +$$

$$\frac{1}{\Gamma(\alpha)} \left\{ \sum_{k=\lceil na \rceil}^{\lfloor nx \rfloor} \Phi(nx-k) \int_{\frac{k}{n}}^{x} \left(J - \frac{k}{n}\right)^{\alpha-1} \left(D_{x-}^{\alpha} f(J) - D_{x-}^{\alpha} f(x)\right) dJ + \right.$$

$$\left. \sum_{k=\lfloor nx \rfloor+1}^{\lfloor nb \rfloor} \Phi(nx-k) \int_{x}^{\frac{k}{n}} \left(\frac{k}{n} - J\right)^{\alpha-1} \left(D_{*x}^{\alpha} f(J) - D_{*x}^{\alpha} f(x)\right) dJ \right\}.$$

So we have derived

$$A_n^*(f,x) - f(x) \left(\sum_{k=\lceil na \rceil}^{\lfloor nb \rfloor} \Phi(nx-k) \right) = \tag{1.105}$$

$$\sum_{j=1}^{N-1} \frac{f^{(j)}(x)}{j!} A_n^*\left((\cdot - x)^j\right) + u_n(x),$$

where

$$u_n(x) := \frac{1}{\Gamma(\alpha)} \left\{ \sum_{k=\lceil na \rceil}^{\lfloor nx \rfloor} \Phi(nx-k) \int_{\frac{k}{n}}^{x} \left(J - \frac{k}{n}\right)^{\alpha-1} \left(D_{x-}^{\alpha} f(J) - D_{x-}^{\alpha} f(x)\right) dJ \right.$$

$$\left. + \sum_{k=\lfloor nx \rfloor+1}^{\lfloor nb \rfloor} \Phi(nx-k) \int_{x}^{\frac{k}{n}} \left(\frac{k}{n} - J\right)^{\alpha-1} \left(D_{*x}^{\alpha} f(J) - D_{*x}^{\alpha} f(x)\right) dJ \right\}. \tag{1.106}$$

We set

$$u_{1n}(x) := \frac{1}{\Gamma(\alpha)} \sum_{k=\lceil na \rceil}^{\lfloor nx \rfloor} \Phi(nx-k) \int_{\frac{k}{n}}^{x} \left(J - \frac{k}{n}\right)^{\alpha-1} \left(D_{x-}^{\alpha} f(J) - D_{x-}^{\alpha} f(x)\right) dJ, \tag{1.107}$$

and

$$u_{2n} := \frac{1}{\Gamma(\alpha)} \sum_{k=\lceil nx \rceil+1}^{\lfloor nb \rfloor} \Phi(nx-k) \int_x^{\frac{k}{n}} \left(\frac{k}{n}-J\right)^{\alpha-1} \left(D_{*x}^{\alpha} f(J) - D_{*x}^{\alpha} f(x)\right) dJ, \tag{1.108}$$

i.e.

$$u_n(x) = u_{1n}(x) + u_{2n}(x). \tag{1.109}$$

We assume $b-a > \frac{1}{n^\beta}$, $0<\beta<1$, which is always the case for large enough $n \in \mathbb{N}$, that is when $n > \left\lceil (b-a)^{-\frac{1}{\beta}} \right\rceil$. It is always true that either $\left|\frac{k}{n}-x\right| \le \frac{1}{n^\beta}$ or $\left|\frac{k}{n}-x\right| > \frac{1}{n^\beta}$.

For $k = \lceil na \rceil, \ldots, \lfloor nx \rfloor$, we consider

$$\gamma_{1k} := \left\| \int_{\frac{k}{n}}^{x} \left(J-\frac{k}{n}\right)^{\alpha-1} \left(D_{x-}^{\alpha} f(J) - D_{x-}^{\alpha} f(x)\right) dJ \right\| = \tag{1.110}$$

$$\left\| \int_{\frac{k}{n}}^{x} \left(J-\frac{k}{n}\right)^{\alpha-1} D_{x-}^{\alpha} f(J)\, dJ \right\| \le \int_{\frac{k}{n}}^{x} \left(J-\frac{k}{n}\right)^{\alpha-1} \left\| D_{x-}^{\alpha} f(J) \right\| dJ \le$$

$$\left\| D_{x-}^{\alpha} f(J) \right\|_{\infty,[a,x]} \frac{\left(x-\frac{k}{n}\right)^{\alpha}}{\alpha} \le \left\| D_{x-}^{\alpha} f \right\|_{\infty,[a,x]} \frac{(x-a)^{\alpha}}{\alpha}. \tag{1.111}$$

That is

$$\gamma_{1k} \le \left\| D_{x-}^{\alpha} f \right\|_{\infty,[a,x]} \frac{(x-a)^{\alpha}}{\alpha}, \tag{1.112}$$

for $k = \lceil na \rceil, \ldots, \lfloor nx \rfloor$.

Also we have in case of $\left|\frac{k}{n}-x\right| \le \frac{1}{n^\beta}$ that

$$\gamma_{1k} \le \int_{\frac{k}{n}}^{x} \left(J-\frac{k}{n}\right)^{\alpha-1} \left\| D_{x-}^{\alpha} f(J) - D_{x-}^{\alpha} f(x) \right\| dJ \le$$

$$\int_{\frac{k}{n}}^{x} \left(J-\frac{k}{n}\right)^{\alpha-1} \omega_1\left(D_{x-}^{\alpha} f, |J-x|\right)_{[a,x]} dJ \le \tag{1.113}$$

$$\omega_1\left(D_{x-}^{\alpha} f, \left|x-\frac{k}{n}\right|\right)_{[a,x]} \int_{\frac{k}{n}}^{x} \left(J-\frac{k}{n}\right)^{\alpha-1} dJ \le$$

$$\omega_1\left(D_{x-}^{\alpha} f, \frac{1}{n^\beta}\right)_{[a,x]} \frac{\left(x-\frac{k}{n}\right)^{\alpha}}{\alpha} \le \omega_1\left(D_{x-}^{\alpha} f, \frac{1}{n^\beta}\right)_{[a,x]} \frac{1}{\alpha n^{\alpha\beta}}.$$

That is when $\left|\frac{k}{n}-x\right| \le \frac{1}{n^\beta}$, then

$$\gamma_{1k} \leq \frac{\omega_1\left(D_{x-}^{\alpha} f, \frac{1}{n^{\beta}}\right)_{[a,x]}}{\alpha n^{\alpha\beta}}. \tag{1.114}$$

Consequently we obtain

$$\|u_{1n}(x)\| \leq \frac{1}{\Gamma(\alpha)} \sum_{k=\lceil na \rceil}^{\lfloor nx \rfloor} \Phi(nx-k)\, \gamma_{1k} = \tag{1.115}$$

$$\frac{1}{\Gamma(\alpha)} \left\{ \sum_{\begin{cases} k = \lceil na \rceil \\ : \left|\frac{k}{n} - x\right| \leq \frac{1}{n^{\beta}} \end{cases}}^{\lfloor nx \rfloor} \Phi(nx-k)\, \gamma_{1k} + \sum_{\begin{cases} k = \lceil na \rceil \\ : \left|\frac{k}{n} - x\right| > \frac{1}{n^{\beta}} \end{cases}}^{\lfloor nx \rfloor} \Phi(nx-k)\, \gamma_{1k} \right\} \leq$$

$$\frac{1}{\Gamma(\alpha)} \left\{ \left(\sum_{\begin{cases} k = \lceil na \rceil \\ : \left|\frac{k}{n} - x\right| \leq \frac{1}{n^{\beta}} \end{cases}}^{\lfloor nx \rfloor} \Phi(nx-k) \right) \frac{\omega_1\left(D_{x-}^{\alpha} f, \frac{1}{n^{\beta}}\right)_{[a,x]}}{\alpha n^{\alpha\beta}} + \right.$$

$$\left. \left(\sum_{\begin{cases} k = \lceil na \rceil \\ : \left|\frac{k}{n} - x\right| > \frac{1}{n^{\beta}} \end{cases}}^{\lfloor nx \rfloor} \Phi(nx-k) \right) \left\| D_{x-}^{\alpha} f \right\|_{\infty,[a,x]} \frac{(x-a)^{\alpha}}{\alpha} \right\} \leq \tag{1.116}$$

$$\frac{1}{\Gamma(\alpha+1)} \left\{ \frac{\omega_1\left(D_{x-}^{\alpha} f, \frac{1}{n^{\beta}}\right)_{[a,x]}}{n^{\alpha\beta}} + \right.$$

$$\left. \left(\sum_{\begin{cases} k = -\infty \\ : |nx-k| > n^{1-\beta} \end{cases}}^{\infty} \Phi(nx-k) \right) \left\| D_{x-}^{\alpha} f \right\|_{\infty,[a,x]} (x-a)^{\alpha} \right\} \leq$$

$$\frac{1}{\Gamma(\alpha+1)} \left\{ \frac{\omega_1\left(D_{x-}^{\alpha} f, \frac{1}{n^{\beta}}\right)_{[a,x]}}{n^{\alpha\beta}} + \frac{\left\| D_{x-}^{\alpha} f \right\|_{\infty,[a,x]} (x-a)^{\alpha}}{4m\left(n^{1-\beta}-2\right)^{2m}} \right\}.$$

So we have proved that

$$\|u_{1n}(x)\| \leq \frac{1}{\Gamma(\alpha+1)} \left\{ \frac{\omega_1\left(D_{x-}^{\alpha}f, \frac{1}{n^{\beta}}\right)_{[a,x]}}{n^{\alpha\beta}} + \frac{\left\|D_{x-}^{\alpha}f\right\|_{\infty,[a,x]}(x-a)^{\alpha}}{4m\left(n^{1-\beta}-2\right)^{2m}} \right\}. \tag{1.117}$$

Next when $k = \lfloor nx \rfloor + 1, \ldots, \lfloor nb \rfloor$ we consider

$$\gamma_{2k} := \left\| \int_x^{\frac{k}{n}} \left(\frac{k}{n} - J\right)^{\alpha-1} \left(D_{*x}^{\alpha}f(J) - D_{*x}^{\alpha}f(x)\right) dJ \right\| \leq \tag{1.118}$$

$$\int_x^{\frac{k}{n}} \left(\frac{k}{n} - J\right)^{\alpha-1} \left\|D_{*x}^{\alpha}f(J) - D_{*x}^{\alpha}f(x)\right\| dJ =$$

$$\int_x^{\frac{k}{n}} \left(\frac{k}{n} - J\right)^{\alpha-1} \left\|D_{*x}^{\alpha}f(J)\right\| dJ \leq$$

$$\left\|D_{*x}^{\alpha}f\right\|_{\infty,[x,b]} \frac{\left(\frac{k}{n}-x\right)^{\alpha}}{\alpha} \leq \left\|D_{*x}^{\alpha}f\right\|_{\infty,[x,b]} \frac{(b-x)^{\alpha}}{\alpha}. \tag{1.119}$$

Therefore when $k = \lfloor nx \rfloor + 1, \ldots, \lfloor nb \rfloor$ we get that

That is

$$\gamma_{2k} \leq \left\|D_{*x}^{\alpha}f\right\|_{\infty,[x,b]} \frac{(b-x)^{\alpha}}{\alpha}. \tag{1.120}$$

In case of $\left|\frac{k}{n} - x\right| \leq \frac{1}{n^{\beta}}$ we have

$$\gamma_{2k} \leq \int_x^{\frac{k}{n}} \left(\frac{k}{n} - J\right)^{\alpha-1} \omega_1\left(D_{*x}^{\alpha}f, |J-x|\right)_{[x,b]} dJ \leq$$

$$\omega_1\left(D_{*x}^{\alpha}f, \left|\frac{k}{n}-x\right|\right)_{[x,b]} \int_x^{\frac{k}{n}} \left(\frac{k}{n} - J\right)^{\alpha-1} dJ \leq \tag{1.121}$$

$$\omega_1\left(D_{*x}^{\alpha}f, \frac{1}{n^{\beta}}\right)_{[x,b]} \frac{\left(\frac{k}{n}-x\right)^{\alpha}}{\alpha} \leq \omega_1\left(D_{*x}^{\alpha}f, \frac{1}{n^{\beta}}\right)_{[x,b]} \frac{1}{\alpha n^{\alpha\beta}}.$$

So when $\left|\frac{k}{n} - x\right| \leq \frac{1}{n^{\beta}}$ we derived that

$$\gamma_{2k} \leq \frac{\omega_1\left(D_{*x}^{\alpha}f, \frac{1}{n^{\beta}}\right)_{[x,b]}}{\alpha n^{\alpha\beta}}. \tag{1.122}$$

Similarly we have that

$$\|u_{2n}(x)\| \leq \frac{1}{\Gamma(\alpha)}\left(\sum_{k=\lfloor nx\rfloor+1}^{\lfloor nb\rfloor}\Phi(nx-k)\gamma_{2k}\right) =$$

$$\frac{1}{\Gamma(\alpha)}\left\{\sum_{\substack{k=\lfloor nx\rfloor+1\\ :\left|\frac{k}{n}-x\right|\leq\frac{1}{n^\beta}}}^{\lfloor nb\rfloor}\Phi(nx-k)\gamma_{2k}+\sum_{\substack{k=\lfloor nx\rfloor+1\\ :\left|\frac{k}{n}-x\right|>\frac{1}{n^\beta}}}^{\lfloor nb\rfloor}\Phi(nx-k)\gamma_{2k}\right\} \leq \tag{1.123}$$

$$\frac{1}{\Gamma(\alpha)}\left\{\left(\sum_{\substack{k=\lfloor nx\rfloor+1\\ :\left|\frac{k}{n}-x\right|\leq\frac{1}{n^\beta}}}^{\lfloor nb\rfloor}\Phi(nx-k)\right)\frac{\omega_1\left(D_{*x}^{\alpha}f,\frac{1}{n^\beta}\right)_{[x,b]}}{\alpha n^{\alpha\beta}}+\right.$$

$$\left.\left(\sum_{\substack{k=\lfloor nx\rfloor+1\\ :\left|\frac{k}{n}-x\right|>\frac{1}{n^\beta}}}^{\lfloor nb\rfloor}\Phi(nx-k)\right)\left\|D_{*x}^{\alpha}f\right\|_{\infty,[x,b]}\frac{(b-x)^\alpha}{\alpha}\right\} \leq$$

$$\frac{1}{\Gamma(\alpha+1)}\left\{\frac{\omega_1\left(D_{*x}^{\alpha}f,\frac{1}{n^\beta}\right)_{[x,b]}}{n^{\alpha\beta}}+\right.$$

$$\left.\left(\sum_{\substack{k=-\infty\\ :\left|\frac{k}{n}-x\right|>\frac{1}{n^\beta}}}^{\infty}\Phi(nx-k)\right)\left\|D_{*x}^{\alpha}f\right\|_{\infty,[x,b]}(b-x)^\alpha\right\} \leq \tag{1.124}$$

$$\frac{1}{\Gamma(\alpha+1)}\left\{\frac{\omega_1\left(D_{*x}^{\alpha}f,\frac{1}{n^\beta}\right)_{[x,b]}}{n^{\alpha\beta}}+\frac{\left\|D_{*x}^{\alpha}f\right\|_{\infty,[x,b]}(b-x)^\alpha}{4m\left(n^{1-\beta}-2\right)^{2m}}\right\}.$$

So we have proved that

$$\|u_{2n}(x)\| \leq \frac{1}{\Gamma(\alpha+1)} \left\{ \frac{\omega_1 \left(D_{*x}^{\alpha} f, \frac{1}{n^{\beta}}\right)_{[x,b]}}{n^{\alpha\beta}} + \frac{\left\|D_{*x}^{\alpha} f\right\|_{\infty,[x,b]} (b-x)^{\alpha}}{4m\left(n^{1-\beta}-2\right)^{2m}} \right\}. \tag{1.125}$$

Therefore

$$\|u_n(x)\| \leq \|u_{1n}(x)\| + \|u_{2n}(x)\| \leq$$

$$\frac{1}{\Gamma(\alpha+1)} \left\{ \frac{\omega_1 \left(D_{x-}^{\alpha} f, \frac{1}{n^{\beta}}\right)_{[a,x]} + \omega_1 \left(D_{*x}^{\alpha} f, \frac{1}{n^{\beta}}\right)_{[x,b]}}{n^{\alpha\beta}} + \right. \tag{1.126}$$

$$\left. \frac{1}{4m\left(n^{1-\beta}-2\right)^{2m}} \left(\left\|D_{x-}^{\alpha} f\right\|_{\infty,[a,x]} (x-a)^{\alpha} + \left\|D_{*x}^{\alpha} f\right\|_{\infty,[x,b]} (b-x)^{\alpha} \right) \right\}.$$

From the proof of Theorem 1.14 we get that

$$\left|A_n^*\left((\cdot - x)^j\right)(x)\right| \leq \frac{1}{n^{\beta j}} + \frac{(b-a)^j}{4m\left(n^{1-\beta}-2\right)^{2m}}, \tag{1.127}$$

for $j = 1, ..., N-1$, $\forall\, x \in [a, b]$.

Putting things together, we have established

$$\left\| A_n^*(f, x) - f(x) \left(\sum_{k=\lceil na \rceil}^{\lfloor nb \rfloor} \Phi(nx-k) \right) \right\| \leq \sum_{j=1}^{N-1} \frac{\left\|f^{(j)}(x)\right\|}{j!} \tag{1.128}$$

$$\left[\frac{1}{n^{\beta j}} + \frac{(b-a)^j}{4m\left(n^{1-\alpha}-2\right)^{2m}} \right] +$$

$$\frac{1}{\Gamma(\alpha+1)} \left\{ \frac{\omega_1 \left(D_{x-}^{\alpha} f, \frac{1}{n^{\beta}}\right)_{[a,x]} + \omega_1 \left(D_{*x}^{\alpha} f, \frac{1}{n^{\beta}}\right)_{[x,b]}}{n^{\alpha\beta}} + \right.$$

$$\left. \frac{1}{4m\left(n^{1-\beta}-2\right)^{2m}} \left(\left\|D_{x-}^{\alpha} f\right\|_{\infty,[a,x]} (x-a)^{\alpha} + \left\|D_{*x}^{\alpha} f\right\|_{\infty,[x,b]} (b-x)^{\alpha} \right) \right\} =: K_n(x). \tag{1.129}$$

As a result we derive (see (1.41))

$$\|A_n(f, x) - f(x)\| \leq 2\left(\sqrt[2m]{1+4^m}\right) K_n(x), \ \forall\, x \in [a, b]. \tag{1.130}$$

We further have that

$$\|K_n\|_\infty \leq \sum_{j=1}^{N-1} \frac{\|f^{(j)}\|_\infty}{j!} \left[\frac{1}{n^{\beta j}} + (b-a)^j \frac{1}{4m\left(n^{1-\alpha}-2\right)^{2m}} \right] + \tag{1.131}$$

$$\frac{1}{\Gamma(\alpha+1)} \left\{ \frac{\left\{ \sup_{x\in[a,b]} \left(\omega_1\left(D_{x-}^{\alpha} f, \frac{1}{n^\beta}\right)_{[a,x]} \right) + \sup_{x\in[a,b]} \left(\omega_1\left(D_{*x}^{\alpha} f, \frac{1}{n^\beta}\right)_{[x,b]} \right) \right\}}{n^{\alpha\beta}} + \right.$$

$$(b-a)^\alpha \frac{1}{4m\left(n^{1-\beta}-2\right)^{2m}}$$

$$\left.\left\{ \left(\sup_{x\in[a,b]} \left(\left\| D_{x-}^{\alpha} f \right\|_{\infty,[a,x]} \right) + \sup_{x\in[a,b]} \left(\left\| D_{*x}^{\alpha} f \right\|_{\infty,[x,b]} \right) \right) \right\} \right\} =: E_n.$$

Hence it holds

$$\|A_n f - f\|_\infty \leq 2\left(\sqrt[2m]{1+4^m}\right) E_n. \tag{1.132}$$

We observe the following:

We have

$$\left(D_{x-}^{\alpha} f\right)(y) = \frac{(-1)^N}{\Gamma(N-\alpha)} \int_y^x (J-y)^{N-\alpha-1} f^{(N)}(J)\, dJ, \ \forall\, y \in [a,x] \tag{1.133}$$

and

$$\left\| \left(D_{x-}^{\alpha} f\right)(y) \right\| \leq \frac{1}{\Gamma(N-\alpha)} \left(\int_y^x (J-y)^{N-\alpha-1} dJ \right) \left\| f^{(N)} \right\|_\infty =$$

$$\frac{1}{\Gamma(N-\alpha)} \frac{(x-y)^{N-\alpha}}{(N-\alpha)} \left\| f^{(N)} \right\|_\infty = \frac{(x-y)^{N-\alpha}}{\Gamma(N-\alpha+1)} \left\| f^{(N)} \right\|_\infty$$

$$\leq \frac{(b-a)^{N-\alpha}}{\Gamma(N-\alpha+1)} \left\| f^{(N)} \right\|_\infty. \tag{1.134}$$

That is

$$\left\| D_{x-}^{\alpha} f \right\|_{\infty,[a,x]} \leq \frac{(b-a)^{N-\alpha}}{\Gamma(N-\alpha+1)} \left\| f^{(N)} \right\|_\infty, \tag{1.135}$$

and

$$\sup_{x\in[a,b]} \left\| D_{x-}^{\alpha} f \right\|_{\infty,[a,x]} \leq \frac{(b-a)^{N-\alpha}}{\Gamma(N-\alpha+1)} \left\| f^{(N)} \right\|_{\infty}. \tag{1.136}$$

Similarly we have

$$\left(D_{*x}^{\alpha} f\right)(y) = \frac{1}{\Gamma(N-\alpha)} \int_{x}^{y} (y-t)^{N-\alpha-1} f^{(N)}(t)\, dt, \ \forall\, y \in [x,b]. \tag{1.137}$$

Thus we get

$$\left\| \left(D_{*x}^{\alpha} f\right)(y) \right\| \leq \frac{1}{\Gamma(N-\alpha)} \left(\int_{x}^{y} (y-t)^{N-\alpha-1} dt \right) \left\| f^{(N)} \right\|_{\infty} \leq \tag{1.138}$$

$$\frac{1}{\Gamma(N-\alpha)} \frac{(y-x)^{N-\alpha}}{(N-\alpha)} \left\| f^{(N)} \right\|_{\infty} \leq \frac{(b-a)^{N-\alpha}}{\Gamma(N-\alpha+1)} \left\| f^{(N)} \right\|_{\infty}.$$

Hence

$$\left\| D_{*x}^{\alpha} f \right\|_{\infty,[x,b]} \leq \frac{(b-a)^{N-\alpha}}{\Gamma(N-\alpha+1)} \left\| f^{(N)} \right\|_{\infty}, \tag{1.139}$$

and

$$\sup_{x\in[a,b]} \left\| D_{*x}^{\alpha} f \right\|_{\infty,[x,b]} \leq \frac{(b-a)^{N-\alpha}}{\Gamma(N-\alpha+1)} \left\| f^{(N)} \right\|_{\infty}. \tag{1.140}$$

From (1.92) and (1.93) we get

$$\sup_{x\in[a,b]} \omega_1 \left(D_{x-}^{\alpha} f, \frac{1}{n^{\beta}} \right)_{[a,x]} \leq \frac{2 \left\| f^{(N)} \right\|_{\infty}}{\Gamma(N-\alpha+1)} (b-a)^{N-\alpha}, \tag{1.141}$$

and

$$\sup_{x\in[a,b]} \omega_1 \left(D_{*x}^{\alpha} f, \frac{1}{n^{\beta}} \right)_{[x,b]} \leq \frac{2 \left\| f^{(N)} \right\|_{\infty}}{\Gamma(N-\alpha+1)} (b-a)^{N-\alpha}. \tag{1.142}$$

That is $E_n < \infty$.

We finally notice that

$$A_n(f,x) - \sum_{j=1}^{N-1} \frac{f^{(j)}(x)}{j!} A_n\left((\cdot - x)^j\right)(x) - f(x) =$$

$$\frac{A_n^*(f,x)}{\left(\sum_{k=\lceil na \rceil}^{\lfloor nb \rfloor} \Phi(nx-k)\right)} - \frac{1}{\left(\sum_{k=\lceil na \rceil}^{\lfloor nb \rfloor} \Phi(nx-k)\right)}.$$

$$\left(\sum_{j=1}^{N-1}\frac{f^{(j)}(x)}{j!}A_n^*\left((\cdot-x)^j\right)(x)\right)-f(x)=$$

$$\frac{1}{\left(\sum_{k=\lceil na\rceil}^{\lfloor nb\rfloor}\Phi(nx-k)\right)}\left[A_n^*(f,x)-\left(\sum_{j=1}^{N-1}\frac{f^{(j)}(x)}{j!}A_n^*\left((\cdot-x)^j\right)(x)\right)\right. \tag{1.143}$$

$$\left.-\left(\sum_{k=\lceil na\rceil}^{\lfloor nb\rfloor}\Phi(nx-k)\right)f(x)\right].$$

Therefore we get

$$\left\|A_n(f,x)-\sum_{j=1}^{N-1}\frac{f^{(j)}(x)}{j!}A_n\left((\cdot-x)^j\right)(x)-f(x)\right\|\leq 2\left(\sqrt[2m]{1+4^m}\right)\cdot$$

$$\left\|A_n^*(f,x)-\left(\sum_{j=1}^{N-1}\frac{f^{(j)}(x)}{j!}A_n^*\left((\cdot-x)^j\right)(x)\right)-\left(\sum_{k=\lceil na\rceil}^{\lfloor nb\rfloor}\Phi(nx-k)\right)f(x)\right\|, \tag{1.144}$$

$\forall\, x\in[a,b]$.

The proof of the theorem is now completed. ∎

Next we apply Theorem 1.30 for $N=1$.

Theorem 1.31 *Let* $0<\alpha,\beta<1$, $f\in C^1([a,b],X)$, $x\in[a,b]$, $n\in\mathbb{N}:n^{1-\beta}>2$, $m\in\mathbb{N}$. *Then*

(i)

$$\|A_n(f,x)-f(x)\|\leq$$

$$\frac{2\left(\sqrt[2m]{1+4^m}\right)}{\Gamma(\alpha+1)}\left\{\frac{\left(\omega_1\left(D_{x-}^{\alpha}f,\frac{1}{n^{\beta}}\right)_{[a,x]}+\omega_1\left(D_{*x}^{\alpha}f,\frac{1}{n^{\beta}}\right)_{[x,b]}\right)}{n^{\alpha\beta}}+\right.$$

$$\left.\frac{1}{4m\left(n^{1-\beta}-2\right)^{2m}}\left(\left\|D_{x-}^{\alpha}f\right\|_{\infty,[a,x]}(x-a)^{\alpha}+\left\|D_{*x}^{\alpha}f\right\|_{\infty,[x,b]}(b-x)^{\alpha}\right)\right\}, \tag{1.145}$$

and

(ii)

$$\|A_nf-f\|_{\infty}\leq\frac{2\left(\sqrt[2m]{1+4^m}\right)}{\Gamma(\alpha+1)}\cdot$$

$$\left\{ \frac{\left(\sup_{x\in[a,b]} \omega_1 \left(D_{x-}^{\alpha} f, \frac{1}{n^{\beta}}\right)_{[a,x]} + \sup_{x\in[a,b]} \omega_1 \left(D_{*x}^{\alpha} f, \frac{1}{n^{\beta}}\right)_{[x,b]} \right)}{n^{\alpha\beta}} + \right.$$

$$\left. \frac{(b-a)^{\alpha}}{4m\left(n^{1-\beta}-2\right)^{2m}} \left(\sup_{x\in[a,b]} \left\| D_{x-}^{\alpha} f \right\|_{\infty,[a,x]} + \sup_{x\in[a,b]} \left\| D_{*x}^{\alpha} f \right\|_{\infty,[x,b]} \right) \right\}. \quad (1.146)$$

When $\alpha = \frac{1}{2}$ we derive

Corollary 1.32 *Let* $0 < \beta < 1$, $f \in C^1([a,b], X)$, $x \in [a,b]$, $n \in \mathbb{N} : n^{1-\beta} > 2$, $m \in \mathbb{N}$. *Then*

(i)

$$\|A_n(f,x) - f(x)\| \leq$$

$$\frac{4\left(\sqrt[2m]{1+4^m}\right)}{\sqrt{\pi}} \left\{ \frac{\left(\omega_1 \left(D_{x-}^{\frac{1}{2}} f, \frac{1}{n^{\beta}}\right)_{[a,x]} + \omega_1 \left(D_{*x}^{\frac{1}{2}} f, \frac{1}{n^{\beta}}\right)_{[x,b]} \right)}{n^{\frac{\beta}{2}}} + \right.$$

$$\left. \frac{1}{4m\left(n^{1-\beta}-2\right)^{2m}} \left(\left\| D_{x-}^{\frac{1}{2}} f \right\|_{\infty,[a,x]} \sqrt{(x-a)} + \left\| D_{*x}^{\frac{1}{2}} f \right\|_{\infty,[x,b]} \sqrt{(b-x)} \right) \right\}, \quad (1.147)$$

and

(ii)

$$\|A_n f - f\|_{\infty} \leq \frac{4\left(\sqrt[2m]{1+4^m}\right)}{\sqrt{\pi}} \cdot$$

$$\left\{ \frac{\left(\sup_{x\in[a,b]} \omega_1 \left(D_{x-}^{\frac{1}{2}} f, \frac{1}{n^{\beta}}\right)_{[a,x]} + \sup_{x\in[a,b]} \omega_1 \left(D_{*x}^{\frac{1}{2}} f, \frac{1}{n^{\beta}}\right)_{[x,b]} \right)}{n^{\frac{\beta}{2}}} + \right.$$

$$\left. \frac{\sqrt{(b-a)}}{4m\left(n^{1-\beta}-2\right)^{2m}} \left(\sup_{x\in[a,b]} \left\| D_{x-}^{\frac{1}{2}} f \right\|_{\infty,[a,x]} + \sup_{x\in[a,b]} \left\| D_{*x}^{\frac{1}{2}} f \right\|_{\infty,[x,b]} \right) \right\} < \infty. \quad (1.148)$$

Next we make

Remark 1.33 Some convergence analysis follows:

Let $0 < \beta < 1$, $f \in C^1([a,b], X)$, $x \in [a,b]$, $n \in \mathbb{N} : n^{1-\beta} > 2$, $m \in \mathbb{N}$. We elaborate on (1.148). Assume that

$$\omega_1\left(D_{x-}^{\frac{1}{2}} f, \frac{1}{n^{\beta}}\right)_{[a,x]} \leq \frac{K_1}{n^{\beta}}, \tag{1.149}$$

and

$$\omega_1\left(D_{*x}^{\frac{1}{2}} f, \frac{1}{n^{\beta}}\right)_{[x,b]} \leq \frac{K_2}{n^{\beta}}, \tag{1.150}$$

$\forall\, x \in [a,b]$, $\forall\, n \in \mathbb{N}$, where $K_1, K_2 > 0$.

Then it holds

$$\frac{\left[\sup_{x\in[a,b]} \omega_1\left(D_{x-}^{\frac{1}{2}} f, \frac{1}{n^{\beta}}\right)_{[a,x]} + \sup_{x\in[a,b]} \omega_1\left(D_{*x}^{\frac{1}{2}} f, \frac{1}{n^{\beta}}\right)_{[x,b]}\right]}{n^{\frac{\beta}{2}}} \leq$$

$$\frac{\frac{(K_1+K_2)}{n^{\beta}}}{n^{\frac{\beta}{2}}} = \frac{(K_1+K_2)}{n^{\frac{3\beta}{2}}} = \frac{K}{n^{\frac{3\beta}{2}}}, \tag{1.151}$$

where $K := K_1 + K_2 > 0$.

The other summand of the right hand side of (1.148), for large enough n, converges to zero at the speed $\frac{1}{n^{2m(1-\beta)}}$, so it is about $\frac{L}{n^{2m(1-\beta)}}$, where $L > 0$ is a constant.

Then, for large enough $n \in \mathbb{N}$, by (1.148), (1.151) and the above comment, we obtain that

$$\|A_n f - f\|_{\infty} \leq \frac{M}{\min\left(n^{\frac{3\beta}{2}}, n^{2m(1-\beta)}\right)}, \tag{1.152}$$

where $M > 0$.

Clearly we have two cases:

(i)

$$\|A_n f - f\|_{\infty} \leq \frac{M}{n^{2m(1-\beta)}}, \text{ when } \frac{4m}{3+4m} \leq \beta < 1, \tag{1.153}$$

with speed of convergence $\frac{1}{n^{2m(1-\beta)}}$,

and

(ii)

$$\|A_n f - f\|_{\infty} \leq \frac{M}{n^{\frac{3\beta}{2}}}, \text{ when } 0 < \beta \leq \frac{4m}{3+4m}, \tag{1.154}$$

with speed of convergence $\frac{1}{n^{\frac{3\beta}{2}}}$.

In Theorem 1.11, for $f \in C([a,b], X)$ and for large enough $n \in \mathbb{N}$, when $0 < \beta \leq \frac{2m}{1+2m}$, the speed is $\frac{1}{n^{\beta}}$. So when $0 < \beta \leq \frac{4m}{3+4m}$ $(< \frac{2m}{1+2m})$, we get by (1.154) that $\|A_n f - f\|_\infty$ converges much faster to zero. The last comes because we assumed differentiability of f.

Notice that in Corollary 1.32 no initial condition is assumed.

Next, we will present an alternative fractional approximation by A_n, $n \in \mathbb{N}$.

Notation 1.34 *Let $\overline{n} \in \mathbb{N}$, we denote the left iterated fractional derivative*

$$D_{*x}^{\overline{n}x} = D_{*x}^{\alpha} D_{*x}^{\alpha} ... D_{*x}^{\alpha}, \ (\overline{n} \text{ - times}), \tag{1.155}$$

$x \in [a,b]$, $0 < \alpha < 1$.

Similarly, we also denote the right iterated fractional derivative

$$D_{x-}^{\overline{n}x} = D_{x-}^{\alpha} D_{x-}^{\alpha} ... D_{x-}^{\alpha}, \ (\overline{n} \text{ - times}), \tag{1.156}$$

$x \in [a,b]$.

We need

Definition 1.35 Let $\overline{n} \in \mathbb{N}$, $D_x^{(\overline{n}+1)\alpha} f$ denote any of $D_{*x}^{(\overline{n}+1)\alpha}$, $D_{x-}^{(\overline{n}+1)\alpha}$, and $\delta > 0$. We set

$$\omega_1 \left(D_x^{(\overline{n}+1)\alpha} f, \delta \right) = \max \left\{ \omega_1 \left(D_{*x}^{(\overline{n}+1)\alpha} f, \delta \right)_{[x,b]}, \omega_1 \left(D_{x-}^{(\overline{n}+1)\alpha} f, \delta \right)_{[a,x]} \right\}, \tag{1.157}$$

where $x \in [a,b]$. Here the moduli of continuity are considered over $[x,b]$ and $[a,x]$, respectively.

We also need

Theorem 1.36 ([13], p. 123) *Let $0 < \alpha < 1$, $f : [a,b] \to \mathbb{R}$, $f' \in L_\infty([a,b])$, $x \in [a,b]$ being fixed. Assume that $D_{*x}^{k\alpha} f \in C([x,b])$, $k = 0, 1, ..., \overline{n}+1$, $\overline{n} \in \mathbb{N}$, and $\left(D_{*x}^{i\alpha} f\right)(x) = 0$, $i = 2, 3, ..., \overline{n}+1$. Also, suppose that $D_{x-}^{k\alpha} f \in C([a,x])$, for $k = 0, 1, ..., \overline{n}+1$, and $\left(D_{x-}^{i\alpha} f\right)(x) = 0$, for $i = 2, 3, ..., \overline{n}+1$. Then*

$$|f(\cdot) - f(x)| \leq \frac{\omega_1 \left(D_x^{(\overline{n}+1)\alpha} f, \delta \right)}{\Gamma((\overline{n}+1)\alpha + 1)} \left[|\cdot - x|^{(\overline{n}+1)\alpha} + \frac{|\cdot - x|^{(\overline{n}+1)\alpha+1}}{\delta((\overline{n}+1)\alpha + 1)} \right], \ \delta > 0. \tag{1.158}$$

We present

Theorem 1.37 *Let* $f \in C([a,b])$ *and all as in Theorem 1.36,* $n \in \mathbb{N} : n^{1-\alpha} > 2$, $m \in \mathbb{N}$.

Then

$$|(A_n f)(x) - f(x)| \leq \frac{2\left(\sqrt[2m]{1+4^m}\right)\omega_1\left(D_x^{(\overline{n}+1)\alpha} f, \delta\right)}{\Gamma((\overline{n}+1)\alpha+1)}$$

$$\left\{\left[\frac{1}{n^{(\overline{n}+1)\alpha^2}} + \frac{(b-a)^{(\overline{n}+1)\alpha}}{4m\left(n^{1-\alpha}-2\right)^{2m}}\right] + \right. \tag{1.159}$$

$$\left.\frac{1}{\delta((\overline{n}+1)\alpha+1)}\left[\frac{1}{n^{\alpha[(\overline{n}+1)\alpha+1]}} + \frac{(b-a)^{(\overline{n}+1)\alpha+1}}{4m\left(n^{1-\alpha}-2\right)^{2m}}\right]\right\}, \quad \delta > 0.$$

Hence $\lim_{n \to +\infty} A_n(f)(x) = f(x)$.

Proof We notice that A_n is a positive linear operator with $A_n(1) = 1$.

Let $f \in C([a,b], \mathbb{R})$, then $|f| \leq |f|$ and $-|f| \leq f \leq |f|$.

Hence $-A_n(|f|) \leq A_n(f) \leq A_n(|f|)$ and $|A_n(f)| \leq A_n(|f|)$.

Therefore

$$|(A_n f)(x) - f(x)| = |(A_n f)(x) - A_n(f(x))(x)| =$$

$$|A_n(f - f(x))(x)| \overset{(1.158)}{\leq} A_n(|f - f(x)|)(x) \leq \tag{1.160}$$

$$\frac{\omega_1\left(D_x^{(\overline{n}+1)\alpha} f, \delta\right)}{\Gamma((\overline{n}+1)\alpha+1)}\left[A_n\left(|\cdot - x|^{(\overline{n}+1)\alpha}\right)(x) + \frac{A_n\left(|\cdot - x|^{(\overline{n}+1)\alpha+1}\right)(x)}{\delta((\overline{n}+1)\alpha+1)}\right] =$$

$$\frac{\omega_1\left(D_x^{(\overline{n}+1)\alpha} f, \delta\right)}{\Gamma((\overline{n}+1)\alpha+1)\sum_{k=\lceil na \rceil}^{\lfloor nb \rfloor}\Phi(nx-k)}$$

$$\left[\sum_{k=\lceil na \rceil}^{\lfloor nb \rfloor}\left|\frac{k}{n} - x\right|^{(\overline{n}+1)\alpha}\Phi(nx-k) + \frac{\sum_{k=\lceil na \rceil}^{\lfloor nb \rfloor}\left|\frac{k}{n} - x\right|^{(\overline{n}+1)\alpha+1}\Phi(nx-k)}{\delta((\overline{n}+1)\alpha+1)}\right] \overset{(1.26)}{\leq} \tag{1.161}$$

$$\frac{2\left(\sqrt[2m]{1+4^m}\right)\omega_1\left(D_x^{(\overline{n}+1)\alpha} f, \delta\right)}{\Gamma((\overline{n}+1)\alpha+1)}$$

$$\Bigg\{\Bigg[\sum_{\substack{k=\lceil na\rceil \\ :\left|\frac{k}{n}-x\right|\leq\frac{1}{n^{\alpha}}}}^{\lfloor nb\rfloor}\left|\frac{k}{n}-x\right|^{(\overline{n}+1)\alpha}\Phi(nx-k)+$$

$$\sum_{\substack{k=\lceil na\rceil \\ :\left|\frac{k}{n}-x\right|>\frac{1}{n^{\alpha}}}}^{\lfloor nb\rfloor}\left|\frac{k}{n}-x\right|^{(\overline{n}+1)\alpha}\Phi(nx-k)\Bigg]+$$

$$\frac{1}{\delta((\overline{n}+1)\alpha+1)}\Bigg[\sum_{\substack{k=\lceil na\rceil \\ :\left|\frac{k}{n}-x\right|\leq\frac{1}{n^{\alpha}}}}^{\lfloor nb\rfloor}\left|\frac{k}{n}-x\right|^{(\overline{n}+1)\alpha+1}\Phi(nx-k)+ \tag{1.162}$$

$$\sum_{\substack{k=\lceil na\rceil \\ :\left|\frac{k}{n}-x\right|>\frac{1}{n^{\alpha}}}}^{\lfloor nb\rfloor}\left|\frac{k}{n}-x\right|^{(\overline{n}+1)\alpha+1}\Phi(nx-k)\Bigg]\Bigg\}\overset{(1.19)}{\leq}$$

$$\frac{2\left(\sqrt[2m]{1+4^{m}}\right)\omega_{1}\left(D_{x}^{(\overline{n}+1)\alpha}f,\delta\right)}{\Gamma((\overline{n}+1)\alpha+1)}\left\{\left[\frac{1}{n^{(\overline{n}+1)\alpha^{2}}}+\frac{(b-a)^{(\overline{n}+1)\alpha}}{4m\left(n^{1-\alpha}-2\right)^{2m}}\right]\right.$$

$$\left.+\frac{1}{\delta((\overline{n}+1)\alpha+1)}\left[\frac{1}{n^{\alpha[(\overline{n}+1)\alpha+1]}}+\frac{(b-a)^{(\overline{n}+1)\alpha+1}}{4m\left(n^{1-\alpha}-2\right)^{2m}}\right]\right\},\ \delta>0, \tag{1.163}$$

proving the claim. ■

We finish with

Corollary 1.38 *All as in Theorem 1.37, with* $\delta=\frac{1}{(\overline{n}+1)\alpha+1}$. *Then*

$$|(A_{n}f)(x)-f(x)|\leq\frac{2\left(\sqrt[2m]{1+4^{m}}\right)\omega_{1}\left(D_{x}^{(\overline{n}+1)\alpha}f,\frac{1}{(\overline{n}+1)\alpha+1}\right)}{\Gamma((\overline{n}+1)\alpha+1)}$$

$$\left\{\left[\frac{1}{n^{(\overline{n}+1)\alpha^2}}+\frac{(b-a)^{(\overline{n}+1)\alpha}}{4m\left(n^{1-\alpha}-2\right)^{2m}}\right]+\left[\frac{1}{n^{\alpha[(\overline{n}+1)\alpha+1]}}+\frac{(b-a)^{(\overline{n}+1)\alpha+1}}{4m\left(n^{1-\alpha}-2\right)^{2m}}\right]\right\}. \tag{1.164}$$

Hence $\lim_{n\to+\infty} A_n(f)(x) = f(x)$.

References

1. G.A. Anastassiou, Rate of convergence of some neural network operators to the unit-univariate case. J. Math. Anal. Appl. **212**, 237–262 (1997)
2. G.A. Anastassiou, *Quantitative Approximations* (Chapman & Hall / CRC, Boca Raton, New York, 2001)
3. G.A. Anastassiou, Univariate hyperbolic tangent neural network approximation. Math. Comput. Model. **53**, 1111–1132 (2011)
4. G.A. Anastassiou, Multivariate hyperbolic tangent neural network approximation. Comput. Math. **61**, 809–821 (2011)
5. G.A. Anastassiou, Multivariate sigmoidal neural network approximation. Neural Netw. **24**, 378–386 (2011)
6. G.A. Anastassiou, *Inteligent Systems: Approximation by Artificial Neural Networks, Intelligent Systems Reference Library*, vol. 19 (Springer, Heidelberg, 2011)
7. G.A. Anastassiou, Univariate sigmoidal neural network approximation. J. Comput. Anal. Appl. **14**(4), 659–690 (2012)
8. G.A. Anastassiou, Fractional neural network approximation. Comput. Math. Appl. **64**, 1655–1676 (2012)
9. G.A. Anastassiou, *Intelligent Systems II: Complete Approximation by Neural Network Operators* (Springer, Heidelberg, New York, 2016)
10. G.A. Anastassiou, Strong right fractional calculus for Banach space valued functions. Revista Proyecciones **36**(1), 149–186 (2017)
11. G.A. Anastassiou, Vector fractional Korovkin type Approximations. Dynam. Syst. Appl. **26**, 81–104 (2017)
12. G.A. Anastassiou, A strong fractional calculus theory for Banach space valued functions. Nonlinear Funct. Anal. Appl. (Korea) **22**(3), 495–524 (2017)
13. G.A. Anastassiou, *Nonlinearity: Ordinary and Fractional Approximations by Sublinear and Max-Product Operators* (Springer, Heidelberg, New York, 2018)
14. G.A. Anastassiou, Algebraic function based Banach space valued ordinary and fractional neural network approximations. New Trends Math. Sci., NTMSCI 10 Special Issue; (1), 100–125 (2022)
15. Z. Chen, F. Cao, The approximation operators with sigmoidal functions. Comput. Math. Appl. **58**, 758–765 (2009)
16. S. Haykin, *Neural Networks: A Comprehensive Foundation*, 2nd edn. (Prentice Hall, New York, 1998)
17. M. Kreuter, Sobolev Spaces of Vector-valued functions. Ulm University, Master Thesis in Mathematics, Ulm, Germany (2015)
18. W. McCulloch, W. Pitts, A logical calculus of the ideas immanent in nervous activity. Bull. Math. Biophys. **7**, 115–133 (1943)
19. J. Mikusinski, *The Bochner Integral* (Academic Press, New York, 1978)
20. T.M. Mitchell, *Machine Learning* (WCB-McGraw-Hill, New York, 1997)
21. G.E. Shilov, *Elementary Functional Analysis* (Dover Publications Inc, New York, 1996)

Chapter 2
Gudermannian Function Induced Banach Space Valued Ordinary and Fractional Neural Network Approximations

Here we investigate the univariate quantitative approximation, ordinary and fractional, of Banach space valued continuous functions on a compact interval or all the real line by quasi-interpolation Banach space valued neural network operators. These approximations are derived by establishing Jackson type inequalities involving the modulus of continuity of the engaged function or its Banach space valued high order derivative or fractional derivatives. Our operators are defined by using a density function generated by the Gudermannian sigmoid function. The approximations are pointwise and of the uniform norm. The related Banach space valued feed-forward neural networks are with one hidden layer. It relies on [14].

2.1 Introduction

The author in [1, 2], see in this chapter and Chaps. 3–5, was the first to establish neural network approximation to continuous functions with rated by very specifically defined neural network operators of Cardaliagnet-Euvrard and "Squashing" types, by employing the modulus of continuity of the engaged function or its high order derivative, and producing very tight Jackson type inequalities. He treats there both the univariate and multivariate cases. The defining these operators "bell-shaped" and "squashing" functions are assumed to be of compact support. Also in [2] he gives the Nth order asymptotic expansion for the error of weak approximation of these two operators to a special natural class of smooth functions, see Chaps. 4 and 5 there.

© The Author(s), under exclusive license to Springer Nature Switzerland AG 2023

G. A. Anastassiou, *Banach Space Valued Neural Network*,
Studies in Computational Intelligence 1062,
https://doi.org/10.1007/978-3-031-16400-2_2

The author inspired by [15], continued his studies on neural networks approximation by introducing and using the proper quasi-interpolation operators of sigmoidal and hyperbolic tangent type which resulted into [3–7], by treating both the univariate and multivariate cases. He did also the corresponding fractional cases [8, 9, 13].

The author here performs Gudermannian sigmoidal activated neural network approximations to continuous functions over compact intervals of the real line or over the whole $\mathbb{R}$ with valued to an arbitrary Banach space $(X, \|\cdot\|)$. Finally he treats completely the related X-valued fractional approximation. All convergences here are with rates expressed via the modulus of continuity of the involved function or its X-valued high order derivative, or X-valued fractional derivatives and given by very tight Jackson type inequalities. Iterated fractional approximation is also included.

Our compact intervals are not necessarily symmetric to the origin. Some of our upper bounds to error quantity are very flexible and general. In preparation to prove our results we establish important properties of the basic density function defining our operators which is induced by the Gudermannian sigmoid function.

Feed-forward X-valued neural networks (FNNs) with one hidden layer, the only type of networks we deal with in this chapter, are mathematically expressed as

$$N_n(x) = \sum_{=0}^{n} c_j \sigma\left(\langle a_j \cdot x \rangle + b_j\right), \quad x \in \mathbb{R}^s, \, s \in \mathbb{N},$$

where for $0 \leq j \leq n$, $b_j \in \mathbb{R}$ are the thresholds, $a_j \in \mathbb{R}^s$ are the connection weights, $c_j \in X$ are the coefficients, $\langle a_j \cdot x \rangle$ is the inner product of a_j and x, and σ is the activation function of the network. About neural networks in general read [16, 18, 20]. See also [9] for a complete study of real valued approximation by neural network operators.

2.2 Basics

Here we consider $gd(x)$ the Gudermannian function [22], which is a sigmoid function, as a generator function:

$$\sigma(x) = 2\arctan\left(\tanh\left(\frac{x}{2}\right)\right) = \int_0^x \frac{dt}{\cosh t} =: gd(x), x \in \mathbb{R}, \tag{2.1}$$

Let the normalized generator sigmoid function

$$f(x) := \frac{4}{\pi}\sigma(x) = \frac{4}{\pi}\int_0^x \frac{dt}{\cosh t} = \frac{8}{\pi}\int_0^x \frac{1}{e^t + e^{-t}} dt, \; x \in \mathbb{R} \tag{2.2}$$

Here $f'(x) = \frac{4}{\pi \cosh x} > 0$, $\forall\, x \in \mathbb{R}$, hence f is strictly increasing on $\mathbb{R}$. Notice that $\tanh(-x) = -\tanh x$ and $\arctan(-x) = -\arctan x$, $x \in \mathbb{R}$.

So, here the neural network activation function will be:

$$\psi(x) = \frac{1}{4}(f(x+1) - f(x-1)),\ x \in \mathbb{R}. \tag{2.3}$$

We notice that

$$\psi(x) = \frac{2}{\pi}\left(\arctan\left(\tanh\left(\frac{x+1}{2}\right)\right) - \arctan\left(\tanh\left(\frac{x-1}{2}\right)\right)\right), \tag{2.4}$$

and

$$\psi(-x) = \frac{2}{\pi}\left(\arctan\left(\tanh\left(\frac{-x+1}{2}\right)\right) - \arctan\left(\tanh\left(\frac{-x-1}{2}\right)\right)\right) =$$

$$\frac{2}{\pi}\left(\arctan\left(\tanh\left(-\frac{(x-1)}{2}\right)\right) - \arctan\left(\tanh\left(-\frac{(x+1)}{2}\right)\right)\right) =$$

$$\frac{2}{\pi}\left(\arctan\left(-\tanh\frac{(x-1)}{2}\right) - \arctan\left(-\tanh\left(\frac{x+1}{2}\right)\right)\right) = \tag{2.5}$$

$$\frac{2}{\pi}\left(-\arctan\left(\tanh\left(\frac{x-1}{2}\right)\right) + \arctan\left(\tanh\left(\frac{x+1}{2}\right)\right)\right) =$$

$$\frac{2}{\pi}\left(\arctan\left(\tanh\left(\frac{x+1}{2}\right)\right) - \arctan\left(\tanh\left(\frac{x-1}{2}\right)\right)\right) = \psi(x).$$

That is

$$\psi(x) = \psi(-x),\ \forall x \in \mathbb{R}, \tag{2.6}$$

i.e. it is even and symmetric with respect to y-axis. Let $x \to +\infty$, then $\tanh\left(\frac{x}{2}\right) \to 1$ and $\arctan\left(\tanh\left(\frac{x}{2}\right)\right) \to \frac{\pi}{4}$. Let $x \to -\infty$, then $\tan\left(\frac{x}{2}\right) \to -1$ and $\arctan\left(\tanh\left(\frac{x}{2}\right)\right) \to -\frac{\pi}{4}$.

So, we deal with the normalized sigmoid function

$$f(x) = \frac{8}{\pi}\arctan\left(\tanh\left(\frac{x}{2}\right)\right) = \frac{4}{\pi}\int_0^x \frac{dt}{\cosh t}. \tag{2.7}$$

Clearly, we obtain that $f(+\infty) = 1$, and $f(-\infty) = -1$, along with $f(0) = 0$.

Notice that

$$f(-x) = \frac{8}{\pi}\arctan\left(\tanh\left(-\frac{x}{2}\right)\right) = \frac{8}{\pi}\arctan\left(-\tanh\frac{x}{2}\right) =$$

$$-\frac{8}{\pi}\arctan\left(\tanh\frac{x}{2}\right) = -f(x),$$

that is

$$f(-x) = -f(x), \ \forall x \in \mathbb{R}, \tag{2.8}$$

and odd function, symmetric with respect to the origin.

Since $x + 1 > x - 1$, and $f(x+1) > f(x-1)$, we obtain $\psi(x) > 0, \forall x \in \mathbb{R}$.

We see that

$$\psi(0) = \frac{1}{4}(f(1) - f(-1)) = \frac{1}{4}(f(1) - (-f(1))) = \frac{f(1)}{2} =$$

$$\frac{2}{\pi}\int_0^1 \frac{dt}{\cosh t} = \frac{2}{\pi} gd(1) \cong \frac{2}{\pi}(0.86577) \cong 0.551. \tag{2.9}$$

Thus, it is

$$\psi(0) \cong 0.551. \tag{2.10}$$

Furthermore we have

$$\psi'(x) = \frac{1}{4}\left(f'(x+1) - f'(x-1)\right) = \frac{1}{\pi}\left(\frac{1}{\cosh(x+1)} - \frac{1}{\cosh(x-1)}\right)$$

$$= \frac{1}{\pi}\left(\frac{\cosh(x-1) - \cosh(x+1)}{\cosh(x+1)\cosh(x-1)}\right). \tag{2.11}$$

If $x > 0$, then $x > -x$, and let $x > 1$, then $x - 1 > 0$, hence it holds $0 < x - 1 < x + 1$ and $\cosh(x-1) < \cosh(x+1)$, thus $\psi'(x) < 0$.

Let now $x < 1$, then $1 - x > 0$ and $0 < 1 - x < 1 + x$, then $\cosh(x-1) = \cosh(1-x) < \cosh(x+1)$, so that again $\psi'(x) < 0$. Consequently ψ is strictly decreasing on $(0, +\infty)$.

Clearly, ψ is strictly increasing on $(-\infty, 0)$, and $\psi'(0) = 0$.

We observe that

$$\lim_{x \to +\infty} \psi(x) = \tfrac{1}{4}(f(+\infty) - f(+\infty)) = 0,$$
$$\lim_{x \to -\infty} \psi(x) = \tfrac{1}{4}(f(-\infty) - f(-\infty)) = 0. \tag{2.12}$$

That is, the x-axis is the horizontal asymptote for ψ.

Conclusion, ψ is a bell symmetric function with maximum

$$\psi(0) \cong \frac{2}{\pi}(0.86577) \cong 0.551. \tag{2.13}$$

We need

Theorem 2.1 *We have that*

$$\sum_{i=-\infty}^{\infty} \psi(x-i) = 1, \ \forall x \in \mathbb{R}. \tag{2.14}$$

Proof We observe that

$$\sum_{i=-\infty}^{\infty} (f(x-i) - f(x-1-i)) =$$

$$\sum_{i=0}^{\infty} (f(x-i) - f(x-1-i)) + \sum_{i=-\infty}^{-1} (f(x-i) - f(x-1-i)).$$

Furthermore ($\lambda \in \mathbb{Z}^+$)

$$\sum_{i=0}^{\infty} (f(x-i) - f(x-1-i)) = \tag{2.15}$$

$$\lim_{\lambda \to \infty} \sum_{i=0}^{\lambda} (f(x-i) - f(x-1-i)) \text{ (telescoping sum)}$$

$$= \lim_{\lambda \to \infty} (f(x) - f(x-(\lambda+1))) = 1 + f(x).$$

Similarly, it holds

$$\sum_{i=-\infty}^{-1} (f(x-i) - f(x-1-i)) = \lim_{\lambda \to \infty} \sum_{i=-\lambda}^{-1} (f(x-i) - f(x-1-i))$$

$$= \lim_{\lambda \to \infty} (f(x+\lambda) - f(x)) = 1 - f(x). \tag{2.16}$$

Therefore we derive

$$\sum_{i=-\infty}^{\infty} (f(x-i) - f(x-1-i)) = 2, \ \forall x \in \mathbb{R}, \tag{2.17}$$

and

$$\sum_{i=-\infty}^{\infty} (f(x+1-i) - f(x-i)) = 2, \ \forall x \in \mathbb{R}. \tag{2.18}$$

Adding (2.17) and (2.18) we get

$$\sum_{i=-\infty}^{\infty} \left(f\left(x+1-i\right) - f\left(x-1-i\right) \right) = 4, \forall x \in \mathbb{R}. \tag{2.19}$$

Since

$$\psi\left(x\right) = \frac{1}{4}\left(f\left(x+1\right) - f\left(x-1\right) \right),$$

we have that

$$\psi\left(x-i\right) = \frac{1}{4}\left[f\left(x+1-i\right) - f\left(x-1-i\right) \right], \tag{2.20}$$

giving $\sum_{i=-\infty}^{\infty} \psi\left(x-i\right) = 1$. ■

Remark 2.2 Because ψ is even it holds

$$\sum_{i=-\infty}^{\infty} \psi\left(i-x\right) = 1, \forall x \in \mathbb{R}.$$

Hence

$$\sum_{i=-\infty}^{\infty} \psi\left(i+x\right) = 1, \forall x \in \mathbb{R},$$

and

$$\sum_{i=-\infty}^{\infty} \psi\left(x+i\right) = 1, \forall x \in \mathbb{R}. \tag{2.21}$$

Theorem 2.3 *It holds*

$$\int_{-\infty}^{\infty} \psi\left(x\right) dx = 1. \tag{2.22}$$

Proof We observe that

$$\int_{-\infty}^{\infty} \psi\left(x\right) dx = \sum_{j=-\infty}^{\infty} \int_{j}^{j+1} \psi\left(x\right) dx = \sum_{j=-\infty}^{\infty} \int_{0}^{1} \psi\left(x+j\right) dx =$$

$$\int_{0}^{1} \left(\sum_{j=-\infty}^{\infty} \psi\left(x+j\right) dx \right) = \int_{0}^{1} 1 dx = 1.$$

So $\psi\left(x\right)$ is a density function. ■

We need

Theorem 2.4 *Let $0 < \alpha < 1$, and $n \in \mathbb{N}$ with $n^{1-\alpha} > 2$. It holds*

$$\sum_{\begin{cases} k=-\infty \\ : |nx-k| \geq n^{1-\alpha} \end{cases}}^{\infty} \psi(nx-k) < \frac{4}{\pi e^{(n^{1-\alpha}-2)}} = \frac{4e^2}{\pi e^{n^{1-\alpha}}}. \tag{2.23}$$

Proof We have that

$$\psi(x) = \frac{1}{4}[f(x+1) - f(x-1)], \ \forall x \in \mathbb{R}.$$

Let $x \geq 1$. That is $0 \leq x-1 < x+1$. Applying the mean value theorem we get

$$\psi(x) = \frac{1}{4} \cdot 2 \cdot f'(\xi) = \frac{2}{4} \frac{4}{\pi \cosh \xi} = \frac{2}{\pi \cosh \xi} > 0,$$

where $0 \leq x-1 < \xi < x+1$.

We have $\cosh(x-1) < \cosh \xi$ and $\frac{1}{\cosh \xi} < \frac{1}{\cosh(x-1)}$.

Hence

$$\psi(x) < \frac{2}{\pi \cosh(x-1)}, \ \forall x \geq 1. \tag{2.24}$$

Thus, we have

$$\sum_{\begin{cases} k=-\infty \\ : |nx-k| \geq n^{1-\alpha} \end{cases}}^{\infty} \psi(nx-k) = \sum_{\begin{cases} k=-\infty \\ : |nx-k| \geq n^{1-\alpha} \end{cases}}^{\infty} \psi(|nx-k|) <$$

$$\frac{2}{\pi} \sum_{\begin{cases} k=-\infty \\ : |nx-k| \geq n^{1-\alpha} \end{cases}}^{\infty} \frac{1}{\cosh(|nx-k|-1)} \leq$$

$$\frac{2}{\pi} \int_{n^{1-\alpha}-1}^{\infty} \frac{1}{\cosh(x-1)} dx = \frac{2}{\pi} \int_{n^{1-\alpha}-2}^{\infty} \frac{1}{\cosh z} dz = \tag{2.25}$$

$$\frac{2}{\pi} \int_{n^{1-\alpha}-2}^{\infty} \frac{2}{e^x + e^{-x}} dx < \frac{4}{\pi} \int_{n^{1-\alpha}-2}^{\infty} e^{-x} dx = \frac{4}{\pi} \left\{ -e^{-x} \Big|_{n^{1-\alpha}-2}^{\infty} \right\} = \tag{2.26}$$

$$\frac{4}{\pi} \left\{ e^{-x} \Big|_{\infty}^{n^{1-\alpha}-2} \right\} = \frac{4}{\pi \left(e^{n^{1-\alpha}-2}\right)}.$$

We have found that

$$\sum_{\begin{cases} k=-\infty \\ : |nx-k| \geq n^{1-\alpha} \end{cases}}^{\infty} \psi(nx-k) < \frac{4}{\pi\left(e^{n^{1-\alpha}-2}\right)}, \tag{2.27}$$

for $n^{1-\alpha} > 2$, $n \in \mathbb{N}$. ∎

Denote by $\lfloor\cdot\rfloor$ the integral part of the number and by $\lceil\cdot\rceil$ the ceiling of the number.

Theorem 2.5 *Let* $[a, b] \subset \mathbb{R}$ *and* $n \in \mathbb{N}$ *so that* $\lceil na \rceil \leq \lfloor nb \rfloor$. *It holds*

$$\frac{1}{\sum_{k=\lceil na\rceil}^{\lfloor nb\rfloor} \psi(nx-k)} < 2.412, \tag{2.28}$$

$\forall\, x \in [a, b]$.

Proof Let $x \in [a, b]$. We see that

$$1 = \sum_{k=-\infty}^{\infty} \psi(nx-k) > \sum_{k=\lceil na\rceil}^{\lfloor nb\rfloor} \psi(nx-k) =$$

$$\sum_{k=\lceil na\rceil}^{\lfloor nb\rfloor} \psi(|nx-k|) > \psi(|nx-k_0|), \tag{2.29}$$

$\forall\, k_0 \in [\lceil na\rceil, \lfloor nb\rfloor] \cap \mathbb{Z}$.

We can choose $k_0 \in [\lceil na\rceil, \lfloor nb\rfloor] \cap \mathbb{Z}$ such that $|nx-k_0| < 1$.

Therefore we get that

$$\psi(|nx-k_0|) > \psi(1) = \frac{1}{4} f(2) = \frac{1}{4}\frac{4}{\pi}\int_0^2 \frac{dt}{\cosh t} = \tag{2.30}$$

$$\frac{1}{\pi}\int_0^2 \frac{dt}{\cosh t} = \frac{1}{\pi} gd(2) \cong \frac{1}{\pi}(1.3018),$$

and

$$\sum_{k=\lceil na\rceil}^{\lfloor nb\rfloor} \psi(|nx-k|) > \frac{1.3018}{\pi}. \tag{2.31}$$

That is

$$\frac{1}{\sum_{k=\lceil na\rceil}^{\lfloor nb\rfloor} \psi\left(|nx-k|\right)} < \frac{\pi}{1.3018} \approx 2.412, \tag{2.32}$$

proving the claim. ∎

We make

Remark 2.6 We also notice that

$$1 - \sum_{k=\lceil na\rceil}^{\lfloor nb\rfloor} \psi\left(nb-k\right) = \sum_{k=-\infty}^{\lceil na\rceil-1} \psi\left(nb-k\right) + \sum_{k=\lfloor nb\rfloor+1}^{\infty} \psi\left(nb-k\right)$$

$$> \psi\left(nb - \lfloor nb\rfloor - 1\right) \tag{2.33}$$

(call $\varepsilon := nb - \lfloor nb\rfloor$, $0 \leq \varepsilon < 1$)

$$= \psi\left(\varepsilon - 1\right) = \psi\left(1-\varepsilon\right) \geq \psi\left(1\right) > 0.$$

Therefore

$$\lim_{n\to\to\infty}\left(1 - \sum_{k=\lceil na\rceil}^{\lfloor nb\rfloor} \psi\left(nb-k\right)\right) > 0. \tag{2.34}$$

Similarly,

$$1 - \sum_{k=\lceil na\rceil}^{\lfloor nb\rfloor} \psi\left(na-k\right) = \sum_{k=-\infty}^{\lceil na\rceil-1} \psi\left(na-k\right) + \sum_{k=\lfloor nb\rfloor+1}^{\infty} \psi\left(na-k\right)$$

$$> \psi\left(na - \lceil na\rceil + 1\right)$$

(call $\eta := \lceil na\rceil - na$, $0 \leq \eta < 1$)

$$= \psi\left(1-\eta\right) \geq \psi\left(1\right) > 0. \tag{2.35}$$

Therefore again

$$\lim_{n\to\infty}\left(1 - \sum_{k=\lceil na\rceil}^{\lfloor nb\rfloor} \psi\left(na-k\right)\right) > 0. \tag{2.36}$$

Here we find that

$$\lim_{n\to\infty} \sum_{k=\lceil na\rceil}^{\lfloor nb\rfloor} \psi\left(nx-k\right) \neq 1, \text{ for at least some } x \in [a,b]. \tag{2.37}$$

Note 2.7 *Let $[a,b] \subset \mathbb{R}$. For large enough n we always obtain $\lceil na \rceil \leq \lfloor nb \rfloor$. Also $a \leq \frac{k}{n} \leq b$, iff $\lceil na \rceil \leq k \leq \lfloor nb \rfloor$.*

In general it holds (by $\sum_{i=-\infty}^{\infty} \psi(x-i) = 1, \forall\, x \in \mathbb{R}$) that is

$$\sum_{k=\lceil na \rceil}^{\lfloor nb \rfloor} \psi(nx-k) \leq 1. \tag{2.38}$$

Let $(X, \|\cdot\|)$ be a Banach space.

Definition 2.8 Let $f \in C([a,b], X)$ and $n \in \mathbb{N} : \lceil na \rceil \leq \lfloor nb \rfloor$. We introduce and define the X-valued linear neural network operators

$$G_n(f,x) := \frac{\sum_{k=\lceil na \rceil}^{\lfloor nb \rfloor} f\left(\frac{k}{n}\right) \psi(nx-k)}{\sum_{k=\lceil na \rceil}^{\lfloor nb \rfloor} \psi(nx-k)}, \quad x \in [a,b]. \tag{2.39}$$

Clearly here $G_n(f,x) \in C([a,b], X)$.

For convenience we use the same G_n for real valued functions when needed. We study here the pointwise and uniform convergence of $G_n(f,x)$ to $f(x)$ with rates.

For convenience, also we call

$$G_n^*(f,x) := \sum_{k=\lceil na \rceil}^{\lfloor nb \rfloor} f\left(\frac{k}{n}\right) \psi(nx-k), \tag{2.40}$$

(similarly, G_n^* can be defined for real valued functions) that is

$$G_n(f,x) := \frac{G_n^*(f,x)}{\sum_{k=\lceil na \rceil}^{\lfloor nb \rfloor} \psi(nx-k)}. \tag{2.41}$$

So that

$$G_n(f,x) - f(x) = \frac{G_n^*(f,x)}{\sum_{k=\lceil na \rceil}^{\lfloor nb \rfloor} \psi(nx-k)} - f(x) = \tag{2.42}$$

$$\frac{G_n^*(f,x) - f(x)\left(\sum_{k=\lceil na \rceil}^{\lfloor nb \rfloor} \psi(nx-k)\right)}{\sum_{k=\lceil na \rceil}^{\lfloor nb \rfloor} \psi(nx-k)}.$$

Consequently, we derive that

$$\|G_n(f,x) - f(x)\| \leq 2.412 \left\| G_n^*(f,x) - f(x)\left(\sum_{k=\lceil na \rceil}^{\lfloor nb \rfloor} \psi(nx-k)\right)\right\| =$$

$$2.412 \left\| \sum_{k=\lceil na \rceil}^{\lfloor nb \rfloor} \left(f\left(\frac{k}{n}\right) - f(x)\right) \psi(nx-k) \right\|. \tag{2.43}$$

We will estimate the right and hand side of (2.43).

For that we need, for $f \in C([a,b], X)$ the first modulus of continuity

$$\omega_1(f,\delta) := \sup_{\substack{x,y \in [a,b] \\ |x-y| \leq \delta}} \|f(x) - f(y)\|, \ \delta > 0. \tag{2.44}$$

Similarly, it is defined ω_1 for $f \in C_{uB}(\mathbb{R}, X)$ (uniformly continuous and bounded functions from $\mathbb{R}$ into X), for $f \in C_B(\mathbb{R}, X)$ (continuous and bounded X-valued), and for $f \in C_u(\mathbb{R}, X)$ (uniformly continuous).

The fact $f \in C([a,b], X)$ or $f \in C_u(\mathbb{R}, X)$, is equivalent to $\lim_{\delta \to 0} \omega_1(f,\delta) = 0$, see [11].

We make

Definition 2.9 When $f \in C_{uB}(\mathbb{R}, X)$, or $f \in C_B(\mathbb{R}, X)$, we define

$$\overline{G_n}(f,x) := \sum_{k=-\infty}^{\infty} f\left(\frac{k}{n}\right) \psi(nx-k), \tag{2.45}$$

$n \in \mathbb{N}$, $x \in \mathbb{R}$,

the X-valued quasi-interpolation neural network operator.

We make

Remark 2.10 We have that

$$\left\| f\left(\frac{k}{n}\right)\right\| \leq \|f\|_{\infty,\mathbb{R}} < +\infty,$$

and

$$\left\| f\left(\frac{k}{n}\right)\right\| \psi(nx-k) \leq \|f\|_{\infty,\mathbb{R}} \psi(nx-k) \tag{2.46}$$

and

$$\sum_{k=-\lambda}^{\lambda} \left\| f\left(\frac{k}{n}\right)\right\| \psi(nx-k) \leq \|f\|_{\infty,\mathbb{R}} \left(\sum_{k=-\lambda}^{\lambda} \psi(nx-k)\right),$$

and finally

$$\sum_{k=-\infty}^{\infty} \left\| f\left(\frac{k}{n}\right) \right\| \psi(nx-k) \leq \|f\|_{\infty,\mathbb{R}}, \tag{2.47}$$

a convergent in $\mathbb{R}$ series.

So, the series $\sum_{k=-\infty}^{\infty} \left\| f\left(\frac{k}{n}\right) \right\| \psi(nx-k)$ is absolutely convergent in X, hence it is convergent in X and $\overline{G_n}(f,x) \in X$. We denote by $\|f\|_{\infty} := \sup_{x\in[a,b]} \|f(x)\|$, for $f \in C([a,b], X)$, similarly it is defined for $f \in C_B(\mathbb{R}, X)$.

2.3 Main Results

We present a set of X-valued neural network approximations to a function given with rates.

Theorem 2.11 *Let $f \in C([a,b], X)$, $0 < \alpha < 1$, $n \in \mathbb{N} : n^{1-\alpha} > 2$, $x \in [a,b]$. Then*

(i)

$$\|G_n(f,x) - f(x)\| \leq 2.412 \left[\omega_1\left(f, \frac{1}{n^{\alpha}}\right) + \frac{8\|f\|_{\infty}}{\pi e^{(n^{1-\alpha}-2)}} \right] =: \rho, \tag{2.48}$$

and

(ii)

$$\|G_n(f) - f\|_{\infty} \leq \rho. \tag{2.49}$$

We get that $\lim_{n\to\infty} G_n(f) = f$, pointwise and uniformly.

Proof We see that

$$\left\| \sum_{k=\lceil na \rceil}^{\lfloor nb \rfloor} \left(f\left(\frac{k}{n}\right) - f(x) \right) \psi(nx-k) \right\| \leq$$

$$\sum_{k=\lceil na \rceil}^{\lfloor nb \rfloor} \left\| f\left(\frac{k}{n}\right) - f(x) \right\| \psi(nx-k) =$$

$$\sum_{\substack{k=\lceil na \rceil \\ \left|\frac{k}{n}-x\right| \leq \frac{1}{n^{\alpha}}}}^{\lfloor nb \rfloor} \left\| f\left(\frac{k}{n}\right) - f(x) \right\| \psi(nx-k) +$$

$$\sum_{\substack{k=\lceil na\rceil \\ \left|\frac{k}{n}-x\right|>\frac{1}{n^\alpha}}}^{\lfloor nb\rfloor} \left\| f\left(\frac{k}{n}\right)-f(x)\right\| \psi(nx-k) \leq \tag{2.50}$$

$$\sum_{\substack{k=\lceil na\rceil \\ \left|\frac{k}{n}-x\right|\leq\frac{1}{n^\alpha}}}^{\lfloor nb\rfloor} \omega_1\left(f,\left|\frac{k}{n}-x\right|\right)\psi(nx-k)+$$

$$2\|f\|_\infty \sum_{\substack{k=\lceil na\rceil \\ |k-nx|>n^{1-\alpha}}}^{\lfloor nb\rfloor} \psi(nx-k) \leq$$

$$\omega_1\left(f,\frac{1}{n^\alpha}\right) \sum_{\substack{k=-\infty \\ \left|\frac{k}{n}-x\right|\leq\frac{1}{n^\alpha}}}^{\infty} \psi(nx-k)+$$

$$2\|f\|_\infty \sum_{\substack{k=-\infty \\ |k-nx|>n^{1-\alpha}}}^{\infty} \psi(nx-k) \underset{\text{(by Theorem 2.4)}}{\leq}$$

$$\omega_1\left(f,\frac{1}{n^\alpha}\right)+\frac{8\|f\|_\infty}{\pi e^{\left(n^{1-\alpha}-2\right)}}. \tag{2.51}$$

That is

$$\left\|\sum_{k=\lceil na\rceil}^{\lfloor nb\rfloor}\left(f\left(\frac{k}{n}\right)-f(x)\right)\psi(nx-k)\right\| \leq$$

$$\omega_1\left(f,\frac{1}{n^\alpha}\right)+\frac{8\|f\|_\infty}{\pi e^{\left(n^{1-\alpha}-2\right)}}. \tag{2.52}$$

Using (2.52) we derive (2.48). ■

Next we give

Theorem 2.12 *Let $f \in C_B(\mathbb{R}, X)$, $0 < \alpha < 1$, $n \in \mathbb{N} : n^{1-\alpha} > 2$, $x \in \mathbb{R}$. Then*

(i)

$$\left\| \overline{G}_n(f, x) - f(x) \right\| \leq \omega_1 \left(f, \frac{1}{n^\alpha} \right) + \frac{8 \| f \|_\infty}{\pi e^{\left(n^{1-\alpha} - 2\right)}} =: \mu, \tag{2.53}$$

and

(ii)

$$\left\| \overline{G}_n(f) - f \right\|_\infty \leq \mu. \tag{2.54}$$

For $f \in C_{uB}(\mathbb{R}, X)$ we get $\lim_{n \to \infty} \overline{G}_n(f) = f$, pointwise and uniformly.

Proof We observe that

$$\left\| \overline{G}_n(f, x) - f(x) \right\| = \left\| \sum_{k=-\infty}^{\infty} f\left(\frac{k}{n}\right) \psi(nx - k) - f(x) \sum_{k=-\infty}^{\infty} \psi(nx - k) \right\| =$$

$$\left\| \sum_{k=-\infty}^{\infty} \left(f\left(\frac{k}{n}\right) - f(x) \right) \psi(nx - k) \right\| \leq$$

$$\sum_{k=-\infty}^{\infty} \left\| f\left(\frac{k}{n}\right) - f(x) \right\| \psi(nx - k) =$$

$$\sum_{\begin{cases} k = -\infty \\ \left| \frac{k}{n} - x \right| \leq \frac{1}{n^\alpha} \end{cases}}^{\infty} \left\| f\left(\frac{k}{n}\right) - f(x) \right\| \psi(nx - k) +$$

$$\sum_{\begin{cases} k = -\infty \\ \left| \frac{k}{n} - x \right| > \frac{1}{n^\alpha} \end{cases}}^{\infty} \left\| f\left(\frac{k}{n}\right) - f(x) \right\| \psi(nx - k) \leq \tag{2.55}$$

$$\sum_{\begin{cases} k = -\infty \\ \left| \frac{k}{n} - x \right| \leq \frac{1}{n^\alpha} \end{cases}}^{\infty} \omega_1 \left(f, \left| \frac{k}{n} - x \right| \right) \psi(nx - k) +$$

$$2 \| f \|_\infty \sum_{\begin{cases} k = -\infty \\ \left| \frac{k}{n} - x \right| > \frac{1}{n^\alpha} \end{cases}}^{\infty} \psi(nx - k) \leq$$

$$\omega_1\left(f, \frac{1}{n^\alpha}\right) \sum_{\substack{k=-\infty \\ \left|\frac{k}{n}-x\right| \le \frac{1}{n^\alpha}}}^{\infty} \psi(nx-k) + \frac{8\|f\|_\infty}{\pi e^{(n^{1-\alpha}-2)}} \le$$

$$\omega_1\left(f, \frac{1}{n^\alpha}\right) + \frac{8\|f\|_\infty}{\pi e^{(n^{1-\alpha}-2)}}, \tag{2.56}$$

proving the claim. ■

We need the X-valued Taylor's formula in an appropriate form:

Theorem 2.13 ([10, 12]) *Let $N \in \mathbb{N}$, and $f \in C^N([a,b], X)$, where $[a,b] \subset \mathbb{R}$ and X is a Banach space. Let any $x, y \in [a,b]$. Then*

$$f(x) = \sum_{i=0}^{N} \frac{(x-y)^i}{i!} f^{(i)}(y) + \frac{1}{(N-1)!} \int_y^x (x-t)^{N-1} \left(f^{(N)}(t) - f^{(N)}(y)\right) dt. \tag{2.57}$$

The derivatives $f^{(i)}$, $i \in \mathbb{N}$, are defined like the numerical ones, see [21], p. 83. The integral $\int_y^x$ in (2.57) is of Bochner type, see [19].

By [12, 17] we have that: if $f \in C([a,b], X)$, then $f \in L_\infty([a,b], X)$ and $f \in L_1([a,b], X)$.

In the next we discuss high order neural network X-valued approximation by using the smoothness of f.

Theorem 2.14 *Let $f \in C^N([a,b], X)$, $n, N \in \mathbb{N}$, $0 < \alpha < 1$, $x \in [a,b]$ and $n^{1-\alpha} > 2$. Then*

(i)

$$\|G_n(f,x) - f(x)\| \le 2.412 \left\{ \sum_{j=1}^{N} \frac{\left\|f^{(j)}(x)\right\|}{j!} \left[\frac{1}{n^{\alpha j}} + \frac{4(b-a)^j}{\pi e^{(n^{1-\alpha}-2)}}\right] + \right. \tag{2.58}$$

$$\left. \left[\omega_1\left(f^{(N)}, \frac{1}{n^\alpha}\right) \frac{1}{n^{\alpha N} N!} + \frac{8\left\|f^{(N)}\right\|_\infty (b-a)^N}{N! \pi e^{(n^{1-\alpha}-2)}}\right] \right\},$$

(ii) assume further $f^{(j)}(x_0) = 0$, $j = 1, \ldots, N$, for some $x_0 \in [a,b]$, it holds

$$\|G_n(f,x_0) - f(x_0)\| \le 2.412 \cdot$$

$$\left\{\omega_1\left(f^{(N)}, \frac{1}{n^\alpha}\right) \frac{1}{n^{\alpha N} N!} + \frac{8\left\|f^{(N)}\right\|_\infty (b-a)^N}{N! \pi e^{(n^{1-\alpha}-2)}}\right\}, \tag{2.59}$$

and

(iii)

$$\|G_n(f)-f\|_\infty \leq 2.412\left\{\sum_{j=1}^{N}\frac{\|f^{(j)}\|_\infty}{j!}\left[\frac{1}{n^{\alpha j}}+\frac{4(b-a)^j}{\pi e^{(n^{1-\alpha}-2)}}\right]+\right.$$

$$\left.\left[\omega_1\left(f^{(N)},\frac{1}{n^\alpha}\right)\frac{1}{n^{\alpha N}N!}+\frac{8\|f^{(N)}\|_\infty (b-a)^N}{N!\pi e^{(n^{1-\alpha}-2)}}\right]\right\}. \tag{2.60}$$

Again we obtain $\lim_{n\to\infty} G_n(f)=f$, *pointwise and uniformly.*

Proof Next we apply the X-valued Taylor's formula with Bochner integral remainder (2.57). We have (here $\frac{k}{n}, x \in [a,b]$)

$$f\left(\frac{k}{n}\right)=\sum_{j=0}^{N}\frac{f^{(j)}(x)}{j!}\left(\frac{k}{n}-x\right)^j+\int_x^{\frac{k}{n}}\left(f^{(N)}(t)-f^{(N)}(x)\right)\frac{\left(\frac{k}{n}-t\right)^{N-1}}{(N-1)!}dt. \tag{2.61}$$

Then

$$f\left(\frac{k}{n}\right)\psi(nx-k)=\sum_{j=0}^{N}\frac{f^{(j)}(x)}{j!}\psi(nx-k)\left(\frac{k}{n}-x\right)^j+ \tag{2.62}$$

$$\psi(nx-k)\int_x^{\frac{k}{n}}\left(f^{(N)}(t)-f^{(N)}(x)\right)\frac{\left(\frac{k}{n}-t\right)^{N-1}}{(N-1)!}dt.$$

Hence

$$\sum_{k=\lceil na\rceil}^{\lfloor nb\rfloor} f\left(\frac{k}{n}\right)\psi(nx-k)-f(x)\sum_{k=\lceil na\rceil}^{\lfloor nb\rfloor}\psi(nx-k)= \tag{2.63}$$

$$\sum_{j=0}^{N}\frac{f^{(j)}(x)}{j!}\sum_{k=\lceil na\rceil}^{\lfloor nb\rfloor}\psi(nx-k)\left(\frac{k}{n}-x\right)^j+$$

$$\sum_{k=\lceil na\rceil}^{\lfloor nb\rfloor}\psi(nx-k)\int_x^{\frac{k}{n}}\left(f^{(N)}(t)-f^{(N)}(x)\right)\frac{\left(\frac{k}{n}-t\right)^{N-1}}{(N-1)!}dt.$$

Thus

$$G_n^*(f,x) - f(x)\left(\sum_{k=\lceil na\rceil}^{\lfloor nb\rfloor}\psi(nx-k)\right) =$$

$$\sum_{j=1}^{N}\frac{f^{(j)}(x)}{j!}G_n^*\left((\cdot - x)^j\right) + \Lambda_n(x), \tag{2.64}$$

where

$$\Lambda_n(x) := \sum_{k=\lceil na\rceil}^{\lfloor nb\rfloor}\psi(nx-k)\int_x^{\frac{k}{n}}\left(f^{(N)}(t) - f^{(N)}(x)\right)\frac{\left(\frac{k}{n}-t\right)^{N-1}}{(N-1)!}dt. \tag{2.65}$$

We assume that $b - a > \frac{1}{n^\alpha}$, which is always the case for large enough $n \in \mathbb{N}$, that is when $n > \left\lceil (b-a)^{-\frac{1}{\alpha}}\right\rceil$.

Thus $\left|\frac{k}{n} - x\right| \le \frac{1}{n^\alpha}$ or $\left|\frac{k}{n} - x\right| > \frac{1}{n^\alpha}$.

Let

$$\gamma := \int_x^{\frac{k}{n}}\left(f^{(N)}(t) - f^{(N)}(x)\right)\frac{\left(\frac{k}{n}-t\right)^{N-1}}{(N-1)!}dt, \tag{2.66}$$

in the case of $\left|\frac{k}{n} - x\right| \le \frac{1}{n^\alpha}$, we find that

$$\|\gamma\| \le \omega_1\left(f^{(N)}, \frac{1}{n^\alpha}\right)\frac{1}{n^{\alpha N}N!} \tag{2.67}$$

for $x \le \frac{k}{n}$ or $x \ge \frac{k}{n}$.

We prove it next.

(i) Indeed, for the case of $x \le \frac{k}{n}$, we have

$$\|\gamma\| = \left\|\int_x^{\frac{k}{n}}\left(f^{(N)}(t) - f^{(N)}(x)\right)\frac{\left(\frac{k}{n}-t\right)^{N-1}}{(N-1)!}dt\right\| \le$$

$$\int_x^{\frac{k}{n}}\left\|f^{(N)}(t) - f^{(N)}(x)\right\|\frac{\left(\frac{k}{n}-t\right)^{N-1}}{(N-1)!}dt \le$$

$$\int_x^{\frac{k}{n}}\omega_1\left(f^{(N)}, |t-x|\right)\frac{\left(\frac{k}{n}-t\right)^{N-1}}{(N-1)!}dt \le \omega_1\left(f^{(N)}, \frac{1}{n^\alpha}\right)\int_x^{\frac{k}{n}}\frac{\left(\frac{k}{n}-t\right)^{N-1}}{(N-1)!}dt = \tag{2.68}$$

$$\omega_1\left(f^{(N)}, \frac{1}{n^\alpha}\right)\frac{\left(\frac{k}{n}-x\right)^{N}}{N!} \le \omega_1\left(f^{(N)}, \frac{1}{n^\alpha}\right)\frac{1}{n^{\alpha N}N!}.$$

(ii) for the case of $x > \frac{k}{n}$, we have

$$\|\gamma\| = \left\| \int_{x}^{\frac{k}{n}} \left(f^{(N)}(t) - f^{(N)}(x)\right) \frac{\left(\frac{k}{n} - t\right)^{N-1}}{(N-1)!} dt \right\| =$$

$$\left\| \int_{\frac{k}{n}}^{x} \left(f^{(N)}(t) - f^{(N)}(x)\right) \frac{\left(t - \frac{k}{n}\right)^{N-1}}{(N-1)!} dt \right\| \leq$$

$$\int_{\frac{k}{n}}^{x} \left\| f^{(N)}(t) - f^{(N)}(x) \right\| \frac{\left(t - \frac{k}{n}\right)^{N-1}}{(N-1)!} dt \leq \tag{2.69}$$

$$\int_{\frac{k}{n}}^{x} \omega_1\left(f^{(N)}, |t - x|\right) \frac{\left(t - \frac{k}{n}\right)^{N-1}}{(N-1)!} dt \leq \omega_1\left(f^{(N)}, \frac{1}{n^\alpha}\right) \int_{\frac{k}{n}}^{x} \frac{\left(t - \frac{k}{n}\right)^{N-1}}{(N-1)!} dt =$$

$$\omega_1\left(f^{(N)}, \frac{1}{n^\alpha}\right) \frac{\left(x - \frac{k}{n}\right)^N}{N!} \leq \omega_1\left(f^{(N)}, \frac{1}{n^\alpha}\right) \frac{1}{n^{\alpha N} N!}.$$

We have proved (2.67).

We treat again γ, see (2.66), but differently:

Notice also for $x \leq \frac{k}{n}$ that

$$\left\| \int_{x}^{\frac{k}{n}} \left(f^{(N)}(t) - f^{(N)}(x)\right) \frac{\left(\frac{k}{n} - t\right)^{N-1}}{(N-1)!} dt \right\| \leq$$

$$\int_{x}^{\frac{k}{n}} \left\| f^{(N)}(t) - f^{(N)}(x) \right\| \frac{\left(\frac{k}{n} - t\right)^{N-1}}{(N-1)!} dt \leq \tag{2.70}$$

$$2 \left\| f^{(N)} \right\|_\infty \int_{x}^{\frac{k}{n}} \frac{\left(\frac{k}{n} - t\right)^{N-1}}{(N-1)!} dt = 2 \left\| f^{(N)} \right\|_\infty \frac{\left(\frac{k}{n} - x\right)^N}{N!}$$

$$\leq 2 \left\| f^{(N)} \right\|_\infty \frac{(b-a)^N}{N!}.$$

Next assume $\frac{k}{n} \leq x$, then

$$\left\| \int_{x}^{\frac{k}{n}} \left(f^{(N)}(t) - f^{(N)}(x)\right) \frac{\left(\frac{k}{n} - t\right)^{N-1}}{(N-1)!} dt \right\| =$$

$$\left\| \int_{\frac{k}{n}}^{x} \left(f^{(N)}(t) - f^{(N)}(x) \right) \frac{\left(t - \frac{k}{n} \right)^{N-1}}{(N-1)!} dt \right\| \leq$$

$$\int_{\frac{k}{n}}^{x} \left\| f^{(N)}(t) - f^{(N)}(x) \right\| \frac{\left(t - \frac{k}{n} \right)^{N-1}}{(N-1)!} dt \leq \tag{2.71}$$

$$2 \left\| f^{(N)} \right\|_{\infty} \int_{\frac{k}{n}}^{x} \frac{\left(t - \frac{k}{n} \right)^{N-1}}{(N-1)!} dt = 2 \left\| f^{(N)} \right\|_{\infty} \frac{\left(x - \frac{k}{n} \right)^{N}}{N!}$$

$$\leq 2 \left\| f^{(N)} \right\|_{\infty} \frac{(b-a)^N}{N!}.$$

Thus

$$\|\gamma\| \leq 2 \left\| f^{(N)} \right\|_{\infty} \frac{(b-a)^N}{N!}. \tag{2.72}$$

in the two cases.

Therefore

$$\Lambda_n(x) = \sum_{\substack{\left\{ \begin{array}{l} k = \lceil na \rceil \\ \left| \frac{k}{n} - x \right| \leq \frac{1}{n^{\alpha}} \end{array} \right.}}^{\lfloor nb \rfloor} \psi(nx - k) \gamma + \sum_{\substack{\left\{ \begin{array}{l} k = \lceil na \rceil \\ \left| \frac{k}{n} - x \right| > \frac{1}{n^{\alpha}} \end{array} \right.}}^{\lfloor nb \rfloor} \psi(nx - k) \gamma. \tag{2.73}$$

Hence

$$\|\Lambda_n(x)\| \leq \sum_{\substack{\left\{ \begin{array}{l} k = \lceil na \rceil \\ \left| \frac{k}{n} - x \right| \leq \frac{1}{n^{\alpha}} \end{array} \right.}}^{\lfloor nb \rfloor} \psi(nx - k) \left(\omega_1 \left(f^{(N)}, \frac{1}{n^{\alpha}} \right) \frac{1}{N! n^{\alpha N}} \right) + \tag{2.74}$$

$$\left(\sum_{\substack{\left\{ \begin{array}{l} k = \lceil na \rceil \\ \left| \frac{k}{n} - x \right| > \frac{1}{n^{\alpha}} \end{array} \right.}}^{\lfloor nb \rfloor} \psi(nx - k) \right) 2 \left\| f^{(N)} \right\|_{\infty} \frac{(b-a)^N}{N!} \leq$$

$$\omega_1 \left(f^{(N)}, \frac{1}{n^{\alpha}} \right) \frac{1}{N! n^{\alpha N}} + \frac{4}{\pi e^{\left(n^{1-\alpha} - 2 \right)}} 2 \left\| f^{(N)} \right\|_{\infty} \frac{(b-a)^N}{N!} =$$

$$\omega_1 \left(f^{(N)}, \frac{1}{n^{\alpha}} \right) \frac{1}{N! n^{\alpha N}} + \frac{8 \left\| f^{(N)} \right\|_{\infty} (b-a)^N}{N! \pi e^{\left(n^{1-\alpha} - 2 \right)}}.$$

That is

$$\|\Lambda_n(x)\| \leq \frac{\omega_1\left(f^{(N)}, \frac{1}{n^\alpha}\right)}{N!n^{\alpha N}} + \frac{8\left\|f^{(N)}\right\|_\infty (b-a)^N}{N!\pi e^{\left(n^{1-\alpha}-2\right)}}, \tag{2.75}$$

$\forall\, x \in [a, b]$.

We further see that

$$G_n^*\left((\cdot - x)^j\right) = \sum_{k=\lceil na \rceil}^{\lfloor nb \rfloor} \psi(nx-k)\left(\frac{k}{n} - x\right)^j, \tag{2.76}$$

where G_n^* is defined similarly for real valued functions.

Therefore

$$\left|G_n^*\left((\cdot - x)^j\right)\right| \leq \sum_{k=\lceil na \rceil}^{\lfloor nb \rfloor} \psi(nx-k)\left|\frac{k}{n} - x\right|^j =$$

$$\sum_{\substack{k=\lceil na \rceil \\ \left|\frac{k}{n}-x\right| \leq \frac{1}{n^\alpha}}}^{\lfloor nb \rfloor} \psi(nx-k)\left|\frac{k}{n} - x\right|^j + \sum_{\substack{k=\lceil na \rceil \\ \left|\frac{k}{n}-x\right| > \frac{1}{n^\alpha}}}^{\lfloor nb \rfloor} \psi(nx-k)\left|\frac{k}{n} - x\right|^j \overset{(2.23.)}{\leq}$$

$$\frac{1}{n^{\alpha j}} + (b-a)^j \frac{4}{\pi e^{\left(n^{1-\alpha}-2\right)}}. \tag{2.77}$$

That is

$$\left|G_n^*\left((\cdot - x)^j\right)\right| \leq \frac{1}{n^{\alpha j}} + (b-a)^j \frac{4}{\pi e^{\left(n^{1-\alpha}-2\right)}}, \tag{2.78}$$

for $j = 1, \ldots, N$.

Putting things together we have proved

$$\left\|G_n^*(f, x) - f(x)\left(\sum_{k=\lceil na \rceil}^{\lfloor nb \rfloor} \psi(nx-k)\right)\right\| \leq \sum_{j=1}^{N} \frac{\left\|f^{(j)}(x)\right\|}{j!} \tag{2.79}$$

$$\left[\frac{1}{n^{\alpha j}} + \frac{4(b-a)^j}{\pi e^{\left(n^{1-\alpha}-2\right)}}\right] + \left[\omega_1\left(f^{(N)}, \frac{1}{n^\alpha}\right)\frac{1}{n^{\alpha N} N!} + \frac{8\left\|f^{(N)}\right\|_\infty (b-a)^N}{N!\pi e^{\left(n^{1-\alpha}-2\right)}}\right],$$

that is establishing the theorem. ■

All integrals from now on are of Bochner type [19].

We need

Definition 2.15 ([12]) Let $[a, b] \subset \mathbb{R}$, X be a Banach space, $\alpha > 0$; $m = \lceil \alpha \rceil \in \mathbb{N}$, ($\lceil \cdot \rceil$ is the ceiling of the number), $f : [a, b] \to X$. We assume that $f^{(m)} \in L_1([a, b], X)$. We call the Caputo-Bochner left fractional derivative of order α:

$$\left(D_{*a}^{\alpha} f\right)(x) := \frac{1}{\Gamma(m-\alpha)} \int_a^x (x-t)^{m-\alpha-1} f^{(m)}(t)\, dt, \ \forall x \in [a, b]. \tag{2.80}$$

If $\alpha \in \mathbb{N}$, we set $D_{*a}^{\alpha} f := f^{(m)}$ the ordinary X-valued derivative (defined similar to numerical one, see [21], p. 83), and also set $D_{*a}^{0} f := f$.

By [12], $\left(D_{*a}^{\alpha} f\right)(x)$ exists almost everywhere in $x \in [a, b]$ and $D_{*a}^{\alpha} f \in L_1([a, b], X)$.

If $\left\| f^{(m)} \right\|_{L_\infty([a,b],X)} < \infty$, then by [12], $D_{*a}^{\alpha} f \in C([a, b], X)$, hence $\left\| D_{*a}^{\alpha} f \right\| \in C([a, b])$.

We mention

Lemma 2.16 ([11]) *Let $\alpha > 0$, $\alpha \notin \mathbb{N}$, $m = \lceil \alpha \rceil$, $f \in C^{m-1}([a, b], X)$ and $f^{(m)} \in L_\infty([a, b], X)$. Then $D_{*a}^{\alpha} f(a) = 0$.*

We mention

Definition 2.17 ([10]) Let $[a, b] \subset \mathbb{R}$, X be a Banach space, $\alpha > 0$, $m := \lceil \alpha \rceil$. We assume that $f^{(m)} \in L_1([a, b], X)$, where $f : [a, b] \to X$. We call the Caputo-Bochner right fractional derivative of order α:

$$\left(D_{b-}^{\alpha} f\right)(x) := \frac{(-1)^m}{\Gamma(m-\alpha)} \int_x^b (z-x)^{m-\alpha-1} f^{(m)}(z)\, dz, \ \forall x \in [a, b]. \tag{2.81}$$

We observe that $\left(D_{b-}^{m} f\right)(x) = (-1)^m f^{(m)}(x)$, for $m \in \mathbb{N}$, and $\left(D_{b-}^{0} f\right)(x) = f(x)$.

By [10], $\left(D_{b-}^{\alpha} f\right)(x)$ exists almost everywhere on $[a, b]$ and $\left(D_{b-}^{\alpha} f\right) \in L_1([a, b], X)$.

If $\left\| f^{(m)} \right\|_{L_\infty([a,b],X)} < \infty$, and $\alpha \notin \mathbb{N}$, by [10], $D_{b-}^{\alpha} f \in C([a, b], X)$, hence $\left\| D_{b-}^{\alpha} f \right\| \in C([a, b])$.

We need

Lemma 2.18 ([11]) *Let $f \in C^{m-1}([a, b], X)$, $f^{(m)} \in L_\infty([a, b], X)$, $m = \lceil \alpha \rceil$, $\alpha > 0$, $\alpha \notin \mathbb{N}$. Then $D_{b-}^{\alpha} f(b) = 0$.*

We mention the left fractional Taylor formula

Theorem 2.19 ([12]) *Let $m \in \mathbb{N}$ and $f \in C^m([a, b], X)$, where $[a, b] \subset \mathbb{R}$ and X is a Banach space, and let $\alpha > 0 : m = \lceil \alpha \rceil$. Then*

$$f(x) = \sum_{i=0}^{m-1} \frac{(x-a)^i}{i!} f^{(i)}(a) + \frac{1}{\Gamma(\alpha)} \int_a^x (x-z)^{\alpha-1} \left(D_{*a}^{\alpha} f\right)(z)\, dz, \tag{2.82}$$

$\forall x \in [a, b]$.

We also mention the right fractional Taylor formula

Theorem 2.20 ([10]) *Let* $[a,b] \subset \mathbb{R}$, *X be a Banach space,* $\alpha > 0$, $m = \lceil \alpha \rceil$, $f \in C^m([a,b], X)$. *Then*

$$f(x) = \sum_{i=0}^{m-1} \frac{(x-b)^i}{i!} f^{(i)}(b) + \frac{1}{\Gamma(\alpha)} \int_x^b (z-x)^{\alpha-1} \left(D_{b-}^{\alpha} f\right)(z)\, dz, \tag{2.83}$$

$\forall\, x \in [a,b]$.

Convention 2.21 *We assume that*

$$D_{*x_0}^{\alpha} f(x) = 0, \text{ for } x < x_0, \tag{2.84}$$

and

$$D_{x_0-}^{\alpha} f(x) = 0, \text{ for } x > x_0, \tag{2.85}$$

for all $x, x_0 \in [a,b]$.

We mention

Proposition 2.22 ([11]) *Let* $f \in C^n([a,b], X)$, $n = \lceil \nu \rceil$, $\nu > 0$. *Then* $D_{*a}^{\nu} f(x)$ *is continuous in* $x \in [a,b]$.

Proposition 2.23 ([11]) *Let* $f \in C^m([a,b], X)$, $m = \lceil \alpha \rceil$, $\alpha > 0$. *Then* $D_{b-}^{\nu} f(x)$ *is continuous in* $x \in [a,b]$.

We also mention

Proposition 2.24 ([11]) *Let* $f \in C^{m-1}([a,b], X)$, $f^{(m)} \in L_{\infty}([a,b], X)$, $m = \lceil \alpha \rceil$, $\alpha > 0$ *and*

$$D_{*x_0}^{\alpha} f(x) = \frac{1}{\Gamma(m-\alpha)} \int_{x_0}^{x} (x-t)^{m-\alpha-1} f^{(m)}(t)\, dt, \tag{2.86}$$

for all $x, x_0 \in [a,b] : x \geq x_0$.

Then $D_{*x_0}^{\alpha} f(x)$ *is continuous in* x_0.

Proposition 2.25 ([11]) *Let* $f \in C^{m-1}([a,b], X)$, $f^{(m)} \in L_{\infty}([a,b], X)$, $m = \lceil \alpha \rceil$, $\alpha > 0$ *and*

$$D_{x_0-}^{\alpha} f(x) = \frac{(-1)^m}{\Gamma(m-\alpha)} \int_{x}^{x_0} (\zeta - x)^{m-\alpha-1} f^{(m)}(\zeta)\, d\zeta, \tag{2.87}$$

for all $x, x_0 \in [a,b] : x_0 \geq x$.

Then $D_{x_0-}^{\alpha} f(x)$ *is continuous in* x_0.

Corollary 2.26 ([11]) *Let* $f \in C^m([a,b], X)$, $m = \lceil \alpha \rceil$, $\alpha > 0$, $x, x_0 \in [a,b]$. *Then* $D^a_{*x_0} f(x)$, $D^a_{x_0-} f(x)$ *are jointly continuous functions in* (x, x_0) *from* $[a,b]^2$ *into* X, X *is a Banach space.*

We need

Theorem 2.27 ([11]) *Let* $f : [a,b]^2 \to X$ *be jointly continuous,* X *is a Banach space. Consider*

$$G(x) = \omega_1(f(\cdot, x), \delta, [x, b]), \tag{2.88}$$

$\delta > 0$, $x \in [a,b]$.

Then G *is continuous on* $[a,b]$.

Theorem 2.28 ([11]) *Let* $f : [a,b]^2 \to X$ *be jointly continuous,* X *is a Banach space. Then*

$$H(x) = \omega_1(f(\cdot, x), \delta, [a, x]), \tag{2.89}$$

$x \in [a,b]$, *is continuous in* $x \in [a,b]$, $\delta > 0$.

We make

Remark 2.29 ([11]) Let $f \in C^{n-1}([a,b])$, $f^{(n)} \in L_\infty([a,b])$, $n = \lceil \nu \rceil$, $\nu > 0$, $\nu \notin \mathbb{N}$. Then

$$\left\| D^\nu_{*a} f(x) \right\| \le \frac{\left\| f^{(n)} \right\|_{L_\infty([a,b],X)}}{\Gamma(n-\nu+1)} (x-a)^{n-\nu}, \ \forall x \in [a,b]. \tag{2.90}$$

Thus we observe

$$\omega_1\left(D^\nu_{*a} f, \delta\right) = \sup_{\substack{x,y \in [a,b] \\ |x-y| \le \delta}} \left\| D^\nu_{*a} f(x) - D^\nu_{*a} f(y) \right\| \le \tag{2.91}$$

$$\sup_{\substack{x,y \in [a,b] \\ |x-y| \le \delta}} \left(\frac{\left\| f^{(n)} \right\|_{L_\infty([a,b],X)}}{\Gamma(n-\nu+1)} (x-a)^{n-\nu} + \frac{\left\| f^{(n)} \right\|_{L_\infty([a,b],X)}}{\Gamma(n-\nu+1)} (y-a)^{n-\nu} \right)$$

$$\le \frac{2 \left\| f^{(n)} \right\|_{L_\infty([a,b],X)}}{\Gamma(n-\nu+1)} (b-a)^{n-\nu}.$$

Consequently

$$\omega_1\left(D^\nu_{*a} f, \delta\right) \le \frac{2 \left\| f^{(n)} \right\|_{L_\infty([a,b],X)}}{\Gamma(n-\nu+1)} (b-a)^{n-\nu}. \tag{2.92}$$

Similarly, let $f \in C^{m-1}([a,b])$, $f^{(m)} \in L_{\infty}([a,b])$, $m = \lceil \alpha \rceil$, $\alpha > 0$, $\alpha \notin \mathbb{N}$, then

$$\omega_1\left(D_{b-}^{\alpha} f, \delta\right) \leq \frac{2\left\|f^{(m)}\right\|_{L_{\infty}([a,b],X)}}{\Gamma(m-\alpha+1)}(b-a)^{m-\alpha}. \tag{2.93}$$

So for $f \in C^{m-1}([a,b])$, $f^{(m)} \in L_{\infty}([a,b])$, $m = \lceil \alpha \rceil$, $\alpha > 0$, $\alpha \notin \mathbb{N}$, we find

$$\sup_{x_0 \in [a,b]} \omega_1\left(D_{*x_0}^{\alpha} f, \delta\right)_{[x_0,b]} \leq \frac{2\left\|f^{(m)}\right\|_{L_{\infty}([a,b],X)}}{\Gamma(m-\alpha+1)}(b-a)^{m-\alpha}, \tag{2.94}$$

and

$$\sup_{x_0 \in [a,b]} \omega_1\left(D_{x_0-}^{\alpha} f, \delta\right)_{[a,x_0]} \leq \frac{2\left\|f^{(m)}\right\|_{L_{\infty}([a,b],X)}}{\Gamma(m-\alpha+1)}(b-a)^{m-\alpha}. \tag{2.95}$$

By [12] we get that $D_{*x_0}^{\alpha} f \in C([x_0,b], X)$, and by [10] we obtain that $D_{x_0-}^{\alpha} f \in C([a,x_0], X)$.

We present the following X-valued fractional approximation result by neural networks.

Theorem 2.30 *Let $\alpha > 0$, $N = \lceil \alpha \rceil$, $\alpha \notin \mathbb{N}$, $f \in C^N([a,b], X)$, $0 < \beta < 1$, $x \in [a,b]$, $n \in \mathbb{N} : n^{1-\beta} > 2$. Then*

(i)

$$\left\|G(f,x) - \sum_{j=1}^{N-1} \frac{f^{(j)}(x)}{j!} G_n\left((\cdot - x)^j\right)(x) - f(x)\right\| \leq$$

$$\frac{2.412}{\Gamma(\alpha+1)} \left\{ \frac{\left(\omega_1\left(D_{x-}^{\alpha} f, \frac{1}{n^{\beta}}\right)_{[a,x]} + \omega_1\left(D_{*x}^{\alpha} f, \frac{1}{n^{\beta}}\right)_{[x,b]}\right)}{n^{\alpha\beta}} + \right.$$

$$\left. \frac{4}{\pi e^{\left(n^{1-\beta}-2\right)}} \left(\left\|D_{x-}^{\alpha} f\right\|_{\infty,[a,x]} (x-a)^{\alpha} + \left\|D_{*x}^{\alpha} f\right\|_{\infty,[x,b]} (b-x)^{\alpha} \right) \right\}, \tag{2.96}$$

(ii) if $f^{(j)}(x) = 0$, for $j = 1, \ldots, N-1$, we have

$$\left\|G_n(f,x) - f(x)\right\| \leq \frac{2.412}{\Gamma(\alpha+1)}$$

$$\left\{ \frac{\left(\omega_1\left(D_{x-}^{\alpha} f, \frac{1}{n^{\beta}}\right)_{[a,x]} + \omega_1\left(D_{*x}^{\alpha} f, \frac{1}{n^{\beta}}\right)_{[x,b]}\right)}{n^{\alpha\beta}} + \right.$$

$$\left. \frac{4}{\pi e^{\left(n^{1-\beta}-2\right)}} \left(\left\|D_{x-}^{\alpha} f\right\|_{\infty,[a,x]} (x-a)^{\alpha} + \left\|D_{*x}^{\alpha} f\right\|_{\infty,[x,b]} (b-x)^{\alpha} \right) \right\}, \tag{2.97}$$

(iii)

$$\|G_n(f,x) - f(x)\| \leq 2.412 \cdot$$

$$\left\{ \sum_{j=1}^{N-1} \frac{\|f^{(j)}(x)\|}{j!} \left\{ \frac{1}{n^{\beta j}} + (b-a)^j \frac{4}{\pi e^{(n^{1-\beta}-2)}} \right\} + \right.$$

$$\frac{1}{\Gamma(\alpha+1)} \left\{ \frac{\left(\omega_1 \left(D^{\alpha}_{x-} f, \frac{1}{n^{\beta}} \right)_{[a,x]} + \omega_1 \left(D^{\alpha}_{*x} f, \frac{1}{n^{\beta}} \right)_{[x,b]} \right)}{n^{\alpha\beta}} + \right.$$

$$\left. \left. \frac{4}{\pi e^{(n^{1-\beta}-2)}} \left(\left\| D^{\alpha}_{x-} f \right\|_{\infty,[a,x]} (x-a)^{\alpha} + \left\| D^{\alpha}_{*x} f \right\|_{\infty,[x,b]} (b-x)^{\alpha} \right) \right\} \right\}, \quad (2.98)$$

$\forall x \in [a,b]$,

and

(iv)

$$\|G_n f - f\|_{\infty} \leq 2.412 \cdot$$

$$\left\{ \sum_{j=1}^{N-1} \frac{\|f^{(j)}\|_{\infty}}{j!} \left\{ \frac{1}{n^{\beta j}} + (b-a)^j \frac{4}{\pi e^{(n^{1-\beta}-2)}} \right\} + \right.$$

$$\frac{1}{\Gamma(\alpha+1)} \left\{ \frac{\left(\sup_{x\in[a,b]} \omega_1 \left(D^{\alpha}_{x-} f, \frac{1}{n^{\beta}} \right)_{[a,x]} + \sup_{x\in[a,b]} \omega_1 \left(D^{\alpha}_{*x} f, \frac{1}{n^{\beta}} \right)_{[x,b]} \right)}{n^{\alpha\beta}} + \right.$$

$$\left. \left. \frac{4}{\pi e^{(n^{1-\beta}-2)}} (b-a)^{\alpha} \left(\sup_{x\in[a,b]} \left\| D^{\alpha}_{x-} f \right\|_{\infty,[a,x]} + \sup_{x\in[a,b]} \left\| D^{\alpha}_{*x} f \right\|_{\infty,[x,b]} \right) \right\} \right\}. \quad (2.99)$$

Above, when $N = 1$ the sum $\sum_{j=1}^{N-1} \cdot = 0$.

As we see here we obtain X-valued fractionally type pointwise and uniform convergence with rates of $G_n \to I$ the unit operator, as $n \to \infty$.

Proof Let $x \in [a,b]$. We have that $D^{\alpha}_{x-} f(x) = D^{\alpha}_{*x} f(x) = 0$.

From Theorem 2.19, we get by the left Caputo fractional Taylor formula that

$$f\left(\frac{k}{n}\right)=\sum_{j=0}^{N-1}\frac{f^{(j)}(x)}{j!}\left(\frac{k}{n}-x\right)^{j}+ \tag{2.100}$$

$$\frac{1}{\Gamma(\alpha)}\int_{x}^{\frac{k}{n}}\left(\frac{k}{n}-J\right)^{\alpha-1}\left(D_{*x}^{\alpha}f(J)-D_{*x}^{\alpha}f(x)\right)dJ,$$

for all $x \leq \frac{k}{n} \leq b$.

Also from Theorem 2.20, using the right Caputo fractional Taylor formula we get

$$f\left(\frac{k}{n}\right)=\sum_{j=0}^{N-1}\frac{f^{(j)}(x)}{j!}\left(\frac{k}{n}-x\right)^{j}+ \tag{2.101}$$

$$\frac{1}{\Gamma(\alpha)}\int_{\frac{k}{n}}^{x}\left(J-\frac{k}{n}\right)^{\alpha-1}\left(D_{x-}^{\alpha}f(J)-D_{x-}^{\alpha}f(x)\right)dJ,$$

for all $a \leq \frac{k}{n} \leq x$.

Hence we have

$$f\left(\frac{k}{n}\right)\psi(nx-k)=\sum_{j=0}^{N-1}\frac{f^{(j)}(x)}{j!}\psi(nx-k)\left(\frac{k}{n}-x\right)^{j}+ \tag{2.102}$$

$$\frac{\psi(nx-k)}{\Gamma(\alpha)}\int_{x}^{\frac{k}{n}}\left(\frac{k}{n}-J\right)^{\alpha-1}\left(D_{*x}^{\alpha}f(J)-D_{*x}^{\alpha}f(x)\right)dJ,$$

for all $x \leq \frac{k}{n} \leq b$, iff $\lceil nx \rceil \leq k \leq \lfloor nb \rfloor$, and

$$f\left(\frac{k}{n}\right)\psi(nx-k)=\sum_{j=0}^{N-1}\frac{f^{(j)}(x)}{j!}\psi(nx-k)\left(\frac{k}{n}-x\right)^{j}+ \tag{2.103}$$

$$\frac{\psi(nx-k)}{\Gamma(\alpha)}\int_{\frac{k}{n}}^{x}\left(J-\frac{k}{n}\right)^{\alpha-1}\left(D_{x-}^{\alpha}f(J)-D_{x-}^{\alpha}f(x)\right)dJ,$$

for all $a \leq \frac{k}{n} \leq x$, iff $\lceil na \rceil \leq k \leq \lfloor nx \rfloor$.

Therefore it holds

$$\sum_{k=\lfloor nx \rfloor+1}^{\lfloor nb \rfloor} f\left(\frac{k}{n}\right)\psi(nx-k)=\sum_{j=0}^{N-1}\frac{f^{(j)}(x)}{j!}\sum_{k=\lfloor nx \rfloor+1}^{\lfloor nb \rfloor}\psi(nx-k)\left(\frac{k}{n}-x\right)^{j}+ \tag{2.104}$$

$$\frac{1}{\Gamma(\alpha)} \sum_{k=\lfloor nx \rfloor+1}^{\lfloor nb \rfloor} \psi(nx-k) \int_x^{\frac{k}{n}} \left(\frac{k}{n}-J\right)^{\alpha-1} \left(D_{*x}^{\alpha} f(J) - D_{*x}^{\alpha} f(x)\right) dJ,$$

and

$$\sum_{k=\lceil na \rceil}^{\lfloor nx \rfloor} f\left(\frac{k}{n}\right) \psi(nx-k) = \sum_{j=0}^{N-1} \frac{f^{(j)}(x)}{j!} \sum_{k=\lceil na \rceil}^{\lfloor nx \rfloor} \psi(nx-k) \left(\frac{k}{n}-x\right)^j + \tag{2.105}$$

$$\frac{1}{\Gamma(\alpha)} \sum_{k=\lceil na \rceil}^{\lfloor nx \rfloor} \psi(nx-k) \int_{\frac{k}{n}}^{x} \left(J-\frac{k}{n}\right)^{\alpha-1} \left(D_{x-}^{\alpha} f(J) - D_{x-}^{\alpha} f(x)\right) dJ.$$

Adding the last two equalities (2.104) and (2.105) obtain

$$G_n^*(f,x) = \sum_{k=\lceil na \rceil}^{\lfloor nb \rfloor} f\left(\frac{k}{n}\right) \psi(nx-k) = \tag{2.106}$$

$$\sum_{j=0}^{N-1} \frac{f^{(j)}(x)}{j!} \sum_{k=\lceil na \rceil}^{\lfloor nb \rfloor} \psi(nx-k) \left(\frac{k}{n}-x\right)^j +$$

$$\frac{1}{\Gamma(\alpha)} \left\{ \sum_{k=\lceil na \rceil}^{\lfloor nx \rfloor} \psi(nx-k) \int_{\frac{k}{n}}^{x} \left(J-\frac{k}{n}\right)^{\alpha-1} \left(D_{x-}^{\alpha} f(J) - D_{x-}^{\alpha} f(x)\right) dJ + \right.$$

$$\left. \sum_{k=\lfloor nx \rfloor+1}^{\lfloor nb \rfloor} \psi(nx-k) \int_x^{\frac{k}{n}} \left(\frac{k}{n}-J\right)^{\alpha-1} \left(D_{*x}^{\alpha} f(J) - D_{*x}^{\alpha} f(x)\right) dJ \right\}.$$

So we have derived

$$G_n^*(f,x) - f(x) \left(\sum_{k=\lceil na \rceil}^{\lfloor nb \rfloor} \psi(nx-k) \right) = \tag{2.107}$$

$$\sum_{j=1}^{N-1} \frac{f^{(j)}(x)}{j!} G_n^*\left((\cdot - x)^j\right) + u_n(x),$$

where

$$u_n(x) := \frac{1}{\Gamma(\alpha)} \left\{ \sum_{k=\lceil na \rceil}^{\lfloor nx \rfloor} \psi(nx - k) \int_{\frac{k}{n}}^{x} \left(J - \frac{k}{n} \right)^{\alpha - 1} \left(D_{x-}^{\alpha} f(J) - D_{x-}^{\alpha} f(x) \right) dJ \right.$$

$$\left. + \sum_{k=\lfloor nx \rfloor + 1}^{\lfloor nb \rfloor} \psi(nx - k) \int_{x}^{\frac{k}{n}} \left(\frac{k}{n} - J \right)^{\alpha - 1} \left(D_{*x}^{\alpha} f(J) - D_{*x}^{\alpha} f(x) \right) dJ \right\}. \quad (2.108)$$

We set

$$u_{1n}(x) := \frac{1}{\Gamma(\alpha)} \sum_{k=\lceil na \rceil}^{\lfloor nx \rfloor} \psi(nx - k) \int_{\frac{k}{n}}^{x} \left(J - \frac{k}{n} \right)^{\alpha - 1} \left(D_{x-}^{\alpha} f(J) - D_{x-}^{\alpha} f(x) \right) dJ, \quad (2.109)$$

and

$$u_{2n} := \frac{1}{\Gamma(\alpha)} \sum_{k=\lceil nx \rceil + 1}^{\lfloor nb \rfloor} \psi(nx - k) \int_{x}^{\frac{k}{n}} \left(\frac{k}{n} - J \right)^{\alpha - 1} \left(D_{*x}^{\alpha} f(J) - D_{*x}^{\alpha} f(x) \right) dJ, \quad (2.110)$$

i.e.

$$u_n(x) = u_{1n}(x) + u_{2n}(x). \quad (2.111)$$

We assume $b - a > \frac{1}{n^{\beta}}$, $0 < \beta < 1$, which is always the case for large enough $n \in \mathbb{N}$, that is when $n > \left\lceil (b - a)^{-\frac{1}{\beta}} \right\rceil$. It is always true that either $\left| \frac{k}{n} - x \right| \leq \frac{1}{n^{\beta}}$ or $\left| \frac{k}{n} - x \right| > \frac{1}{n^{\beta}}$.

For $k = \lceil na \rceil, \ldots, \lfloor nx \rfloor$, we consider

$$\gamma_{1k} := \left\| \int_{\frac{k}{n}}^{x} \left(J - \frac{k}{n} \right)^{\alpha - 1} \left(D_{x-}^{\alpha} f(J) - D_{x-}^{\alpha} f(x) \right) dJ \right\| = \quad (2.112)$$

$$\left\| \int_{\frac{k}{n}}^{x} \left(J - \frac{k}{n} \right)^{\alpha - 1} D_{x-}^{\alpha} f(J) \, dJ \right\| \leq \int_{\frac{k}{n}}^{x} \left(J - \frac{k}{n} \right)^{\alpha - 1} \left\| D_{x-}^{\alpha} f(J) \right\| dJ \leq$$

$$\left\| D_{x-}^{\alpha} f(J) \right\|_{\infty, [a, x]} \frac{\left(x - \frac{k}{n} \right)^{\alpha}}{\alpha} \leq \left\| D_{x-}^{\alpha} f \right\|_{\infty, [a, x]} \frac{(x - a)^{\alpha}}{\alpha}. \quad (2.113)$$

That is

$$\gamma_{1k} \leq \left\| D_{x-}^{\alpha} f \right\|_{\infty, [a, x]} \frac{(x - a)^{\alpha}}{\alpha}, \quad (2.114)$$

for $k = \lceil na \rceil, \ldots, \lfloor nx \rfloor$.

Also we have in case of $\left| \frac{k}{n} - x \right| \leq \frac{1}{n^{\beta}}$ that

$$\gamma_{1k} \leq \int_{\frac{k}{n}}^{x} \left(J - \frac{k}{n}\right)^{\alpha-1} \left\| D_{x-}^{\alpha} f(J) - D_{x-}^{\alpha} f(x) \right\| dJ \leq$$

$$\int_{\frac{k}{n}}^{x} \left(J - \frac{k}{n}\right)^{\alpha-1} \omega_1 \left(D_{x-}^{\alpha} f, |J - x|\right)_{[a,x]} dJ \leq \tag{2.115}$$

$$\omega_1 \left(D_{x-}^{\alpha} f, \left|x - \frac{k}{n}\right|\right)_{[a,x]} \int_{\frac{k}{n}}^{x} \left(J - \frac{k}{n}\right)^{\alpha-1} dJ \leq$$

$$\omega_1 \left(D_{x-}^{\alpha} f, \frac{1}{n^{\beta}}\right)_{[a,x]} \frac{\left(x - \frac{k}{n}\right)^{\alpha}}{\alpha} \leq \omega_1 \left(D_{x-}^{\alpha} f, \frac{1}{n^{\beta}}\right)_{[a,x]} \frac{1}{\alpha n^{\alpha\beta}}.$$

That is when $\left|\frac{k}{n} - x\right| \leq \frac{1}{n^{\beta}}$, then

$$\gamma_{1k} \leq \frac{\omega_1 \left(D_{x-}^{\alpha} f, \frac{1}{n^{\beta}}\right)_{[a,x]}}{\alpha n^{\alpha\beta}}. \tag{2.116}$$

Consequently we obtain

$$\|u_{1n}(x)\| \leq \frac{1}{\Gamma(\alpha)} \sum_{k=\lceil na \rceil}^{\lfloor nx \rfloor} \psi(nx - k) \gamma_{1k} = \tag{2.117}$$

$$\frac{1}{\Gamma(\alpha)} \left\{ \sum_{\substack{k = \lceil na \rceil \\ : \left|\frac{k}{n} - x\right| \leq \frac{1}{n^{\beta}}}}^{\lfloor nx \rfloor} \psi(nx - k) \gamma_{1k} + \sum_{\substack{k = \lceil na \rceil \\ : \left|\frac{k}{n} - x\right| > \frac{1}{n^{\beta}}}}^{\lfloor nx \rfloor} \psi(nx - k) \gamma_{1k} \right\} \leq$$

$$\frac{1}{\Gamma(\alpha)} \left\{ \left(\sum_{\substack{k = \lceil na \rceil \\ : \left|\frac{k}{n} - x\right| \leq \frac{1}{n^{\beta}}}}^{\lfloor nx \rfloor} \psi(nx - k) \right) \frac{\omega_1 \left(D_{x-}^{\alpha} f, \frac{1}{n^{\beta}}\right)_{[a,x]}}{\alpha n^{\alpha\beta}} + \right.$$

$$\left. \left(\sum_{\substack{k = \lceil na \rceil \\ : \left|\frac{k}{n} - x\right| > \frac{1}{n^{\beta}}}}^{\lfloor nx \rfloor} \psi(nx - k) \right) \left\| D_{x-}^{\alpha} f \right\|_{\infty,[a,x]} \frac{(x - a)^{\alpha}}{\alpha} \right\} \leq \tag{2.118}$$

$$\frac{1}{\Gamma(\alpha+1)}\left\{\frac{\omega_1\left(D_{x-}^{\alpha}f,\frac{1}{n^{\beta}}\right)_{[a,x]}}{n^{\alpha\beta}}+\right.$$

$$\left.\left(\sum_{\begin{cases}k=-\infty\\ :|nx-k|>n^{1-\beta}\end{cases}}^{\infty}\psi(nx-k)\right)\left\|D_{x-}^{\alpha}f\right\|_{\infty,[a,x]}(x-a)^{\alpha}\right\}\leq$$

$$\frac{1}{\Gamma(\alpha+1)}\left\{\frac{\omega_1\left(D_{x-}^{\alpha}f,\frac{1}{n^{\beta}}\right)_{[a,x]}}{n^{\alpha\beta}}+\frac{4\left\|D_{x-}^{\alpha}f\right\|_{\infty,[a,x]}(x-a)^{\alpha}}{\pi e^{\left(n^{1-\beta}-2\right)}}\right\}.$$

So we have proved that

$$\|u_{1n}(x)\|\leq\frac{1}{\Gamma(\alpha+1)}\left\{\frac{\omega_1\left(D_{x-}^{\alpha}f,\frac{1}{n^{\beta}}\right)_{[a,x]}}{n^{\alpha\beta}}+\right.\tag{2.119}$$

$$\left.\frac{4}{\pi e^{\left(n^{1-\beta}-2\right)}}\left\|D_{x-}^{\alpha}f\right\|_{\infty,[a,x]}(x-a)^{\alpha}\right\}.$$

Next when $k=\lfloor nx\rfloor+1,\ldots,\lfloor nb\rfloor$ we consider

$$\gamma_{2k}:=\left\|\int_x^{\frac{k}{n}}\left(\frac{k}{n}-J\right)^{\alpha-1}\left(D_{*x}^{\alpha}f(J)-D_{*x}^{\alpha}f(x)\right)dJ\right\|\leq\tag{2.120}$$

$$\int_x^{\frac{k}{n}}\left(\frac{k}{n}-J\right)^{\alpha-1}\left\|D_{*x}^{\alpha}f(J)-D_{*x}^{\alpha}f(x)\right\|dJ=$$

$$\int_x^{\frac{k}{n}}\left(\frac{k}{n}-J\right)^{\alpha-1}\left\|D_{*x}^{\alpha}f(J)\right\|dJ\leq$$

$$\left\|D_{*x}^{\alpha}f\right\|_{\infty,[x,b]}\frac{\left(\frac{k}{n}-x\right)^{\alpha}}{\alpha}\leq\left\|D_{*x}^{\alpha}f\right\|_{\infty,[x,b]}\frac{(b-x)^{\alpha}}{\alpha}.\tag{2.121}$$

Therefore when $k=\lfloor nx\rfloor+1,\ldots,\lfloor nb\rfloor$ we get that

That is

$$\gamma_{2k}\leq\left\|D_{*x}^{\alpha}f\right\|_{\infty,[x,b]}\frac{(b-x)^{\alpha}}{\alpha}.\tag{2.122}$$

In case of $\left|\frac{k}{n}-x\right|\leq\frac{1}{n^{\beta}}$ we have

$$\gamma_{2k}\leq\int_{x}^{\frac{k}{n}}\left(\frac{k}{n}-J\right)^{\alpha-1}\omega_{1}\left(D_{*x}^{\alpha}f,|J-x|\right)_{[x,b]}dJ\leq$$

$$\omega_{1}\left(D_{*x}^{\alpha}f,\left|\frac{k}{n}-x\right|\right)_{[x,b]}\int_{x}^{\frac{k}{n}}\left(\frac{k}{n}-J\right)^{\alpha-1}dJ\leq \tag{2.123}$$

$$\omega_{1}\left(D_{*x}^{\alpha}f,\frac{1}{n^{\beta}}\right)_{[x,b]}\frac{\left(\frac{k}{n}-x\right)^{\alpha}}{\alpha}\leq\omega_{1}\left(D_{*x}^{\alpha}f,\frac{1}{n^{\beta}}\right)_{[x,b]}\frac{1}{\alpha n^{\alpha\beta}}.$$

So when $\left|\frac{k}{n}-x\right|\leq\frac{1}{n^{\beta}}$ we derived that

$$\gamma_{2k}\leq\frac{\omega_{1}\left(D_{*x}^{\alpha}f,\frac{1}{n^{\beta}}\right)_{[x,b]}}{\alpha n^{\alpha\beta}}. \tag{2.124}$$

Similarly we have that

$$\|u_{2n}(x)\|\leq\frac{1}{\Gamma(\alpha)}\left(\sum_{k=\lfloor nx\rfloor+1}^{\lfloor nb\rfloor}\psi(nx-k)\,\gamma_{2k}\right)=$$

$$\frac{1}{\Gamma(\alpha)}\left\{\sum_{\begin{cases}k=\lfloor nx\rfloor+1\\ :\left|\frac{k}{n}-x\right|\leq\frac{1}{n^{\beta}}\end{cases}}^{\lfloor nb\rfloor}\psi(nx-k)\,\gamma_{2k}+\sum_{\begin{cases}k=\lfloor nx\rfloor+1\\ :\left|\frac{k}{n}-x\right|>\frac{1}{n^{\beta}}\end{cases}}^{\lfloor nb\rfloor}\psi(nx-k)\,\gamma_{2k}\right\}\leq \tag{2.125}$$

$$\frac{1}{\Gamma(\alpha)}\left\{\left(\sum_{\begin{cases}k=\lfloor nx\rfloor+1\\ :\left|\frac{k}{n}-x\right|\leq\frac{1}{n^{\beta}}\end{cases}}^{\lfloor nb\rfloor}\psi(nx-k)\right)\frac{\omega_{1}\left(D_{*x}^{\alpha}f,\frac{1}{n^{\beta}}\right)_{[x,b]}}{\alpha n^{\alpha\beta}}+\right.$$

$$\left.\left(\sum_{\begin{cases}k=\lfloor nx\rfloor+1\\ :\left|\frac{k}{n}-x\right|>\frac{1}{n^{\beta}}\end{cases}}^{\lfloor nb\rfloor}\psi(nx-k)\right)\left\|D_{*x}^{\alpha}f\right\|_{\infty,[x,b]}\frac{(b-x)^{\alpha}}{\alpha}\right\}\leq$$

$$\frac{1}{\Gamma(\alpha+1)}\left\{\frac{\omega_1\left(D_{*x}^{\alpha}f,\frac{1}{n^{\beta}}\right)_{[x,b]}}{n^{\alpha\beta}}+\right.$$

$$\left.\left(\sum_{\substack{\left\{k=-\infty\right.\\ \left.:\left|\frac{k}{n}-x\right|>\frac{1}{n^{\beta}}\right.}}^{\infty}\psi(nx-k)\right)\left\|D_{*x}^{\alpha}f\right\|_{\infty,[x,b]}(b-x)^{\alpha}\right\}\leq \tag{2.126}$$

$$\frac{1}{\Gamma(\alpha+1)}\left\{\frac{\omega_1\left(D_{*x}^{\alpha}f,\frac{1}{n^{\beta}}\right)_{[x,b]}}{n^{\alpha\beta}}+\frac{4}{\pi e^{\left(n^{1-\beta}-2\right)}}\left\|D_{*x}^{\alpha}f\right\|_{\infty,[x,b]}(b-x)^{\alpha}\right\}.$$

So we have proved that

$$\|u_{2n}(x)\|\leq\frac{1}{\Gamma(\alpha+1)}\left\{\frac{\omega_1\left(D_{*x}^{\alpha}f,\frac{1}{n^{\beta}}\right)_{[x,b]}}{n^{\alpha\beta}}+\right. \tag{2.127}$$

$$\left.\frac{4}{\pi e^{\left(n^{1-\beta}-2\right)}}\left\|D_{*x}^{\alpha}f\right\|_{\infty,[x,b]}(b-x)^{\alpha}\right\}.$$

Therefore

$$\|u_n(x)\|\leq\|u_{1n}(x)\|+\|u_{2n}(x)\|\leq$$

$$\frac{1}{\Gamma(\alpha+1)}\left\{\frac{\omega_1\left(D_{x-}^{\alpha}f,\frac{1}{n^{\beta}}\right)_{[a,x]}+\omega_1\left(D_{*x}^{\alpha}f,\frac{1}{n^{\beta}}\right)_{[x,b]}}{n^{\alpha\beta}}+\right. \tag{2.128}$$

$$\left.\frac{4}{\pi e^{\left(n^{1-\beta}-2\right)}}\left(\left\|D_{x-}^{\alpha}f\right\|_{\infty,[a,x]}(x-a)^{\alpha}+\left\|D_{*x}^{\alpha}f\right\|_{\infty,[x,b]}(b-x)^{\alpha}\right)\right\}.$$

From the proof of Theorem 2.14 we get that

$$\left|G_n^*\left((\cdot-x)^j\right)(x)\right|\leq\frac{1}{n^{\beta j}}+(b-a)^j\frac{4}{\pi e^{\left(n^{1-\beta}-2\right)}}, \tag{2.129}$$

for $j=1,\ldots,N-1$, $\forall\, x\in[a,b]$.

Putting things together, we have established

$$\left\|G_n^*(f,x)-f(x)\left(\sum_{k=\lceil na\rceil}^{\lfloor nb\rfloor}\psi(nx-k)\right)\right\|\leq\sum_{j=1}^{N-1}\frac{\left\|f^{(j)}(x)\right\|}{j!} \tag{2.130}$$

$$\left[\frac{1}{n^{\beta j}} + (b-a)^j \frac{4}{\pi e^{(n^{1-\beta}-2)}}\right] +$$

$$\frac{1}{\Gamma(\alpha+1)} \left\{ \frac{\omega_1\left(D_{x-}^{\alpha} f, \frac{1}{n^{\beta}}\right)_{[a,x]} + \omega_1\left(D_{*x}^{\alpha} f, \frac{1}{n^{\beta}}\right)_{[x,b]}}{n^{\alpha\beta}} + \right.$$

$$\left. \frac{4}{\pi e^{(n^{1-\beta}-2)}} \left(\left\| D_{x-}^{\alpha} f \right\|_{\infty,[a,x]} (x-a)^{\alpha} + \left\| D_{*x}^{\alpha} f \right\|_{\infty,[x,b]} (b-x)^{\alpha} \right) \right\} =: K_n(x). \tag{2.131}$$

As a result we derive

$$\| G_n(f, x) - f(x) \| \le 2.412 K_n(x), \ \forall x \in [a, b]. \tag{2.132}$$

We further have that

$$\| K_n \|_{\infty} \le \sum_{j=1}^{N-1} \frac{\| f^{(j)} \|_{\infty}}{j!} \left[\frac{1}{n^{\beta j}} + (b-a)^j \frac{4}{\pi e^{(n^{1-\beta}-2)}}\right] + \tag{2.133}$$

$$\frac{1}{\Gamma(\alpha+1)} \left\{ \frac{\left\{ \sup_{x\in[a,b]} \left(\omega_1\left(D_{x-}^{\alpha} f, \frac{1}{n^{\beta}}\right)_{[a,x]} \right) + \sup_{x\in[a,b]} \left(\omega_1\left(D_{*x}^{\alpha} f, \frac{1}{n^{\beta}}\right)_{[x,b]} \right) \right\}}{n^{\alpha\beta}} + \right.$$

$$\frac{4}{\pi e^{(n^{1-\beta}-2)}} (b-a)^{\alpha} \cdot$$

$$\left. \left\{ \left(\sup_{x\in[a,b]} \left(\| D_{x-}^{\alpha} f \| \right) + \sup_{x\in[a,b]} \left(\| D_{*x}^{\alpha} f \|_{\infty,[x,b]} \right) \right) \right\} \right\} =: E_n.$$

Hence it holds

$$\| G_n f - f \|_{\infty} \le 2.412 E_n. \tag{2.134}$$

We observe the following:

We have

$$\left(D_{x-}^{\alpha} f\right)(y) = \frac{(-1)^N}{\Gamma(N-\alpha)} \int_y^x (J-y)^{N-\alpha-1} f^{(N)}(J) \, dJ, \ \forall y \in [a, x] \tag{2.135}$$

and

$$\left\| \left(D_{x-}^{\alpha} f \right)(y) \right\| \leq \frac{1}{\Gamma(N-\alpha)} \left(\int_y^x (J-y)^{N-\alpha-1} dJ \right) \left\| f^{(N)} \right\|_{\infty} =$$

$$\frac{1}{\Gamma(N-\alpha)} \frac{(x-y)^{N-\alpha}}{(N-\alpha)} \left\| f^{(N)} \right\|_{\infty} = \frac{(x-y)^{N-\alpha}}{\Gamma(N-\alpha+1)} \left\| f^{(N)} \right\|_{\infty}$$

$$\leq \frac{(b-a)^{N-\alpha}}{\Gamma(N-\alpha+1)} \left\| f^{(N)} \right\|_{\infty}. \tag{2.136}$$

That is

$$\left\| D_{x-}^{\alpha} f \right\|_{\infty,[a,x]} \leq \frac{(b-a)^{N-\alpha}}{\Gamma(N-\alpha+1)} \left\| f^{(N)} \right\|_{\infty}, \tag{2.137}$$

and

$$\sup_{x\in[a,b]} \left\| D_{x-}^{\alpha} f \right\|_{\infty,[a,x]} \leq \frac{(b-a)^{N-\alpha}}{\Gamma(N-\alpha+1)} \left\| f^{(N)} \right\|_{\infty}. \tag{2.138}$$

Similarly we have

$$\left(D_{*x}^{\alpha} f \right)(y) = \frac{1}{\Gamma(N-\alpha)} \int_x^y (y-t)^{N-\alpha-1} f^{(N)}(t)\, dt, \ \forall\, y \in [x,b]. \tag{2.139}$$

Thus we get

$$\left\| \left(D_{*x}^{\alpha} f \right)(y) \right\| \leq \frac{1}{\Gamma(N-\alpha)} \left(\int_x^y (y-t)^{N-\alpha-1} dt \right) \left\| f^{(N)} \right\|_{\infty} \leq \tag{2.140}$$

$$\frac{1}{\Gamma(N-\alpha)} \frac{(y-x)^{N-\alpha}}{(N-\alpha)} \left\| f^{(N)} \right\|_{\infty} \leq \frac{(b-a)^{N-\alpha}}{\Gamma(N-\alpha+1)} \left\| f^{(N)} \right\|_{\infty}.$$

Hence

$$\left\| D_{*x}^{\alpha} f \right\|_{\infty,[x,b]} \leq \frac{(b-a)^{N-\alpha}}{\Gamma(N-\alpha+1)} \left\| f^{(N)} \right\|_{\infty}, \tag{2.141}$$

and

$$\sup_{x\in[a,b]} \left\| D_{*x}^{\alpha} f \right\|_{\infty,[x,b]} \leq \frac{(b-a)^{N-\alpha}}{\Gamma(N-\alpha+1)} \left\| f^{(N)} \right\|_{\infty}. \tag{2.142}$$

From (2.94) and (2.95) we get

$$\sup_{x\in[a,b]} \omega_1 \left(D_{x-}^{\alpha} f, \frac{1}{n^{\beta}} \right)_{[a,x]} \leq \frac{2 \left\| f^{(N)} \right\|_{\infty}}{\Gamma(N-\alpha+1)} (b-a)^{N-\alpha}, \tag{2.143}$$

and

$$\sup_{x\in[a,b]} \omega_1 \left(D^{\alpha}_{*x} f, \frac{1}{n^{\beta}} \right)_{[x,b]} \leq \frac{2 \left\| f^{(N)} \right\|_{\infty}}{\Gamma(N-\alpha+1)} (b-a)^{N-\alpha}. \tag{2.144}$$

That is $E_n < \infty$.

We finally notice that

$$G_n(f,x) - \sum_{j=1}^{N-1} \frac{f^{(j)}(x)}{j!} G_n\left((\cdot - x)^j \right)(x) - f(x) =$$

$$\frac{G_n^*(f,x)}{\left(\sum_{k=\lceil na \rceil}^{\lfloor nb \rfloor} \psi(nx-k) \right)} - \frac{1}{\left(\sum_{k=\lceil na \rceil}^{\lfloor nb \rfloor} \psi(nx-k) \right)} \cdot$$

$$\left(\sum_{j=1}^{N-1} \frac{f^{(j)}(x)}{j!} G_n^*\left((\cdot - x)^j \right)(x) \right) - f(x) =$$

$$\frac{1}{\left(\sum_{k=\lceil na \rceil}^{\lfloor nb \rfloor} \psi(nx-k) \right)} \left[G_n^*(f,x) - \left(\sum_{j=1}^{N-1} \frac{f^{(j)}(x)}{j!} G_n^*\left((\cdot - x)^j \right)(x) \right) \right. \tag{2.145}$$

$$\left. - \left(\sum_{k=\lceil na \rceil}^{\lfloor nb \rfloor} \psi(nx-k) \right) f(x) \right].$$

Therefore we get

$$\left\| G_n(f,x) - \sum_{j=1}^{N-1} \frac{f^{(j)}(x)}{j!} G_n\left((\cdot - x)^j \right)(x) - f(x) \right\| \leq 2.412 \cdot$$

$$\left\| G_n^*(f,x) - \left(\sum_{j=1}^{N-1} \frac{f^{(j)}(x)}{j!} G_n^*\left((\cdot - x)^j \right)(x) \right) - \left(\sum_{k=\lceil na \rceil}^{\lfloor nb \rfloor} \psi(nx-k) \right) f(x) \right\|, \tag{2.146}$$

$\forall\, x \in [a,b]$.

The proof of the theorem is now finished. ■

Next we apply Theorem 2.30 for $N = 1$.

Theorem 2.31 *Let* $0 < \alpha, \beta < 1$, $f \in C^1([a,b], X)$, $x \in [a,b]$, $n \in \mathbb{N} : n^{1-\beta} > 2$. *Then (i)*

$$\|G_n(f,x) - f(x)\| \leq$$

$$\frac{2.412}{\Gamma(\alpha+1)} \left\{ \frac{\left(\omega_1 \left(D_{x-}^{\alpha} f, \frac{1}{n^{\beta}} \right)_{[a,x]} + \omega_1 \left(D_{*x}^{\alpha} f, \frac{1}{n^{\beta}} \right)_{[x,b]} \right)}{n^{\alpha\beta}} + \right.$$

$$\left. \frac{4}{\pi e^{\left(n^{1-\beta}-2\right)}} \left(\left\| D_{x-}^{\alpha} f \right\|_{\infty,[a,x]} (x-a)^{\alpha} + \left\| D_{*x}^{\alpha} f \right\|_{\infty,[x,b]} (b-x)^{\alpha} \right) \right\}, \tag{2.147}$$

and

(ii)

$$\|G_n f - f\|_{\infty} \leq \frac{2.412}{\Gamma(\alpha+1)} \cdot$$

$$\left\{ \frac{\left(\sup_{x\in[a,b]} \omega_1 \left(D_{x-}^{\alpha} f, \frac{1}{n^{\beta}} \right)_{[a,x]} + \sup_{x\in[a,b]} \omega_1 \left(D_{*x}^{\alpha} f, \frac{1}{n^{\beta}} \right)_{[x,b]} \right)}{n^{\alpha\beta}} + \right.$$

$$\left. \frac{4(b-a)^{\alpha}}{\pi e^{\left(n^{1-\beta}-2\right)}} \left(\sup_{x\in[a,b]} \left\| D_{x-}^{\alpha} f \right\|_{\infty,[a,x]} + \sup_{x\in[a,b]} \left\| D_{*x}^{\alpha} f \right\|_{\infty,[x,b]} \right) \right\}. \tag{2.148}$$

When $\alpha = \frac{1}{2}$ we derive

Corollary 2.32 *Let* $0 < \beta < 1$, $f \in C^1([a,b], X)$, $x \in [a,b]$, $n \in \mathbb{N} : n^{1-\beta} > 2$. *Then*

(i)

$$\|G_n(f,x) - f(x)\| \leq$$

$$\frac{4.824}{\sqrt{\pi}} \left\{ \frac{\left(\omega_1 \left(D_{x-}^{\frac{1}{2}} f, \frac{1}{n^{\beta}} \right)_{[a,x]} + \omega_1 \left(D_{*x}^{\frac{1}{2}} f, \frac{1}{n^{\beta}} \right)_{[x,b]} \right)}{n^{\frac{\beta}{2}}} + \right.$$

$$\left. \frac{4}{\pi e^{\left(n^{1-\beta}-2\right)}} \left(\left\| D_{x-}^{\frac{1}{2}} f \right\|_{\infty,[a,x]} \sqrt{(x-a)} + \left\| D_{*x}^{\frac{1}{2}} f \right\|_{\infty,[x,b]} \sqrt{(b-x)} \right) \right\}, \tag{2.149}$$

and

(ii)

$$\|G_n f - f\|_\infty \le \frac{4.824}{\sqrt{\pi}} \cdot$$

$$\left\{ \frac{\left(\sup_{x\in[a,b]} \omega_1 \left(D_{x-}^{\frac{1}{2}} f, \frac{1}{n^\beta} \right)_{[a,x]} + \sup_{x\in[a,b]} \omega_1 \left(D_{*x}^{\frac{1}{2}} f, \frac{1}{n^\beta} \right)_{[x,b]} \right)}{n^{\frac{\beta}{2}}} + \right.$$

$$\left. \frac{4\sqrt{(b-a)}}{\pi e^{(n^{1-\beta}-2)}} \left(\sup_{x\in[a,b]} \left\| D_{x-}^{\frac{1}{2}} f \right\|_{\infty,[a,x]} + \sup_{x\in[a,b]} \left\| D_{*x}^{\frac{1}{2}} f \right\|_{\infty,[x,b]} \right) \right\} < \infty. \quad (2.150)$$

We make

Remark 2.33 Some convergence analysis follows based on Corollary 2.32.

Let $0 < \beta < 1$, $f \in C^1([a,b], X)$, $x \in [a,b]$, $n \in \mathbb{N} : n^{1-\beta} > 2$. We elaborate on (2.150). Assume that

$$\omega_1 \left(D_{x-}^{\frac{1}{2}} f, \frac{1}{n^\beta} \right)_{[a,x]} \le \frac{K_1}{n^\beta}, \quad (2.151)$$

and

$$\omega_1 \left(D_{*x}^{\frac{1}{2}} f, \frac{1}{n^\beta} \right)_{[x,b]} \le \frac{K_2}{n^\beta}, \quad (2.152)$$

$\forall\, x \in [a,b]$, $\forall\, n \in \mathbb{N}$, where $K_1, K_2 > 0$.

Then it holds

$$\frac{\left[\sup_{x\in[a,b]} \omega_1 \left(D_{x-}^{\frac{1}{2}} f, \frac{1}{n^\beta} \right)_{[a,x]} + \sup_{x\in[a,b]} \omega_1 \left(D_{*x}^{\frac{1}{2}} f, \frac{1}{n^\beta} \right)_{[x,b]} \right]}{n^{\frac{\beta}{2}}} \le$$

$$\frac{\frac{(K_1+K_2)}{n^\beta}}{n^{\frac{\beta}{2}}} = \frac{(K_1+K_2)}{n^{\frac{3\beta}{2}}} = \frac{K}{n^{\frac{3\beta}{2}}}, \quad (2.153)$$

where $K := K_1 + K_2 > 0$.

The other summand of the right hand side of (2.150), for large enough n, converges to zero at the speed $\frac{1}{e^{(n^{1-\beta}-2)}}$, so it is about $\frac{L}{e^{(n^{1-\beta}-2)}}$, where $L > 0$ is a constant.

Then, for large enough $n \in \mathbb{N}$, by (2.150), (2.153) and the above comment, we obtain that

$$\|G_n f - f\|_\infty \leq \frac{M}{n^{\frac{3\beta}{2}}}, \tag{2.154}$$

where $M > 0$, converging to zero at the high speed of $\frac{1}{n^{\frac{3\beta}{2}}}$.

In Theorem 2.11, for $f \in C([a, b], X)$ and for large enough $n \in \mathbb{N}$, the speed is $\frac{1}{n^\beta}$. So by (2.154), $\|G_n f - f\|_\infty$ converges much faster to zero. The last comes because we assumed differentiability of f. Notice that in Corollary 2.32 no initial condition is assumed.

Next, we will present an alternative fractional approximation by G_n, $n \in \mathbb{N}$.

Notation 2.34 *Let $\overline{n} \in \mathbb{N}$, we denote the left iterated fractional derivative*

$$D_{*x}^{\overline{n}x} = D_{*x}^{\alpha} D_{*x}^{\alpha} ... D_{*x}^{\alpha}, \ (\overline{n} \text{ - times}), \tag{2.155}$$

$x \in [a, b]$, $0 < \alpha < 1$.

Similarly, we also denote the right iterated fractional derivative

$$D_{x-}^{\overline{n}x} = D_{x-}^{\alpha} D_{x-}^{\alpha} ... D_{x-}^{\alpha}, \ (\overline{n} \text{ - times}), \tag{2.156}$$

$x \in [a, b]$.

We need

Definition 2.35 Let $\overline{n} \in \mathbb{N}$, $D_x^{(\overline{n}+1)\alpha} f$ denote any of $D_{*x}^{(\overline{n}+1)\alpha}$, $D_{x-}^{(\overline{n}+1)\alpha}$, and $\delta > 0$. We set

$$\omega_1 \left(D_x^{(\overline{n}+1)\alpha} f, \delta \right) = \max \left\{ \omega_1 \left(D_{*x}^{(\overline{n}+1)\alpha} f, \delta \right)_{[x,b]}, \omega_1 \left(D_{x-}^{(\overline{n}+1)\alpha} f, \delta \right)_{[a,x]} \right\}, \tag{2.157}$$

where $x \in [a, b]$. Here the moduli of continuity are considered over $[x, b]$ and $[a, x]$, respectively.

We also need

Theorem 2.36 ([13], p. 123) *Let $0 < \alpha < 1$, $f : [a, b] \to \mathbb{R}$, $f' \in L_\infty([a, b])$, $x \in [a, b]$ being fixed. Assume that $D_{*x}^{k\alpha} f \in C([x, b])$, $k = 0, 1, \ldots, \overline{n}+1$, $\overline{n} \in \mathbb{N}$, and $\left(D_{*x}^{i\alpha} f\right)(x) = 0$, $i = 2, 3, \ldots, \overline{n}+1$. Also, suppose that $D_{x-}^{k\alpha} f \in C([a, x])$, for $k = 0, 1, \ldots, \overline{n}+1$, and $\left(D_{x-}^{i\alpha} f\right)(x) = 0$, for $i = 2, 3, \ldots, \overline{n}+1$. Then*

$$|f(\cdot) - f(x)| \leq \frac{\omega_1 \left(D_x^{(\overline{n}+1)\alpha} f, \delta \right)}{\Gamma((\overline{n}+1)\alpha + 1)} \left[|\cdot - x|^{(\overline{n}+1)\alpha} + \frac{|\cdot - x|^{(\overline{n}+1)\alpha+1}}{\delta((\overline{n}+1)\alpha + 1)} \right], \ \delta > 0. \tag{2.158}$$

We present

Theorem 2.37 *Let $f \in C([a,b])$ and all as in Theorem 2.36, $n \in \mathbb{N} : n^{1-\alpha} > 2$. Then*

$$|(G_n f)(x) - f(x)| \leq \frac{2.412\omega_1\left(D_x^{(\overline{n}+1)\alpha} f, \delta\right)}{\Gamma((\overline{n}+1)\alpha+1)}$$

$$\left\{\left[\frac{1}{n^{(\overline{n}+1)\alpha^2}} + \frac{4(b-a)^{(\overline{n}+1)\alpha}}{\pi e^{(n^{1-\alpha}-2)}}\right] + \right. \tag{2.159}$$

$$\left.\frac{1}{\delta((\overline{n}+1)\alpha+1)}\left[\frac{1}{n^{\alpha[(\overline{n}+1)\alpha+1]}} + \frac{4(b-a)^{(\overline{n}+1)\alpha+1}}{\pi e^{(n^{1-\alpha}-2)}}\right]\right\}, \ \delta > 0.$$

Hence $\lim_{n \to +\infty} G_n(f)(x) = f(x)$.

Proof We notice that G_n is a positive linear operator with $G_n(1) = 1$.
Let $f \in C([a,b], \mathbb{R})$, then $|f| \leq |f|$ and $-|f| \leq f \leq |f|$.
Hence $-G_n(|f|) \leq G_n(f) \leq G_n(|f|)$ and $|G_n(f)| \leq G_n(|f|)$.
Therefore

$$|(G_n f)(x) - f(x)| = |(G_n f)(x) - G_n(f(x))(x)| =$$

$$|G_n(f - f(x))(x)| \leq G_n(|f - f(x)|)(x) \text{ (by (2.158))} \leq \tag{2.160}$$

$$\frac{\omega_1\left(D_x^{(\overline{n}+1)\alpha} f, \delta\right)}{\Gamma((\overline{n}+1)\alpha+1)}\left[G_n\left(|\cdot - x|^{(\overline{n}+1)\alpha}\right)(x) + \frac{G\left(|\cdot - x|^{(\overline{n}+1)\alpha+1}\right)(x)}{\delta((\overline{n}+1)\alpha+1)}\right] =$$

$$\frac{\omega_1\left(D_x^{(\overline{n}+1)\alpha} f, \delta\right)}{\Gamma((\overline{n}+1)\alpha+1)\sum_{k=\lceil na \rceil}^{\lfloor nb \rfloor}\psi(nx-k)}$$

$$\left[\sum_{k=\lceil na \rceil}^{\lfloor nb \rfloor}\left|\frac{k}{n} - x\right|^{(\overline{n}+1)\alpha}\psi(nx-k) + \frac{\sum_{k=\lceil na \rceil}^{\lfloor nb \rfloor}\left|\frac{k}{n} - x\right|^{(\overline{n}+1)\alpha+1}\psi(nx-k)}{\delta((\overline{n}+1)\alpha+1)}\right] \overset{(2.28.)}{\leq} \tag{2.161}$$

$$\frac{2.412\omega_1\left(D_x^{(\overline{n}+1)\alpha} f, \delta\right)}{\Gamma((\overline{n}+1)\alpha+1)}$$

$$\left\{\left[\sum_{\substack{k=\lceil na\rceil \\ :\left|\frac{k}{n}-x\right|\leq\frac{1}{n^{\alpha}}}}^{\lfloor nb\rfloor}\left|\frac{k}{n}-x\right|^{(\overline{n}+1)\alpha}\psi(nx-k)+\right.\right.$$

$$\left.\sum_{\substack{k=\lceil na\rceil \\ :\left|\frac{k}{n}-x\right|>\frac{1}{n^{\alpha}}}}^{\lfloor nb\rfloor}\left|\frac{k}{n}-x\right|^{(\overline{n}+1)\alpha}\psi(nx-k)\right]+$$

$$\frac{1}{\delta((\overline{n}+1)\alpha+1)}\left[\sum_{\substack{k=\lceil na\rceil \\ :\left|\frac{k}{n}-x\right|\leq\frac{1}{n^{\alpha}}}}^{\lfloor nb\rfloor}\left|\frac{k}{n}-x\right|^{(\overline{n}+1)\alpha+1}\psi(nx-k)+\right. \tag{2.162}$$

$$\left.\left.\sum_{\substack{k=\lceil na\rceil \\ :\left|\frac{k}{n}-x\right|>\frac{1}{n^{\alpha}}}}^{\lfloor nb\rfloor}\left|\frac{k}{n}-x\right|^{(\overline{n}+1)\alpha+1}\psi(nx-k)\right]\right\}\overset{(2.23.)}{\leq}$$

$$\frac{2.412\omega_1\left(D_x^{(\overline{n}+1)\alpha}f,\delta\right)}{\Gamma((\overline{n}+1)\alpha+1)}\left\{\left[\frac{1}{n^{(\overline{n}+1)\alpha^2}}+\frac{4(b-a)^{(\overline{n}+1)\alpha}}{\pi e^{(n^{1-\alpha}-2)}}\right]\right.$$

$$\left.+\frac{1}{\delta((\overline{n}+1)\alpha+1)}\left[\frac{1}{n^{\alpha[(\overline{n}+1)\alpha+1]}}+\frac{4(b-a)^{(\overline{n}+1)\alpha+1}}{\pi e^{(n^{1-\alpha}-2)}}\right]\right\},\ \delta>0, \tag{2.163}$$

proving the claim. ■

We finish with

Corollary 2.38 *All as in Theorem 2.37, with* $\delta=\frac{1}{(\overline{n}+1)\alpha+1}$. *Then*

$$|(G_nf)(x)-f(x)|\leq\frac{2.412\omega_1\left(D_x^{(\overline{n}+1)\alpha}f,\frac{1}{(\overline{n}+1)\alpha+1}\right)}{\Gamma((\overline{n}+1)\alpha+1)}$$

$$\left\{\left[\frac{1}{n^{(\overline{n}+1)\alpha^2}}+\frac{4(b-a)^{(\overline{n}+1)\alpha}}{\pi e^{(n^{1-\alpha}-2)}}\right]+\left[\frac{1}{n^{\alpha[(\overline{n}+1)\alpha+1]}}+\frac{4(b-a)^{(\overline{n}+1)\alpha+1}}{\pi e^{(n^{1-\alpha}-2)}}\right]\right\}. \tag{2.164}$$

Hence $\lim_{n\to+\infty} G_n(f)(x) = f(x)$.

References

1. G.A. Anastassiou, Rate of convergence of some neural network operators to the unit-univariate case. J. Math. Anal. Appl. **212**, 237–262 (1997)
2. G.A. Anastassiou, *Quantitative Approximations* (Chapman & Hall / CRC, Boca Raton, New York, 2001)
3. G.A. Anastassiou, Univariate hyperbolic tangent neural network approximation. Math. Comput. Model. **53**, 1111–1132 (2011)
4. G.A. Anastassiou, Multivariate hyperbolic tangent neural network approximation. Comput. Math. **61**, 809–821 (2011)
5. G.A. Anastassiou, Multivariate sigmoidal neural network approximation. Neural Netw. **24**, 378–386 (2011)
6. G.A. Anastassiou, *Inteligent Systems: Approximation by Artificial Neural Networks, Intelligent Systems Reference Library*, vol. 19 (Springer, Heidelberg, 2011)
7. G.A. Anastassiou, Univariate sigmoidal neural network approximation. J. Comput. Anal. Appl. **14**(4), 659–690 (2012)
8. G.A. Anastassiou, Fractional neural network approximation. Comput. Math. Appl. **64**, 1655–1676 (2012)
9. G.A. Anastassiou, *Intelligent Systems II: Complete Approximation by Neural Network Operators* (Springer, Heidelberg, New York, 2016)
10. G.A. Anastassiou, Strong right fractional calculus for Banach space valued functions. Revista Proyecciones **36**(1), 149–186 (2017)
11. G.A. Anastassiou, Vector fractional Korovkin type Approximations. Dyn. Syst. Appl. **26**, 81–104 (2017)
12. G.A. Anastassiou, A strong fractional calculus theory for Banach space valued functions. Nonlinear Funct. Anal. Appl. (Korea) **22**(3), 495–524 (2017)
13. G.A. Anastassiou, *Nonlinearity: Ordinary and Fractional Approximations by Sublinear and Max-Product Operators* (Springer, Heidelberg, New York, 2018)
14. G.A. Anastassiou, Gudermannian function activated Banach space valued ordinary and fractional neural network approximations. Adv. Nonlinear Var. Inequalities (2022)
15. Z. Chen, F. Cao, The approximation operators with sigmoidal functions. Comput. Math. Appl. **58**, 758–765 (2009)
16. S. Haykin, *Neural Networks: A Comprehensive Foundation*, 2nd edn. (Prentice Hall, New York, 1998)
17. M. Kreuter, Sobolev spaces of vector-valued functions. Ulm University, Master Thesis in Mathematics, Ulm, Germany (2015)
18. W. McCulloch, W. Pitts, A logical calculus of the ideas immanent in nervous activity. Bull. Math. Biophys. **7**, 115–133 (1943)
19. J. Mikusinski, *The Bochner Integral* (Academic Press, New York, 1978)
20. T.M. Mitchell, *Machine Learning* (WCB-McGraw-Hill, New York, 1997)
21. G.E. Shilov, *Elementary Functional Analysis* (Dover Publications Inc, New York, 1996)
22. E.W. Weisstein, *Gudermannian*, MathWorld

Chapter 3
Generalized Symmetrical Sigmoid Function Induced Banach Space Valued Ordinary and Fractional Neural Network Approximations

Here we research the univariate quantitative approximation, ordinary and fractional, of Banach space valued continuous functions on a compact interval or all the real line by quasi-interpolation Banach space valued neural network operators. These approximations are derived by establishing Jackson type inequalities involving the modulus of continuity of the engaged function or its Banach space valued high order derivative or fractional derivatives. Our operators are defined by using a density function generated by a generalized symmetrical sigmoid function. The approximations are pointwise and of the uniform norm. The related Banach space valued feed-forward neural networks are with one hidden layer. It relies on [15].

3.1 Introduction

The author in [1, 2], see Chaps. 2–5, was the first to establish neural network approximation to continuous functions with rates by very specifically defined neural network operators of Cardaliagnet-Euvrard and "Squashing" types, by employing the modulus of continuity of the engaged function or its high order derivative, and producing very tight Jackson type inequalities. He treats there both the univariate and multivariate cases. The defining these operators "bell-shaped" and "squashing" functions are assumed to be of compact support. Also in [2] he gives the Nth order asymptotic expansion for the error of weak approximation of these two operators to a special natural class of smooth functions, see Chaps. 4–5 there.

The author inspired by [16], continued his studies on neural networks approximation by introducing and using the proper quasi-interpolation operators of sigmoidal and hyperbolic tangent type which resulted into [3–7], by treating both the univariate and multivariate cases. He did also the corresponding fractional cases [8, 9, 13].

© The Author(s), under exclusive license to Springer Nature Switzerland AG 2023
G. A. Anastassiou, *Banach Space Valued Neural Network*,
Studies in Computational Intelligence 1062,
https://doi.org/10.1007/978-3-031-16400-2_3

The author here performs generalized symmetrical sigmoid function based neural network approximations to continuous functions over compact intervals of the real line or over the whole $\mathbb{R}$ with valued to an arbitrary Banach space $(X, \|\cdot\|)$. Finally he treats completely the related X-valued fractional approximation. All convergences here are with rates expressed via the modulus of continuity of the involved function or its X-valued high order derivative, or X-valued fractional derivatives and given by very tight Jackson type inequalities.

Our compact intervals are not necessarily symmetric to the origin. Some of our upper bounds to error quantity are very flexible and general. In preparation to prove our results we establish important properties of the basic density function defining our operators which is induced by generalized symmetrical sigmoid function.

Feed-forward X-valued neural networks (FNNs) with one hidden layer, the only type of networks we deal with in this chapter, are mathematically expressed as

$$N_n(x) = \sum_{=0}^{n} c_j \sigma\left(\langle a_j \cdot x \rangle + b_j\right), \quad x \in \mathbb{R}^s, \ s \in \mathbb{N},$$

where for $0 \leq j \leq n$, $b_j \in \mathbb{R}$ are the thresholds, $a_j \in \mathbb{R}^s$ are the connection weights, $c_j \in X$ are the coefficients, $\langle a_j \cdot x \rangle$ is the inner product of a_j and x, and σ is the activation function of the network. About neural networks in general read [18, 19, 21]. See also [9] for a complete study of real valued approximation by neural network operators.

3.2 Auxiliary Results

Here we consider the generalized symmetrical sigmoid function [17]

$$f_1(x) = \frac{x}{(1 + |x|^\mu)^{\frac{1}{\mu}}}, \quad \mu > 0, x \in \mathbb{R}. \tag{3.1}$$

This has applications in immunology and protection from disease together with probability theory. It is also called a symmetrical protection curve.

The parameter μ is a shape parameter controling how fast the curve approaches the asymptotes for a given slope at the inflection point. When $\mu = 1$ f_1 is the absolute sigmoid function, and when $\mu = 2$ f_1 is the square root sigmoid function. When $\mu = 1.5$ the function approximates the arctangent function, when $\mu = 2.9$ it approximates the logistic function, and when $\mu = 3.4$ it approximates the error function. Parameter μ is estimated in the likelihood maximization [17]. For more see [17].

Next we study the particular generator sigmoid function

$$f_2(x) = \frac{x}{\left(1 + |x|^\lambda\right)^{\frac{1}{\lambda}}}, \quad \lambda \text{ is an odd number, } x \in \mathbb{R}. \tag{3.2}$$

We have that $f_2(0) = 0$, and

$$f_2(-x) = -\frac{x}{\left(1+|x|^\lambda\right)^{\frac{1}{\lambda}}} = -f_2(x),$$

i.e.

$$f_2(-x) = -f_2(x), \tag{3.3}$$

so f_2 is symmetric with respect to zero.

Let $x \geq 0$, then $f_2(x) = \frac{x}{(1+|x|^\lambda)^{\frac{1}{\lambda}}}$, and

$$f_2{}'(x) = \frac{\left(1+x^\lambda\right)^{\frac{1}{\lambda}} - x\frac{1}{\lambda}\left(1+x^\lambda\right)^{\frac{1}{\lambda}-1}\lambda x^{\lambda-1}}{\left(1+x^\lambda\right)^{\frac{2}{\lambda}}} =$$

$$\frac{\left(1+x^\lambda\right)^{\frac{1}{\lambda}}\left[1-\left(1+x^\lambda\right)^{-1}x^\lambda\right]}{\left(1+x^\lambda\right)^{\frac{2}{\lambda}}} = \frac{\left(1+x^\lambda\right)^{\frac{1}{\lambda}}\left[1-\frac{x^\lambda}{1+x^\lambda}\right]}{\left(1+x^\lambda\right)^{\frac{2}{\lambda}}} =$$

$$\frac{\left(1+x^\lambda\right)^{\frac{1}{\lambda}}\left[1+x^\lambda-x^\lambda\right]}{\left(1+x^\lambda\right)\left(1+x^\lambda\right)^{\frac{2}{\lambda}}} = \frac{\left(1+x^\lambda\right)^{\frac{1}{\lambda}}}{\left(1+x^\lambda\right)\left(1+x^\lambda\right)^{\frac{2}{\lambda}}} = \frac{1}{\left(1+x^\lambda\right)^{\frac{\lambda+1}{\lambda}}}. \tag{3.4}$$

That is when $x \geq 0$, we get that

$$f_2'(x) = \frac{1}{\left(1+x^\lambda\right)^{\frac{\lambda+1}{\lambda}}} > 0, \tag{3.5}$$

and f_2 is strictly increasing on $[0, +\infty)$. Let $x_1 < x_2 \leq 0$, then $-x_1 > -x_2$ and $f_2(-x_1) > f_2(-x_2)$, hence $-f_2(x_1) > -f_2(x_2)$ and $f_2(x_1) < f_2(x_2)$. Therefore f_2 is strictly increasing on $(-\infty, 0]$. Consequently f_2 is strictly increasing on $\mathbb{R}$.

Let $x > 0$, then $f_2(x) = \frac{x}{(1+x^\lambda)^{\frac{1}{\lambda}}} = \frac{1}{\frac{(1+x^\lambda)^{\frac{1}{\lambda}}}{x^{\frac{\lambda}{\lambda}}}} = \frac{1}{\left(\frac{1}{x^\lambda}+1\right)^{\frac{1}{\lambda}}} \underset{x\to+\infty}{\to} 1$. I.e.

$$f_2(+\infty) = 1. \tag{3.6}$$

Let $x < 0$, then $f_2(x) = \frac{x}{(1+(-1)^\lambda x^\lambda)^{\frac{1}{\lambda}}} = \frac{x}{(1-x^\lambda)^{\frac{1}{\lambda}}} = \frac{1}{\left(\frac{1}{x^\lambda}-1\right)^{\frac{1}{\lambda}}} \underset{x\to-\infty}{\to} \frac{1}{(-1)^{\frac{1}{\lambda}}} = -1.$

I.e.

$$f_2(-\infty) = 1. \tag{3.7}$$

Let us consider the activation function:

$$\chi(x) = \frac{1}{4}\left[f_2(x+1) - f_2(x-1)\right] = \tag{3.8}$$

$$\frac{1}{4}\left[\frac{(x+1)}{\left(1+|x+1|^{\lambda}\right)^{\frac{1}{\lambda}}} - \frac{(x-1)}{\left(1+|x-1|^{\lambda}\right)^{\frac{1}{\lambda}}}\right].$$

We have that

$$\chi(-x) = \frac{1}{4}\left[\frac{-x+1}{\left(1+|-x+1|^{\lambda}\right)^{\frac{1}{\lambda}}} - \frac{(-x-1)}{\left(1+|-x-1|^{\lambda}\right)^{\frac{1}{\lambda}}}\right] =$$

$$\frac{1}{4}\left[\frac{-(x-1)}{\left(1+|x-1|^{\lambda}\right)^{\frac{1}{\lambda}}} + \frac{(x+1)}{\left(1+|x+1|^{\lambda}\right)^{\frac{1}{\lambda}}}\right] =$$

$$\frac{1}{4}\left[\frac{(x+1)}{\left(1+|x+1|^{\lambda}\right)^{\frac{1}{\lambda}}} - \frac{(x-1)}{\left(1+|x-1|^{\lambda}\right)^{\frac{1}{\lambda}}}\right] = \chi(x).$$

I.e.

$$\chi(x) = \chi(-x), \quad \forall\, x \in \mathbb{R}. \tag{3.9}$$

We see that

$$\chi(0) = \frac{1}{4}\left[f_2(1) - f_2(-1)\right] = \frac{1}{4}\left[f_2(1) - (-f_2(1))\right] =$$

$$\frac{1}{4}\left[f_2(1) + f_2(1)\right] = \frac{1}{2} f_2(1) = \frac{1}{2}\frac{1}{2^{\frac{1}{\lambda}}} = \frac{1}{2\sqrt[\lambda]{2}}.$$

That is

$$\chi(0) = \frac{1}{2\sqrt[\lambda]{2}}. \tag{3.10}$$

Since $x+1 > x-1$, we get $\chi(x) > 0$, $\forall\, x \in \mathbb{R}$.

Let $x \geq 1$, then $x-1 \geq 0$, hence $x+1 > x-1 (\geq 0)$. Thus

$$\chi'(x) = \frac{1}{4}\left[f_2'(x+1) - f_2'(x-1)\right] =$$

$$\frac{1}{4}\left[\frac{1}{\left(1+(x+1)^{\lambda}\right)^{\frac{\lambda+1}{\lambda}}} - \frac{1}{\left(1+(x-1)^{\lambda}\right)^{\frac{\lambda+1}{\lambda}}}\right] = \tag{3.11}$$

$$\frac{1}{4}\left[\frac{\left(1+(x-1)^{\lambda}\right)^{\frac{\lambda+1}{\lambda}}-\left(1+(x+1)^{\lambda}\right)^{\frac{\lambda+1}{\lambda}}}{\left(1+(x+1)^{\lambda}\right)^{\frac{\lambda+1}{\lambda}}\left(1+(x-1)^{\lambda}\right)^{\frac{\lambda+1}{\lambda}}}\right]=:(\xi).$$

(from $0 \leq x-1 < x+1$, then $(x-1)^{\lambda} < (x+1)^{\lambda}$ and $1+(x-1)^{\lambda} < 1+(x+1)^{\lambda}$, and $\left(1+(x-1)^{\lambda}\right)^{\frac{\lambda+1}{\lambda}} < \left(1+(x+1)^{\lambda}\right)^{\frac{\lambda+1}{\lambda}}$).

Hence $\chi'(x) = (\xi) < 0$.

Therefore χ is strictly decreasing over $[1, +\infty)$.

Let now $0 \leq x \leq 1$, then $1-x \geq 0$, and $f_2(x-1) = f_2(-(1-x)) = -f_2(1-x)$, and $(f_2(x-1))' = -(f_2(1-x))' = -f_2'(1-x)(1-x)' = -f_2'(1-x)(-1) = f_2'(1-x)$.

That is

$$(f_2(x-1))' = f_2'(1-x). \tag{3.12}$$

Again we have

$$\chi'(x) = \frac{1}{4}\left[f_2'(x+1) - f_2'(x-1)\right] = \frac{1}{4}\left[f_2'(x+1) - f_2'(1-x)\right] =$$

$$\frac{1}{4}\left[\frac{1}{\left(1+(x+1)^{\lambda}\right)^{\frac{\lambda+1}{\lambda}}} - \frac{1}{\left(1+(1-x)^{\lambda}\right)^{\frac{\lambda+1}{\lambda}}}\right] =$$

$$\frac{1}{4}\left[\frac{\left(1+(1-x)^{\lambda}\right)^{\frac{\lambda+1}{\lambda}}-\left(1+(1+x)^{\lambda}\right)^{\frac{\lambda+1}{\lambda}}}{\left(1+(x+1)^{\lambda}\right)^{\frac{\lambda+1}{\lambda}}\left(1+(1-x)^{\lambda}\right)^{\frac{\lambda+1}{\lambda}}}\right] < 0, \tag{3.13}$$

(by $-x < x$ and $0 \leq 1-x < 1+x$, $(1-x)^{\lambda} < (1+x)^{\lambda}$ and $\left(1+(1-x)^{\lambda}\right)^{\frac{\lambda+1}{\lambda}} < \left(1+(1+x)^{\lambda}\right)^{\frac{\lambda+1}{\lambda}}$)

when $0 < x \leq 1$, therefore $\chi'(x) < 0$, when $0 < x \leq 1$.

That is χ is strictly decreasing on $(0, 1]$.

By continuity of χ it is strictly decreasing on $[0, \infty)$.

Since χ is symmetric with respect to y-axis, χ is stricly increasing on $(-\infty, 0]$.

Next we would like to find $\chi'(0)$:

Let now $-1 \leq x \leq 0 (< 1)$, then $x+1 \geq 0$, and $1-x > 0$, and again $(f_2(x-1))' = f_2'(1-x)$.

Again it holds

$$\chi'(x) = \frac{1}{4}\left[f_2'(x+1) - (f_2(x-1))'\right] = \frac{1}{4}\left[f_2'(x+1) - f_2'(1-x)\right] =$$

$$\frac{1}{4}\left[\frac{\left(1+(1-x)^{\lambda}\right)^{\frac{\lambda+1}{\lambda}}-\left(1+(1+x)^{\lambda}\right)^{\frac{\lambda+1}{\lambda}}}{\left(1+(x+1)^{\lambda}\right)^{\frac{\lambda+1}{\lambda}}\left(1+(1-x)^{\lambda}\right)^{\frac{\lambda+1}{\lambda}}}\right], \tag{3.14}$$

true for $-1 \leq x \leq 0$.

In fact $\chi'(x)$ is given by the same formula (see (3.13), (3.14)) over $[-1, 1]$, which is a continuous function.

We have that

$$\chi'(0+) = \lim_{\substack{x \to 0+ \\ (x>0)}} \chi'(x) = 0. \tag{3.15}$$

Clearly it holds

$$\chi'(0-) = \lim_{\substack{x \to 0- \\ (x<0)}} \chi'(x) = 0. \tag{3.16}$$

Therefore it is

$$\chi'(0) = 0,$$

by continuity of χ' over $[-1, 1]$.

Clearly it is

$$\lim_{x \to +\infty} \chi(x) = 0, \quad \lim_{x \to -\infty} \chi(x) = 0, \tag{3.17}$$

therefore the x-axis is the horizontal asymptote of $\chi(x)$.

The value

$$\chi(0) = \frac{1}{2\sqrt[\lambda]{2}}$$

is the maximum of χ, which is a bell shaped function.

We give

Theorem 3.1 *We have that*

$$\sum_{i=-\infty}^{\infty} \chi(x-i) = 1, \ \forall x \in \mathbb{R}. \tag{3.18}$$

Proof We observe that

$$\sum_{i=-\infty}^{\infty} (f_2(x-i) - f_2(x-1-i)) =$$

$$\sum_{i=0}^{\infty} (f_2(x-i) - f_2(x-1-i)) + \sum_{i=-\infty}^{-1} (f_2(x-i) - f_2(x-1-i)).$$

Furthermore ($\overline{\lambda} \in \mathbb{Z}^+$) we have

$$\sum_{i=0}^{\infty}\left(f_2(x-i)-f_2(x-1-i)\right)=\lim_{\overline{\lambda}\to\infty}\sum_{i=0}^{\overline{\lambda}}\left(f_2(x-i)-f_2(x-1-i)\right) \tag{3.19}$$

(telescoping sum)

$$=\lim_{\overline{\lambda}\to\infty}\left(f_2(x)-f_2\left(x-\left(\overline{\lambda}+1\right)\right)\right)=1+f_2(x).$$

Similarly

$$\sum_{i=-\infty}^{-1}\left(f_2(x-i)-f_2(x-1-i)\right)=\lim_{\overline{\lambda}\to\infty}\sum_{i=-\overline{\lambda}}^{-1}\left(f_2(x-i)-f_2(x-1-i)\right)$$

$$=\lim_{\overline{\lambda}\to\infty}\left(f_2\left(x+\overline{\lambda}\right)-f_2(x)\right)=1-f_2(x). \tag{3.20}$$

So, by adding the last two limits we obtain

$$\sum_{i=-\infty}^{\infty}\left(f_2(x-i)-f_2(x-1-i)\right)=2,\ \forall\, x\in\mathbb{R}. \tag{3.21}$$

Therefore

$$\sum_{i=-\infty}^{\infty}\left(f_2(x+1-i)-f_2(x-i)\right)=2,\ \forall\, x\in\mathbb{R}. \tag{3.22}$$

Consequently, by adding (3.21), (3.22) we get

$$\sum_{i=-\infty}^{\infty}\left(f_2(x+1-i)-f_2(x-1-i)\right)=4,\ \forall\, x\in\mathbb{R}, \tag{3.23}$$

proving the claim. □

Furthemore we give:
Because χ is even it holds

$$\sum_{i=-\infty}^{\infty}\chi(i-x)=1,\ \forall\, x\in\mathbb{R}, \tag{3.24}$$

and

$$\sum_{i=-\infty}^{\infty}\chi(i+x)=1,\ \forall\, x\in\mathbb{R}. \tag{3.25}$$

That is

$$\sum_{i=-\infty}^{\infty} \chi(x+i) = 1, \ \forall x \in \mathbb{R}. \tag{3.26}$$

We give

Theorem 3.2 *It holds*

$$\int_{-\infty}^{\infty} \chi(x)\, dx = 1.$$

Proof We observe that

$$\int_{-\infty}^{\infty} \chi(x)\, dx = \sum_{j=-\infty}^{\infty} \int_{j}^{j+1} \chi(x)\, dx = \sum_{j=-\infty}^{\infty} \int_{0}^{1} \chi(x+j)\, dx = \tag{3.27}$$

$$\int_{0}^{1} \left(\sum_{j=-\infty}^{\infty} \chi(x+j)\, dx \right) = \int_{0}^{1} 1dx = 1.$$

□

So that $\chi(x)$ is a density function on $\mathbb{R}$.
We need

Theorem 3.3 *Let* $0 < \alpha < 1$, *and* $n \in \mathbb{N}$ *with* $n^{1-\alpha} > 2$. *It holds*

$$\sum_{\substack{j=-\infty \\ : |nx-j| \geq n^{1-\alpha}}}^{\infty} \chi(nx-j) < \frac{1}{2\lambda\left(n^{1-\alpha}-2\right)^{\lambda}}, \tag{3.28}$$

where $\lambda \in \mathbb{N}$ *is an odd number.*

Proof We have that

$$\chi(x) = \frac{1}{4}\left[f_2(x+1) - f_2(x-1)\right], \ \forall x \in \mathbb{R}.$$

Let $x \geq 1$. That is $0 \leq x-1 < x+1$. Applying the mean value theorem we get

$$\chi(x) = \frac{1}{4} \cdot 2 \cdot f_2'(\xi) = \frac{1}{2\left(1+\xi^{\lambda}\right)^{\frac{\lambda+1}{\lambda}}} > 0, \tag{3.29}$$

where $0 \leq x-1 < \xi < x+1$.
Then,

$$(x-1)^{\lambda} < \xi^{\lambda} < (x+1)^{\lambda},$$

and

$$1+(x-1)^{\lambda} < 1+\xi^{\lambda} < 1+(x+1)^{\lambda}, \tag{3.30}$$

and

$$2\left(1+(x-1)^{\lambda}\right)^{\frac{\lambda+1}{\lambda}} < 2\left(1+\xi^{\lambda}\right)^{\frac{\lambda+1}{\lambda}} < 2\left(1+(x+1)^{\lambda}\right)^{\frac{\lambda+1}{\lambda}}.$$

Hence it holds

$$\frac{1}{2\left(1+\xi^{\lambda}\right)^{\frac{\lambda+1}{\lambda}}} < \frac{1}{2\left(1+(x-1)^{\lambda}\right)^{\frac{\lambda+1}{\lambda}}}, \quad \forall\, x \geq 1.$$

We have found that

$$\chi(x) < \frac{1}{2\left(1+(x-1)^{\lambda}\right)^{\frac{\lambda+1}{\lambda}}}, \quad \forall\, x \geq 1. \tag{3.31}$$

Thus, we have

$$\sum_{\substack{j=-\infty \\ : |nx-j| \geq n^{1-\alpha}}}^{\infty} \chi(nx-j) = \sum_{\substack{j=-\infty \\ : |nx-j| \geq n^{1-\alpha}}}^{\infty} \chi(|nx-j|) <$$

$$\sum_{\substack{j=-\infty \\ : |nx-j| \geq n^{1-\alpha}}}^{\infty} \frac{1}{2\left(1+(|nx-j|-1)^{\lambda}\right)^{\frac{\lambda+1}{\lambda}}} \leq$$

$$\frac{1}{2}\int_{(n^{1-\alpha}-1)}^{\infty} \frac{1}{\left(1+(x-1)^{\lambda}\right)^{\frac{\lambda+1}{\lambda}}}dx = \frac{1}{2}\int_{(n^{1-\alpha}-2)}^{\infty} \frac{1}{\left(1+z^{\lambda}\right)^{\frac{\lambda+1}{\lambda}}}dz =: (*). \tag{3.32}$$

We see that

$$1+z^{\lambda} > z^{\lambda},$$

$$\left(1+z^{\lambda}\right)^{\frac{\lambda+1}{\lambda}} > z^{\lambda+1}, \tag{3.33}$$

$$\frac{1}{z^{\lambda+1}} > \frac{1}{\left(1+z^{\lambda}\right)^{\frac{\lambda+1}{\lambda}}}.$$

Therefore it holds

$$(*) < \frac{1}{2}\int_{(n^{1-\alpha}-2)}^{\infty}\frac{1}{z^{\lambda+1}}dz = \frac{1}{2}\int_{(n^{1-\alpha}-2)}^{\infty} z^{-(\lambda+1)}dz =$$

$$\frac{1}{2}\left(\frac{z^{-(\lambda+1)+1}}{-(\lambda+1)+1}\right)\Bigg|_{n^{1-\alpha}-2}^{\infty} = \frac{1}{2}\left(-\frac{z^{-\lambda}}{\lambda}\right)\Bigg|_{n^{1-\alpha}-2}^{\infty} =$$

$$\frac{1}{2\lambda}\left(z^{-\lambda}\right)\Bigg|_{\infty}^{n^{1-\alpha}-2} = \frac{1}{2\lambda\left(n^{1-\alpha}-2\right)^{\lambda}}, \tag{3.34}$$

proving (3.28). □

Denote by $\lfloor\cdot\rfloor$ the integral part of the number and by $\lceil\cdot\rceil$ the ceiling of the number.

Theorem 3.4 *Let $[a,b] \subset \mathbb{R}$ and $n \in \mathbb{N}$ so that $\lceil na\rceil \leq \lfloor nb\rfloor$. Then*

$$\frac{1}{\sum_{k=\lceil na\rceil}^{\lfloor nb\rfloor}\chi\left(|nx-k|\right)} < 2\sqrt[\lambda]{1+2^{\lambda}}, \tag{3.35}$$

where λ is an odd number, $\forall\, x \in [a,b]$.

Proof Let $x \in [a,b]$. We see that

$$1 = \sum_{k=-\infty}^{\infty}\chi\left(nx-k\right) > \sum_{k=\lceil na\rceil}^{\lfloor nb\rfloor}\chi\left(nx-k\right) =$$

$$\sum_{k=\lceil na\rceil}^{\lfloor nb\rfloor}\chi\left(|nx-k|\right) > \chi\left(|nx-k_0|\right), \tag{3.36}$$

$\forall\, k_0 \in [\lceil na\rceil, \lfloor nb\rfloor] \cap \mathbb{Z}$.

We can choose $k_0 \in [\lceil na\rceil, \lfloor nb\rfloor] \cap \mathbb{Z}$ such that $|nx-k_0| < 1$.

Therefore we get

$$\chi\left(|nx-k_0|\right) > \chi(1) = \frac{1}{4}\left(f_2(2) - f_2(0)\right) = \frac{1}{4}f_2(2) = \tag{3.37}$$

$$\frac{1}{4}\frac{2}{\left(1+2^{\lambda}\right)^{\frac{1}{\lambda}}} = \frac{1}{2\left(1+2^{\lambda}\right)^{\frac{1}{\lambda}}},$$

and

$$\sum_{k=\lceil na\rceil}^{\lfloor nb\rfloor}\chi\left(|nx-k|\right) > \frac{1}{2\left(1+2^{\lambda}\right)^{\frac{1}{\lambda}}}. \tag{3.38}$$

That is

$$\frac{1}{\sum_{k=\lceil na\rceil}^{\lfloor nb\rfloor}\chi\left(|nx-k|\right)}<2\left(1+2^{\lambda}\right)^{\frac{1}{\lambda}}, \tag{3.39}$$

proving the claim. □

We make

Remark 3.5 We also notice that

$$1-\sum_{k=\lceil na\rceil}^{\lfloor nb\rfloor}\chi\left(nb-k\right)=\sum_{k=-\infty}^{\lceil na\rceil-1}\chi\left(nb-k\right)+\sum_{k=\lfloor nb\rfloor+1}^{\infty}\chi\left(nb-k\right)$$

$$>\chi\left(nb-\lfloor nb\rfloor-1\right) \tag{3.40}$$

(call $\varepsilon:=nb-\lfloor nb\rfloor$, $0\leq\varepsilon<1$)

$$=\chi\left(\varepsilon-1\right)=\chi\left(1-\varepsilon\right)\geq\chi\left(1\right)>0.$$

Therefore

$$\lim_{n\to\to\infty}\left(1-\sum_{k=\lceil na\rceil}^{\lfloor nb\rfloor}\chi\left(nb-k\right)\right)>0. \tag{3.41}$$

Similarly, it holds

$$1-\sum_{k=\lceil na\rceil}^{\lfloor nb\rfloor}\chi\left(na-k\right)=\sum_{k=-\infty}^{\lceil na\rceil-1}\chi\left(na-k\right)+\sum_{k=\lfloor nb\rfloor+1}^{\infty}\chi\left(na-k\right)$$

$$>\chi\left(na-\lceil na\rceil+1\right) \tag{3.42}$$

(call $\eta:=\lceil na\rceil-na$, $0\leq\eta<1$)

$$=\chi\left(1-\eta\right)\geq\chi\left(1\right)>0.$$

Therefore again

$$\lim_{n\to\infty}\left(1-\sum_{k=\lceil na\rceil}^{\lfloor nb\rfloor}\chi\left(na-k\right)\right)>0. \tag{3.43}$$

Here we find that

$$\lim_{n\to\infty} \sum_{k=\lceil na\rceil}^{\lfloor nb\rfloor} \chi(nx-k) \neq 1, \quad \text{for at least some } x \in [a,b]. \tag{3.44}$$

Note 3.6 Let $[a,b] \subset \mathbb{R}$. For large enough n we always obtain $\lceil na\rceil \leq \lfloor nb\rfloor$. Also $a \leq \frac{k}{n} \leq b$, iff $\lceil na\rceil \leq k \leq \lfloor nb\rfloor$.

In general it holds (by $\sum_{i=-\infty}^{\infty} \chi(x-i) = 1, \forall\, x \in \mathbb{R}$) that

$$\sum_{k=\lceil na\rceil}^{\lfloor nb\rfloor} \chi(nx-k) \leq 1. \tag{3.45}$$

Let $(X, \|\cdot\|)$ be a Banach space.

Definition 3.7 Let $f \in C([a,b], X)$ and $n \in \mathbb{N} : \lceil na\rceil \leq \lfloor nb\rfloor$. We introduce and define the X-valued linear neural network operators

$$H_n(f,x) := \frac{\sum_{k=\lceil na\rceil}^{\lfloor nb\rfloor} f\left(\frac{k}{n}\right)\chi(nx-k)}{\sum_{k=\lceil na\rceil}^{\lfloor nb\rfloor} \chi(nx-k)}, \quad x \in [a,b]. \tag{3.46}$$

Clearly here $H_n(f,x) \in C([a,b], X)$.

For convenience we use the same H_n for real valued functions when needed. We study here the pointwise and uniform convergence of $H_n(f,x)$ to $f(x)$ with rates.

For convenience, also we call

$$H_n^*(f,x) := \sum_{k=\lceil na\rceil}^{\lfloor nb\rfloor} f\left(\frac{k}{n}\right)\chi(nx-k), \tag{3.47}$$

(similarly, H_n^* can be defined for real valued functions) that is

$$H_n(f,x) := \frac{H_n^*(f,x)}{\sum_{k=\lceil na\rceil}^{\lfloor nb\rfloor} \chi(nx-k)}. \tag{3.48}$$

So that

$$H_n(f,x) - f(x) = \frac{H_n^*(f,x)}{\sum_{k=\lceil na \rceil}^{\lfloor nb \rfloor} \chi(nx-k)} - f(x) =$$

$$\frac{H_n^*(f,x) - f(x)\left(\sum_{k=\lceil na \rceil}^{\lfloor nb \rfloor} \chi(nx-k)\right)}{\sum_{k=\lceil na \rceil}^{\lfloor nb \rfloor} \chi(nx-k)}. \tag{3.49}$$

Consequently, we derive that

$$\|H_n(f,x) - f(x)\| \leq 2\left(\sqrt[\lambda]{1+2^\lambda}\right) \left\| H_n^*(f,x) - f(x)\left(\sum_{k=\lceil na \rceil}^{\lfloor nb \rfloor} \chi(nx-k)\right)\right\| =$$

$$2\left(\sqrt[\lambda]{1+2^\lambda}\right) \left\| \sum_{k=\lceil na \rceil}^{\lfloor nb \rfloor} \left(f\left(\frac{k}{n}\right) - f(x)\right)\chi(nx-k)\right\|. \tag{3.50}$$

We will estimate the right and hand side of (3.50).

For that we need, for $f \in C([a,b], X)$ the first modulus of continuity

$$\omega_1(f,\delta) := \sup_{\substack{x,y \in [a,b] \\ |x-y| \leq \delta}} \|f(x) - f(y)\|, \quad \delta > 0. \tag{3.51}$$

Similarly, it is defined ω_1 for $f \in C_{uB}(\mathbb{R}, X)$ (uniformly continuous and bounded functions from $\mathbb{R}$ into X), for $f \in C_B(\mathbb{R}, X)$ (continuous and bounded X-valued), and for $f \in C_u(\mathbb{R}, X)$ (uniformly continuous).

The fact $f \in C([a,b], X)$ or $f \in C_u(\mathbb{R}, X)$, is equivalent to $\lim_{\delta\to 0} \omega_1(f,\delta) = 0$, see [11].

We make

Definition 3.8 When $f \in C_{uB}(\mathbb{R}, X)$, or $f \in C_B(\mathbb{R}, X)$, we define

$$\overline{H_n}(f,x) := \sum_{k=-\infty}^{\infty} f\left(\frac{k}{n}\right)\chi(nx-k), \tag{3.52}$$

$n \in \mathbb{N}$, $x \in \mathbb{R}$,

the X-valued quasi-interpolation neural network operator.

We make

Remark 3.9 We have that

$$\left\| f\left(\frac{k}{n}\right)\right\| \leq \|f\|_{\infty,\mathbb{R}} < +\infty, \tag{3.53}$$

and

$$\left\| f\left(\frac{k}{n}\right)\right\| \chi(nx-k) \leq \|f\|_{\infty,\mathbb{R}}\, \chi(nx-k) \tag{3.54}$$

and

$$\sum_{k=-\overline{\lambda}}^{\overline{\lambda}} \left\| f\left(\frac{k}{n}\right)\right\| \chi(nx-k) \leq \|f\|_{\infty,\mathbb{R}} \left(\sum_{k=-\overline{\lambda}}^{\overline{\lambda}} \chi(nx-k)\right), \tag{3.55}$$

and finally

$$\sum_{k=-\infty}^{\infty} \left\| f\left(\frac{k}{n}\right)\right\| \chi(nx-k) \leq \|f\|_{\infty,\mathbb{R}}, \tag{3.56}$$

a convergent in $\mathbb{R}$ series.

So, the series $\sum_{k=-\infty}^{\infty} \left\| f\left(\frac{k}{n}\right)\right\| \chi(nx-k)$ is absolutely convergent in X, hence it is convergent in X and $\overline{H_n}(f,x) \in X$. We denote by $\|f\|_{\infty} := \sup_{x\in[a,b]} \|f(x)\|$, for $f \in C([a,b], X)$, similarly it is defined for $f \in C_B(\mathbb{R}, X)$.

3.3 Main Results

We present a set of X-valued neural network approximations to a function given with rates.

Theorem 3.10 *Let* $f \in C([a,b], X)$, $0<\alpha<1$, $n \in \mathbb{N} : n^{1-\alpha} > 2$, $x \in [a,b]$, $\lambda \in \mathbb{N}$ *is odd. Then*

(i)

$$\|H_n(f,x) - f(x)\| \leq 2\sqrt[\lambda]{1+2^{\lambda}} \left[\omega_1\left(f, \frac{1}{n^{\alpha}}\right) + \frac{\|f\|_{\infty}}{\lambda\left(n^{1-\alpha}-2\right)^{\lambda}}\right] =: \rho_1, \tag{3.57}$$

and

(ii)

$$\|H_n(f) - f\|_{\infty} \leq \rho_1. \tag{3.58}$$

We get that $\lim_{n\to\infty} H_n(f) = f$, *pointwise and uniformly.*

Proof We see that

$$\left\| \sum_{k=\lceil na \rceil}^{\lfloor nb \rfloor} \left(f\left(\frac{k}{n}\right) - f(x) \right) \chi(nx-k) \right\| \leq$$

$$\sum_{k=\lceil na \rceil}^{\lfloor nb \rfloor} \left\| f\left(\frac{k}{n}\right) - f(x) \right\| \chi(nx-k) =$$

$$\sum_{\begin{cases} k=\lceil na \rceil \\ : \left|\frac{k}{n}-x\right| \leq \frac{1}{n^{\alpha}} \end{cases}}^{\lfloor nb \rfloor} \left\| f\left(\frac{k}{n}\right) - f(x) \right\| \chi(nx-k) +$$

$$\sum_{\begin{cases} k=\lceil na \rceil \\ : \left|\frac{k}{n}-x\right| > \frac{1}{n^{\alpha}} \end{cases}}^{\lfloor nb \rfloor} \left\| f\left(\frac{k}{n}\right) - f(x) \right\| \chi(nx-k) \leq \tag{3.59}$$

$$\sum_{\begin{cases} k=\lceil na \rceil \\ : \left|\frac{k}{n}-x\right| \leq \frac{1}{n^{\alpha}} \end{cases}}^{\lfloor nb \rfloor} \omega_1\left(f, \left|\frac{k}{n}-x\right|\right) \chi(nx-k) +$$

$$2\|f\|_{\infty} \sum_{\begin{cases} k=-\infty \\ : |k-nx| > n^{1-\alpha} \end{cases}}^{\infty} \chi(nx-k) \leq$$

$$\omega_1\left(f, \frac{1}{n^{\alpha}}\right) \sum_{\begin{cases} k=-\infty \\ : \left|\frac{k}{n}-x\right| \leq \frac{1}{n^{\alpha}} \end{cases}}^{\infty} \chi(nx-k) +$$

$$2\|f\|_{\infty} \sum_{\begin{cases} k=-\infty \\ : |k-nx| > n^{1-\alpha} \end{cases}}^{\infty} \chi(nx-k) \underset{\text{(by Theorem 3.3)}}{\leq}$$

$$\omega_1\left(f, \frac{1}{n^{\alpha}}\right) + \frac{\|f\|_{\infty}}{\lambda\left(n^{1-\alpha}-2\right)^{\lambda}}.$$

That is

$$\left\| \sum_{k=\lceil na \rceil}^{\lfloor nb \rfloor} \left(f\left(\frac{k}{n}\right) - f(x) \right) \chi(nx - k) \right\| \leq$$

$$\omega_1\left(f, \frac{1}{n^\alpha}\right) + \frac{\|f\|_\infty}{\lambda\left(n^{1-\alpha} - 2\right)^\lambda}. \tag{3.60}$$

Using (3.50) we derive (3.57). □

It follows

Theorem 3.11 *Let $f \in C_B(\mathbb{R}, X)$, $0 < \alpha < 1$, $n \in \mathbb{N} : n^{1-\alpha} > 2$, $x \in \mathbb{R}$, $\lambda \in \mathbb{N}$ is odd. Then*

(i)

$$\left\| \overline{H_n}(f, x) - f(x) \right\| \leq \omega_1\left(f, \frac{1}{n^\alpha}\right) + \frac{\|f\|_\infty}{\lambda\left(n^{1-\alpha} - 2\right)^\lambda} =: \rho_2, \tag{3.61}$$

and

(ii)

$$\left\| \overline{H_n}(f) - f \right\|_\infty \leq \rho_2. \tag{3.62}$$

For $f \in C_{uB}(\mathbb{R}, X)$ we get $\lim_{n \to \infty} \overline{H_n}(f) = f$, pointwise and uniformly.

Proof We observe that

$$\left\| \overline{H_n}(f, x) - f(x) \right\| = \left\| \sum_{k=-\infty}^{\infty} f\left(\frac{k}{n}\right) \chi(nx - k) - f(x) \sum_{k=-\infty}^{\infty} \chi(nx - k) \right\| =$$

$$\left\| \sum_{k=-\infty}^{\infty} \left(f\left(\frac{k}{n}\right) - f(x) \right) \chi(nx - k) \right\| \leq$$

$$\sum_{k=-\infty}^{\infty} \left\| f\left(\frac{k}{n}\right) - f(x) \right\| \chi(nx - k) = \tag{3.63}$$

$$\sum_{\begin{cases} k = -\infty \\ : \left|\frac{k}{n} - x\right| \leq \frac{1}{n^\alpha} \end{cases}}^{\infty} \left\| f\left(\frac{k}{n}\right) - f(x) \right\| \chi(nx - k) +$$

$$\sum_{\begin{cases} k=-\infty \\ : \left|\frac{k}{n}-x\right| > \frac{1}{n^{\alpha}} \end{cases}}^{\infty} \left\| f\left(\frac{k}{n}\right) - f(x) \right\| \chi(nx-k) \leq$$

$$\sum_{\begin{cases} k=-\infty \\ : \left|\frac{k}{n}-x\right| \leq \frac{1}{n^{\alpha}} \end{cases}}^{\infty} \omega_1\left(f, \left|\frac{k}{n}-x\right|\right) \chi(nx-k) +$$

$$2\|f\|_{\infty} \sum_{\begin{cases} k=-\infty \\ : \left|\frac{k}{n}-x\right| > \frac{1}{n^{\alpha}} \end{cases}}^{\infty} \chi(nx-k) \leq$$

$$\omega_1\left(f, \frac{1}{n^{\alpha}}\right) \sum_{\begin{cases} k=-\infty \\ : \left|\frac{k}{n}-x\right| \leq \frac{1}{n^{\alpha}} \end{cases}}^{\infty} \chi(nx-k) + \frac{2\|f\|_{\infty}}{2\lambda\left(n^{1-\alpha}-2\right)^{\lambda}} \leq \tag{3.64}$$

$$\omega_1\left(f, \frac{1}{n^{\alpha}}\right) + \frac{\|f\|_{\infty}}{\lambda\left(n^{1-\alpha}-2\right)^{\lambda}},$$

proving the claim. □

In the next we study the high order neural network X-valued approximation by using the smoothness of f. Derivarives are defined similar to numerical ones, see [22].

Theorem 3.12 *Let $f \in C^N([a,b], X)$, $n, N \in \mathbb{N}$, λ is odd, $0 < \alpha < 1$, $x \in [a,b]$ and $n^{1-\alpha} > 2$. Then*

(i)

$$\|H_n(f,x) - f(x)\| \leq \sqrt[\lambda]{1+2^{\lambda}} \left\{ \sum_{j=1}^{N} \frac{\left\|f^{(j)}(x)\right\|}{j!} \left[\frac{2}{n^{\alpha j}} + \frac{(b-a)^j}{\lambda\left(n^{1-\alpha}-2\right)^{\lambda}} \right] + \right.$$

$$\left. \left[\omega_1\left(f^{(N)}, \frac{1}{n^{\alpha}}\right) \frac{2}{n^{\alpha N} N!} + \frac{2\left\|f^{(N)}\right\|_{\infty} (b-a)^N}{N!\lambda\left(n^{1-\alpha}-2\right)^{\lambda}} \right] \right\}, \tag{3.65}$$

(ii) assume further $f^{(j)}(x_0) = 0$, $j = 1, ..., N$, for some $x_0 \in [a, b]$, it holds

$$\|H_n(f, x_0) - f(x_0)\| \leq \sqrt[\lambda]{1 + 2^\lambda} \cdot$$

$$\left[\omega_1 \left(f^{(N)}, \frac{1}{n^\alpha} \right) \frac{2}{n^{\alpha N} N!} + \frac{2 \left\| f^{(N)} \right\|_\infty (b-a)^N}{N! \lambda \left(n^{1-\alpha} - 2 \right)^\lambda} \right], \tag{3.66}$$

and

(iii)

$$\|H_n(f) - f\|_\infty \leq \sqrt[\lambda]{1 + 2^\lambda} \left\{ \sum_{j=1}^{N} \frac{\left\| f^{(j)} \right\|_\infty}{j!} \left[\frac{2}{n^{\alpha j}} + \frac{(b-a)^j}{\lambda \left(n^{1-\alpha} - 2 \right)^\lambda} \right] + \right.$$

$$\left. \left[\omega_1 \left(f^{(N)}, \frac{1}{n^\alpha} \right) \frac{2}{n^{\alpha N} N!} + \frac{2 \left\| f^{(N)} \right\|_\infty (b-a)^N}{N! \lambda \left(n^{1-\alpha} - 2 \right)^\lambda} \right] \right\}. \tag{3.67}$$

We derive that $\lim_{n \to \infty} H_n(f) = f$, pointwise and uniformly.

Proof The proof is lengthy and very similar to the proof of the corresponding result in [14]. As such is omitted. □

The integrals from now on are of Bochner type [20].

We need

Definition 3.13 ([12]) Let $[a, b] \subset \mathbb{R}$, X be a Banach space, $\alpha > 0$; $m = \lceil \alpha \rceil \in \mathbb{N}$, ($\lceil \cdot \rceil$ is the ceiling of the number), $f : [a, b] \to X$. We assume that $f^{(m)} \in L_1([a, b], X)$. We call the Caputo-Bochner left fractional derivative of order α:

$$\left(D_{*a}^\alpha f \right)(x) := \frac{1}{\Gamma(m - \alpha)} \int_a^x (x - t)^{m-\alpha-1} f^{(m)}(t) \, dt, \quad \forall \, x \in [a, b]. \tag{3.68}$$

If $\alpha \in \mathbb{N}$, we set $D_{*a}^\alpha f := f^{(m)}$ the ordinary X-valued derivative (defined similar to numerical one, see [22], p. 83), and also set $D_{*a}^0 f := f$.

By [12], $\left(D_{*a}^\alpha f \right)(x)$ exists almost everywhere in $x \in [a, b]$ and $D_{*a}^\alpha f \in L_1([a, b], X)$.

If $\left\| f^{(m)} \right\|_{L_\infty([a,b],X)} < \infty$, then by [12], $D_{*a}^\alpha f \in C([a, b], X)$, hence $\left\| D_{*a}^\alpha f \right\| \in C([a, b])$.

We mention

Lemma 3.14 ([11]) *Let $\alpha > 0$, $\alpha \notin \mathbb{N}$, $m = \lceil \alpha \rceil$, $f \in C^{m-1}([a, b], X)$ and $f^{(m)} \in L_\infty([a, b], X)$. Then $D_{*a}^\alpha f(a) = 0$.*

We mention

Definition 3.15 ([10]) Let $[a, b] \subset \mathbb{R}$, X be a Banach space, $\alpha > 0$, $m := \lceil \alpha \rceil$. We assume that $f^{(m)} \in L_1([a, b], X)$, where $f : [a, b] \to X$. We call the Caputo-Bochner right fractional derivative of order α:

$$\left(D_{b-}^{\alpha} f\right)(x) := \frac{(-1)^m}{\Gamma(m-\alpha)} \int_x^b (z-x)^{m-\alpha-1} f^{(m)}(z)\,dz, \quad \forall\, x \in [a, b]. \tag{3.69}$$

We observe that $\left(D_{b-}^{m} f\right)(x) = (-1)^m f^{(m)}(x)$, for $m \in \mathbb{N}$, and $\left(D_{b-}^{0} f\right)(x) = f(x)$.

By [10], $\left(D_{b-}^{\alpha} f\right)(x)$ exists almost everywhere on $[a, b]$ and $\left(D_{b-}^{\alpha} f\right) \in L_1([a, b], X)$.

If $\left\| f^{(m)} \right\|_{L_\infty([a,b],X)} < \infty$, and $\alpha \notin \mathbb{N}$, by [10], $D_{b-}^{\alpha} f \in C([a, b], X)$, hence $\left\| D_{b-}^{\alpha} f \right\| \in C([a, b])$.

We need

Lemma 3.16 ([11]) *Let $f \in C^{m-1}([a, b], X)$, $f^{(m)} \in L_\infty([a, b], X)$, $m = \lceil \alpha \rceil$, $\alpha > 0$, $\alpha \notin \mathbb{N}$. Then $D_{b-}^{\alpha} f(b) = 0$.*

Convention 3.17 *We assume that*

$$D_{*x_0}^{\alpha} f(x) = 0, \text{ for } x < x_0, \tag{3.70}$$

and

$$D_{x_0-}^{\alpha} f(x) = 0, \text{ for } x > x_0, \tag{3.71}$$

for all $x, x_0 \in [a, b]$.

Next we present the corresponding X-valued fractional approximation by neural networks.

Theorem 3.18 *Let $\alpha > 0$, $N = \lceil \alpha \rceil$, $\alpha \notin \mathbb{N}$, $f \in C^N([a, b], X)$, $0 < \beta < 1$, λ is odd, $x \in [a, b]$, $n \in \mathbb{N} : n^{1-\beta} > 2$. Then*

(i)

$$\left\| H_n(f, x) - \sum_{j=1}^{N-1} \frac{f^{(j)}(x)}{j!} H_n\left((\cdot - x)^j\right)(x) - f(x) \right\| \leq$$

$$\frac{2\sqrt[\lambda]{1+2^\lambda}}{\Gamma(\alpha+1)} \left\{ \frac{\left(\omega_1\left(D_{x-}^{\alpha} f, \frac{1}{n^\beta}\right)_{[a,x]} + \omega_1\left(D_{*x}^{\alpha} f, \frac{1}{n^\beta}\right)_{[x,b]}\right)}{n^{\alpha\beta}} + \right.$$

$$\left. \frac{1}{2\lambda\left(n^{1-\beta}-2\right)^\lambda} \left(\left\| D_{x-}^{\alpha} f \right\|_{\infty,[a,x]} (x-a)^\alpha + \left\| D_{*x}^{\alpha} f \right\|_{\infty,[x,b]} (b-x)^\alpha \right) \right\}, \tag{3.72}$$

(ii) if $f^{(j)}(x) = 0$, *for* $j = 1, ..., N-1$, *we have*

$$\|H_n(f,x) - f(x)\| \leq \frac{2\sqrt[\lambda]{1+2^\lambda}}{\Gamma(\alpha+1)}$$

$$\left\{ \frac{\left(\omega_1\left(D_{x-}^{\alpha} f, \frac{1}{n^\beta}\right)_{[a,x]} + \omega_1\left(D_{*x}^{\alpha} f, \frac{1}{n^\beta}\right)_{[x,b]}\right)}{n^{\alpha\beta}} + \right.$$

$$\left. \frac{1}{2\lambda\left(n^{1-\beta}-2\right)^\lambda}\left(\left\|D_{x-}^{\alpha} f\right\|_{\infty,[a,x]}(x-a)^\alpha + \left\|D_{*x}^{\alpha} f\right\|_{\infty,[x,b]}(b-x)^\alpha\right)\right\}, \tag{3.73}$$

(iii)

$$\|H_n(f,x) - f(x)\| \leq 2\sqrt[\lambda]{1+2^\lambda}\cdot$$

$$\left\{ \sum_{j=1}^{N-1} \frac{\left\|f^{(j)}(x)\right\|}{j!}\left\{\frac{1}{n^{\beta j}} + \frac{(b-a)^j}{2\lambda\left(n^{1-\beta}-2\right)^\lambda}\right\} + \right.$$

$$\frac{1}{\Gamma(\alpha+1)}\left\{ \frac{\left(\omega_1\left(D_{x-}^{\alpha} f, \frac{1}{n^\beta}\right)_{[a,x]} + \omega_1\left(D_{*x}^{\alpha} f, \frac{1}{n^\beta}\right)_{[x,b]}\right)}{n^{\alpha\beta}} + \right.$$

$$\left.\left. \frac{1}{2\lambda\left(n^{1-\beta}-2\right)^\lambda}\left(\left\|D_{x-}^{\alpha} f\right\|_{\infty,[a,x]}(x-a)^\alpha + \left\|D_{*x}^{\alpha} f\right\|_{\infty,[x,b]}(b-x)^\alpha\right)\right\}\right\}, \tag{3.74}$$

$\forall\, x \in [a,b]$,

and

(iv)

$$\|H_n f - f\|_\infty \leq 2\sqrt[\lambda]{1+2^\lambda}\cdot$$

$$\left\{ \sum_{j=1}^{N-1} \frac{\left\|f^{(j)}\right\|_\infty}{j!}\left\{\frac{1}{n^{\beta j}} + \frac{(b-a)^j}{2\lambda\left(n^{1-\beta}-2\right)^\lambda}\right\} + \right.$$

$$\frac{1}{\Gamma(\alpha+1)}\left\{ \frac{\left(\sup_{x\in[a,b]}\omega_1\left(D_{x-}^{\alpha} f, \frac{1}{n^\beta}\right)_{[a,x]} + \sup_{x\in[a,b]}\omega_1\left(D_{*x}^{\alpha} f, \frac{1}{n^\beta}\right)_{[x,b]}\right)}{n^{\alpha\beta}} + \right.$$

$$\left.\left. \frac{(b-a)^\alpha}{2\lambda\left(n^{1-\beta}-2\right)^\lambda}\left(\sup_{x\in[a,b]}\left\|D_{x-}^{\alpha} f\right\|_{\infty,[a,x]} + \sup_{x\in[a,b]}\left\|D_{*x}^{\alpha} f\right\|_{\infty,[x,b]}\right)\right\}\right\}. \tag{3.75}$$

Above, when $N = 1$ the sum $\sum_{j=1}^{N-1} \cdot = 0$.

As we see here we obtain X-valued fractionally type pointwise and uniform convergence with rates of $H_n \to I$ the unit operator, as $n \to \infty$.

Proof The proof is very lengthy and very similar to the proof of the corresponding result in [14]. As such is omitted. □

Next we apply Theorem 3.18 for $N = 1$.

Theorem 3.19 *Let $0 < \alpha, \beta < 1$, $f \in C^1([a,b], X)$, $x \in [a,b]$, $n \in \mathbb{N} : n^{1-\beta} > 2$, λ is odd. Then*

(i)

$$\|H_n(f,x) - f(x)\| \leq$$

$$\frac{2\sqrt[\lambda]{1+2^\lambda}}{\Gamma(\alpha+1)} \left\{ \frac{\left(\omega_1\left(D_{x-}^\alpha f, \frac{1}{n^\beta}\right)_{[a,x]} + \omega_1\left(D_{*x}^\alpha f, \frac{1}{n^\beta}\right)_{[x,b]}\right)}{n^{\alpha\beta}} + \right.$$

$$\left. \frac{1}{2\lambda\left(n^{1-\beta}-2\right)^\lambda} \left(\left\| D_{x-}^\alpha f \right\|_{\infty,[a,x]} (x-a)^\alpha + \left\| D_{*x}^\alpha f \right\|_{\infty,[x,b]} (b-x)^\alpha \right) \right\}, \quad (3.76)$$

and

(ii)

$$\|H_n f - f\|_\infty \leq \frac{2\sqrt[\lambda]{1+2^\lambda}}{\Gamma(\alpha+1)} \cdot$$

$$\left\{ \frac{\left(\sup_{x\in[a,b]} \omega_1\left(D_{x-}^\alpha f, \frac{1}{n^\beta}\right)_{[a,x]} + \sup_{x\in[a,b]} \omega_1\left(D_{*x}^\alpha f, \frac{1}{n^\beta}\right)_{[x,b]} \right)}{n^{\alpha\beta}} + \right.$$

$$\left. \frac{(b-a)^\alpha}{2\lambda\left(n^{1-\beta}-2\right)^\lambda} \left(\sup_{x\in[a,b]} \left\| D_{x-}^\alpha f \right\|_{\infty,[a,x]} + \sup_{x\in[a,b]} \left\| D_{*x}^\alpha f \right\|_{\infty,[x,b]} \right) \right\}. \quad (3.77)$$

When $\alpha = \frac{1}{2}$ we derive

Corollary 3.20 *Let $0 < \beta < 1$, $f \in C^1([a,b], X)$, $x \in [a,b]$, $n \in \mathbb{N} : n^{1-\beta} > 2$, λ is odd. Then*

(i)

$$\|H_n(f,x) - f(x)\| \leq$$

$$\frac{4\sqrt[\lambda]{1+2^{\lambda}}}{\sqrt{\pi}}\left\{\frac{\left(\omega_1\left(D_{x-}^{\frac{1}{2}}f,\frac{1}{n^{\beta}}\right)_{[a,x]}+\omega_1\left(D_{*x}^{\frac{1}{2}}f,\frac{1}{n^{\beta}}\right)_{[x,b]}\right)}{n^{\frac{\beta}{2}}}+\right.$$

$$\left.\frac{1}{2\lambda\left(n^{1-\beta}-2\right)^{\lambda}}\left(\left\|D_{x-}^{\frac{1}{2}}f\right\|_{\infty,[a,x]}\sqrt{(x-a)}+\left\|D_{*x}^{\frac{1}{2}}f\right\|_{\infty,[x,b]}\sqrt{(b-x)}\right)\right\},\tag{3.78}$$

and
(ii)

$$\|H_n f-f\|_{\infty}\leq\frac{4\sqrt[\lambda]{1+2^{\lambda}}}{\sqrt{\pi}}\cdot$$

$$\left\{\frac{\left(\sup_{x\in[a,b]}\omega_1\left(D_{x-}^{\frac{1}{2}}f,\frac{1}{n^{\beta}}\right)_{[a,x]}+\sup_{x\in[a,b]}\omega_1\left(D_{*x}^{\frac{1}{2}}f,\frac{1}{n^{\beta}}\right)_{[x,b]}\right)}{n^{\frac{\beta}{2}}}+\right.$$

$$\left.\frac{\sqrt{(b-a)}}{2\lambda\left(n^{1-\beta}-2\right)^{\lambda}}\left(\sup_{x\in[a,b]}\left\|D_{x-}^{\frac{1}{2}}f\right\|_{\infty,[a,x]}+\sup_{x\in[a,b]}\left\|D_{*x}^{\frac{1}{2}}f\right\|_{\infty,[x,b]}\right)\right\}<\infty.\tag{3.79}$$

Next we make

Conclusion 3.21 *Some convergence analysis follows:*

Let $0<\beta<1$, $f\in C^1([a,b],X)$, $x\in[a,b]$, $n\in\mathbb{N}: n^{1-\beta}>2$, λ *is odd. We elaborate on (3.79). Assume that*

$$\omega_1\left(D_{x-}^{\frac{1}{2}}f,\frac{1}{n^{\beta}}\right)_{[a,x]}\leq\frac{K_1}{n^{\beta}},\tag{3.80}$$

and

$$\omega_1\left(D_{*x}^{\frac{1}{2}}f,\frac{1}{n^{\beta}}\right)_{[x,b]}\leq\frac{K_2}{n^{\beta}},\tag{3.81}$$

$\forall\, x\in[a,b]$, $\forall\, n\in\mathbb{N}$, *where* $K_1, K_2>0$.

Then it holds

$$\frac{\left[\sup_{x\in[a,b]}\omega_1\left(D_{x-}^{\frac{1}{2}}f,\frac{1}{n^{\beta}}\right)_{[a,x]}+\sup_{x\in[a,b]}\omega_1\left(D_{*x}^{\frac{1}{2}}f,\frac{1}{n^{\beta}}\right)_{[x,b]}\right]}{n^{\frac{\beta}{2}}}\leq$$

$$\frac{\frac{(K_1+K_2)}{n^{\beta}}}{n^{\frac{\beta}{2}}} = \frac{(K_1+K_2)}{n^{\frac{3\beta}{2}}} = \frac{K}{n^{\frac{3\beta}{2}}}, \tag{3.82}$$

where $K := K_1 + K_2 > 0$.

The other summand of the right hand side of (3.79), for large enough n, converges to zero at the speed $\frac{1}{n^{\lambda(1-\beta)}}$*, so it is about* $\frac{L}{n^{\lambda(1-\beta)}}$*, where* $L > 0$ *is a constant.*

Then, for large enough $n \in \mathbb{N}$*, by (3.79), (3.82) and the above comment, we obtain that*

$$\|H_n f - f\|_\infty \le \frac{M}{\min\left(n^{\frac{3\beta}{2}}, n^{\lambda(1-\beta)}\right)}, \tag{3.83}$$

where $M > 0$.

Clearly we have two cases:

(i)

$$\|H_n f - f\|_\infty \le \frac{M}{n^{\lambda(1-\beta)}}, \text{ when } \frac{2\lambda}{3+2\lambda} \le \beta < 1, \tag{3.84}$$

with speed of convergence $\frac{1}{n^{\lambda(1-\beta)}}$,

and

(ii)

$$\|H_n f - f\|_\infty \le \frac{M}{n^{\frac{3\beta}{2}}}, \text{ when } 0 < \beta \le \frac{2\lambda}{3+2\lambda}, \tag{3.85}$$

with speed of convergence $\frac{1}{n^{\frac{3\beta}{2}}}$.

In Theorem 3.10, for $f \in C([a,b], X)$ *and for large enough* $n \in \mathbb{N}$*, when* $0 < \beta \le \frac{\lambda}{1+\lambda}$*, the speed is* $\frac{1}{n^{\beta}}$*. So when* $0 < \beta \le \frac{2\lambda}{3+2\lambda}$ $(< \frac{\lambda}{1+\lambda})$*, we get by (3.85) that* $\|H_n f - f\|_\infty$ *converges much faster to zero. The last comes because we assumed differentiability of* f.

Notice that in Corollary 3.20 no initial condition is assumed.

Conclusion 3.22 *When* $(X, \|\cdot\|) = (\mathbb{C}, |\cdot|)$*, all of our main results give nice and great applications, etc.*

References

1. G.A. Anastassiou, Rate of convergence of some neural network operators to the unit-univariate case. J. Math. Anal. Appl. **212**, 237–262 (1997)
2. G.A. Anastassiou, *Quantitative Approximations* (Chapman & Hall / CRC, Boca Raton, New York, 2001)
3. G.A. Anastassiou, Univariate hyperbolic tangent neural network approximation. Math. Comput. Model. **53**, 1111–1132 (2011)
4. G.A. Anastassiou, Multivariate hyperbolic tangent neural network approximation. Comput. Math. **61**, 809–821 (2011)

5. G.A. Anastassiou, Multivariate sigmoidal neural network approximation. Neural Netw. **24**, 378–386 (2011)
6. G.A. Anastassiou, *Inteligent Systems: Approximation by Artificial Neural Networks, Intelligent Systems Reference Library*, vol. 19 (Springer, Heidelberg, 2011)
7. G.A. Anastassiou, Univariate sigmoidal neural network approximation. J. Comput. Anal. Appl. **14**(4), 659–690 (2012)
8. G.A. Anastassiou, Fractional neural network approximation. Comput. Math. Appl. **64**, 1655–1676 (2012)
9. G.A. Anastassiou, *Intelligent Systems II: Complete Approximation by Neural Network Operators* (Springer, Heidelberg, New York, 2016)
10. G.A. Anastassiou, Strong right fractional calculus for Banach space valued functions. Revista Proyecciones **36**(1), 149–186 (2017)
11. G.A. Anastassiou, Vector fractional Korovkin type Approximations. Dyn. Syst. Appl. **26**, 81–104 (2017)
12. G.A. Anastassiou, A strong fractional calculus theory for Banach space valued functions. Nonlinear Funct. Anal. Appl. (Korea) **22**(3), 495–524 (2017)
13. G.A. Anastassiou, *Nonlinearity: Ordinary and Fractional Approximations by Sublinear and Max-Product Operators* (Springer, Heidelberg, New York, 2018)
14. G.A. Anastassiou, Algebraic function based Banach space valued ordinary and fractional neural network approximations. New Trends Math. Sci., NTMSCI 10 Special Issue, (1), 100–125 (2022)
15. G.A. Anastassiou, Generalized symmetrical sigmoid function activated Banach space valued ordinary and fractional neural network approximation (Analele Universitatii Oradea, Fascicula Matematica, 2022)
16. Z. Chen, F. Cao, The approximation operators with sigmoidal functions. Comput. Math. Appl. **58**, 758–765 (2009)
17. A.J. Dunning, J. Kensler, L. Goudeville, F. Bailleux, Some extensions in continuous methods for immunological correlates of protection. BMC Med. Res. Methodol. **15**(107) (28 Dec 2015). https://doi.org/10.1186/s12874-015-0096-9
18. S. Haykin, *Neural Networks: A Comprehensive Foundation*, 2nd edn. (Prentice Hall, New York, 1998)
19. W. McCulloch, W. Pitts, A logical calculus of the ideas immanent in nervous activity. Bull. Math. Biophys. **7**, 115–133 (1943)
20. J. Mikusinski, *The Bochner Integral* (Academic Press, New York, 1978)
21. T.M. Mitchell, *Machine Learning* (WCB-McGraw-Hill, New York, 1997)
22. G.E. Shilov, *Elementary Functional Analysis* (Dover Publications Inc., New York, 1996)

Chapter 4
Abstract Multivariate Algebraic Function Induced Neural Network Approximations

Here we exhibit multivariate quantitative approximations of Banach space valued continuous multivariate functions on a box or $\mathbb{R}^N$, $N \in \mathbb{N}$, by the multivariate normalized, quasi-interpolation, Kantorovich type and quadrature type neural network operators. We study also the case of approximation by iterated operators of the last four types. These approximations are achieved by establishing multidimensional Jackson type inequalities involving the multivariate modulus of continuity of the engaged function or its high order Fréchet derivatives. Our multivariate operators are defined by using a multidimensional density function induced by the algebraic sigmoid function. The approximations are pointwise and uniform. The related feed-forward neural network is with one hidden layer. It follows [14].

4.1 Introduction

G. A. Anastassiou in [2, 3], see Chaps. 2–5, was the first to establish neural network approximations to continuous functions with rates by very specifically defined neural network operators of Cardaliagnet–Euvrard and "Squashing" types, by employing the modulus of continuity of the engaged function or its high order derivative, and producing very tight Jackson type inequalities. He treats there both the univariate and multivariate cases. The defining these operators "bell-shaped" and "squashing" functions are assumed to be of compact support. Also in [3] he gives the Nth order asymptotic expansion for the error of weak approximation of these two operators to a special natural class of smooth functions, see Chaps. 4 and 5 there.

Motivations for this chapter are the articles [16] of Z. Chen and F. Cao, and [4–13, 17, 18].

© The Author(s), under exclusive license to Springer Nature Switzerland AG 2023
G. A. Anastassiou, *Banach Space Valued Neural Network*,
Studies in Computational Intelligence 1062,
https://doi.org/10.1007/978-3-031-16400-2_4

Here we perform multivariate algebraic sigmoid function based neural network approximations to continuous functions over boxes or over the whole $\mathbb{R}^N$, $N \in \mathbb{N}$, and also iterated approximations. All convergences here are with rates expressed via the multivariate modulus of continuity of the involved function or its high order Fréchet derivative and given by very tight multidimensional Jackson type inequalities.

We come up with the "right" precisely defined multivariate normalized, quasi-interpolation neural network operators related to boxes or $\mathbb{R}^N$, as well as Kantorovich type and quadrature type related operators on $\mathbb{R}^N$. Our boxes are not necessarily symmetric to the origin. In preparation to prove our results we establish important properties of the basic multivariate density function induced by algebraic sigmoid function and defining our operators.

Feed-forward neural networks (FNNs) with one hidden layer, the only type of networks we deal with in this chapter, are mathematically expressed as

$$N_n(x) = \sum_{j=0}^{n} c_j \sigma\left(\langle a_j \cdot x \rangle + b_j\right), \quad x \in \mathbb{R}^s, \quad s \in \mathbb{N},$$

where for $0 \leq j \leq n$, $b_j \in \mathbb{R}$ are the thresholds, $a_j \in \mathbb{R}^s$ are the connection weights, $c_j \in \mathbb{R}$ are the coefficients, $\langle a_j \cdot x \rangle$ is the inner product of a_j and x, and σ is the activation function of the network. In many fundamental network models, the activation function is the algebraic sigmoid function. About neural networks see [19–21].

4.2 Basic

Here see also [12].

We consider the generator algebraic function

$$\varphi(x) = \frac{x}{\sqrt[2m]{1 + x^{2m}}}, \quad m \in \mathbb{N}, x \in \mathbb{R}, \tag{4.1}$$

which is a sigmoid type of function and is a strictly increasing function.

We see that $\varphi(-x) = -\varphi(x)$ with $\varphi(0) = 0$. We get that

$$\varphi'(x) = \frac{1}{\left(1 + x^{2m}\right)^{\frac{2m+1}{m}}} > 0, \ \forall\, x \in \mathbb{R}, \tag{4.2}$$

proving φ as strictly increasing over $\mathbb{R}$, $\varphi'(x) = \varphi'(-x)$. We easily find that $\lim_{x \to +\infty} \varphi(x) = 1$, $\varphi(+\infty) = 1$, and $\lim_{x \to -\infty} \varphi(x) = -1$, $\varphi(-\infty) = -1$.

We consider the activation function

$$\Phi(x) = \frac{1}{4}\left[\varphi(x+1) - \varphi(x-1)\right]. \tag{4.3}$$

Clearly it is $\Phi(x) = \Phi(-x)$, $\forall\, x \in \mathbb{R}$, so that Φ is an even function and symmetric with respect to the y-axis. Clearly $\Phi(x) > 0$, $\forall\, x \in \mathbb{R}$.

Also it is

$$\Phi(0) = \frac{1}{2\sqrt[2m]{2}}. \tag{4.4}$$

By [12], we have that $\Phi'(x) < 0$ for $x > 0$. That is Φ is strictly decreasing over $(0, +\infty)$.

Clearly, Φ is strictly increasing over $(-\infty, 0)$ and $\Phi'(0) = 0$.

Furthermore we obtain that

$$\lim_{x \to +\infty} \Phi(x) = \frac{1}{4}\left[\varphi(+\infty) - \varphi(+\infty)\right] = 0, \tag{4.5}$$

and

$$\lim_{x \to -\infty} \Phi(x) = \frac{1}{4}\left[\varphi(-\infty) - \varphi(-\infty)\right] = 0. \tag{4.6}$$

That is the x-axis is the horizontal asymptote of Φ.

Conclusion, Φ is a bell shape symmetric function with maximum

$$\Phi(0) = \frac{1}{2\sqrt[2m]{2}}, \quad m \in \mathbb{N}. \tag{4.7}$$

We need

Theorem 4.1 ([12]) *We have that*

$$\sum_{i=-\infty}^{\infty} \Phi(x - i) = 1, \ \forall\, x \in \mathbb{R}. \tag{4.8}$$

Theorem 4.2 ([12]) *It holds*

$$\int_{-\infty}^{\infty} \Phi(x)\, dx = 1. \tag{4.9}$$

Theorem 4.3 ([12]) *Let* $0 < \alpha < 1$, *and* $n \in \mathbb{N}$ *with* $n^{1-\alpha} > 2$. *It holds*

$$\sum_{\substack{k = -\infty \\ : |nx - k| \geq n^{1-\alpha}}}^{\infty} \Phi(nx - k) < \frac{1}{4m\left(n^{1-\alpha} - 2\right)^{2m}}, \quad m \in \mathbb{N}. \tag{4.10}$$

Denote by $\lfloor \cdot \rfloor$ the integral part of the number and by $\lceil \cdot \rceil$ the ceiling of the number. We need

Theorem 4.4 ([12]) *Let* $[a, b] \subset \mathbb{R}$ *and* $n \in \mathbb{N}$ *so that* $\lceil na \rceil \leq \lfloor nb \rfloor$. *It holds*

$$\frac{1}{\sum_{k=\lceil na \rceil}^{\lfloor nb \rfloor} \Phi(nx-k)} < 2\left(\sqrt[2m]{1+4^m}\right), \tag{4.11}$$

$\forall\, x \in [a, b]$, $m \in \mathbb{N}$.

Note 4.5 (1) By [12] we have that

$$\lim_{n \to \infty} \sum_{k=\lceil na \rceil}^{\lfloor nb \rfloor} \Phi(nx-k) \neq 1, \tag{4.12}$$

for at least some $x \in [a, b]$.

(2) Let $[a, b] \subset \mathbb{R}$. For large $n \in \mathbb{N}$ we always have $\lceil na \rceil \leq \lfloor nb \rfloor$. Also $a \leq \frac{k}{n} \leq b$, iff $\lceil na \rceil \leq k \leq \lfloor nb \rfloor$.

In general it holds that

$$\sum_{k=\lceil na \rceil}^{\lfloor nb \rfloor} \Phi(nx-k) \leq 1. \tag{4.13}$$

We introduce

$$Z(x_1, \ldots, x_N) := Z(x) := \prod_{i=1}^{N} \Phi(x_i), \quad x = (x_1, \ldots, x_N) \in \mathbb{R}^N, \ N \in \mathbb{N}. \tag{4.14}$$

It has the properties:

(i) $Z(x) > 0$, $\forall\, x \in \mathbb{R}^N$,

(ii)

$$\sum_{k=-\infty}^{\infty} Z(x-k) := \sum_{k_1=-\infty}^{\infty} \sum_{k_2=-\infty}^{\infty} \ldots \sum_{k_N=-\infty}^{\infty} Z(x_1-k_1, \ldots, x_N-k_N) = 1, \tag{4.15}$$

where $k := (k_1, \ldots, k_n) \in \mathbb{Z}^N$, $\forall\, x \in \mathbb{R}^N$,

hence

(iii)

$$\sum_{k=-\infty}^{\infty} Z(nx-k) = 1, \tag{4.16}$$

$\forall\, x \in \mathbb{R}^N$; $n \in \mathbb{N}$,

and

(iv)

$$\int_{\mathbb{R}^N} Z(x)\,dx = 1, \tag{4.17}$$

that is Z is a multivariate density function.

Here denote $\|x\|_\infty := \max\{|x_1|, ..., |x_N|\}$, $x \in \mathbb{R}^N$, also set $\infty := (\infty, ..., \infty)$, $-\infty := (-\infty, ..., -\infty)$ upon the multivariate context, and

$$\begin{aligned} \lceil na \rceil &:= (\lceil na_1 \rceil, ..., \lceil na_N \rceil), \\ \lfloor nb \rfloor &:= (\lfloor nb_1 \rfloor, ..., \lfloor nb_N \rfloor), \end{aligned} \tag{4.18}$$

where $a := (a_1, ..., a_N)$, $b := (b_1, ..., b_N)$.

We obviously see that

$$\sum_{k=\lceil na \rceil}^{\lfloor nb \rfloor} Z(nx - k) = \sum_{k=\lceil na \rceil}^{\lfloor nb \rfloor} \left(\prod_{i=1}^{N} \Phi(nx_i - k_i) \right) =$$

$$\sum_{k_1=\lceil na_1 \rceil}^{\lfloor nb_1 \rfloor} \cdots \sum_{k_N=\lceil na_N \rceil}^{\lfloor nb_N \rfloor} \left(\prod_{i=1}^{N} \Phi(nx_i - k_i) \right) = \prod_{i=1}^{N} \left(\sum_{k_i=\lceil na_i \rceil}^{\lfloor nb_i \rfloor} \Phi(nx_i - k_i) \right). \tag{4.19}$$

For $0 < \beta < 1$ and $n \in \mathbb{N}$, a fixed $x \in \mathbb{R}^N$, we have that

$$\sum_{k=\lceil na \rceil}^{\lfloor nb \rfloor} Z(nx - k) =$$

$$\sum_{\begin{cases} k = \lceil na \rceil \\ \left\| \frac{k}{n} - x \right\|_\infty \leq \frac{1}{n^\beta} \end{cases}}^{\lfloor nb \rfloor} Z(nx - k) + \sum_{\begin{cases} k = \lceil na \rceil \\ \left\| \frac{k}{n} - x \right\|_\infty > \frac{1}{n^\beta} \end{cases}}^{\lfloor nb \rfloor} Z(nx - k). \tag{4.20}$$

In the last two sums the counting is over disjoint vector sets of k's, because the condition $\left\| \frac{k}{n} - x \right\|_\infty > \frac{1}{n^\beta}$ implies that there exists at least one $\left| \frac{k_r}{n} - x_r \right| > \frac{1}{n^\beta}$, where $r \in \{1, ..., N\}$.

(v) As in [10], pp. 379–380, we derive that

$$\sum_{\begin{cases} k = \lceil na \rceil \\ \left\| \frac{k}{n} - x \right\|_\infty > \frac{1}{n^\beta} \end{cases}}^{\lfloor nb \rfloor} Z(nx - k) \overset{4.10}{<} \frac{1}{4m \left(n^{1-\beta} - 2\right)^{2m}}, \quad 0 < \beta < 1, \ m \in \mathbb{N}, \tag{4.21}$$

with $n \in \mathbb{N} : n^{1-\beta} > 2$, $x \in \prod_{i=1}^{N} [a_i, b_i]$.

(vi) By Theorem 4.4 we get that

$$0 < \frac{1}{\sum_{k=\lceil na \rceil}^{\lfloor nb \rfloor} Z(nx-k)} < \frac{1}{(\Phi(1))^N} \cong \left[2\left(\sqrt[2m]{1+4^m}\right)\right]^N, \tag{4.22}$$

$\forall\, x \in \left(\prod_{i=1}^N [a_i, b_i]\right)$, $n \in \mathbb{N}$.

It is also clear that

(vii)

$$\sum_{\substack{k=-\infty \\ \left\| \frac{k}{n} - x \right\|_\infty > \frac{1}{n^\beta}}}^{\infty} Z(nx-k) < \frac{1}{4m\left(n^{1-\beta}-2\right)^{2m}}, \tag{4.23}$$

$0 < \beta < 1$, $n \in \mathbb{N} : n^{1-\beta} > 2$, $x \in \mathbb{R}^N$, $m \in \mathbb{N}$.

Furthermore it holds

$$\lim_{n\to\infty} \sum_{k=\lceil na \rceil}^{\lfloor nb \rfloor} Z(nx-k) \neq 1, \tag{4.24}$$

for at least some $x \in \left(\prod_{i=1}^N [a_i, b_i]\right)$.

Here $\left(X, \|\cdot\|_\gamma\right)$ is a Banach space.

Let $f \in C\left(\prod_{i=1}^N [a_i, b_i], X\right)$, $x = (x_1, ..., x_N) \in \prod_{i=1}^N [a_i, b_i]$, $n \in \mathbb{N}$ such that $\lceil na_i \rceil \leq \lfloor nb_i \rfloor$, $i = 1, ..., N$.

We introduce and define the following multivariate linear normalized neural network operator ($x := (x_1, ..., x_N) \in \left(\prod_{i=1}^N [a_i, b_i]\right)$):

$$A_n(f, x_1, ..., x_N) := A_n(f, x) := \frac{\sum_{k=\lceil na \rceil}^{\lfloor nb \rfloor} f\left(\frac{k}{n}\right) Z(nx-k)}{\sum_{k=\lceil na \rceil}^{\lfloor nb \rfloor} Z(nx-k)} =$$

$$\frac{\sum_{k_1=\lceil na_1 \rceil}^{\lfloor nb_1 \rfloor} \sum_{k_2=\lceil na_2 \rceil}^{\lfloor nb_2 \rfloor} \cdots \sum_{k_N=\lceil na_N \rceil}^{\lfloor nb_N \rfloor} f\left(\frac{k_1}{n}, ..., \frac{k_N}{n}\right)\left(\prod_{i=1}^N \Phi(nx_i - k_i)\right)}{\prod_{i=1}^N \left(\sum_{k_i=\lceil na_i \rceil}^{\lfloor nb_i \rfloor} \Phi(nx_i - k_i)\right)}. \tag{4.25}$$

For large enough $n \in \mathbb{N}$ we always obtain $\lceil na_i \rceil \leq \lfloor nb_i \rfloor$, $i = 1, ..., N$. Also $a_i \leq \frac{k_i}{n} \leq b_i$, iff $\lceil na_i \rceil \leq k_i \leq \lfloor nb_i \rfloor$, $i = 1, ..., N$.

When $g \in C\left(\prod_{i=1}^N [a_i, b_i]\right)$ we define the companion operator

$$\widetilde{A}_n(g, x) := \frac{\sum_{k=\lceil na \rceil}^{\lfloor nb \rfloor} g\left(\frac{k}{n}\right) Z(nx-k)}{\sum_{k=\lceil na \rceil}^{\lfloor nb \rfloor} Z(nx-k)}. \tag{4.26}$$

Clearly $\widetilde{A}_n$ is a positive linear operator. We have that

$$\widetilde{A}_n(1,x)=1,\ \forall\, x\in\left(\prod_{i=1}^{N}[a_i,b_i]\right).$$

Notice that $A_n(f)\in C\left(\prod_{i=1}^{N}[a_i,b_i],X\right)$ and $\widetilde{A}_n(g)\in C\left(\prod_{i=1}^{N}[a_i,b_i]\right)$.

Furthermore it holds

$$\|A_n(f,x)\|_\gamma\le\frac{\sum_{k=\lceil na\rceil}^{\lfloor nb\rfloor}\left\|f\left(\frac{k}{n}\right)\right\|_\gamma Z(nx-k)}{\sum_{k=\lceil na\rceil}^{\lfloor nb\rfloor}Z(nx-k)}=\widetilde{A}_n\left(\|f\|_\gamma,x\right),\tag{4.27}$$

$\forall\, x\in\prod_{i=1}^{N}[a_i,b_i]$.

Clearly $\|f\|_\gamma\in C\left(\prod_{i=1}^{N}[a_i,b_i]\right)$.

So, we have that

$$\|A_n(f,x)\|_\gamma\le\widetilde{A}_n\left(\|f\|_\gamma,x\right),\tag{4.28}$$

$\forall\, x\in\prod_{i=1}^{N}[a_i,b_i]$, $\forall\, n\in\mathbb{N}$, $\forall\, f\in C\left(\prod_{i=1}^{N}[a_i,b_i],X\right)$.

Let $c\in X$ and $g\in C\left(\prod_{i=1}^{N}[a_i,b_i]\right)$, then $cg\in C\left(\prod_{i=1}^{N}[a_i,b_i],X\right)$.

Furthermore it holds

$$A_n(cg,x)=c\widetilde{A}_n(g,x),\ \forall\, x\in\prod_{i=1}^{N}[a_i,b_i].\tag{4.29}$$

Since $\widetilde{A}_n(1)=1$, we get that

$$A_n(c)=c,\ \forall\, c\in X.\tag{4.30}$$

We call $\widetilde{A}_n$ the companion operator of A_n.

For convenience we call

$$A_n^*(f,x):=\sum_{k=\lceil na\rceil}^{\lfloor nb\rfloor}f\left(\frac{k}{n}\right)Z(nx-k)=$$

$$\sum_{k_1=\lceil na_1\rceil}^{\lfloor nb_1\rfloor}\sum_{k_2=\lceil na_2\rceil}^{\lfloor nb_2\rfloor}\cdots\sum_{k_N=\lceil na_N\rceil}^{\lfloor nb_N\rfloor}f\left(\frac{k_1}{n},\ldots,\frac{k_N}{n}\right)\left(\prod_{i=1}^{N}\Phi(nx_i-k_i)\right),\tag{4.31}$$

$\forall\, x\in\left(\prod_{i=1}^{N}[a_i,b_i]\right)$.

That is

$$A_n(f,x) := \frac{A_n^*(f,x)}{\sum_{k=\lceil na \rceil}^{\lfloor nb \rfloor} Z(nx-k)}, \tag{4.32}$$

$\forall\, x \in \left(\prod_{i=1}^{N} [a_i, b_i]\right)$, $n \in \mathbb{N}$.

Hence

$$A_n(f,x) - f(x) = \frac{A_n^*(f,x) - f(x)\left(\sum_{k=\lceil na \rceil}^{\lfloor nb \rfloor} Z(nx-k)\right)}{\sum_{k=\lceil na \rceil}^{\lfloor nb \rfloor} Z(nx-k)}. \tag{4.33}$$

Consequently we derive

$$\|A_n(f,x) - f(x)\|_\gamma \overset{(4.22)}{\leq} \left[2\left(\sqrt[2m]{1+4^m}\right)\right]^N \left\| A_n^*(f,x) - f(x) \sum_{k=\lceil na \rceil}^{\lfloor nb \rfloor} Z(nx-k) \right\|_\gamma, \tag{4.34}$$

$\forall\, x \in \left(\prod_{i=1}^{N} [a_i, b_i]\right)$.

We will estimate the right hand side of (4.34).

For the last and others we need

Definition 4.6 ([11], p. 274) Let M be a convex and compact subset of $\left(\mathbb{R}^N, \|\cdot\|_p\right)$, $p \in [1, \infty]$, and $\left(X, \|\cdot\|_\gamma\right)$ be a Banach space. Let $f \in C(M, X)$. We define the first modulus of continuity of f as

$$\omega_1(f,\delta) := \sup_{\substack{x,\, y \in M: \\ \|x-y\|_p \leq \delta}} \|f(x) - f(y)\|_\gamma, \quad 0 < \delta \leq diam(M). \tag{4.35}$$

If $\delta > diam(M)$, then

$$\omega_1(f,\delta) = \omega_1(f, diam(M)). \tag{4.36}$$

Notice $\omega_1(f,\delta)$ is increasing in $\delta > 0$. For $f \in C_B(M, X)$ (continuous and bounded functions) $\omega_1(f,\delta)$ is defined similarly.

Lemma 4.7 ([11], p. 274) *We have* $\omega_1(f,\delta) \to 0$ *as* $\delta \downarrow 0$, *iff* $f \in C(M, X)$, *where* M *is a convex compact subset of* $\left(\mathbb{R}^N, \|\cdot\|_p\right)$, $p \in [1, \infty]$.

Clearly we have also: $f \in C_U\left(\mathbb{R}^N, X\right)$ (uniformly continuous functions), iff $\omega_1(f,\delta) \to 0$ as $\delta \downarrow 0$, where ω_1 is defined similarly to (4.35). The space $C_B\left(\mathbb{R}^N, X\right)$ denotes the continuous and bounded functions on $\mathbb{R}^N$.

When $f \in C_B\left(\mathbb{R}^N, X\right)$ we define,

$$B_n(f,x) := B_n(f, x_1, ..., x_N) := \sum_{k=-\infty}^{\infty} f\left(\frac{k}{n}\right) Z(nx-k) :=$$

$$\sum_{k_1=-\infty}^{\infty}\sum_{k_2=-\infty}^{\infty}...\sum_{k_N=-\infty}^{\infty} f\left(\frac{k_1}{n},\frac{k_2}{n},...,\frac{k_N}{n}\right)\left(\prod_{i=1}^{N}\Phi\left(nx_i-k_i\right)\right), \tag{4.37}$$

$n \in \mathbb{N}, \forall\, x \in \mathbb{R}^N$, $N \in \mathbb{N}$, the multivariate quasi-interpolation neural network operator.

Also for $f \in C_B\left(\mathbb{R}^N, X\right)$ we define the multivariate Kantorovich type neural network operator

$$C_n\left(f,x\right) := C_n\left(f,x_1,...,x_N\right) := \sum_{k=-\infty}^{\infty}\left(n^N\int_{\frac{k}{n}}^{\frac{k+1}{n}} f\left(t\right)dt\right) Z\left(nx-k\right) =$$

$$\sum_{k_1=-\infty}^{\infty}\sum_{k_2=-\infty}^{\infty}...\sum_{k_N=-\infty}^{\infty}\left(n^N\int_{\frac{k_1}{n}}^{\frac{k_1+1}{n}}\int_{\frac{k_2}{n}}^{\frac{k_2+1}{n}}...\int_{\frac{k_N}{n}}^{\frac{k_N+1}{n}} f\left(t_1,...,t_N\right)dt_1...dt_N\right)$$

$$\cdot\left(\prod_{i=1}^{N}\Phi\left(nx_i-k_i\right)\right), \tag{4.38}$$

$n \in \mathbb{N}, \ \forall\, x \in \mathbb{R}^N$.

Again for $f \in C_B\left(\mathbb{R}^N, X\right)$, $N \in \mathbb{N}$, we define the multivariate neural network operator of quadrature type $D_n\left(f,x\right)$, $n \in \mathbb{N}$, as follows.

Let $\theta = \left(\theta_1,...,\theta_N\right) \in \mathbb{N}^N$, $r = \left(r_1,...,r_N\right) \in \mathbb{Z}_+^N$, $w_r = w_{r_1,r_2,...r_N} \geq 0$, such that $\sum_{r=0}^{\theta} w_r = \sum_{r_1=0}^{\theta_1}\sum_{r_2=0}^{\theta_2}...\sum_{r_N=0}^{\theta_N} w_{r_1,r_2,...,r_N} = 1$; $k \in \mathbb{Z}^N$ and

$$\delta_{nk}\left(f\right) := \delta_{n,k_1,k_2,...,k_N}\left(f\right) := \sum_{r=0}^{\theta} w_r f\left(\frac{k}{n}+\frac{r}{n\theta}\right) =$$

$$\sum_{r_1=0}^{\theta_1}\sum_{r_2=0}^{\theta_2}...\sum_{r_N=0}^{\theta_N} w_{r_1,r_2,...r_N} f\left(\frac{k_1}{n}+\frac{r_1}{n\theta_1},\frac{k_2}{n}+\frac{r_2}{n\theta_2},...,\frac{k_N}{n}+\frac{r_N}{n\theta_N}\right), \tag{4.39}$$

where $\frac{r}{\theta} := \left(\frac{r_1}{\theta_1},\frac{r_2}{\theta_2},...,\frac{r_N}{\theta_N}\right)$.

We set

$$D_n\left(f,x\right) := D_n\left(f,x_1,...,x_N\right) := \sum_{k=-\infty}^{\infty}\delta_{nk}\left(f\right) Z\left(nx-k\right) = \tag{4.40}$$

$$\sum_{k_1=-\infty}^{\infty}\sum_{k_2=-\infty}^{\infty}...\sum_{k_N=-\infty}^{\infty}\delta_{n,k_1,k_2,...,k_N}\left(f\right)\left(\prod_{i=1}^{N}\Phi\left(nx_i-k_i\right)\right),$$

$\forall\, x \in \mathbb{R}^N$.

In this chapter we study the approximation properties of A_n, B_n, C_n, D_n neural network operators and as well of their iterates. That is, the quantitative pointwise and uniform convergence of these operators to the unit operator I.

4.3 Multivariate General Neural Network Approximations

Here we present several vectorial neural network approximations to Banach space valued functions given with rates.

We give

Theorem 4.8 *Let $f \in C\left(\prod_{i=1}^N [a_i, b_i], X\right)$, $0 < \beta < 1$, $x \in \left(\prod_{i=1}^N [a_i, b_i]\right)$, m, $N, n \in \mathbb{N}$ with $n^{1-\beta} > 2$. Then*

(1)

$$\|A_n(f, x) - f(x)\|_\gamma \leq \left[2\left(\sqrt[2m]{1+4^m}\right)\right]^N \left[\omega_1\left(f, \frac{1}{n^\beta}\right) + \frac{\| \|f\|_\gamma \|_\infty}{2m\left(n^{1-\beta}-2\right)^{2m}}\right] =: \lambda_1(n), \tag{4.41}$$

and

(2)

$$\left\| \|A_n(f) - f\|_\gamma \right\|_\infty \leq \lambda_1(n). \tag{4.42}$$

We notice that $\lim_{n\to\infty} A_n(f) \overset{\|\cdot\|_\gamma}{=} f$, pointwise and uniformly.

Above ω_1 is with respect to $p = \infty$.

Proof We observe that

$$\Delta(x) := A_n^*(f, x) - f(x) \sum_{k=\lceil na \rceil}^{\lfloor nb \rfloor} Z(nx - k) =$$

$$\sum_{k=\lceil na \rceil}^{\lfloor nb \rfloor} f\left(\frac{k}{n}\right) Z(nx-k) - \sum_{k=\lceil na \rceil}^{\lfloor nb \rfloor} f(x) Z(nx-k) =$$

$$\sum_{k=\lceil na \rceil}^{\lfloor nb \rfloor} \left(f\left(\frac{k}{n}\right) - f(x)\right) Z(nx-k). \tag{4.43}$$

Thus

$$\|\Delta(x)\|_\gamma \le \sum_{k=\lceil na\rceil}^{\lfloor nb\rfloor} \left\| f\left(\frac{k}{n}\right) - f(x)\right\|_\gamma Z(nx-k) =$$

$$\sum_{\substack{k=\lceil na\rceil \\ \left\|\frac{k}{n}-x\right\|_\infty \le \frac{1}{n^\beta}}}^{\lfloor nb\rfloor} \left\| f\left(\frac{k}{n}\right) - f(x)\right\|_\gamma Z(nx-k) +$$

$$\sum_{\substack{k=\lceil na\rceil \\ \left\|\frac{k}{n}-x\right\|_\infty > \frac{1}{n^\beta}}}^{\lfloor nb\rfloor} \left\| f\left(\frac{k}{n}\right) - f(x)\right\|_\gamma Z(nx-k) \overset{(4.16)}{\le}$$

$$\omega_1\left(f, \frac{1}{n^\beta}\right) + 2\left\|\|f\|_\gamma\right\|_\infty \sum_{\substack{k=\lceil na\rceil \\ \left\|\frac{k}{n}-x\right\|_\infty > \frac{1}{n^\beta}}}^{\lfloor nb\rfloor} Z(nx-k) \overset{(4.21)}{\le}$$

$$\omega_1\left(f, \frac{1}{n^\beta}\right) + \frac{2\left\|\|f\|_\gamma\right\|_\infty}{4m\left(n^{1-\beta}-2\right)^{2m}}. \tag{4.44}$$

So that

$$\|\Delta(x)\|_\gamma \le \omega_1\left(f, \frac{1}{n^\beta}\right) + \frac{\left\|\|f\|_\gamma\right\|_\infty}{2m\left(n^{1-\beta}-2\right)^{2m}}. \tag{4.45}$$

Now using (4.34) we finish the proof. ■

We make

Remark 4.9 ([11], pp. 263–266) Let $\left(\mathbb{R}^N, \|\cdot\|_p\right)$, $N \in \mathbb{N}$; where $\|\cdot\|_p$ is the L_p-norm, $1 \le p \le \infty$. $\mathbb{R}^N$ is a Banach space, and $\left(\mathbb{R}^N\right)^j$ denotes the j-fold product space $\mathbb{R}^N \times \ldots \times \mathbb{R}^N$ endowed with the max-norm $\|x\|_{\left(\mathbb{R}^N\right)^j} := \max_{1\le\lambda\le j} \|x_\lambda\|_p$, where $x := (x_1, \ldots, x_j) \in \left(\mathbb{R}^N\right)^j$.

Let $\left(X, \|\cdot\|_\gamma\right)$ be a general Banach space. Then the space $L_j := L_j\left(\left(\mathbb{R}^N\right)^j; X\right)$ of all j-multilinear continuous maps $g : \left(\mathbb{R}^N\right)^j \to X$, $j = 1, \ldots, m$, is a Banach space with norm

$$\|g\| := \|g\|_{L_j} := \sup_{\left(\|x\|_{\left(\mathbb{R}^N\right)^j}=1\right)} \|g(x)\|_\gamma = \sup \frac{\|g(x)\|_\gamma}{\|x_1\|_p \ldots \|x_j\|_p}. \tag{4.46}$$

Let M be a non-empty convex and compact subset of $\mathbb{R}^N$ and $x_0 \in M$ is fixed.

Let O be an open subset of $\mathbb{R}^N : M \subset O$. Let $f : O \to X$ be a continuous function, whose Fréchet derivatives (see [22]) $f^{(j)} : O \to L_j = L_j\left(\left(\mathbb{R}^N\right)^j ; X\right)$ exist and are continuous for $1 \leq j \leq \overline{m}, \overline{m} \in \mathbb{N}$.

Call $(x - x_0)^j := (x - x_0, ..., x - x_0) \in \left(\mathbb{R}^N\right)^j$, $x \in M$.

We will work with $f|_M$.

Then, by Taylor's formula [15], ([22], p. 124), we get

$$f(x) = \sum_{j=0}^{\overline{m}} \frac{f^{(j)}(x_0)(x - x_0)^j}{j!} + R_{\overline{m}}(x, x_0), \text{ all } x \in M, \tag{4.47}$$

where the remainder is the Riemann integral

$$R_{\overline{m}}(x, x_0) := \int_0^1 \frac{(1-u)^{\overline{m}-1}}{(\overline{m}-1)!} \left(f^{(\overline{m})}(x_0 + u(x - x_0)) - f^{(\overline{m})}(x_0)\right)(x - x_0)^{\overline{m}} du, \tag{4.48}$$

here we set $f^{(0)}(x_0)(x - x_0)^0 = f(x_0)$.

We consider

$$w := \omega_1\left(f^{(\overline{m})}, h\right) := \sup_{\substack{x, y \in M: \\ \|x-y\|_p \leq h}} \left\| f^{(\overline{m})}(x) - f^{(\overline{m})}(y) \right\|, \tag{4.49}$$

$h > 0$.

We obtain

$$\begin{aligned} &\left\| \left(f^{(\overline{m})}(x_0 + u(x - x_0)) - f^{(\overline{m})}(x_0)\right)(x - x_0)^{\overline{m}} \right\|_\gamma \leq \\ &\left\| f^{(\overline{m})}(x_0 + u(x - x_0)) - f^{(\overline{m})}(x_0) \right\| \cdot \|x - x_0\|_p^{\overline{m}} \leq \\ &w \|x - x_0\|_p^{\overline{m}} \left\lceil \frac{u \|x - x_0\|_p}{h} \right\rceil, \end{aligned} \tag{4.50}$$

by Lemma 7.1.1, [1], p. 208, where $\lceil \cdot \rceil$ is the ceiling.

Therefore for all $x \in M$ (see [1], pp. 121–122):

$$\begin{aligned} \|R_{\overline{m}}(x, x_0)\|_\gamma &\leq w \|x - x_0\|_p^{\overline{m}} \int_0^1 \left\lceil \frac{u \|x - x_0\|_p}{h} \right\rceil \frac{(1-u)^{\overline{m}-1}}{(\overline{m}-1)!} du \\ &= \overline{w} \Phi_{\overline{m}}\left(\|x - x_0\|_p\right) \end{aligned} \tag{4.51}$$

by a change of variable, where

$$\Phi_{\overline{m}}(t) := \int_0^{|t|} \left\lceil \frac{s}{h} \right\rceil \frac{(|t|-s)^{\overline{m}-1}}{(\overline{m}-1)!} ds = \frac{1}{\overline{m}!} \left(\sum_{j=0}^{\infty} (|t| - jh)_+^{\overline{m}} \right), \ \forall \, t \in \mathbb{R}, \quad (4.52)$$

is a (polynomial) spline function, see [1], pp. 210–211.

Also from there we get

$$\Phi_{\overline{m}}(t) \leq \left(\frac{|t|^{\overline{m}+1}}{(\overline{m}+1)!h} + \frac{|t|^{\overline{m}}}{2\overline{m}!} + \frac{h\,|t|^{\overline{m}-1}}{8\,(\overline{m}-1)!} \right), \ \forall \, t \in \mathbb{R}, \quad (4.53)$$

with equality true only at $t = 0$.

Therefore it holds

$$\| R_{\overline{m}}(x, x_0) \|_{\gamma} \leq w \left(\frac{\|x - x_0\|_p^{\overline{m}+1}}{(\overline{m}+1)!h} + \frac{\|x - x_0\|_p^{\overline{m}}}{2\overline{m}!} + \frac{h\,\|x - x_0\|_p^{\overline{m}-1}}{8\,(\overline{m}-1)!} \right), \ \forall \, x \in M. \quad (4.54)$$

We have found that

$$\left\| f(x) - \sum_{j=0}^{\overline{m}} \frac{f^{(j)}(x_0)(x - x_0)^j}{j!} \right\|_{\gamma} \leq$$

$$\omega_1\left(f^{(\overline{m})}, h\right) \left(\frac{\|x - x_0\|_p^{\overline{m}+1}}{(\overline{m}+1)!h} + \frac{\|x - x_0\|_p^{\overline{m}}}{2\overline{m}!} + \frac{h\,\|x - x_0\|_p^{\overline{m}-1}}{8\,(\overline{m}-1)!} \right) < \infty, \quad (4.55)$$

$\forall \, x, x_0 \in M$.

Here $0 < \omega_1\left(f^{(\overline{m})}, h\right) < \infty$, by M being compact and $f^{(\overline{m})}$ being continuous on M.

One can rewrite (4.55) as follows:

$$\left\| f(\cdot) - \sum_{j=0}^{\overline{m}} \frac{f^{(j)}(x_0)(\cdot - x_0)^j}{j!} \right\|_{\gamma} \leq$$

$$\omega_1\left(f^{(\overline{m})}, h\right) \left(\frac{\|\cdot - x_0\|_p^{\overline{m}+1}}{(\overline{m}+1)!h} + \frac{\|\cdot - x_0\|_p^{\overline{m}}}{2\overline{m}!} + \frac{h\,\|\cdot - x_0\|_p^{\overline{m}-1}}{8\,(\overline{m}-1)!} \right), \ \forall \, x_0 \in M, \quad (4.56)$$

a pointwise functional inequality on M.

Here $(\cdot - x_0)^j$ maps M into $\left(\mathbb{R}^N\right)^j$ and it is continuous, also $f^{(j)}(x_0)$ maps $\left(\mathbb{R}^N\right)^j$ into X and it is continuous. Hence their composition $f^{(j)}(x_0)(\cdot - x_0)^j$ is continuous from M into X.

Clearly $f(\cdot) - \sum_{j=0}^{\overline{m}} \frac{f^{(j)}(x_0)(\cdot - x_0)^j}{j!} \in C(M, X)$, hence $\left\| f(\cdot) - \sum_{j=0}^{\overline{m}} \frac{f^{(j)}(x_0)(\cdot - x_0)^j}{j!} \right\|_\gamma \in C(M)$.

Let $\{\widetilde{L}_N\}_{N \in \mathbb{N}}$ be a sequence of positive linear operators mapping $C(M)$ into $C(M)$.

Therefore we obtain

$$\left(\widetilde{L}_N \left(\left\| f(\cdot) - \sum_{j=0}^{\overline{m}} \frac{f^{(j)}(x_0)(\cdot - x_0)^j}{j!} \right\|_\gamma \right) \right)(x_0) \leq$$

$$\omega_1\left(f^{(\overline{m})}, h\right) \left[\frac{\left(\widetilde{L}_N\left(\|\cdot - x_0\|_p^{\overline{m}+1}\right)\right)(x_0)}{(\overline{m}+1)!h} + \frac{\left(\widetilde{L}_N\left(\|\cdot - x_0\|_p^{\overline{m}}\right)\right)(x_0)}{2\overline{m}!} \right.$$

$$\left. + \frac{h\left(\widetilde{L}_N\left(\|\cdot - x_0\|_p^{\overline{m}-1}\right)\right)(x_0)}{8(\overline{m}-1)!} \right], \tag{4.57}$$

$\forall\, N \in \mathbb{N}, \forall\, x_0 \in M$.

Clearly (4.57) is valid when $M = \prod_{i=1}^{N} [a_i, b_i]$ and $\widetilde{L}_n = \widetilde{A}_n$, see (4.26).

All the above is preparation for the following theorem, where we assume Fr échet differentiability of functions.

This will be a direct application of Theorem 10.2, [11], pp. 268–270. The operators A_n, $\widetilde{A}_n$ fulfill its assumptions, see (4.25), (4.26), (4.28), (4.29) and (4.30).

We present the following high order approximation results.

Theorem 4.10 *Let O open subset of $\left(\mathbb{R}^N, \|\cdot\|_p\right)$, $p \in [1, \infty]$, such that $\prod_{i=1}^{N} [a_i, b_i] \subset O \subseteq \mathbb{R}^N$, and let $\left(X, \|\cdot\|_\gamma\right)$ be a general Banach space. Let $\overline{m} \in \mathbb{N}$ and $f \in C^{\overline{m}}(O, X)$, the space of $\overline{m}$-times continuously Fréchet differentiable functions from O into X. We study the approximation of $f|_{\prod_{i=1}^{N}[a_i, b_i]}$. Let $x_0 \in \left(\prod_{i=1}^{N} [a_i, b_i]\right)$ and $r > 0$. Then*

(1)

$$\left\| (A_n(f))(x_0) - \sum_{j=0}^{\overline{m}} \frac{1}{j!} \left(A_n\left(f^{(j)}(x_0)(\cdot - x_0)^j\right)\right)(x_0) \right\|_\gamma \leq$$

$$\frac{\omega_1\left(f^{(\overline{m})}, r\left(\left(\widetilde{A}_n\left(\|\cdot - x_0\|_p^{\overline{m}+1}\right)\right)(x_0)\right)^{\frac{1}{\overline{m}+1}}\right)}{r\overline{m}!} \left(\left(\widetilde{A}_n\left(\|\cdot - x_0\|_p^{\overline{m}+1}\right)\right)(x_0)\right)^{\left(\frac{\overline{m}}{\overline{m}+1}\right)}$$

$$\left[\frac{1}{(\overline{m}+1)}+\frac{r}{2}+\frac{\overline{m}r^{2}}{8}\right], \tag{4.58}$$

(2) additionally if $f^{(j)}(x_0)=0,\ j=1,\dots,\overline{m}$, *we have*

$$\|(A_n(f))(x_0)-f(x_0)\|_{\gamma}\leq$$

$$\frac{\omega_1\left(f^{(\overline{m})},r\left(\left(\widetilde{A}_n\left(\|\cdot-x_0\|_p^{\overline{m}+1}\right)\right)(x_0)\right)^{\frac{1}{\overline{m}+1}}\right)}{r\overline{m}!}\left(\left(\widetilde{A}_n\left(\|\cdot-x_0\|_p^{\overline{m}+1}\right)\right)(x_0)\right)^{\left(\frac{\overline{m}}{\overline{m}+1}\right)}$$
$$\left[\frac{1}{(\overline{m}+1)}+\frac{r}{2}+\frac{\overline{m}r^{2}}{8}\right], \tag{4.59}$$

(3)

$$\|(A_n(f))(x_0)-f(x_0)\|_{\gamma}\leq\sum_{j=1}^{\overline{m}}\frac{1}{j!}\left\|\left(A_n\left(f^{(j)}(x_0)(\cdot-x_0)^j\right)\right)(x_0)\right\|_{\gamma}+$$

$$\frac{\omega_1\left(f^{(\overline{m})},r\left(\left(\widetilde{A}_n\left(\|\cdot-x_0\|_p^{\overline{m}+1}\right)\right)(x_0)\right)^{\frac{1}{\overline{m}+1}}\right)}{r\overline{m}!}\left(\left(\widetilde{A}_n\left(\|\cdot-x_0\|_p^{\overline{m}+1}\right)\right)(x_0)\right)^{\left(\frac{\overline{m}}{\overline{m}+1}\right)}$$
$$\left[\frac{1}{(\overline{m}+1)}+\frac{r}{2}+\frac{\overline{m}r^{2}}{8}\right], \tag{4.60}$$

and
(4)

$$\left\|\|A_n(f)-f\|_{\gamma}\right\|_{\infty,\prod_{i=1}^{N}[a_i,b_i]}\leq$$

$$\sum_{j=1}^{\overline{m}}\frac{1}{j!}\left\|\left\|\left(A_n\left(f^{(j)}(x_0)(\cdot-x_0)^j\right)\right)(x_0)\right\|_{\gamma}\right\|_{\infty,x_0\in\prod_{i=1}^{N}[a_i,b_i]}+$$

$$\frac{\omega_1\left(f^{(\overline{m})},r\left\|\left(\widetilde{A}_n\left(\|\cdot-x_0\|_p^{\overline{m}+1}\right)\right)(x_0)\right\|_{\infty,x_0\in\prod_{i=1}^{N}[a_i,b_i]}^{\frac{1}{\overline{m}+1}}\right)}{r\overline{m}!}$$

$$\left\|\left(\widetilde{A}_n\left(\|\cdot-x_0\|_p^{\overline{m}+1}\right)\right)(x_0)\right\|_{\infty,x_0\in\prod_{i=1}^{N}[a_i,b_i]}^{\left(\frac{\overline{m}}{\overline{m}+1}\right)} \tag{4.61}$$

$$\left[\frac{1}{(\overline{m}+1)}+\frac{r}{2}+\frac{\overline{m}r^2}{8}\right].$$

We need

Lemma 4.11 *The function* $\left(\widetilde{A}_n\left(\|\cdot - x_0\|_p^{\overline{m}}\right)\right)(x_0)$ *is continuous in* $x_0 \in \left(\prod_{i=1}^{N}[a_i, b_i]\right)$, $\overline{m} \in \mathbb{N}$.

Proof By Lemma 10.3, [11], p. 272. ■

We give

Corollary 4.12 (to Theorem 4.10, case of $\overline{m} = 1$) *Then*

(1)

$$\|(A_n(f))(x_0) - f(x_0)\|_\gamma \leq \left\|\left(A_n\left(f^{(1)}(x_0)(\cdot - x_0)\right)\right)(x_0)\right\|_\gamma +$$

$$\frac{1}{2r}\omega_1\left(f^{(1)}, r\left(\left(\widetilde{A}_n\left(\|\cdot - x_0\|_p^2\right)\right)(x_0)\right)^{\frac{1}{2}}\right)\left(\left(\widetilde{A}_n\left(\|\cdot - x_0\|_p^2\right)\right)(x_0)\right)^{\frac{1}{2}} \tag{4.62}$$

$$\left[1+r+\frac{r^2}{4}\right],$$

and

(2)

$$\left\|\|(A_n(f)) - f\|_\gamma\right\|_{\infty, \prod_{i=1}^{N}[a_i,b_i]} \leq$$

$$\left\|\left\|\left(A_n\left(f^{(1)}(x_0)(\cdot - x_0)\right)\right)(x_0)\right\|_\gamma\right\|_{\infty, x_0 \in \prod_{i=1}^{N}[a_i,b_i]} +$$

$$\frac{1}{2r}\omega_1\left(f^{(1)}, r\left\|\left(\widetilde{A}_n\left(\|\cdot - x_0\|_p^2\right)\right)(x_0)\right\|^{\frac{1}{2}}_{\infty, x_0 \in \prod_{i=1}^{N}[a_i,b_i]}\right)$$

$$\left\|\left(\widetilde{A}_n\left(\|\cdot - x_0\|_p^2\right)\right)(x_0)\right\|^{\frac{1}{2}}_{\infty, x_0 \in \prod_{i=1}^{N}[a_i,b_i]}\left[1+r+\frac{r^2}{4}\right], \tag{4.63}$$

$r > 0$.

We make

Remark 4.13 We estimate $0 < \alpha < 1, m, \overline{m}, n \in \mathbb{N} : n^{1-\alpha} > 2$,

$$\widetilde{A}_n \left(\|\cdot - x_0\|_\infty^{\overline{m}+1} \right)(x_0) = \frac{\sum_{k=\lceil na \rceil}^{\lfloor nb \rfloor} \left\| \frac{k}{n} - x_0 \right\|_\infty^{\overline{m}+1} Z(nx_0 - k)}{\sum_{k=\lceil na \rceil}^{\lfloor nb \rfloor} Z(nx_0 - k)} \overset{(4.22)}{<}$$

$$\left[2 \left(\sqrt[2m]{1+4^m} \right) \right]^N \sum_{k=\lceil na \rceil}^{\lfloor nb \rfloor} \left\| \frac{k}{n} - x_0 \right\|_\infty^{\overline{m}+1} Z(nx_0 - k) = \tag{4.64}$$

$$\left[2 \left(\sqrt[2m]{1+4^m} \right) \right]^N \left\{ \sum_{\substack{k = \lceil na \rceil \\ : \left\| \frac{k}{n} - x_0 \right\|_\infty \leq \frac{1}{n^\alpha}}}^{\lfloor nb \rfloor} \left\| \frac{k}{n} - x_0 \right\|_\infty^{\overline{m}+1} Z(nx_0 - k) + \right.$$

$$\left. \sum_{\substack{k = \lceil na \rceil \\ : \left\| \frac{k}{n} - x_0 \right\|_\infty > \frac{1}{n^\alpha}}}^{\lfloor nb \rfloor} \left\| \frac{k}{n} - x_0 \right\|_\infty^{\overline{m}+1} Z(nx_0 - k) \right\} \overset{(4.23)}{\leq}$$

$$\left[2 \left(\sqrt[2m]{1+4^m} \right) \right]^N \left\{ \frac{1}{n^{\alpha(\overline{m}+1)}} + \frac{\|b-a\|_\infty^{\overline{m}+1}}{4m \left(n^{1-\alpha} - 2 \right)^{2m}} \right\}, \tag{4.65}$$

(where $b - a = (b_1 - a_1, ..., b_N - a_N)$).

We have proved that $\left(\forall \, x_0 \in \prod_{i=1}^N [a_i, b_i] \right)$

$$\widetilde{A}_n \left(\|\cdot - x_0\|_\infty^{\overline{m}+1} \right)(x_0) < \left[2 \left(\sqrt[2m]{1+4^m} \right) \right]^N \left\{ \frac{1}{n^{\alpha(\overline{m}+1)}} + \frac{\|b-a\|_\infty^{\overline{m}+1}}{4m \left(n^{1-\alpha} - 2 \right)^{2m}} \right\} =: \varphi_1(n) \tag{4.66}$$

$(0 < \alpha < 1, m, \overline{m}, n \in \mathbb{N} : n^{1-\alpha} > 2)$.

And, consequently it holds

$$\left\| \widetilde{A}_n \left(\|\cdot - x_0\|_\infty^{\overline{m}+1} \right)(x_0) \right\|_{\infty, x_0 \in \prod_{i=1}^N [a_i, b_i]} <$$

$$\left[2 \left(\sqrt[2m]{1+4^m} \right) \right]^N \left\{ \frac{1}{n^{\alpha(\overline{m}+1)}} + \frac{\|b-a\|_\infty^{\overline{m}+1}}{4m \left(n^{1-\alpha} - 2 \right)^{2m}} \right\} = \varphi_1(n) \to 0, \text{ as } n \to +\infty. \tag{4.67}$$

So, we have that $\varphi_1(n) \to 0$, as $n \to +\infty$. Thus, when $p \in [1, \infty]$, from Theorem 4.10 we have the convergence to zero in the right hand sides of parts (1), (2).

Next we estimate $\left\| \left(\widetilde{A}_n \left(f^{(j)}(x_0) (\cdot - x_0)^j \right) \right)(x_0) \right\|_\gamma$.

We have that

$$\left(\widetilde{A}_n \left(f^{(j)}(x_0) (\cdot - x_0)^j \right) \right)(x_0) = \frac{\sum_{k=\lceil na \rceil}^{\lfloor nb \rfloor} f^{(j)}(x_0) \left(\frac{k}{n} - x_0 \right)^j Z(nx_0 - k)}{\sum_{k=\lceil na \rceil}^{\lfloor nb \rfloor} Z(nx_0 - k)}. \tag{4.68}$$

When $p = \infty$, $j = 1, ..., \overline{m}$, we obtain

$$\left\| f^{(j)}(x_0) \left(\frac{k}{n} - x_0 \right)^j \right\|_\gamma \leq \left\| f^{(j)}(x_0) \right\| \left\| \frac{k}{n} - x_0 \right\|_\infty^j. \tag{4.69}$$

We further have that

$$\left\| \left(\widetilde{A}_n \left(f^{(j)}(x_0) (\cdot - x_0)^j \right) \right)(x_0) \right\|_\gamma \overset{(4.22)}{<}$$

$$\left[2 \left(\sqrt[2m]{1 + 4^m} \right) \right]^N \left(\sum_{k=\lceil na \rceil}^{\lfloor nb \rfloor} \left\| f^{(j)}(x_0) \left(\frac{k}{n} - x_0 \right)^j \right\|_\gamma Z(nx_0 - k) \right) \leq$$

$$\left[2 \left(\sqrt[2m]{1 + 4^m} \right) \right]^N \left(\sum_{k=\lceil na \rceil}^{\lfloor nb \rfloor} \left\| f^{(j)}(x_0) \right\| \left\| \frac{k}{n} - x_0 \right\|_\infty^j Z(nx_0 - k) \right) = \tag{4.70}$$

$$\left[2 \left(\sqrt[2m]{1 + 4^m} \right) \right]^N \left\| f^{(j)}(x_0) \right\| \left(\sum_{k=\lceil na \rceil}^{\lfloor nb \rfloor} \left\| \frac{k}{n} - x_0 \right\|_\infty^j Z(nx_0 - k) \right) =$$

$$\left[2 \left(\sqrt[2m]{1 + 4^m} \right) \right]^N \left\| f^{(j)}(x_0) \right\| \left\{ \sum_{\substack{k = \lceil na \rceil \\ : \left\| \frac{k}{n} - x_0 \right\|_\infty \leq \frac{1}{n^\alpha}}}^{\lfloor nb \rfloor} \left\| \frac{k}{n} - x_0 \right\|_\infty^j Z(nx_0 - k) \right.$$

$$\left. + \sum_{\substack{k = \lceil na \rceil \\ : \left\| \frac{k}{n} - x_0 \right\|_\infty > \frac{1}{n^\alpha}}}^{\lfloor nb \rfloor} \left\| \frac{k}{n} - x_0 \right\|_\infty^j Z(nx_0 - k) \right\} \overset{(4.21)}{\leq} \tag{4.71}$$

$$\left[2\left(\sqrt[2m]{1+4^m}\right)\right]^N \left\|f^{(j)}(x_0)\right\| \left\{\frac{1}{n^{\alpha j}} + \frac{\|b-a\|_\infty^j}{4m\left(n^{1-\alpha}-2\right)^{2m}}\right\} \to 0, \text{ as } n \to \infty.$$

That is

$$\left\|\left(\widetilde{A}_n\left(f^{(j)}(x_0)(\cdot - x_0)^j\right)\right)(x_0)\right\|_\gamma \to 0, \text{ as } n \to \infty.$$

Therefore when $p = \infty$, for $j = 1, ..., \overline{m}$, we have proved:

$$\left\|\left(\widetilde{A}_n\left(f^{(j)}(x_0)(\cdot - x_0)^j\right)\right)(x_0)\right\|_\gamma <$$

$$\left[2\left(\sqrt[2m]{1+4^m}\right)\right]^N \left\|f^{(j)}(x_0)\right\| \left\{\frac{1}{n^{\alpha j}} + \frac{\|b-a\|_\infty^j}{4m\left(n^{1-\alpha}-2\right)^{2m}}\right\} \leq \tag{4.72}$$

$$\left[2\left(\sqrt[2m]{1+4^m}\right)\right]^N \left\|f^{(j)}\right\|_\infty \left\{\frac{1}{n^{\alpha j}} + \frac{\|b-a\|_\infty^j}{4m\left(n^{1-\alpha}-2\right)^{2m}}\right\} =: \varphi_{2j}(n) < \infty,$$

and converges to zero, as $n \to \infty$.

We conclude:

In Theorem 4.10, the right hand sides of (4.60) and (4.61) converge to zero as $n \to \infty$, for any $p \in [1, \infty]$.

Also in Corollary 4.12, the right hand sides of (4.62) and (4.63) converge to zero as $n \to \infty$, for any $p \in [1, \infty]$.

Conclusion 4.14 *We have proved that the left hand sides of (4.58), (4.59), (4.60), (4.61) and (4.62), (4.63) converge to zero as $n \to \infty$, for $p \in [1, \infty]$. Consequently $A_n \to I$ (unit operator) pointwise and uniformly, as $n \to \infty$, where $p \in [1, \infty]$. In the presence of initial conditions we achieve a higher speed of convergence, see (4.59). Higher speed of convergence happens also to the left hand side of (4.58).*

We further give

Corollary 4.15 (to Theorem 4.10) *Let O open subset of $\left(\mathbb{R}^N, \|\cdot\|_\infty\right)$, such that $\prod_{i=1}^N [a_i, b_i] \subset O \subseteq \mathbb{R}^N$, and let $\left(X, \|\cdot\|_\gamma\right)$ be a general Banach space. Let $\overline{m} \in \mathbb{N}$ and $f \in C^{\overline{m}}(O, X)$, the space of $\overline{m}$-times continuously Fréchet differentiable functions from O into X. We study the approximation of $f|_{\prod_{i=1}^N [a_i, b_i]}$. Let $x_0 \in \left(\prod_{i=1}^N [a_i, b_i]\right)$ and $r > 0$. Here $\varphi_1(n)$ as in (4.66) and $\varphi_{2j}(n)$ as in (4.72), where $n \in \mathbb{N} : n^{1-\alpha} > 2$, $0 < \alpha < 1$, $j = 1, ..., \overline{m}$. Then*

(1)

$$\left\| (A_n(f))(x_0) - \sum_{j=0}^{\overline{m}} \frac{1}{j!} \left(A_n\left(f^{(j)}(x_0)(\cdot - x_0)^j\right)\right)(x_0) \right\|_\gamma \leq$$

$$\frac{\omega_1\left(f^{(\overline{m})}, r(\varphi_1(n))^{\frac{1}{\overline{m}+1}}\right)}{r\overline{m}!} (\varphi_1(n))^{\left(\frac{\overline{m}}{\overline{m}+1}\right)} \left[\frac{1}{(\overline{m}+1)} + \frac{r}{2} + \frac{\overline{m}r^2}{8}\right], \tag{4.73}$$

(2) additionally, if $f^{(j)}(x_0) = 0$, $j = 1, ..., \overline{m}$, we have

$$\|(A_n(f))(x_0) - f(x_0)\|_\gamma \leq$$

$$\frac{\omega_1\left(f^{(\overline{m})}, r(\varphi_1(n))^{\frac{1}{\overline{m}+1}}\right)}{r\overline{m}!} (\varphi_1(n))^{\left(\frac{\overline{m}}{\overline{m}+1}\right)} \left[\frac{1}{(\overline{m}+1)} + \frac{r}{2} + \frac{\overline{m}r^2}{8}\right], \tag{4.74}$$

(3)

$$\left\| \|A_n(f) - f\|_\gamma \right\|_{\infty, \prod_{i=1}^{N}[a_i,b_i]} \leq \sum_{j=1}^{\overline{m}} \frac{\varphi_{2j}(n)}{j!} +$$

$$\frac{\omega_1\left(f^{(\overline{m})}, r(\varphi_1(n))^{\frac{1}{\overline{m}+1}}\right)}{r\overline{m}!} (\varphi_1(n))^{\left(\frac{\overline{m}}{\overline{m}+1}\right)} \tag{4.75}$$

$$\left[\frac{1}{(\overline{m}+1)} + \frac{r}{2} + \frac{\overline{m}r^2}{8}\right] =: \varphi_3(n) \to 0, \text{ as } n \to \infty.$$

We continue with

Theorem 4.16 *Let $f \in C_B\left(\mathbb{R}^N, X\right)$, $0 < \beta < 1$, $x \in \mathbb{R}^N$, $m, N, n \in \mathbb{N}$ with $n^{1-\beta} > 2$, ω_1 is for $p = \infty$. Then*
(1)

$$\|B_n(f,x) - f(x)\|_\gamma \leq \omega_1\left(f, \frac{1}{n^\beta}\right) + \frac{\left\| \|f\|_\gamma \right\|_\infty}{2m\left(n^{1-\beta} - 2\right)^{2m}} =: \lambda_2(n), \tag{4.76}$$

(2)

$$\left\| \|B_n(f) - f\|_\gamma \right\|_\infty \leq \lambda_2(n). \tag{4.77}$$

Given that $f \in \left(C_U\left(\mathbb{R}^N, X\right) \cap C_B\left(\mathbb{R}^N, X\right)\right)$, we obtain $\lim_{n\to\infty} B_n(f) = f$, uniformly.

Proof We have that

$$B_n(f,x) - f(x) \overset{(4.16)}{=} \sum_{k=-\infty}^{\infty} f\left(\frac{k}{n}\right) Z(nx-k) - f(x) \sum_{k=-\infty}^{\infty} Z(nx-k) =$$
$$\sum_{k=-\infty}^{\infty} \left(f\left(\frac{k}{n}\right) - f(x) \right) Z(nx-k). \tag{4.78}$$

Hence

$$\|B_n(f,x) - f(x)\|_\gamma \leq \sum_{k=-\infty}^{\infty} \left\| f\left(\frac{k}{n}\right) - f(x) \right\|_\gamma Z(nx-k) =$$

$$\sum_{\begin{cases} k=-\infty \\ \left\|\frac{k}{n}-x\right\|_\infty \leq \frac{1}{n^\beta} \end{cases}}^{\infty} \left\| f\left(\frac{k}{n}\right) - f(x) \right\|_\gamma Z(nx-k) +$$

$$\sum_{\begin{cases} k=-\infty \\ \left\|\frac{k}{n}-x\right\|_\infty > \frac{1}{n^\beta} \end{cases}}^{\infty} \left\| f\left(\frac{k}{n}\right) - f(x) \right\|_\gamma Z(nx-k) \overset{(4.16)}{\leq}$$

$$\omega_1\left(f, \frac{1}{n^\beta}\right) + 2\,\| \|f\|_\gamma \|_\infty \sum_{\begin{cases} k=-\infty \\ \left\|\frac{k}{n}-x\right\|_\infty > \frac{1}{n^\beta} \end{cases}}^{\infty} Z(nx-k) \overset{(4.23)}{\leq}$$

$$\omega_1\left(f, \frac{1}{n^\beta}\right) + \frac{\| \|f\|_\gamma \|_\infty}{2m\left(n^{1-\beta}-2\right)^{2m}}, \tag{4.79}$$

proving the claim. ■

We give

Theorem 4.17 *Let* $f \in C_B\left(\mathbb{R}^N, X\right)$, $0 < \beta < 1$, $x \in \mathbb{R}^N$, $m, N, n \in \mathbb{N}$ *with* $n^{1-\beta} > 2$, ω_1 *is for* $p = \infty$. *Then*

(1)

$$\|C_n(f,x) - f(x)\|_\gamma \leq \omega_1\left(f, \frac{1}{n} + \frac{1}{n^\beta}\right) + \frac{\| \|f\|_\gamma \|_\infty}{2m\left(n^{1-\beta}-2\right)^{2m}} =: \lambda_3(n), \tag{4.80}$$

(2)

$$\| \|C_n(f) - f\|_\gamma \|_\infty \leq \lambda_3(n). \tag{4.81}$$

Given that $f \in \left(C_U\left(\mathbb{R}^N, X\right) \cap C_B\left(\mathbb{R}^N, X\right)\right)$, *we obtain* $\lim_{n\to\infty} C_n(f) = f$, *uniformly.*

Proof We notice that

$$\int_{\frac{k}{n}}^{\frac{k+1}{n}} f(t)\,dt = \int_{\frac{k_1}{n}}^{\frac{k_1+1}{n}} \int_{\frac{k_2}{n}}^{\frac{k_2+1}{n}} \dots \int_{\frac{k_N}{n}}^{\frac{k_N+1}{n}} f(t_1, t_2, \dots, t_N)\,dt_1 dt_2 \dots dt_N =$$

$$\int_0^{\frac{1}{n}} \int_0^{\frac{1}{n}} \dots \int_0^{\frac{1}{n}} f\left(t_1 + \frac{k_1}{n}, t_2 + \frac{k_2}{n}, \dots, t_N + \frac{k_N}{n}\right) dt_1 \dots dt_N = \int_0^{\frac{1}{n}} f\left(t + \frac{k}{n}\right) dt. \tag{4.82}$$

Thus it holds (by (4.38))

$$C_n(f, x) = \sum_{k=-\infty}^{\infty} \left(n^N \int_0^{\frac{1}{n}} f\left(t + \frac{k}{n}\right) dt\right) Z(nx - k). \tag{4.83}$$

We observe that

$$\|C_n(f, x) - f(x)\|_\gamma =$$

$$\left\| \sum_{k=-\infty}^{\infty} \left(n^N \int_0^{\frac{1}{n}} f\left(t + \frac{k}{n}\right) dt\right) Z(nx - k) - \sum_{k=-\infty}^{\infty} f(x) Z(nx - k) \right\|_\gamma =$$

$$\left\| \sum_{k=-\infty}^{\infty} \left(\left(n^N \int_0^{\frac{1}{n}} f\left(t + \frac{k}{n}\right) dt\right) - f(x)\right) Z(nx - k) \right\|_\gamma =$$

$$\left\| \sum_{k=-\infty}^{\infty} \left(n^N \int_0^{\frac{1}{n}} \left(f\left(t + \frac{k}{n}\right) - f(x)\right) dt\right) Z(nx - k) \right\|_\gamma \leq \tag{4.84}$$

$$\sum_{k=-\infty}^{\infty} \left(n^N \int_0^{\frac{1}{n}} \left\| f\left(t + \frac{k}{n}\right) - f(x)\right\|_\gamma dt\right) Z(nx - k) =$$

$$\sum_{\substack{k=-\infty \\ \left\|\frac{k}{n} - x\right\|_\infty \leq \frac{1}{n^\beta}}}^{\infty} \left(n^N \int_0^{\frac{1}{n}} \left\| f\left(t + \frac{k}{n}\right) - f(x)\right\|_\gamma dt\right) Z(nx - k) +$$

$$\sum_{\substack{k=-\infty \\ \left\|\frac{k}{n} - x\right\|_\infty > \frac{1}{n^\beta}}}^{\infty} \left(n^N \int_0^{\frac{1}{n}} \left\| f\left(t + \frac{k}{n}\right) - f(x)\right\|_\gamma dt\right) Z(nx - k) \leq$$

$$\sum_{\substack{k=-\infty \\ \left\| \frac{k}{n} - x \right\|_{\infty} \leq \frac{1}{n^{\beta}}}}^{\infty} \left(n^N \int_0^{\frac{1}{n}} \omega_1 \left(f, \|t\|_{\infty} + \left\| \frac{k}{n} - x \right\|_{\infty} \right) dt \right) Z(nx-k) +$$

$$2 \left\| \|f\|_{\gamma} \right\|_{\infty} \left(\sum_{\substack{k=-\infty \\ \left\| \frac{k}{n} - x \right\|_{\infty} > \frac{1}{n^{\beta}}}}^{\infty} Z(|nx-k|) \right) \leq$$

$$\omega_1 \left(f, \frac{1}{n} + \frac{1}{n^{\beta}} \right) + \frac{\left\| \|f\|_{\gamma} \right\|_{\infty}}{2m \left(n^{1-\beta} - 2 \right)^{2m}}, \tag{4.85}$$

proving the claim. ■

We also present

Theorem 4.18 *Let $f \in C_B\left(\mathbb{R}^N, X\right)$, $0 < \beta < 1$, $x \in \mathbb{R}^N$, $m, N, n \in \mathbb{N}$ with $n^{1-\beta} > 2$, ω_1 is for $p = \infty$. Then*

(1)

$$\|D_n(f,x) - f(x)\|_{\gamma} \leq \omega_1 \left(f, \frac{1}{n} + \frac{1}{n^{\beta}} \right) + \frac{\left\| \|f\|_{\gamma} \right\|_{\infty}}{2m \left(n^{1-\beta} - 2 \right)^{2m}} = \lambda_4(n), \tag{4.86}$$

(2)

$$\left\| \|D_n(f) - f\|_{\gamma} \right\|_{\infty} \leq \lambda_4(n). \tag{4.87}$$

Given that $f \in \left(C_U\left(\mathbb{R}^N, X\right) \cap C_B\left(\mathbb{R}^N, X\right) \right)$, we obtain $\lim_{n \to \infty} D_n(f) = f$, uniformly.

Proof We have that (by (4.40))

$$\|D_n(f,x) - f(x)\|_{\gamma} = \left\| \sum_{k=-\infty}^{\infty} \delta_{nk}(f) Z(nx-k) - \sum_{k=-\infty}^{\infty} f(x) Z(nx-k) \right\|_{\gamma} = \tag{4.88}$$

$$\left\| \sum_{k=-\infty}^{\infty} \left(\delta_{nk}(f) - f(x) \right) Z(nx-k) \right\|_{\gamma} =$$

$$\left\| \sum_{k=-\infty}^{\infty} \left(\sum_{r=0}^{\theta} w_r \left(f \left(\frac{k}{n} + \frac{r}{n\theta} \right) - f(x) \right) \right) Z(nx-k) \right\|_{\gamma} \leq$$

$$\sum_{k=-\infty}^{\infty}\left(\sum_{r=0}^{\theta} w_r \left\| f\left(\frac{k}{n}+\frac{r}{n\theta}\right)-f(x)\right\|_{\gamma}\right) Z(nx-k) =$$

$$\sum_{\substack{k=-\infty \\ \left\| \frac{k}{n}-x \right\|_{\infty} \leq \frac{1}{n^{\beta}}}}^{\infty}\left(\sum_{r=0}^{\theta} w_r \left\| f\left(\frac{k}{n}+\frac{r}{n\theta}\right)-f(x)\right\|_{\gamma}\right) Z(nx-k) +$$

$$\sum_{\substack{k=-\infty \\ \left\| \frac{k}{n}-x \right\|_{\infty} > \frac{1}{n^{\beta}}}}^{\infty}\left(\sum_{r=0}^{\theta} w_r \left\| f\left(\frac{k}{n}+\frac{r}{n\theta}\right)-f(x)\right\|_{\gamma}\right) Z(nx-k) \leq$$

$$\sum_{\substack{k=-\infty \\ \left\| \frac{k}{n}-x \right\|_{\infty} \leq \frac{1}{n^{\beta}}}}^{\infty}\left(\sum_{r=0}^{\theta} w_r \omega_1\left(f, \left\|\frac{k}{n}-x\right\|_{\infty}+\left\|\frac{r}{n\theta}\right\|_{\infty}\right)\right) Z(nx-k) +$$

$$2\left\| \|f\|_{\gamma} \right\|_{\infty}\left(\sum_{\substack{k=-\infty \\ \left\| \frac{k}{n}-x \right\|_{\infty} > \frac{1}{n^{\beta}}}}^{\infty} Z(nx-k)\right) \leq$$

$$\omega_1\left(f, \frac{1}{n}+\frac{1}{n^{\beta}}\right)+\frac{\left\| \|f\|_{\gamma} \right\|_{\infty}}{2m\left(n^{1-\beta}-2\right)^{2m}}, \tag{4.89}$$

proving the claim. ∎

We make

Definition 4.19 Let $f \in C_B\left(\mathbb{R}^N, X\right)$, $N \in \mathbb{N}$, where $\left(X, \|\cdot\|_{\gamma}\right)$ is a Banach space. We define the general neural network operator

$$F_n(f, x) := \sum_{k=-\infty}^{\infty} l_{nk}(f) Z(nx-k) =$$

$$\begin{cases} B_n(f, x), \text{ if } l_{nk}(f) = f\left(\frac{k}{n}\right), \\ C_n(f, x), \text{ if } l_{nk}(f) = n^N \int_{\frac{k}{n}}^{\frac{k+1}{n}} f(t)\, dt, \\ D_n(f, x), \text{ if } l_{nk}(f) = \delta_{nk}(f). \end{cases} \tag{4.90}$$

Clearly $l_{nk}(f)$ is an X-valued bounded linear functional such that $\|l_{nk}(f)\|_\gamma \leq \left\| \|f\|_\gamma \right\|_\infty$.

Hence $F_n(f)$ is a bounded linear operator with $\left\| \|F_n(f)\|_\gamma \right\|_\infty \leq \left\| \|f\|_\gamma \right\|_\infty$.

We need

Theorem 4.20 *Let $f \in C_B(\mathbb{R}^N, X)$, $N \geq 1$. Then $F_n(f) \in C_B(\mathbb{R}^N, X)$.*

Proof Clearly $F_n(f)$ is a bounded function.

Next we prove the continuity of $F_n(f)$. Notice for $N = 1$, $Z = \Phi$ by (4.14).

We will use the generalized Weierstrass M test: If a sequence of positive constants $M_1, M_2, M_3, \ldots$, can be found such that in some interval

(a) $\|u_n(x)\|_\gamma \leq M_n$, $n = 1, 2, 3, \ldots$

(b) $\sum M_n$ converges,

then $\sum u_n(x)$ is uniformly and absolutely convergent in the interval.

Also we will use:

If $\{u_n(x)\}$, $n = 1, 2, 3, \ldots$ are continuous in $[a, b]$ and if $\sum u_n(x)$ converges uniformly to the sum $S(x)$ in $[a, b]$, then $S(x)$ is continuous in $[a, b]$. I.e. a uniformly convergent series of continuous functions is a continuous function. First we prove claim for $N = 1$.

We will prove that $\sum_{k=-\infty}^{\infty} l_{nk}(f)\,\Phi(nx - k)$ is continuous in $x \in \mathbb{R}$.

There always exists $\lambda \in \mathbb{N}$ such that $nx \in [-\lambda, \lambda]$.

Since $nx \leq \lambda$, then $-nx \geq -\lambda$ and $k - nx \geq k - \lambda \geq 0$, when $k \geq \lambda$. Therefore

$$\sum_{k=\lambda}^{\infty} \Phi(nx - k) = \sum_{k=\lambda}^{\infty} \Phi(k - nx) \leq \sum_{k=\lambda}^{\infty} \Phi(k - \lambda) = \sum_{k'=0}^{\infty} \Phi(k') \leq 1. \tag{4.91}$$

So for $k \geq \lambda$ we get

$$\|l_{nk}(f)\|_\gamma\, \Phi(nx - k) \leq \left\| \|f\|_\gamma \right\|_\infty \Phi(k - \lambda),$$

and

$$\left\| \|f\|_\gamma \right\|_\infty \sum_{k=\lambda}^{\infty} \Phi(k - \lambda) \leq \left\| \|f\|_\gamma \right\|_\infty.$$

Hence by the generalized Weierstrass M test we obtain that $\sum_{k=\lambda}^{\infty} l_{nk}(f)\,\Phi(nx - k)$ is uniformly and absolutely convergent on $\left[-\frac{\lambda}{n}, \frac{\lambda}{n}\right]$.

Since $l_{nk}(f)\,\Phi(nx - k)$ is continuous in x, then $\sum_{k=\lambda}^{\infty} l_{nk}(f)\,\Phi(nx - k)$ is continuous on $\left[-\frac{\lambda}{n}, \frac{\lambda}{n}\right]$.

Because $nx \geq -\lambda$, then $-nx \leq \lambda$, and $k - nx \leq k + \lambda \leq 0$, when $k \leq -\lambda$. Therefore

$$\sum_{k=-\infty}^{-\lambda} \Phi(nx - k) = \sum_{k=-\infty}^{-\lambda} \Phi(k - nx) \leq \sum_{k=-\infty}^{-\lambda} \Phi(k + \lambda) = \sum_{k'=-\infty}^{0} \Phi(k') \leq 1.$$

So for $k \leq -\lambda$ we get

$$\|l_{nk}(f)\|_{\gamma} \Phi(nx-k) \leq \left\| \|f\|_{\gamma} \right\|_{\infty} \Phi(k+\lambda), \tag{4.92}$$

and

$$\left\| \|f\|_{\gamma} \right\|_{\infty} \sum_{k=-\infty}^{-\lambda} \Phi(k+\lambda) \leq \left\| \|f\|_{\gamma} \right\|_{\infty}.$$

Hence by Weierstrass M test we obtain that $\sum_{k=-\infty}^{-\lambda} l_{nk}(f) \Phi(nx-k)$ is uniformly and absolutely convergent on $\left[-\frac{\lambda}{n}, \frac{\lambda}{n}\right]$.

Since $l_{nk}(f) \Phi(nx-k)$ is continuous in x, then $\sum_{k=-\infty}^{-\lambda} l_{nk}(f) \Phi(nx-k)$ is continuous on $\left[-\frac{\lambda}{n}, \frac{\lambda}{n}\right]$.

So we proved that $\sum_{k=\lambda}^{\infty} l_{nk}(f) \Phi(nx-k)$ and $\sum_{k=-\infty}^{-\lambda} l_{nk}(f) \Phi(nx-k)$ are continuous on $\mathbb{R}$. Since $\sum_{k=-\lambda+1}^{\lambda-1} l_{nk}(f) \Phi(nx-k)$ is a finite sum of continuous functions on $\mathbb{R}$, it is also a continuous function on $\mathbb{R}$.

Writing

$$\sum_{k=-\infty}^{\infty} l_{nk}(f) \Phi(nx-k) = \sum_{k=-\infty}^{-\lambda} l_{nk}(f) \Phi(nx-k) +$$

$$\sum_{k=-\lambda+1}^{\lambda-1} l_{nk}(f) \Phi(nx-k) + \sum_{k=\lambda}^{\infty} l_{nk}(f) \Phi(nx-k) \tag{4.93}$$

we have it as a continuous function on $\mathbb{R}$. Therefore $F_n(f)$, when $N=1$, is a continuous function on $\mathbb{R}$.

When $N=2$ we have

$$F_n(f, x_1, x_2) = \sum_{k_1=-\infty}^{\infty} \sum_{k_2=-\infty}^{\infty} l_{nk}(f) \Phi(nx_1-k_1) \Phi(nx_2-k_2) =$$

$$\sum_{k_1=-\infty}^{\infty} \Phi(nx_1-k_1) \left(\sum_{k_2=-\infty}^{\infty} l_{nk}(f) \Phi(nx_2-k_2) \right)$$

(there always exist $\lambda_1, \lambda_2 \in \mathbb{N}$ such that $nx_1 \in [-\lambda_1, \lambda_1]$ and $nx_2 \in [-\lambda_2, \lambda_2]$)

$$= \sum_{k_1=-\infty}^{\infty} \Phi(nx_1-k_1) \left[\sum_{k_2=-\infty}^{-\lambda_2} l_{nk}(f) \Phi(nx_2-k_2) + \right.$$

$$\left.\sum_{k_2=-\lambda_2+1}^{\lambda_2-1} l_{nk}(f)\,\Phi(nx_2-k_2)+\sum_{k_2=\lambda_2}^{\infty} l_{nk}(f)\,\Phi(nx_2-k_2)\right]=$$

$$=\sum_{k_1=-\infty}^{\infty}\sum_{k_2=-\infty}^{-\lambda_2} l_{nk}(f)\,\Phi(nx_1-k_1)\,\Phi(nx_2-k_2)+$$

$$\sum_{k_1=-\infty}^{\infty}\sum_{k_2=-\lambda_2+1}^{\lambda_2-1} l_{nk}(f)\,\Phi(nx_1-k_1)\,\Phi(nx_2-k_2)+$$

$$\sum_{k_1=-\infty}^{\infty}\sum_{k_2=\lambda_2}^{\infty} l_{nk}(f)\,\Phi(nx_1-k_1)\,\Phi(nx_2-k_2)=:(*).$$

(For convenience call

$$F(k_1,k_2,x_1,x_2):=l_{nk}(f)\,\Phi(nx_1-k_1)\,\Phi(nx_2-k_2).)$$

Thus

$$(*)=\sum_{k_1=-\infty}^{-\lambda_1}\sum_{k_2=-\infty}^{-\lambda_2}F(k_1,k_2,x_1,x_2)+\sum_{k_1=-\lambda_1+1}^{\lambda_1-1}\sum_{k_2=-\infty}^{-\lambda_2}F(k_1,k_2,x_1,x_2)+$$

$$\sum_{k_1=\lambda_1}^{\infty}\sum_{k_2=-\infty}^{-\lambda_2}F(k_1,k_2,x_1,x_2)+\sum_{k_1=-\infty}^{-\lambda_1}\sum_{k_2=-\lambda_2+1}^{\lambda_2-1}F(k_1,k_2,x_1,x_2)+$$

$$\sum_{k_1=-\lambda_1+1}^{\lambda_1-1}\sum_{k_2=-\lambda_2+1}^{\lambda_2-1}F(k_1,k_2,x_1,x_2)+\sum_{k_1=\lambda_1}^{\infty}\sum_{k_2=-\lambda_2+1}^{\lambda_2-1}F(k_1,k_2,x_1,x_2)+$$

$$\sum_{k_1=-\infty}^{-\lambda_1}\sum_{k_2=\lambda_2}^{\infty}F(k_1,k_2,x_1,x_2)+\sum_{k_1=-\lambda_1+1}^{\lambda_1-1}\sum_{k_2=\lambda_2}^{\infty}F(k_1,k_2,x_1,x_2)+\tag{4.94}$$

$$\sum_{k_1=\lambda_1}^{\infty}\sum_{k_2=\lambda_2}^{\infty}F(k_1,k_2,x_1,x_2).$$

Notice that the finite sum of continuous functions $F(k_1,k_2,x_1,x_2)$, $\sum_{k_1=-\lambda_1+1}^{\lambda_1-1}\sum_{k_2=-\lambda_2+1}^{\lambda_2-1}F(k_1,k_2,x_1,x_2)$ is a continuous function.

The rest of the summands of $F_n(f,x_1,x_2)$ are treated all the same way and similarly to the case of $N=1$. The method is demonstrated as follows.

We will prove that $\sum_{k_1=\lambda_1}^{\infty}\sum_{k_2=-\infty}^{-\lambda_2} l_{nk}(f)\,\Phi(nx_1-k_1)\,\Phi(nx_2-k_2)$ is continuous in $(x_1,x_2)\in\mathbb{R}^2$.

The continuous function

$$\|l_{nk}(f)\|_{\gamma}\,\Phi(nx_1-k_1)\,\Phi(nx_2-k_2)\le\left\|\|f\|_{\gamma}\right\|_{\infty}\Phi(k_1-\lambda_1)\,\Phi(k_2+\lambda_2),$$

and

$$\left\|\|f\|_{\gamma}\right\|_{\infty}\sum_{k_1=\lambda_1}^{\infty}\sum_{k_2=-\infty}^{-\lambda_2}\Phi(k_1-\lambda_1)\,\Phi(k_2+\lambda_2)=$$

$$\left\|\|f\|_{\gamma}\right\|_{\infty}\left(\sum_{k_1=\lambda_1}^{\infty}\Phi(k_1-\lambda_1)\right)\left(\sum_{k_2=-\infty}^{-\lambda_2}\Phi(k_2+\lambda_2)\right)\le$$

$$\left\|\|f\|_{\gamma}\right\|_{\infty}\left(\sum_{k_1'=0}^{\infty}\Phi(k_1')\right)\left(\sum_{k_2'=-\infty}^{0}\Phi(k_2')\right)\le\left\|\|f\|_{\gamma}\right\|_{\infty}.$$

So by the Weierstrass M test we get that $\sum_{k_1=\lambda_1}^{\infty}\sum_{k_2=-\infty}^{-\lambda_2} l_{nk}(f)\,\Phi(nx_1-k_1)\,\Phi(nx_2-k_2)$ is uniformly and absolutely convergent. Therefore it is continuous on $\mathbb{R}^2$.

Next we prove continuity on $\mathbb{R}^2$ of $\sum_{k_1=-\lambda_1+1}^{\lambda_1-1}\sum_{k_2=-\infty}^{-\lambda_2} l_{nk}(f)\,\Phi(nx_1-k_1)\,\Phi(nx_2-k_2)$.

Notice here that

$$\|l_{nk}(f)\|_{\gamma}\,\Phi(nx_1-k_1)\,\Phi(nx_2-k_2)\le\left\|\|f\|_{\gamma}\right\|_{\infty}\Phi(nx_1-k_1)\,\Phi(k_2+\lambda_2)$$

$$\le\left\|\|f\|_{\gamma}\right\|_{\infty}\Phi(0)\,\Phi(k_2+\lambda_2)=\frac{1}{2\sqrt[2m]{2}}\cdot\left\|\|f\|_{\gamma}\right\|_{\infty}\Phi(k_2+\lambda_2),$$

and

$$\frac{1}{2\sqrt[2m]{2}}\cdot\left\|\|f\|_{\gamma}\right\|_{\infty}\left(\sum_{k_1=-\lambda_1+1}^{\lambda_1-1}1\right)\left(\sum_{k_2=-\infty}^{-\lambda_2}\Phi(k_2+\lambda_2)\right)=$$

$$\frac{1}{2\sqrt[2m]{2}}\cdot\left\|\|f\|_{\gamma}\right\|_{\infty}(2\lambda_1-1)\left(\sum_{k_2'=-\infty}^{0}\Phi(k_2')\right)\le\frac{1}{2\sqrt[2m]{2}}\cdot(2\lambda_1-1)\left\|\|f\|_{\gamma}\right\|_{\infty}. \tag{4.95}$$

So the double series under consideration is uniformly convergent and continuous. Clearly $F_n(f,x_1,x_2)$ is proved to be continuous on $\mathbb{R}^2$.

Similarly reasoning one can prove easily now, but with more tedious work, that $F_n(f,x_1,\ldots,x_N)$ is continuous on $\mathbb{R}^N$, for any $N\ge 1$. We choose to omit this similar extra work. ■

Remark 4.21 By (4.25) it is obvious that $\left\| \left\| A_n(f) \right\|_\gamma \right\|_\infty \leq \left\| \left\| f \right\|_\gamma \right\|_\infty < \infty$, and $A_n(f) \in C\left(\prod_{i=1}^{N} [a_i, b_i], X\right)$, given that $f \in C\left(\prod_{i=1}^{N} [a_i, b_i], X\right)$.

Call L_n any of the operators A_n, B_n, C_n, D_n.

Clearly then

$$\left\| \left\| L_n^2(f) \right\|_\gamma \right\|_\infty = \left\| \left\| L_n(L_n(f)) \right\|_\gamma \right\|_\infty \leq \left\| \left\| L_n(f) \right\|_\gamma \right\|_\infty \leq \left\| \left\| f \right\|_\gamma \right\|_\infty, \tag{4.96}$$

etc.

Therefore we get

$$\left\| \left\| L_n^k(f) \right\|_\gamma \right\|_\infty \leq \left\| \left\| f \right\|_\gamma \right\|_\infty, \ \forall\, k \in \mathbb{N}, \tag{4.97}$$

the contraction property.

Also we see that

$$\left\| \left\| L_n^k(f) \right\|_\gamma \right\|_\infty \leq \left\| \left\| L_n^{k-1}(f) \right\|_\gamma \right\|_\infty \leq \ldots \leq \left\| \left\| L_n(f) \right\|_\gamma \right\|_\infty \leq \left\| \left\| f \right\|_\gamma \right\|_\infty. \tag{4.98}$$

Here L_n^k are bounded linear operators.

Notation 4.22 Here $N \in \mathbb{N}$, $0 < \beta < 1$. Denote by

$$c_N := \begin{cases} \left[2\left(\sqrt[2m]{1+4^m}\right)\right]^N, & \text{if } L_n = A_n, \\ 1, & \text{if } L_n = B_n, C_n, D_n, \end{cases} \tag{4.99}$$

$$\varphi(n) := \begin{cases} \frac{1}{n^\beta}, & \text{if } L_n = A_n, B_n, \\ \frac{1}{n} + \frac{1}{n^\beta}, & \text{if } L_n = C_n, D_n, \end{cases} \tag{4.100}$$

$$\Omega := \begin{cases} C\left(\prod_{i=1}^{N} [a_i, b_i], X\right), & \text{if } L_n = A_n, \\ C_B\left(\mathbb{R}^N, X\right), & \text{if } L_n = B_n, C_n, D_n, \end{cases} \tag{4.101}$$

and

$$Y := \begin{cases} \prod_{i=1}^{N} [a_i, b_i], & \text{if } L_n = A_n, \\ \mathbb{R}^N, & \text{if } L_n = B_n, C_n, D_n. \end{cases} \tag{4.102}$$

We give the condensed

Theorem 4.23 *Let $f \in \Omega$, $0 < \beta < 1$, $x \in Y$; $n, m, N \in \mathbb{N}$ with $n^{1-\beta} > 2$. Then*

(i)

$$\left\| L_n(f,x) - f(x) \right\|_\gamma \leq c_N \left[\omega_1(f,\varphi(n)) + \frac{\left\| \|f\|_\gamma \right\|_\infty}{2m\left(n^{1-\beta}-2\right)^{2m}} \right] =: \tau(n), \tag{4.103}$$

where ω_1 is for $p = \infty$,
and
(ii)

$$\left\| \|L_n(f) - f\|_\gamma \right\|_\infty \leq \tau(n) \to 0, \text{ as } n \to \infty. \tag{4.104}$$

For f uniformly continuous and in Ω we obtain

$$\lim_{n\to\infty} L_n(f) = f,$$

pointwise and uniformly.

Proof By Theorems 4.8, 4.16, 4.17, 4.18. ∎

Next we do iterated neural network approximation (see also [9]).
We make

Remark 4.24 Let $r \in \mathbb{N}$ and L_n as above. We observe that

$$L_n^r f - f = \left(L_n^r f - L_n^{r-1} f\right) + \left(L_n^{r-1} f - L_n^{r-2} f\right) +$$
$$\left(L_n^{r-2} f - L_n^{r-3} f\right) + \ldots + \left(L_n^2 f - L_n f\right) + \left(L_n f - f\right).$$

Then

$$\left\| \left\| L_n^r f - f \right\|_\gamma \right\|_\infty \leq \left\| \left\| L_n^r f - L_n^{r-1} f \right\|_\gamma \right\|_\infty + \left\| \left\| L_n^{r-1} f - L_n^{r-2} f \right\|_\gamma \right\|_\infty +$$
$$\left\| \left\| L_n^{r-2} f - L_n^{r-3} f \right\|_\gamma \right\|_\infty + \ldots + \left\| \left\| L_n^2 f - L_n f \right\|_\gamma \right\|_\infty + \left\| \left\| L_n f - f \right\|_\gamma \right\|_\infty =$$
$$\left\| \left\| L_n^{r-1}(L_n f - f) \right\|_\gamma \right\|_\infty + \left\| \left\| L_n^{r-2}(L_n f - f) \right\|_\gamma \right\|_\infty + \left\| \left\| L_n^{r-3}(L_n f - f) \right\|_\gamma \right\|_\infty$$
$$+ \ldots + \left\| \left\| L_n(L_n f - f) \right\|_\gamma \right\|_\infty + \left\| \left\| L_n f - f \right\|_\gamma \right\|_\infty \leq r \left\| \left\| L_n f - f \right\|_\gamma \right\|_\infty. \tag{4.105}$$

That is

$$\left\| \left\| L_n^r f - f \right\|_\gamma \right\|_\infty \leq r \left\| \left\| L_n f - f \right\|_\gamma \right\|_\infty. \tag{4.106}$$

We give

Theorem 4.25 *All here as in Theorem 4.23 and $r \in \mathbb{N}$, $\tau(n)$ as in (4.104). Then*

$$\left\| \left\| L_n^r f - f \right\|_\gamma \right\|_\infty \leq r\tau(n). \tag{4.107}$$

So that the speed of convergence to the unit operator of L_n^r is not worse than of L_n.

Proof By (4.106) and (4.104). ∎

We make

Remark 4.26 Let $m, m_1, ..., m_r \in \mathbb{N} : m_1 \leq m_2 \leq ... \leq m_r,\ 0 < \beta < 1,\ f \in \Omega$. Then $\varphi(m_1) \geq \varphi(m_2) \geq ... \geq \varphi(m_r)$, φ as in (4.100).

Therefore

$$\omega_1(f, \varphi(m_1)) \geq \omega_1(f, \varphi(m_2)) \geq ... \geq \omega_1(f, \varphi(m_r)). \tag{4.108}$$

Assume further that $m_i^{1-\beta} > 2$, $i = 1, ..., r$. Then

$$\frac{1}{4m\left(m_1^{1-\beta} - 2\right)^{2m}} \geq \frac{1}{4m\left(m_2^{1-\beta} - 2\right)^{2m}} \geq ... \geq \frac{1}{4m\left(m_r^{1-\beta} - 2\right)^{2m}}. \tag{4.109}$$

Let L_{m_i} as above, $i = 1, ..., r$, all of the same kind.

We write

$$\begin{gathered} L_{m_r}\left(L_{m_{r-1}}\left(...L_{m_2}\left(L_{m_1} f\right)\right)\right) - f = \\ L_{m_r}\left(L_{m_{r-1}}\left(...L_{m_2}\left(L_{m_1} f\right)\right)\right) - L_{m_r}\left(L_{m_{r-1}}\left(...L_{m_2} f\right)\right) + \\ L_{m_r}\left(L_{m_{r-1}}\left(...L_{m_2} f\right)\right) - L_{m_r}\left(L_{m_{r-1}}\left(...L_{m_3} f\right)\right) + \\ L_{m_r}\left(L_{m_{r-1}}\left(...L_{m_3} f\right)\right) - L_{m_r}\left(L_{m_{r-1}}\left(...L_{m_4} f\right)\right) + ... + \\ L_{m_r}\left(L_{m_{r-1}} f\right) - L_{m_r} f + L_{m_r} f - f = \\ L_{m_r}\left(L_{m_{r-1}}\left(...L_{m_2}\right)\right)\left(L_{m_1} f - f\right) + L_{m_r}\left(L_{m_{r-1}}\left(...L_{m_3}\right)\right)\left(L_{m_2} f - f\right) + \\ L_{m_r}\left(L_{m_{r-1}}\left(...L_{m_4}\right)\right)\left(L_{m_3} f - f\right) + ... + L_{m_r}\left(L_{m_{r-1}} f - f\right) + L_{m_r} f - f. \end{gathered} \tag{4.110}$$

Hence by the triangle inequality property of $\left\| \|\cdot\|_\gamma \right\|_\infty$ we get

$$\begin{gathered} \left\| \left\| L_{m_r}\left(L_{m_{r-1}}\left(...L_{m_2}\left(L_{m_1} f\right)\right)\right) - f \right\|_\gamma \right\|_\infty \leq \\ \left\| \left\| L_{m_r}\left(L_{m_{r-1}}\left(...L_{m_2}\right)\right)\left(L_{m_1} f - f\right) \right\|_\gamma \right\|_\infty + \\ \left\| \left\| L_{m_r}\left(L_{m_{r-1}}\left(...L_{m_3}\right)\right)\left(L_{m_2} f - f\right) \right\|_\gamma \right\|_\infty + \end{gathered}$$

$$\left\| \left\| L_{m_r}\left(L_{m_{r-1}}\left(...L_{m_4}\right)\right)\left(L_{m_3} f - f\right) \right\|_{\gamma} \right\|_{\infty} + ...+$$

$$\left\| \left\| L_{m_r}\left(L_{m_{r-1}} f - f\right) \right\|_{\gamma} \right\|_{\infty} + \left\| \left\| L_{m_r} f - f \right\|_{\gamma} \right\|_{\infty}$$

(repeatedly applying (4.96))

$$\leq \left\| \left\| L_{m_1} f - f \right\|_{\gamma} \right\|_{\infty} + \left\| \left\| L_{m_2} f - f \right\|_{\gamma} \right\|_{\infty} + \left\| \left\| L_{m_3} f - f \right\|_{\gamma} \right\|_{\infty} + ...+$$

$$\left\| \left\| L_{m_{r-1}} f - f \right\|_{\gamma} \right\|_{\infty} + \left\| \left\| L_{m_r} f - f \right\|_{\gamma} \right\|_{\infty} = \sum_{i=1}^{r} \left\| \left\| L_{m_i} f - f \right\|_{\gamma} \right\|_{\infty}. \tag{4.111}$$

That is, we proved

$$\left\| \left\| L_{m_r}\left(L_{m_{r-1}}\left(...L_{m_2}\left(L_{m_1} f\right)\right)\right) - f \right\|_{\gamma} \right\|_{\infty} \leq \sum_{i=1}^{r} \left\| \left\| L_{m_i} f - f \right\|_{\gamma} \right\|_{\infty}. \tag{4.112}$$

We give

Theorem 4.27 *Let $f \in \Omega$; $m, N, m_1, m_2, ..., m_r \in \mathbb{N} : m_1 \leq m_2 \leq ... \leq m_r$, $0 < \beta < 1$; $m_i^{1-\beta} > 2$, $i = 1, ..., r$, $x \in Y$, and let $\left(L_{m_1}, ..., L_{m_r}\right)$ as $\left(A_{m_1}, ..., A_{m_r}\right)$ or $\left(B_{m_1}, ..., B_{m_r}\right)$ or $\left(C_{m_1}, ..., C_{m_r}\right)$ or $\left(D_{m_1}, ..., D_{m_r}\right)$, $p = \infty$. Then*

$$\left\| L_{m_r}\left(L_{m_{r-1}}\left(...L_{m_2}\left(L_{m_1} f\right)\right)\right)(x) - f(x) \right\|_{\gamma} \leq$$

$$\left\| \left\| L_{m_r}\left(L_{m_{r-1}}\left(...L_{m_2}\left(L_{m_1} f\right)\right)\right) - f \right\|_{\gamma} \right\|_{\infty} \leq$$

$$\sum_{i=1}^{r} \left\| \left\| L_{m_i} f - f \right\|_{\gamma} \right\|_{\infty} \leq$$

$$c_N \sum_{i=1}^{r} \left[\omega_1\left(f, \varphi\left(m_i\right)\right) + \frac{\left\| \left\| f \right\|_{\gamma} \right\|_{\infty}}{2m\left(m_i^{1-\beta} - 2\right)^{2m}} \right] \leq$$

$$r c_N \left[\omega_1\left(f, \varphi\left(m_1\right)\right) + \frac{\left\| \left\| f \right\|_{\gamma} \right\|_{\infty}}{2m\left(m_1^{1-\beta} - 2\right)^{2m}} \right]. \tag{4.113}$$

Clearly, we notice that the speed of convergence to the unit operator of the multiply iterated operator is not worse than the speed of L_{m_1}.

Proof Using (4.112), (4.108), (4.109) and (4.103), (4.104). ■

We continue with

Theorem 4.28 *Let all as in Corollary 4.15, and $r \in \mathbb{N}$. Here $\varphi_3(n)$ is as in (4.75). Then*

$$\left\| \left\| A_n^r f - f \right\|_\gamma \right\|_\infty \leq r \left\| \left\| A_n f - f \right\|_\gamma \right\|_\infty \leq r\varphi_3(n). \qquad (4.114)$$

Proof By (4.106) and (4.75). ■

Application 4.29 *A typical application of all of our results is when $\left(X, \|\cdot\|_\gamma\right) = (\mathbb{C}, |\cdot|)$, where $\mathbb{C}$ are the complex numbers.*

References

1. G.A. Anastassiou, *Moments in Probability and Approximation Theory, Pitman Research Notes in Math*, vol. 287 (Longman Sci. & Tech, Harlow, U.K., 1993)
2. G.A. Anastassiou, Rate of convergence of some neural network operators to the unit-univariate case. J. Math. Anal. Appl. **212**, 237–262 (1997)
3. G.A. Anastassiou, *Quantitative Approximations* (Chapman & Hall/CRC, Boca Raton, New York, 2001)
4. G.A. Anastassiou, *Inteligent Systems: Approximation by Artificial Neural Networks, Intelligent Systems Reference Library*, vol. 19 (Springer, Heidelberg, 2011)
5. G.A. Anastassiou, Univariate hyperbolic tangent neural network approximation. Math. Comput. Modell. **53**, 1111–1132 (2011)
6. G.A. Anastassiou, Multivariate hyperbolic tangent neural network approximation. Comput. Math. **61**, 809–821 (2011)
7. G.A. Anastassiou, Multivariate sigmoidal neural network approximation. Neural Netw. **24**, 378–386 (2011)
8. G.A. Anastassiou, Univariate sigmoidal neural network approximation. J. Comput. Anal. Appl. **14**(4), 659–690 (2012)
9. G.A. Anastassiou, Approximation by neural networks iterates, in *Advances in Applied Mathematics and Approximation Theory*, ed. by G. Anastassiou, O. Duman. Springer Proceedings in Mathematics & Statistics (Springer, New York, 2013), pp. 1–20
10. G.A. Anastassiou, *Intelligent Systems II: Complete Approximation by Neural Network Operators* (Springer, Heidelberg, New York, 2016)
11. G.A. Anastassiou, *Intelligent Computations: Abstract Fractional Calculus, Inequalities, Approximations* (Springer, Heidelberg, New York, 2018)
12. G.A. Anastassiou, Algebraic function based Banach space valued ordinary and fractional neural network approximations. New Trends in Mathematical Sciences, 10 special issue (1) (2022), pp. 100–125
13. G.A. Anastassiou, General multimariate arctangent function activated neural network approximations (submitted) (2022)
14. G.A. Anastassiou, R. Kozma, Abstract multivariate algebraic function activated neural network approximations (submitted) (2022)
15. H. Cartan, *Differential Calculus* (Hermann, Paris, 1971)
16. Z. Chen, F. Cao, The approximation operators with sigmoidal functions. Comput. Math. Appl. **58**, 758–765 (2009)
17. D. Costarelli, R. Spigler, Approximation results for neural network operators activated by sigmoidal functions. Neural Netw. **44**, 101–106 (2013)

18. D. Costarelli, R. Spigler, Multivariate neural network operators with sigmoidal activation functions. Neural Netw. **48**, 72–77 (2013)
19. S. Haykin, *Neural Networks: A Comprehensive Foundation*, 2nd edn. (Prentice Hall, New York, 1998)
20. W. McCulloch, W. Pitts, A logical calculus of the ideas immanent in nervous activity. Bull. Math. Biophys. **7**, 115–133 (1943)
21. T.M. Mitchell, *Machine Learning* (WCB-McGraw-Hill, New York, 1997)
22. L.B. Rall, *Computational Solution of Nonlinear Operator Equations* (Wiley, New York, 1969)

Chapter 5
General Multivariate Arctangent Function Induced Neural Network Approximations

Here we expose multivariate quantitative approximations of Banach space valued continuous multivariate functions on a box or $\mathbb{R}^N$, $N \in \mathbb{N}$, by the multivariate normalized, quasi-interpolation, Kantorovich type and quadrature type neural network operators. We treat also the case of approximation by iterated operators of the last four types. These approximations are derived by establishing multidimensional Jackson type inequalities involving the multivariate modulus of continuity of the engaged function or its high order Fréchet derivatives. Our multivariate operators are defined by using a multidimensional density function induced by the arctangent function. The approximations are pointwise and uniform. The related feed-forward neural network is with one hidden layer. It follows [12].

5.1 Introduction

The author in [2, 3], see Chaps. 2–5, was the first to establish neural network approximations to continuous functions with rates by very specifically defined neural network operators of Cardaliagnet–Euvrard and "Squashing" types, by employing the modulus of continuity of the engaged function or its high order derivative, and producing very tight Jackson type inequalities. He treats there both the univariate and multivariate cases. The defining these operators "bell-shaped" and "squashing" functions are assumed to be of compact support. Also in [3] he gives the Nth order asymptotic expansion for the error of weak approximation of these two operators to a special natural class of smooth functions, see Chaps. 4 and 5 there.

For this chapter the author is motivated by the articles [14] of Z. Chen and F. Cao, also by [4–11, 15, 16].

© The Author(s), under exclusive license to Springer Nature Switzerland AG 2023
G. A. Anastassiou, *Banach Space Valued Neural Network*,
Studies in Computational Intelligence 1062,
https://doi.org/10.1007/978-3-031-16400-2_5

The author here performs multivariate arctangent function based neural network approximations to continuous functions over boxes or over the whole $\mathbb{R}^N$, $N \in \mathbb{N}$. Also he does iterated approximation. All convergences here are with rates expressed via the multivariate modulus of continuity of the involved function or its high order Fréchet derivative and given by very tight multidimensional Jackson type inequalities.

The author here comes up with the "right" precisely defined multivariate normalized, quasi-interpolation neural network operators related to boxes or $\mathbb{R}^N$, as well as Kantorovich type and quadrature type related operators on $\mathbb{R}^N$. Our boxes are not necessarily symmetric to the origin. In preparation to prove our results we establish important properties of the basic multivariate density function induced by arctangent function and defining our operators.

Feed-forward neural networks (FNNs) with one hidden layer, the only type of networks we deal with in this chapter, are mathematically expressed as

$$N_n(x) = \sum_{j=0}^{n} c_j \sigma\left(\langle a_j \cdot x \rangle + b_j\right), \quad x \in \mathbb{R}^s, \quad s \in \mathbb{N},$$

where for $0 \leq j \leq n$, $b_j \in \mathbb{R}$ are the thresholds, $a_j \in \mathbb{R}^s$ are the connection weights, $c_j \in \mathbb{R}$ are the coefficients, $\langle a_j \cdot x \rangle$ is the inner product of a_j and x, and σ is the activation function of the network. In many fundamental network models, the activation function is the arctangent function. About neural networks read [17–19].

5.2 Auxiliary Notions

We consider the

$$\arctan x = \int_0^x \frac{dz}{1+z^2}, \quad x \in \mathbb{R}. \tag{5.1}$$

We will be using

$$h(x) := \frac{2}{\pi} \arctan\left(\frac{\pi}{2} x\right) = \frac{2}{\pi} \int_0^{\frac{\pi x}{2}} \frac{dz}{1+z^2}, \quad x \in \mathbb{R}, \tag{5.2}$$

which is a sigmoid type function and it is strictly increasing. We have that

$$h(0) = 0, \ h(-x) = -h(x), \ h(+\infty) = 1, \ h(-\infty) = -1,$$

and

$$h'(x) = \frac{4}{4+\pi^2 x^2} > 0, \ \text{all } x \in \mathbb{R}. \tag{5.3}$$

We consider the activation function

$$\psi(x) := \frac{1}{4}(h(x+1) - h(x-1)), \ x \in \mathbb{R}, \tag{5.4}$$

and we notice that

$$\psi(-x) = \psi(x), \tag{5.5}$$

it is an even function.

Since $x+1 > x-1$, then $h(x+1) > h(x-1)$, and $\psi(x) > 0$, all $x \in \mathbb{R}$.

We see that

$$\psi(0) = \frac{1}{\pi}\arctan\frac{\pi}{2} \cong 0.319. \tag{5.6}$$

Let $x > 0$, we have that

$$\psi'(x) = \frac{1}{4}\left(h'(x+1) - h'(x-1)\right) =$$

$$\frac{-4\pi^2 x}{\left(4+\pi^2(x+1)^2\right)\left(4+\pi^2(x-1)^2\right)} < 0. \tag{5.7}$$

That is

$$\psi'(x) < 0, \text{ for } x > 0. \tag{5.8}$$

That is ψ is strictly decreasing on $[0, \infty)$ and clearly is strictly increasing on $(-\infty, 0]$, and $\psi'(0) = 0$.

Observe that

$$\begin{aligned} &\lim_{x\to+\infty}\psi(x) = \tfrac{1}{4}(h(+\infty) - h(+\infty)) = 0, \\ &\text{and} \\ &\lim_{x\to-\infty}\psi(x) = \tfrac{1}{4}(h(-\infty) - h(-\infty)) = 0. \end{aligned} \tag{5.9}$$

That is the x-axis is the horizontal asymptote on ψ.

All in all, ψ is a bell symmetric function with maximum $\psi(0) \cong 0.319$.

We need

Theorem 5.1 ([11], p. 286) *We have that*

$$\sum_{i=-\infty}^{\infty}\psi(x-i) = 1, \ \forall x \in \mathbb{R}. \tag{5.10}$$

Theorem 5.2 ([11], p. 287) *It holds*

$$\int_{-\infty}^{\infty}\psi(x)\,dx = 1. \tag{5.11}$$

So that $\psi(x)$ is a density function on $\mathbb{R}$.

We mention

Theorem 5.3 ([11], p. 288) *Let $0 < \alpha < 1$, and $n \in \mathbb{N}$ with $n^{1-\alpha} > 2$. It holds*

$$\sum_{\begin{cases} k = -\infty \\ : |nx - k| \geq n^{1-\alpha} \end{cases}}^{\infty} \psi(nx - k) < \frac{2}{\pi^2 \left(n^{1-\alpha} - 2\right)}. \tag{5.12}$$

Denote by $\lfloor \cdot \rfloor$ the integral part of the number and by $\lceil \cdot \rceil$ the ceiling of the number.

We need

Theorem 5.4 ([11], p. 289) *Let $x \in [a, b] \subset \mathbb{R}$ and $n \in \mathbb{N}$ so that $\lceil na \rceil \leq \lfloor nb \rfloor$. It holds*

$$\frac{1}{\sum_{k=\lceil na \rceil}^{\lfloor nb \rfloor} \psi(nx - k)} < \frac{1}{\psi(1)} \cong 4.9737, \quad \forall\, x \in [a, b]. \tag{5.13}$$

Note 5.5 ([11], pp. 290–291)

(i) We have that

$$\lim_{n \to \infty} \sum_{k=\lceil na \rceil}^{\lfloor nb \rfloor} \psi(nx - k) \neq 1, \tag{5.14}$$

for at least some $x \in [a, b]$.

(ii) For large enough $n \in \mathbb{N}$ we always obtain $\lceil na \rceil \leq \lfloor nb \rfloor$. Also $a \leq \frac{k}{n} \leq b$, iff $\lceil na \rceil \leq k \leq \lfloor nb \rfloor$.

In general, by Theorem 5.1, it holds

$$\sum_{k=\lceil na \rceil}^{\lfloor nb \rfloor} \psi(nx - k) \leq 1. \tag{5.15}$$

We introduce

$$Z(x_1, ..., x_N) := Z(x) := \prod_{i=1}^{N} \psi(x_i), \quad x = (x_1, ..., x_N) \in \mathbb{R}^N, \ N \in \mathbb{N}. \tag{5.16}$$

It has the properties:

(i) $Z(x) > 0, \ \forall\, x \in \mathbb{R}^N$,

(ii)

$$\sum_{k=-\infty}^{\infty} Z(x - k) := \sum_{k_1=-\infty}^{\infty} \sum_{k_2=-\infty}^{\infty} ... \sum_{k_N=-\infty}^{\infty} Z(x_1 - k_1, ..., x_N - k_N) = 1, \tag{5.17}$$

where $k := (k_1, ..., k_n) \in \mathbb{Z}^N, \forall\, x \in \mathbb{R}^N$,

hence

(iii)

$$\sum_{k=-\infty}^{\infty} Z(nx-k) = 1, \tag{5.18}$$

$\forall\, x \in \mathbb{R}^N;\ n \in \mathbb{N}$,

and

(iv)

$$\int_{\mathbb{R}^N} Z(x)\,dx = 1, \tag{5.19}$$

that is Z is a multivariate density function.

Here denote $\|x\|_\infty := \max\{|x_1|, ..., |x_N|\}$, $x \in \mathbb{R}^N$, also set $\infty := (\infty, ..., \infty)$, $-\infty := (-\infty, ..., -\infty)$ upon the multivariate context, and

$$\begin{aligned} \lceil na \rceil &:= (\lceil na_1 \rceil, ..., \lceil na_N \rceil), \\ \lfloor nb \rfloor &:= (\lfloor nb_1 \rfloor, ..., \lfloor nb_N \rfloor), \end{aligned} \tag{5.20}$$

where $a := (a_1, ..., a_N)$, $b := (b_1, ..., b_N)$.

We obviously see that

$$\sum_{k=\lceil na \rceil}^{\lfloor nb \rfloor} Z(nx-k) = \sum_{k=\lceil na \rceil}^{\lfloor nb \rfloor} \left(\prod_{i=1}^{N} \psi(nx_i - k_i) \right) =$$

$$\sum_{k_1=\lceil na_1 \rceil}^{\lfloor nb_1 \rfloor} \cdots \sum_{k_N=\lceil na_N \rceil}^{\lfloor nb_N \rfloor} \left(\prod_{i=1}^{N} \psi(nx_i - k_i) \right) = \prod_{i=1}^{N} \left(\sum_{k_i=\lceil na_i \rceil}^{\lfloor nb_i \rfloor} \psi(nx_i - k_i) \right). \tag{5.21}$$

For $0 < \beta < 1$ and $n \in \mathbb{N}$, a fixed $x \in \mathbb{R}^N$, we have that

$$\sum_{k=\lceil na \rceil}^{\lfloor nb \rfloor} \psi(nx-k) =$$

$$\sum_{\begin{cases} k = \lceil na \rceil \\ \left\| \frac{k}{n} - x \right\|_\infty \le \frac{1}{n^\beta} \end{cases}}^{\lfloor nb \rfloor} \psi(nx-k) + \sum_{\begin{cases} k = \lceil na \rceil \\ \left\| \frac{k}{n} - x \right\|_\infty > \frac{1}{n^\beta} \end{cases}}^{\lfloor nb \rfloor} \psi(nx-k). \tag{5.22}$$

In the last two sums the counting is over disjoint vector sets of k's, because the condition $\left\| \frac{k}{n} - x \right\|_\infty > \frac{1}{n^\beta}$ implies that there exists at least one $\left| \frac{k_r}{n} - x_r \right| > \frac{1}{n^\beta}$, where $r \in \{1, ..., N\}$.

(v) As in [10], pp. 379–380, we derive that

$$\sum_{\left\{\begin{array}{c} k = \lceil na \rceil \\ \left\| \frac{k}{n} - x \right\|_{\infty} > \frac{1}{n^{\beta}} \end{array}\right.}^{\lfloor nb \rfloor} Z(nx - k) \overset{(5.12)}{<} \frac{2}{\pi^2 \left(n^{1-\beta} - 2\right)}, \quad 0 < \beta < 1, \tag{5.23}$$

with $n \in \mathbb{N} : n^{1-\beta} > 2$, $x \in \prod_{i=1}^{N} [a_i, b_i]$.

(vi) By Theorem 5.4 we get that

$$0 < \frac{1}{\sum_{k=\lceil na \rceil}^{\lfloor nb \rfloor} Z(nx - k)} < \frac{1}{(\psi(1))^N} \cong (4.9737)^N, \tag{5.24}$$

$\forall\, x \in \left(\prod_{i=1}^{N} [a_i, b_i]\right)$, $n \in \mathbb{N}$.

It is also clear that

(vii)

$$\sum_{\left\{\begin{array}{c} k = -\infty \\ \left\| \frac{k}{n} - x \right\|_{\infty} > \frac{1}{n^{\beta}} \end{array}\right.}^{\infty} Z(nx - k) < \frac{2}{\pi^2 \left(n^{1-\beta} - 2\right)}, \tag{5.25}$$

$0 < \beta < 1$, $n \in \mathbb{N} : n^{1-\beta} > 2$, $x \in \mathbb{R}^N$.

Furthermore it holds

$$\lim_{n \to \infty} \sum_{k=\lceil na \rceil}^{\lfloor nb \rfloor} Z(nx - k) \neq 1, \tag{5.26}$$

for at least some $x \in \left(\prod_{i=1}^{N} [a_i, b_i]\right)$.

Here $\left(X, \|\cdot\|_{\gamma}\right)$ is a Banach space.

Let $f \in C\left(\prod_{i=1}^{N} [a_i, b_i], X\right)$, $x = (x_1, ..., x_N) \in \prod_{i=1}^{N} [a_i, b_i]$, $n \in \mathbb{N}$ such that $\lceil na_i \rceil \leq \lfloor nb_i \rfloor$, $i = 1, ..., N$.

We introduce and define the following multivariate linear normalized neural network operator ($x := (x_1, ..., x_N) \in \left(\prod_{i=1}^{N} [a_i, b_i]\right)$):

$$A_n(f, x_1, ..., x_N) := A_n(f, x) := \frac{\sum_{k=\lceil na \rceil}^{\lfloor nb \rfloor} f\left(\frac{k}{n}\right) Z(nx - k)}{\sum_{k=\lceil na \rceil}^{\lfloor nb \rfloor} Z(nx - k)} =$$

$$\frac{\sum_{k_1=\lceil na_1\rceil}^{\lfloor nb_1\rfloor}\sum_{k_2=\lceil na_2\rceil}^{\lfloor nb_2\rfloor}\cdots\sum_{k_N=\lceil na_N\rceil}^{\lfloor nb_N\rfloor} f\left(\frac{k_1}{n},\ldots,\frac{k_N}{n}\right)\left(\prod_{i=1}^{N}\psi\left(nx_i-k_i\right)\right)}{\prod_{i=1}^{N}\left(\sum_{k_i=\lceil na_i\rceil}^{\lfloor nb_i\rfloor}\psi\left(nx_i-k_i\right)\right)}. \quad (5.27)$$

For large enough $n \in \mathbb{N}$ we always obtain $\lceil na_i\rceil \leq \lfloor nb_i\rfloor$, $i = 1, \ldots, N$. Also $a_i \leq \frac{k_i}{n} \leq b_i$, iff $\lceil na_i\rceil \leq k_i \leq \lfloor nb_i\rfloor$, $i = 1, \ldots, N$.

When $g \in C\left(\prod_{i=1}^{N}[a_i, b_i]\right)$ we define the companion operator

$$\widetilde{A}_n(g, x) := \frac{\sum_{k=\lceil na\rceil}^{\lfloor nb\rfloor} g\left(\frac{k}{n}\right) Z(nx-k)}{\sum_{k=\lceil na\rceil}^{\lfloor nb\rfloor} Z(nx-k)}. \quad (5.28)$$

Clearly $\widetilde{A}_n$ is a positive linear operator. We have that

$$\widetilde{A}_n(1, x) = 1, \ \forall\, x \in \left(\prod_{i=1}^{N}[a_i, b_i]\right).$$

Notice that $A_n(f) \in C\left(\prod_{i=1}^{N}[a_i, b_i], X\right)$ and $\widetilde{A}_n(g) \in C\left(\prod_{i=1}^{N}[a_i, b_i]\right)$.

Furthermore it holds

$$\|A_n(f, x)\|_\gamma \leq \frac{\sum_{k=\lceil na\rceil}^{\lfloor nb\rfloor}\left\|f\left(\frac{k}{n}\right)\right\|_\gamma Z(nx-k)}{\sum_{k=\lceil na\rceil}^{\lfloor nb\rfloor} Z(nx-k)} = \widetilde{A}_n\left(\|f\|_\gamma, x\right), \quad (5.29)$$

$\forall\, x \in \prod_{i=1}^{N}[a_i, b_i]$.

Clearly $\|f\|_\gamma \in C\left(\prod_{i=1}^{N}[a_i, b_i]\right)$.

So, we have that

$$\|A_n(f, x)\|_\gamma \leq \widetilde{A}_n\left(\|f\|_\gamma, x\right), \quad (5.30)$$

$\forall\, x \in \prod_{i=1}^{N}[a_i, b_i]$, $\forall\, n \in \mathbb{N}$, $\forall\, f \in C\left(\prod_{i=1}^{N}[a_i, b_i], X\right)$.

Let $c \in X$ and $g \in C\left(\prod_{i=1}^{N}[a_i, b_i]\right)$, then $cg \in C\left(\prod_{i=1}^{N}[a_i, b_i], X\right)$.

Furthermore it holds

$$A_n(cg, x) = c\widetilde{A}_n(g, x), \ \forall\, x \in \prod_{i=1}^{N}[a_i, b_i]. \quad (5.31)$$

Since $\widetilde{A}_n(1) = 1$, we get that

$$A_n(c) = c, \ \forall\, c \in X. \quad (5.32)$$

We call $\widetilde{A}_n$ the companion operator of A_n.

For convenience we call

$$A_n^*(f,x) := \sum_{k=\lceil na \rceil}^{\lfloor nb \rfloor} f\left(\frac{k}{n}\right) Z(nx-k) =$$

$$\sum_{k_1=\lceil na_1 \rceil}^{\lfloor nb_1 \rfloor} \sum_{k_2=\lceil na_2 \rceil}^{\lfloor nb_2 \rfloor} \cdots \sum_{k_N=\lceil na_N \rceil}^{\lfloor nb_N \rfloor} f\left(\frac{k_1}{n}, \ldots, \frac{k_N}{n}\right) \left(\prod_{i=1}^{N} \psi(nx_i - k_i)\right), \tag{5.33}$$

$\forall\, x \in \left(\prod_{i=1}^{N} [a_i, b_i]\right)$.

That is

$$A_n(f,x) := \frac{A_n^*(f,x)}{\sum_{k=\lceil na \rceil}^{\lfloor nb \rfloor} Z(nx-k)}, \tag{5.34}$$

$\forall\, x \in \left(\prod_{i=1}^{N} [a_i, b_i]\right)$, $n \in \mathbb{N}$.

Hence

$$A_n(f,x) - f(x) = \frac{A_n^*(f,x) - f(x)\left(\sum_{k=\lceil na \rceil}^{\lfloor nb \rfloor} Z(nx-k)\right)}{\sum_{k=\lceil na \rceil}^{\lfloor nb \rfloor} Z(nx-k)}. \tag{5.35}$$

Consequently we derive

$$\|A_n(f,x) - f(x)\|_\gamma \overset{(5.24)}{\leq} (0.0868)^N \left\| A_n^*(f,x) - f(x) \sum_{k=\lceil na \rceil}^{\lfloor nb \rfloor} Z(nx-k) \right\|_\gamma, \tag{5.36}$$

$\forall\, x \in \left(\prod_{i=1}^{N} [a_i, b_i]\right)$.

We will estimate the right hand side of (5.36).

For the last and others we need

Definition 5.6 ([11], p. 274) Let M be a convex and compact subset of $\left(\mathbb{R}^N, \|\cdot\|_p\right)$, $p \in [1, \infty]$, and $\left(X, \|\cdot\|_\gamma\right)$ be a Banach space. Let $f \in C(M, X)$. We define the first modulus of continuity of f as

$$\omega_1(f,\delta) := \sup_{\substack{x, y \in M: \\ \|x-y\|_p \leq \delta}} \|f(x) - f(y)\|_\gamma, \quad 0 < \delta \leq diam(M). \tag{5.37}$$

If $\delta > diam(M)$, then

$$\omega_1(f,\delta) = \omega_1(f, diam(M)). \tag{5.38}$$

Notice $\omega_1(f,\delta)$ is increasing in $\delta>0$. For $f\in C_B(M,X)$ (continuous and bounded functions) $\omega_1(f,\delta)$ is defined similarly.

Lemma 5.7 ([11], p. 274) *We have* $\omega_1(f,\delta)\to 0$ *as* $\delta\downarrow 0$, *iff* $f\in C(M,X)$, *where* M *is a convex compact subset of* $\left(\mathbb{R}^N,\|\cdot\|_p\right)$, $p\in[1,\infty]$.

Clearly we have also: $f\in C_U\left(\mathbb{R}^N,X\right)$ (uniformly continuous functions), iff $\omega_1(f,\delta)\to 0$ as $\delta\downarrow 0$, where ω_1 is defined similarly to (5.37). The space $C_B\left(\mathbb{R}^N,X\right)$ denotes the continuous and bounded functions on $\mathbb{R}^N$.

When $f\in C_B\left(\mathbb{R}^N,X\right)$ we define,

$$B_n(f,x):=B_n(f,x_1,...,x_N):=\sum_{k=-\infty}^{\infty}f\left(\frac{k}{n}\right)Z(nx-k):=$$

$$\sum_{k_1=-\infty}^{\infty}\sum_{k_2=-\infty}^{\infty}...\sum_{k_N=-\infty}^{\infty}f\left(\frac{k_1}{n},\frac{k_2}{n},...,\frac{k_N}{n}\right)\left(\prod_{i=1}^{N}\psi(nx_i-k_i)\right), \tag{5.39}$$

$n\in\mathbb{N}$, $\forall\,x\in\mathbb{R}^N$, $N\in\mathbb{N}$, the multivariate quasi-interpolation neural network operator.

Also for $f\in C_B\left(\mathbb{R}^N,X\right)$ we define the multivariate Kantorovich type neural network operator

$$C_n(f,x):=C_n(f,x_1,...,x_N):=\sum_{k=-\infty}^{\infty}\left(n^N\int_{\frac{k}{n}}^{\frac{k+1}{n}}f(t)\,dt\right)Z(nx-k)=$$

$$\sum_{k_1=-\infty}^{\infty}\sum_{k_2=-\infty}^{\infty}...\sum_{k_N=-\infty}^{\infty}\left(n^N\int_{\frac{k_1}{n}}^{\frac{k_1+1}{n}}\int_{\frac{k_2}{n}}^{\frac{k_2+1}{n}}...\int_{\frac{k_N}{n}}^{\frac{k_N+1}{n}}f(t_1,...,t_N)\,dt_1...dt_N\right)$$

$$\cdot\left(\prod_{i=1}^{N}\psi(nx_i-k_i)\right), \tag{5.40}$$

$n\in\mathbb{N}$, $\forall\,x\in\mathbb{R}^N$.

Again for $f\in C_B\left(\mathbb{R}^N,X\right)$, $N\in\mathbb{N}$, we define the multivariate neural network operator of quadrature type $D_n(f,x)$, $n\in\mathbb{N}$, as follows.

Let $\theta=(\theta_1,...,\theta_N)\in\mathbb{N}^N$, $r=(r_1,...,r_N)\in\mathbb{Z}_+^N$, $w_r=w_{r_1,r_2,...r_N}\geq 0$, such that $\sum_{r=0}^{\theta}w_r=\sum_{r_1=0}^{\theta_1}\sum_{r_2=0}^{\theta_2}...\sum_{r_N=0}^{\theta_N}w_{r_1,r_2,...r_N}=1$; $k\in\mathbb{Z}^N$ and

$$\delta_{nk}(f):=\delta_{n,k_1,k_2,...,k_N}(f):=\sum_{r=0}^{\theta}w_r f\left(\frac{k}{n}+\frac{r}{n\theta}\right)=$$

$$\sum_{r_1=0}^{\theta_1}\sum_{r_2=0}^{\theta_2}\cdots\sum_{r_N=0}^{\theta_N} w_{r_1,r_2,\ldots,r_N} f\left(\frac{k_1}{n}+\frac{r_1}{n\theta_1},\frac{k_2}{n}+\frac{r_2}{n\theta_2},\ldots,\frac{k_N}{n}+\frac{r_N}{n\theta_N}\right), \tag{5.41}$$

where $\frac{r}{\theta} := \left(\frac{r_1}{\theta_1}, \frac{r_2}{\theta_2}, \ldots, \frac{r_N}{\theta_N}\right)$.

We set

$$D_n(f,x) := D_n(f,x_1,\ldots,x_N) := \sum_{k=-\infty}^{\infty} \delta_{nk}(f)\, Z(nx-k) = \tag{5.42}$$

$$\sum_{k_1=-\infty}^{\infty}\sum_{k_2=-\infty}^{\infty}\cdots\sum_{k_N=-\infty}^{\infty} \delta_{n,k_1,k_2,\ldots,k_N}(f)\left(\prod_{i=1}^{N}\psi(nx_i-k_i)\right),$$

$\forall\, x \in \mathbb{R}^N$.

In this chapter we study the approximation properties of A_n, B_n, C_n, D_n neural network operators and as well of their iterates. That is, the quantitative pointwise and uniform convergence of these operators to the unit operator I.

5.3 Multivariate General Neural Network Approximations

Here we present several vectorial neural network approximations to Banach space valued functions given with rates.

We give

Theorem 5.8 *Let $f \in C\left(\prod_{i=1}^{N}[a_i,b_i], X\right)$, $0<\beta<1$, $x \in \left(\prod_{i=1}^{N}[a_i,b_i]\right)$, N, $n \in \mathbb{N}$ with $n^{1-\beta} > 2$. Then*

(1)

$$\|A_n(f,x) - f(x)\|_\gamma \le (4.9737)^N \left[\omega_1\left(f,\frac{1}{n^\beta}\right) + \frac{4\,\big\|\|f\|_\gamma\big\|_\infty}{\pi^2\left(n^{1-\beta}-2\right)}\right] =: \lambda_1(n), \tag{5.43}$$

and

(2)

$$\big\|\|A_n(f) - f\|_\gamma\big\|_\infty \le \lambda_1(n). \tag{5.44}$$

We notice that $\lim_{n\to\infty} A_n(f) \overset{\|\cdot\|_\gamma}{=} f$, pointwise and uniformly.

Above ω_1 is with respect to $p=\infty$.

Proof We observe that

$$\Delta(x) := A_n^*(f,x) - f(x)\sum_{k=\lceil na\rceil}^{\lfloor nb\rfloor} Z(nx-k) =$$

$$\sum_{k=\lceil na\rceil}^{\lfloor nb\rfloor} f\left(\frac{k}{n}\right) Z(nx-k) - \sum_{k=\lceil na\rceil}^{\lfloor nb\rfloor} f(x) Z(nx-k) =$$

$$\sum_{k=\lceil na\rceil}^{\lfloor nb\rfloor} \left(f\left(\frac{k}{n}\right) - f(x)\right) Z(nx-k). \tag{5.45}$$

Thus

$$\|\Delta(x)\|_\gamma \leq \sum_{k=\lceil na\rceil}^{\lfloor nb\rfloor} \left\| f\left(\frac{k}{n}\right) - f(x)\right\|_\gamma Z(nx-k) =$$

$$\sum_{\substack{k=\lceil na\rceil \\ \left\|\frac{k}{n}-x\right\|_\infty \leq \frac{1}{n^\beta}}}^{\lfloor nb\rfloor} \left\| f\left(\frac{k}{n}\right) - f(x)\right\|_\gamma Z(nx-k) +$$

$$\sum_{\substack{k=\lceil na\rceil \\ \left\|\frac{k}{n}-x\right\|_\infty > \frac{1}{n^\beta}}}^{\lfloor nb\rfloor} \left\| f\left(\frac{k}{n}\right) - f(x)\right\|_\gamma Z(nx-k) \overset{(5.18)}{\leq}$$

$$\omega_1\left(f, \frac{1}{n^\beta}\right) + 2\left\|\|f\|_\gamma\right\|_\infty \sum_{\substack{k=\lceil na\rceil \\ \left\|\frac{k}{n}-x\right\|_\infty > \frac{1}{n^\beta}}}^{\lfloor nb\rfloor} Z(nx-k) \overset{(5.23)}{\leq}$$

$$\omega_1\left(f, \frac{1}{n^\beta}\right) + \frac{4\left\|\|f\|_\gamma\right\|_\infty}{\pi^2\left(n^{1-\beta}-2\right)}. \tag{5.46}$$

So that

$$\|\Delta(x)\|_\gamma \leq \omega_1\left(f, \frac{1}{n^\beta}\right) + \frac{4\left\|\|f\|_\gamma\right\|_\infty}{\pi^2\left(n^{1-\beta}-2\right)}. \tag{5.47}$$

Now using (5.36) we finish the proof. ■

We make

Remark 5.9 ([11], pp. 263–266) Let $\left(\mathbb{R}^N, \|\cdot\|_p\right)$, $N \in \mathbb{N}$; where $\|\cdot\|_p$ is the L_p-norm, $1 \leq p \leq \infty$. $\mathbb{R}^N$ is a Banach space, and $\left(\mathbb{R}^N\right)^j$ denotes the j-fold product space $\mathbb{R}^N \times ... \times \mathbb{R}^N$ endowed with the max-norm $\|x\|_{(\mathbb{R}^N)^j} := \max_{1 \leq \lambda \leq j} \|x_\lambda\|_p$, where $x := (x_1, ..., x_j) \in (\mathbb{R}^N)^j$.

Let $(X, \|\cdot\|_\gamma)$ be a general Banach space. Then the space $L_j := L_j\left((\mathbb{R}^N)^j ; X\right)$ of all j-multilinear continuous maps $g : (\mathbb{R}^N)^j \to X$, $j = 1, ..., m$, is a Banach space with norm

$$\|g\| := \|g\|_{L_j} := \sup_{\left(\|x\|_{(\mathbb{R}^N)^j}=1\right)} \|g(x)\|_\gamma = \sup \frac{\|g(x)\|_\gamma}{\|x_1\|_p \ldots \|x_j\|_p}. \tag{5.48}$$

Let M be a non-empty convex and compact subset of $\mathbb{R}^k$ and $x_0 \in M$ is fixed.

Let O be an open subset of $\mathbb{R}^N : M \subset O$. Let $f : O \to X$ be a continuous function, whose Fréchet derivatives (see [20]) $f^{(j)} : O \to L_j = L_j\left((\mathbb{R}^N)^j ; X\right)$ exist and are continuous for $1 \leq j \leq m$, $m \in \mathbb{N}$.

Call $(x - x_0)^j := (x - x_0, ..., x - x_0) \in (\mathbb{R}^N)^j$, $x \in M$.

We will work with $f|_M$.

Then, by Taylor's formula [13], ([20], p. 124), we get

$$f(x) = \sum_{j=0}^{m} \frac{f^{(j)}(x_0)(x - x_0)^j}{j!} + R_m(x, x_0), \text{ all } x \in M, \tag{5.49}$$

where the remainder is the Riemann integral

$$R_m(x, x_0) := \int_0^1 \frac{(1-u)^{m-1}}{(m-1)!} \left(f^{(m)}(x_0 + u(x - x_0)) - f^{(m)}(x_0)\right)(x - x_0)^m du, \tag{5.50}$$

here we set $f^{(0)}(x_0)(x - x_0)^0 = f(x_0)$.

We consider

$$w := \omega_1\left(f^{(m)}, h\right) := \sup_{\substack{x, y \in M: \\ \|x-y\|_p \leq h}} \left\|f^{(m)}(x) - f^{(m)}(y)\right\|, \tag{5.51}$$

$h > 0$.

We obtain

$$\left\|\left(f^{(m)}(x_0 + u(x - x_0)) - f^{(m)}(x_0)\right)(x - x_0)^m\right\|_\gamma \leq$$

$$\left\|f^{(m)}(x_0 + u(x - x_0)) - f^{(m)}(x_0)\right\| \cdot \|x - x_0\|_p^m \leq$$

$$w \left\| x - x_0 \right\|_p^m \left\lceil \frac{u \left\| x - x_0 \right\|_p}{h} \right\rceil, \tag{5.52}$$

by Lemma 7.1.1, [1], p. 208, where $\lceil \cdot \rceil$ is the ceiling.

Therefore for all $x \in M$ (see [1], pp. 121–122):

$$\left\| R_m (x, x_0) \right\|_\gamma \leq w \left\| x - x_0 \right\|_p^m \int_0^1 \left\lceil \frac{u \left\| x - x_0 \right\|_p}{h} \right\rceil \frac{(1-u)^{m-1}}{(m-1)!} du$$

$$= w \Phi_m \left(\left\| x - x_0 \right\|_p \right) \tag{5.53}$$

by a change of variable, where

$$\Phi_m (t) := \int_0^{|t|} \left\lceil \frac{s}{h} \right\rceil \frac{(|t| - s)^{m-1}}{(m-1)!} ds = \frac{1}{m!} \left(\sum_{j=0}^{\infty} (|t| - jh)_+^m \right), \ \forall \, t \in \mathbb{R}, \tag{5.54}$$

is a (polynomial) spline function, see [1], pp. 210–211.

Also from there we get

$$\Phi_m (t) \leq \left(\frac{|t|^{m+1}}{(m+1)!h} + \frac{|t|^m}{2m!} + \frac{h \, |t|^{m-1}}{8 \, (m-1)!} \right), \ \ \forall \, t \in \mathbb{R}, \tag{5.55}$$

with equality true only at $t = 0$.

Therefore it holds

$$\left\| R_m (x, x_0) \right\|_\gamma \leq w \left(\frac{\left\| x - x_0 \right\|_p^{m+1}}{(m+1)!h} + \frac{\left\| x - x_0 \right\|_p^m}{2m!} + \frac{h \left\| x - x_0 \right\|_p^{m-1}}{8 \, (m-1)!} \right), \ \ \forall \, x \in M. \tag{5.56}$$

We have found that

$$\left\| f (x) - \sum_{j=0}^{m} \frac{f^{(j)} (x_0) (x - x_0)^j}{j!} \right\|_\gamma \leq$$

$$\omega_1 \left(f^{(m)}, h \right) \left(\frac{\left\| x - x_0 \right\|_p^{m+1}}{(m+1)!h} + \frac{\left\| x - x_0 \right\|_p^m}{2m!} + \frac{h \left\| x - x_0 \right\|_p^{m-1}}{8 \, (m-1)!} \right) < \infty, \tag{5.57}$$

$\forall \, x, x_0 \in M$.

Here $0 < \omega_1 \left(f^{(m)}, h \right) < \infty$, by M being compact and $f^{(m)}$ being continuous on M.

One can rewrite (5.57) as follows:

$$\left\| f(\cdot) - \sum_{j=0}^{m} \frac{f^{(j)}(x_0)(\cdot - x_0)^j}{j!} \right\|_\gamma \leq$$

$$\omega_1\left(f^{(m)}, h\right)\left(\frac{\|\cdot - x_0\|_p^{m+1}}{(m+1)!h} + \frac{\|\cdot - x_0\|_p^{m}}{2m!} + \frac{h\|\cdot - x_0\|_p^{m-1}}{8(m-1)!}\right), \ \forall\, x_0 \in M, \quad (5.58)$$

a pointwise functional inequality on M.

Here $(\cdot - x_0)^j$ maps M into $\left(\mathbb{R}^N\right)^j$ and it is continuous, also $f^{(j)}(x_0)$ maps $\left(\mathbb{R}^N\right)^j$ into X and it is continuous. Hence their composition $f^{(j)}(x_0)(\cdot - x_0)^j$ is continuous from M into X.

Clearly $f(\cdot) - \sum_{j=0}^{m} \frac{f^{(j)}(x_0)(\cdot - x_0)^j}{j!} \in C(M, X)$, hence $\left\| f(\cdot) - \sum_{j=0}^{m} \frac{f^{(j)}(x_0)(\cdot - x_0)^j}{j!} \right\|_\gamma \in C(M)$.

Let $\left\{\widetilde{L}_N\right\}_{N \in \mathbb{N}}$ be a sequence of positive linear operators mapping $C(M)$ into $C(M)$.

Therefore we obtain

$$\left(\widetilde{L}_N\left(\left\| f(\cdot) - \sum_{j=0}^{m} \frac{f^{(j)}(x_0)(\cdot - x_0)^j}{j!} \right\|_\gamma\right)\right)(x_0) \leq$$

$$\omega_1\left(f^{(m)}, h\right)\left[\frac{\left(\widetilde{L}_N\left(\|\cdot - x_0\|_p^{m+1}\right)\right)(x_0)}{(m+1)!h} + \frac{\left(\widetilde{L}_N\left(\|\cdot - x_0\|_p^{m}\right)\right)(x_0)}{2m!}\right.$$

$$\left. + \frac{h\left(\widetilde{L}_N\left(\|\cdot - x_0\|_p^{m-1}\right)\right)(x_0)}{8(m-1)!}\right], \quad (5.59)$$

$\forall\, N \in \mathbb{N}$, $\forall\, x_0 \in M$.

Clearly (5.59) is valid when $M = \prod_{i=1}^{N} [a_i, b_i]$ and $\widetilde{L}_n = \widetilde{A}_n$, see (5.28).

All the above is preparation for the following theorem, where we assume Fréchet differentiability of functions.

This will be a direct application of Theorem 10.2, [11], pp. 268–270. The operators A_n, $\widetilde{A}_n$ fulfill its assumptions, see (5.27), (5.28), (5.30), (5.31) and (5.32).

We present the following high order approximation results.

Theorem 5.10 *Let O open subset of $\left(\mathbb{R}^N, \|\cdot\|_p\right)$, $p \in [1, \infty]$, such that $\prod_{i=1}^{N} [a_i, b_i] \subset O \subseteq \mathbb{R}^N$, and let $\left(X, \|\cdot\|_\gamma\right)$ be a general Banach space. Let $m \in \mathbb{N}$ and $f \in C^m(O, X)$, the space of m-times continuously Fréchet differentiable functions from*

O into X. We study the approximation of $f|_{\prod_{i=1}^{N}[a_i,b_i]}$. *Let* $x_0 \in \left(\prod_{i=1}^{N}[a_i,b_i]\right)$ *and* $r > 0$.

Then

(1)

$$\left\| (A_n(f))(x_0) - \sum_{j=0}^{m} \frac{1}{j!} \left(A_n \left(f^{(j)}(x_0)(\cdot - x_0)^j \right) \right)(x_0) \right\|_\gamma \leq$$

$$\frac{\omega_1\left(f^{(m)}, r\left(\left(\widetilde{A}_n \left(\|\cdot - x_0\|_p^{m+1} \right) \right)(x_0) \right)^{\frac{1}{m+1}} \right)}{rm!} \left(\left(\widetilde{A}_n \left(\|\cdot - x_0\|_p^{m+1} \right) \right)(x_0) \right)^{\left(\frac{m}{m+1}\right)}$$

$$\left[\frac{1}{(m+1)} + \frac{r}{2} + \frac{mr^2}{8} \right], \tag{5.60}$$

(2) additionally if $f^{(j)}(x_0) = 0$, $j = 1, \ldots, m$, *we have*

$$\|(A_n(f))(x_0) - f(x_0)\|_\gamma \leq$$

$$\frac{\omega_1\left(f^{(m)}, r\left(\left(\widetilde{A}_n \left(\|\cdot - x_0\|_p^{m+1} \right) \right)(x_0) \right)^{\frac{1}{m+1}} \right)}{rm!} \left(\left(\widetilde{A}_n \left(\|\cdot - x_0\|_p^{m+1} \right) \right)(x_0) \right)^{\left(\frac{m}{m+1}\right)}$$

$$\left[\frac{1}{(m+1)} + \frac{r}{2} + \frac{mr^2}{8} \right], \tag{5.61}$$

(3)

$$\|(A_n(f))(x_0) - f(x_0)\|_\gamma \leq \sum_{j=1}^{m} \frac{1}{j!} \left\| \left(A_n \left(f^{(j)}(x_0)(\cdot - x_0)^j \right) \right)(x_0) \right\|_\gamma +$$

$$\frac{\omega_1\left(f^{(m)}, r\left(\left(\widetilde{A}_n \left(\|\cdot - x_0\|_p^{m+1} \right) \right)(x_0) \right)^{\frac{1}{m+1}} \right)}{rm!} \left(\left(\widetilde{A}_n \left(\|\cdot - x_0\|_p^{m+1} \right) \right)(x_0) \right)^{\left(\frac{m}{m+1}\right)}$$

$$\left[\frac{1}{(m+1)} + \frac{r}{2} + \frac{mr^2}{8} \right], \tag{5.62}$$

and

(4)

$$\left\| \|A_n(f) - f\|_\gamma \right\|_{\infty, \prod_{i=1}^{N}[a_i,b_i]} \leq$$

$$\sum_{j=1}^{m} \frac{1}{j!} \left\| \left\| \left(A_n \left(f^{(j)} (x_0) (\cdot - x_0)^j \right) \right) (x_0) \right\|_{\gamma} \right\|_{\infty, x_0 \in \prod_{i=1}^{N} [a_i, b_i]} +$$

$$\frac{\omega_1 \left(f^{(m)}, r \left\| \left(\widetilde{A}_n \left(\|\cdot - x_0\|_p^{m+1} \right) \right) (x_0) \right\|_{\infty, x_0 \in \prod_{i=1}^{N} [a_i, b_i]}^{\frac{1}{m+1}} \right)}{rm!}$$

$$\left\| \left(\widetilde{A}_n \left(\|\cdot - x_0\|_p^{m+1} \right) \right) (x_0) \right\|_{\infty, x_0 \in \prod_{i=1}^{N} [a_i, b_i]}^{\left(\frac{m}{m+1} \right)} \tag{5.63}$$

$$\left[\frac{1}{(m+1)} + \frac{r}{2} + \frac{mr^2}{8} \right].$$

We need

Lemma 5.11 *The function* $\left(\widetilde{A}_n \left(\|\cdot - x_0\|_p^m \right) \right) (x_0)$ *is continuous in* $x_0 \in \left(\prod_{i=1}^{N} [a_i, b_i] \right)$, $m \in \mathbb{N}$.

Proof By Lemma 10.3, [11], p. 272. ■

We make

Remark 5.12 By Remark 10.4, [11], p. 273, we get that

$$\left\| \left(\widetilde{A}_n \left(\|\cdot - x_0\|_p^k \right) \right) (x_0) \right\|_{\infty, x_0 \in \prod_{i=1}^{N} [a_i, b_i]} \leq$$

$$\left\| \left(\widetilde{A}_n \left(\|\cdot - x_0\|_p^{m+1} \right) \right) (x_0) \right\|_{\infty, x_0 \in \prod_{i=1}^{N} [a_i, b_i]}^{\left(\frac{k}{m+1} \right)}, \tag{5.64}$$

for all $k = 1, ..., m$.

We give

Corollary 5.13 (to Theorem 5.10, case of $m = 1$) *Then*
(1)

$$\| (A_n (f)) (x_0) - f (x_0) \|_{\gamma} \leq \left\| \left(A_n \left(f^{(1)} (x_0) (\cdot - x_0) \right) \right) (x_0) \right\|_{\gamma} +$$

$$\frac{1}{2r} \omega_1 \left(f^{(1)}, r \left(\left(\widetilde{A}_n \left(\|\cdot - x_0\|_p^2 \right) \right) (x_0) \right)^{\frac{1}{2}} \right) \left(\left(\widetilde{A}_n \left(\|\cdot - x_0\|_p^2 \right) \right) (x_0) \right)^{\frac{1}{2}} \tag{5.65}$$

$$\left[1 + r + \frac{r^2}{4}\right],$$

and
(2)

$$\left\| \left\| (A_n(f)) - f \right\|_\gamma \right\|_{\infty, \prod_{i=1}^{N} [a_i, b_i]} \leq$$

$$\left\| \left\| \left(A_n\left(f^{(1)}(x_0)(\cdot - x_0)\right)\right)(x_0) \right\|_\gamma \right\|_{\infty, x_0 \in \prod_{i=1}^{N} [a_i, b_i]} +$$

$$\frac{1}{2r} \omega_1 \left(f^{(1)}, r \left\| \left(\widetilde{A}_n\left(\|\cdot - x_0\|_p^2\right)\right)(x_0) \right\|^{\frac{1}{2}}_{\infty, x_0 \in \prod_{i=1}^{N} [a_i, b_i]} \right)$$

$$\left\| \left(\widetilde{A}_n\left(\|\cdot - x_0\|_p^2\right)\right)(x_0) \right\|^{\frac{1}{2}}_{\infty, x_0 \in \prod_{i=1}^{N} [a_i, b_i]} \left[1 + r + \frac{r^2}{4}\right], \tag{5.66}$$

$r > 0$.

We make

Remark 5.14 We estimate $0 < \alpha < 1$, $m, n \in \mathbb{N} : n^{1-\alpha} > 2$,

$$\widetilde{A}_n\left(\|\cdot - x_0\|_\infty^{m+1}\right)(x_0) = \frac{\sum_{k=\lceil na \rceil}^{\lfloor nb \rfloor} \left\| \frac{k}{n} - x_0 \right\|_\infty^{m+1} Z(nx_0 - k)}{\sum_{k=\lceil na \rceil}^{\lfloor nb \rfloor} Z(nx_0 - k)} \overset{(5.24)}{<}$$

$$(4.9737)^N \sum_{k=\lceil na \rceil}^{\lfloor nb \rfloor} \left\| \frac{k}{n} - x_0 \right\|_\infty^{m+1} Z(nx_0 - k) = \tag{5.67}$$

$$(4.9737)^N \left\{ \sum_{\substack{k = \lceil na \rceil \\ : \left\| \frac{k}{n} - x_0 \right\|_\infty \leq \frac{1}{n^\alpha}}}^{\lfloor nb \rfloor} \left\| \frac{k}{n} - x_0 \right\|_\infty^{m+1} Z(nx_0 - k) + \right.$$

$$\left. \sum_{\substack{k = \lceil na \rceil \\ : \left\| \frac{k}{n} - x_0 \right\|_\infty > \frac{1}{n^\alpha}}}^{\lfloor nb \rfloor} \left\| \frac{k}{n} - x_0 \right\|_\infty^{m+1} Z(nx_0 - k) \right\} \overset{(5.25)}{\leq}$$

$$(4.9737)^N \left\{ \frac{1}{n^{\alpha(m+1)}} + \frac{2\|b-a\|_\infty^{m+1}}{\pi^2 \left(n^{1-\alpha}-2\right)} \right\}, \tag{5.68}$$

(where $b - a = (b_1 - a_1, ..., b_N - a_N)$).

We have proved that ($\forall\, x_0 \in \prod_{i=1}^{N} [a_i, b_i]$)

$$\widetilde{A}_n \left(\|\cdot - x_0\|_\infty^{m+1}\right)(x_0) < (4.9737)^N \left\{ \frac{1}{n^{\alpha(m+1)}} + \frac{2\|b-a\|_\infty^{m+1}}{\pi^2 \left(n^{1-\alpha}-2\right)} \right\} =: \varphi_1(n) \tag{5.69}$$

($0 < \alpha < 1, m, n \in \mathbb{N} : n^{1-\alpha} > 2$).

And, consequently it holds

$$\left\| \widetilde{A}_n \left(\|\cdot - x_0\|_\infty^{m+1}\right)(x_0) \right\|_{\infty, x_0 \in \prod_{i=1}^{N} [a_i, b_i]} <$$

$$(4.9737)^N \left\{ \frac{1}{n^{\alpha(m+1)}} + \frac{2\|b-a\|_\infty^{m+1}}{\pi^2 \left(n^{1-\alpha}-2\right)} \right\} = \varphi_1(n) \to 0, \quad \text{as } n \to +\infty. \tag{5.70}$$

So, we have that $\varphi_1(n) \to 0$, as $n \to +\infty$. Thus, when $p \in [1, \infty]$, from Theorem 5.10 we have the convergence to zero in the right hand sides of parts (1), (2).

Next we estimate $\left\| \left(\widetilde{A}_n \left(f^{(j)}(x_0)(\cdot - x_0)^j\right)\right)(x_0) \right\|_\gamma$.

We have that

$$\left(\widetilde{A}_n \left(f^{(j)}(x_0)(\cdot - x_0)^j\right)\right)(x_0) = \frac{\sum_{k=\lceil na \rceil}^{\lfloor nb \rfloor} f^{(j)}(x_0) \left(\frac{k}{n} - x_0\right)^j Z(nx_0 - k)}{\sum_{k=\lceil na \rceil}^{\lfloor nb \rfloor} Z(nx_0 - k)}. \tag{5.71}$$

When $p = \infty$, $j = 1, ..., m$, we obtain

$$\left\| f^{(j)}(x_0) \left(\frac{k}{n} - x_0\right)^j \right\|_\gamma \leq \left\| f^{(j)}(x_0) \right\| \left\| \frac{k}{n} - x_0 \right\|_\infty^j. \tag{5.72}$$

We further have that

$$\left\| \left(\widetilde{A}_n \left(f^{(j)}(x_0)(\cdot - x_0)^j\right)\right)(x_0) \right\|_\gamma \overset{(5.24)}{<}$$

$$(4.9737)^N \left(\sum_{k=\lceil na \rceil}^{\lfloor nb \rfloor} \left\| f^{(j)}(x_0) \left(\frac{k}{n} - x_0\right)^j \right\|_\gamma Z(nx_0 - k) \right) \leq$$

$$(4.9737)^N \left(\sum_{k=\lceil na \rceil}^{\lfloor nb \rfloor} \left\| f^{(j)}(x_0) \right\| \left\| \frac{k}{n} - x_0 \right\|_\infty^j Z(nx_0 - k) \right) = \tag{5.73}$$

$$(4.9737)^N \left\| f^{(j)}(x_0) \right\| \left(\sum_{k=\lceil na \rceil}^{\lfloor nb \rfloor} \left\| \frac{k}{n} - x_0 \right\|_\infty^j Z(nx_0 - k) \right) =$$

$$(4.9737)^N \left\| f^{(j)}(x_0) \right\| \left\{ \sum_{\substack{k = \lceil na \rceil \\ : \left\| \frac{k}{n} - x_0 \right\|_\infty \le \frac{1}{n^\alpha}}}^{\lfloor nb \rfloor} \left\| \frac{k}{n} - x_0 \right\|_\infty^j Z(nx_0 - k) \right.$$

$$\left. + \sum_{\substack{k = \lceil na \rceil \\ : \left\| \frac{k}{n} - x_0 \right\|_\infty > \frac{1}{n^\alpha}}}^{\lfloor nb \rfloor} \left\| \frac{k}{n} - x_0 \right\|_\infty^j Z(nx_0 - k) \right\} \overset{(5.25)}{\le} \tag{5.74}$$

$$(4.9737)^N \left\| f^{(j)}(x_0) \right\| \left\{ \frac{1}{n^{\alpha j}} + \frac{2 \|b - a\|_\infty^j}{\pi^2 \left(n^{1-\alpha} - 2\right)} \right\} \to 0, \text{ as } n \to \infty.$$

That is

$$\left\| \left(\widetilde{A}_n \left(f^{(j)}(x_0) (\cdot - x_0)^j \right) \right)(x_0) \right\|_\gamma \to 0, \text{ as } n \to \infty.$$

Therefore when $p = \infty$, for $j = 1, ..., m$, we have proved:

$$\left\| \left(\widetilde{A}_n \left(f^{(j)}(x_0) (\cdot - x_0)^j \right) \right)(x_0) \right\|_\gamma <$$

$$(4.9737)^N \left\| f^{(j)}(x_0) \right\| \left\{ \frac{1}{n^{\alpha j}} + \frac{2 \|b - a\|_\infty^j}{\pi^2 \left(n^{1-\alpha} - 2\right)} \right\} \le \tag{5.75}$$

$$(4.9737)^N \left\| f^{(j)} \right\|_\infty \left\{ \frac{1}{n^{\alpha j}} + \frac{2 \|b - a\|_\infty^j}{\pi^2 \left(n^{1-\alpha} - 2\right)} \right\} =: \varphi_{2j}(n) < \infty,$$

and converges to zero, as $n \to \infty$.

We conclude:

In Theorem 5.10, the right hand sides of (5.62) and (5.63) converge to zero as $n \to \infty$, for any $p \in [1, \infty]$.

Also in Corollary 5.13, the right hand sides of (5.65) and (5.66) converge to zero as $n \to \infty$, for any $p \in [1, \infty]$.

Conclusion 5.15 *We have proved that the left hand sides of (5.60), (5.61), (5.62), (5.63) and (5.65), (5.66) converge to zero as $n \to \infty$, for $p \in [1, \infty]$. Consequently $A_n \to I$ (unit operator) pointwise and uniformly, as $n \to \infty$, where $p \in [1, \infty]$. In the presence of initial conditions we achieve a higher speed of convergence, see (5.61). Higher speed of convergence happens also to the left hand side of (5.60).*

We give

Corollary 5.16 (to Theorem 5.10) *Let O open subset of $\left(\mathbb{R}^N, \|\cdot\|_\infty\right)$, such that $\prod_{i=1}^{N} [a_i, b_i] \subset O \subseteq \mathbb{R}^N$, and let $\left(X, \|\cdot\|_\gamma\right)$ be a general Banach space. Let $m \in \mathbb{N}$ and $f \in C^m (O, X)$, the space of m-times cont uously Fréchet differentiable functions from O into X. We study the approximation of $f|_{\prod_{i=1}^{N}[a_i, b_i]}$. Let $x_0 \in \left(\prod_{i=1}^{N} [a_i, b_i]\right)$ and $r > 0$. Here $\varphi_1 (n)$ as in (5.69) and $\varphi_{2j} (n)$ as in (5.75), where $n \in \mathbb{N} : n^{1-\alpha} > 2$, $0 < \alpha < 1$, $j = 1, ..., m$. Then*

(1)

$$\left\| (A_n (f)) (x_0) - \sum_{j=0}^{m} \frac{1}{j!} \left(A_n \left(f^{(j)} (x_0) (\cdot - x_0)^j\right)\right) (x_0) \right\|_\gamma \leq$$

$$\frac{\omega_1 \left(f^{(m)}, r (\varphi_1 (n))^{\frac{1}{m+1}}\right)}{rm!} (\varphi_1 (n))^{\left(\frac{m}{m+1}\right)} \left[\frac{1}{(m+1)} + \frac{r}{2} + \frac{mr^2}{8}\right], \tag{5.76}$$

(2) additionally, if $f^{(j)} (x_0) = 0$, $j = 1, ..., m$, we have

$$\|(A_n (f)) (x_0) - f (x_0)\|_\gamma \leq$$

$$\frac{\omega_1 \left(f^{(m)}, r (\varphi_1 (n))^{\frac{1}{m+1}}\right)}{rm!} (\varphi_1 (n))^{\left(\frac{m}{m+1}\right)} \left[\frac{1}{(m+1)} + \frac{r}{2} + \frac{mr^2}{8}\right], \tag{5.77}$$

(3)

$$\left\| \|A_n (f) - f\|_\gamma \right\|_{\infty, \prod_{i=1}^{N}[a_i, b_i]} \leq \sum_{j=1}^{m} \frac{\varphi_{2j} (n)}{j!} +$$

$$\frac{\omega_1 \left(f^{(m)}, r (\varphi_1 (n))^{\frac{1}{m+1}}\right)}{rm!} (\varphi_1 (n))^{\left(\frac{m}{m+1}\right)} \tag{5.78}$$

$$\left[\frac{1}{(m+1)} + \frac{r}{2} + \frac{mr^2}{8}\right] =: \varphi_3(n) \to 0, \text{ as } n \to \infty.$$

We continue with

Theorem 5.17 *Let* $f \in C_B(\mathbb{R}^N, X)$, $0 < \beta < 1$, $x \in \mathbb{R}^N$, $N, n \in \mathbb{N}$ *with* $n^{1-\beta} > 2$, ω_1 *is for* $p = \infty$. *Then*
(1)

$$\|B_n(f,x) - f(x)\|_\gamma \leq \omega_1\left(f, \frac{1}{n^\beta}\right) + \frac{4\left\|\|f\|_\gamma\right\|_\infty}{\pi^2\left(n^{1-\beta}-2\right)} =: \lambda_2(n), \tag{5.79}$$

(2)

$$\left\|\|B_n(f) - f\|_\gamma\right\|_\infty \leq \lambda_2(n). \tag{5.80}$$

Given that $f \in \left(C_U(\mathbb{R}^N, X) \cap C_B(\mathbb{R}^N, X)\right)$, *we obtain* $\lim_{n\to\infty} B_n(f) = f$, *uniformly.*

Proof We have that

$$B_n(f,x) - f(x) \overset{(5.18)}{=} \sum_{k=-\infty}^{\infty} f\left(\frac{k}{n}\right) Z(nx-k) - f(x) \sum_{k=-\infty}^{\infty} Z(nx-k) =$$

$$\sum_{k=-\infty}^{\infty} \left(f\left(\frac{k}{n}\right) - f(x)\right) Z(nx-k). \tag{5.81}$$

Hence

$$\|B_n(f,x) - f(x)\|_\gamma \leq \sum_{k=-\infty}^{\infty} \left\|f\left(\frac{k}{n}\right) - f(x)\right\|_\gamma Z(nx-k) =$$

$$\sum_{\substack{k=-\infty \\ \left\|\frac{k}{n}-x\right\|_\infty \leq \frac{1}{n^\beta}}}^{\infty} \left\|f\left(\frac{k}{n}\right) - f(x)\right\|_\gamma Z(nx-k) +$$

$$\sum_{\substack{k=-\infty \\ \left\|\frac{k}{n}-x\right\|_\infty > \frac{1}{n^\beta}}}^{\infty} \left\|f\left(\frac{k}{n}\right) - f(x)\right\|_\gamma Z(nx-k) \overset{(5.18)}{\leq}$$

$$\omega_1\left(f,\frac{1}{n^\beta}\right)+2\left\|\|f\|_\gamma\right\|_\infty \sum_{\substack{k=-\infty \\ \left\|\frac{k}{n}-x\right\|_\infty>\frac{1}{n^\beta}}}^{\infty} Z(nx-k) \overset{(5.25)}{\leq}$$

$$\omega_1\left(f,\frac{1}{n^\beta}\right)+\frac{4\left\|\|f\|_\gamma\right\|_\infty}{\pi^2\left(n^{1-\beta}-2\right)}, \tag{5.82}$$

proving the claim. ■

We give

Theorem 5.18 *Let $f \in C_B\left(\mathbb{R}^N, X\right)$, $0<\beta<1$, $x \in \mathbb{R}^N$, $N, n \in \mathbb{N}$ with $n^{1-\beta}>2$, ω_1 is for $p=\infty$. Then*
(1)

$$\|C_n(f,x)-f(x)\|_\gamma \leq \omega_1\left(f,\frac{1}{n}+\frac{1}{n^\beta}\right)+\frac{4\left\|\|f\|_\gamma\right\|_\infty}{\pi^2\left(n^{1-\beta}-2\right)}=:\lambda_3(n), \tag{5.83}$$

(2)

$$\left\|\|C_n(f)-f\|_\gamma\right\|_\infty \leq \lambda_3(n). \tag{5.84}$$

Given that $f \in \left(C_U\left(\mathbb{R}^N, X\right) \cap C_B\left(\mathbb{R}^N, X\right)\right)$, we obtain $\lim_{n\to\infty} C_n(f)=f$, uniformly.

Proof We notice that

$$\int_{\frac{k}{n}}^{\frac{k+1}{n}} f(t)\,dt=\int_{\frac{k_1}{n}}^{\frac{k_1+1}{n}}\int_{\frac{k_2}{n}}^{\frac{k_2+1}{n}}\ldots\int_{\frac{k_N}{n}}^{\frac{k_N+1}{n}} f(t_1,t_2,\ldots,t_N)\,dt_1dt_2\ldots dt_N=$$

$$\int_0^{\frac{1}{n}}\int_0^{\frac{1}{n}}\ldots\int_0^{\frac{1}{n}} f\left(t_1+\frac{k_1}{n},t_2+\frac{k_2}{n},\ldots,t_N+\frac{k_N}{n}\right)dt_1\ldots dt_N=\int_0^{\frac{1}{n}} f\left(t+\frac{k}{n}\right)dt. \tag{5.85}$$

Thus it holds (by (5.40))

$$C_n(f,x)=\sum_{k=-\infty}^{\infty}\left(n^N\int_0^{\frac{1}{n}} f\left(t+\frac{k}{n}\right)dt\right)Z(nx-k). \tag{5.86}$$

We observe that

$$\left\| C_n(f,x) - f(x) \right\|_\gamma =$$

$$\left\| \sum_{k=-\infty}^{\infty} \left(n^N \int_0^{\frac{1}{n}} f\left(t + \frac{k}{n}\right) dt \right) Z(nx - k) - \sum_{k=-\infty}^{\infty} f(x) Z(nx - k) \right\|_\gamma =$$

$$\left\| \sum_{k=-\infty}^{\infty} \left(\left(n^N \int_0^{\frac{1}{n}} f\left(t + \frac{k}{n}\right) dt \right) - f(x) \right) Z(nx - k) \right\|_\gamma =$$

$$\left\| \sum_{k=-\infty}^{\infty} \left(n^N \int_0^{\frac{1}{n}} \left(f\left(t + \frac{k}{n}\right) - f(x) \right) dt \right) Z(nx - k) \right\|_\gamma \leq \tag{5.87}$$

$$\sum_{k=-\infty}^{\infty} \left(n^N \int_0^{\frac{1}{n}} \left\| f\left(t + \frac{k}{n}\right) - f(x) \right\|_\gamma dt \right) Z(nx - k) =$$

$$\sum_{\substack{k=-\infty \\ \left\| \frac{k}{n} - x \right\|_\infty \leq \frac{1}{n^\beta}}}^{\infty} \left(n^N \int_0^{\frac{1}{n}} \left\| f\left(t + \frac{k}{n}\right) - f(x) \right\|_\gamma dt \right) Z(nx - k) +$$

$$\sum_{\substack{k=-\infty \\ \left\| \frac{k}{n} - x \right\|_\infty > \frac{1}{n^\beta}}}^{\infty} \left(n^N \int_0^{\frac{1}{n}} \left\| f\left(t + \frac{k}{n}\right) - f(x) \right\|_\gamma dt \right) Z(nx - k) \leq$$

$$\sum_{\substack{k=-\infty \\ \left\| \frac{k}{n} - x \right\|_\infty \leq \frac{1}{n^\beta}}}^{\infty} \left(n^N \int_0^{\frac{1}{n}} \omega_1 \left(f, \|t\|_\infty + \left\| \frac{k}{n} - x \right\|_\infty \right) dt \right) Z(nx - k) +$$

$$2 \left\| \|f\|_\gamma \right\|_\infty \left(\sum_{\substack{k=-\infty \\ \left\| \frac{k}{n} - x \right\|_\infty > \frac{1}{n^\beta}}}^{\infty} Z(|nx - k|) \right) \leq$$

$$\omega_1 \left(f, \frac{1}{n} + \frac{1}{n^\beta} \right) + \frac{4 \left\| \|f\|_\gamma \right\|_\infty}{\pi^2 \left(n^{1-\beta} - 2 \right)}, \tag{5.88}$$

proving the claim. ■

We also present

Theorem 5.19 *Let* $f \in C_B\left(\mathbb{R}^N, X\right)$, $0 < \beta < 1$, $x \in \mathbb{R}^N$, $N, n \in \mathbb{N}$ *with* $n^{1-\beta} > 2$, ω_1 *is for* $p = \infty$. *Then*
(1)

$$\|D_n(f, x) - f(x)\|_\gamma \leq \omega_1\left(f, \frac{1}{n} + \frac{1}{n^\beta}\right) + \frac{4\left\| \|f\|_\gamma \right\|_\infty}{\pi^2\left(n^{1-\beta} - 2\right)} = \lambda_4(n), \quad (5.89)$$

(2)

$$\left\| \|D_n(f) - f\|_\gamma \right\|_\infty \leq \lambda_4(n). \quad (5.90)$$

Given that $f \in \left(C_U\left(\mathbb{R}^N, X\right) \cap C_B\left(\mathbb{R}^N, X\right)\right)$, *we obtain* $\lim_{n\to\infty} D_n(f) = f$, *uniformly.*

Proof We have that (by (5.42))

$$\|D_n(f, x) - f(x)\|_\gamma = \left\| \sum_{k=-\infty}^{\infty} \delta_{nk}(f) Z(nx - k) - \sum_{k=-\infty}^{\infty} f(x) Z(nx - k) \right\|_\gamma = \quad (5.91)$$

$$\left\| \sum_{k=-\infty}^{\infty} (\delta_{nk}(f) - f(x)) Z(nx - k) \right\|_\gamma =$$

$$\left\| \sum_{k=-\infty}^{\infty} \left(\sum_{r=0}^{\theta} w_r \left(f\left(\frac{k}{n} + \frac{r}{n\theta}\right) - f(x) \right) \right) Z(nx - k) \right\|_\gamma \leq$$

$$\sum_{k=-\infty}^{\infty} \left(\sum_{r=0}^{\theta} w_r \left\| f\left(\frac{k}{n} + \frac{r}{n\theta}\right) - f(x) \right\|_\gamma \right) Z(nx - k) =$$

$$\sum_{\substack{k=-\infty \\ \left\|\frac{k}{n} - x\right\|_\infty \leq \frac{1}{n^\beta}}}^{\infty} \left(\sum_{r=0}^{\theta} w_r \left\| f\left(\frac{k}{n} + \frac{r}{n\theta}\right) - f(x) \right\|_\gamma \right) Z(nx - k) +$$

$$\sum_{\substack{k=-\infty \\ \left\|\frac{k}{n} - x\right\|_\infty > \frac{1}{n^\beta}}}^{\infty} \left(\sum_{r=0}^{\theta} w_r \left\| f\left(\frac{k}{n} + \frac{r}{n\theta}\right) - f(x) \right\|_\gamma \right) Z(nx - k) \leq$$

$$\sum_{\substack{k=-\infty \\ \left\| \frac{k}{n} - x \right\|_{\infty} \leq \frac{1}{n^{\beta}}}}^{\infty} \left(\sum_{r=0}^{\theta} w_r \omega_1 \left(f, \left\| \frac{k}{n} - x \right\|_{\infty} + \left\| \frac{r}{n\theta} \right\|_{\infty} \right) \right) Z(nx - k) +$$

$$2 \left\| \|f\|_{\gamma} \right\|_{\infty} \left(\sum_{\substack{k=-\infty \\ \left\| \frac{k}{n} - x \right\|_{\infty} > \frac{1}{n^{\beta}}}}^{\infty} Z(nx - k) \right) \leq$$

$$\omega_1 \left(f, \frac{1}{n} + \frac{1}{n^{\beta}} \right) + \frac{4 \left\| \|f\|_{\gamma} \right\|_{\infty}}{\pi^2 \left(n^{1-\beta} - 2 \right)}, \tag{5.92}$$

proving the claim. ■

We make

Definition 5.20 Let $f \in C_B(\mathbb{R}^N, X)$, $N \in \mathbb{N}$, where $(X, \|\cdot\|_{\gamma})$ is a Banach space. We define the general neural network operator

$$F_n(f, x) := \sum_{k=-\infty}^{\infty} l_{nk}(f) Z(nx - k) =$$

$$\begin{cases} B_n(f, x), \text{ if } l_{nk}(f) = f\left(\frac{k}{n}\right), \\ C_n(f, x), \text{ if } l_{nk}(f) = n^N \int_{\frac{k}{n}}^{\frac{k+1}{n}} f(t)\, dt, \\ D_n(f, x), \text{ if } l_{nk}(f) = \delta_{nk}(f). \end{cases} \tag{5.93}$$

Clearly $l_{nk}(f)$ is an X-valued bounded linear functional such that $\|l_{nk}(f)\|_{\gamma} \leq \left\| \|f\|_{\gamma} \right\|_{\infty}$.

Hence $F_n(f)$ is a bounded linear operator with $\left\| \|F_n(f)\|_{\gamma} \right\|_{\infty} \leq \left\| \|f\|_{\gamma} \right\|_{\infty}$.

We need

Theorem 5.21 *Let $f \in C_B(\mathbb{R}^N, X)$, $N \geq 1$. Then $F_n(f) \in C_B(\mathbb{R}^N, X)$.*

Proof Clearly $F_n(f)$ is a bounded function.

Next we prove the continuity of $F_n(f)$. Notice for $N = 1$, $Z = \psi$ by (5.16).

We will use the generalized Weierstrass M test: If a sequence of positive constants $M_1, M_2, M_3, \ldots$, can be found such that in some interval

(a) $\|u_n(x)\|_{\gamma} \leq M_n$, $n = 1, 2, 3, \ldots$

(b) $\sum M_n$ converges,

then $\sum u_n(x)$ is uniformly and absolutely convergent in the interval.

Also we will use:

If $\{u_n(x)\}$, $n = 1, 2, 3, ...$ are continuous in $[a, b]$ and if $\sum u_n(x)$ converges uniformly to the sum $S(x)$ in $[a, b]$, then $S(x)$ is continuous in $[a, b]$. I.e. a uniformly convergent series of continuous functions is a continuous function. First we prove claim for $N = 1$.

We will prove that $\sum_{k=-\infty}^{\infty} l_{nk}(f)\psi(nx-k)$ is continuous in $x \in \mathbb{R}$.

There always exists $\lambda \in \mathbb{N}$ such that $nx \in [-\lambda, \lambda]$.

Since $nx \leq \lambda$, then $-nx \geq -\lambda$ and $k - nx \geq k - \lambda \geq 0$, when $k \geq \lambda$. Therefore

$$\sum_{k=\lambda}^{\infty} \psi(nx-k) = \sum_{k=\lambda}^{\infty} \psi(k-nx) \leq \sum_{k=\lambda}^{\infty} \psi(k-\lambda) = \sum_{k'=0}^{\infty} \psi(k') \leq 1. \tag{5.94}$$

So for $k \geq \lambda$ we get

$$\|l_{nk}(f)\|_{\gamma}\psi(nx-k) \leq \left\| \|f\|_{\gamma} \right\|_{\infty} \psi(k-\lambda),$$

and

$$\left\| \|f\|_{\gamma} \right\|_{\infty} \sum_{k=\lambda}^{\infty} \psi(k-\lambda) \leq \left\| \|f\|_{\gamma} \right\|_{\infty}.$$

Hence by the generalized Weierstrass M test we obtain that $\sum_{k=\lambda}^{\infty} l_{nk}(f)\psi(nx-k)$ is uniformly and absolutely convergent on $\left[-\frac{\lambda}{n}, \frac{\lambda}{n}\right]$.

Since $l_{nk}(f)\psi(nx-k)$ is continuous in x, then $\sum_{k=\lambda}^{\infty} l_{nk}(f)\psi(nx-k)$ is continuous on $\left[-\frac{\lambda}{n}, \frac{\lambda}{n}\right]$.

Because $nx \geq -\lambda$, then $-nx \leq \lambda$, and $k - nx \leq k + \lambda \leq 0$, when $k \leq -\lambda$. Therefore

$$\sum_{k=-\infty}^{-\lambda} \psi(nx-k) = \sum_{k=-\infty}^{-\lambda} \psi(k-nx) \leq \sum_{k=-\infty}^{-\lambda} \psi(k+\lambda) = \sum_{k'=-\infty}^{0} \psi(k') \leq 1.$$

So for $k \leq -\lambda$ we get

$$\|l_{nk}(f)\|_{\gamma}\psi(nx-k) \leq \left\| \|f\|_{\gamma} \right\|_{\infty} \psi(k+\lambda), \tag{5.95}$$

and

$$\left\| \|f\|_{\gamma} \right\|_{\infty} \sum_{k=-\infty}^{-\lambda} \psi(k+\lambda) \leq \left\| \|f\|_{\gamma} \right\|_{\infty}.$$

Hence by Weierstrass M test we obtain that $\sum_{k=-\infty}^{-\lambda} l_{nk}(f)\psi(nx-k)$ is uniformly and absolutely convergent on $\left[-\frac{\lambda}{n}, \frac{\lambda}{n}\right]$.

Since $l_{nk}(f)\psi(nx-k)$ is continuous in x, then $\sum_{k=-\infty}^{-\lambda} l_{nk}(f)\psi(nx-k)$ is continuous on $\left[-\frac{\lambda}{n}, \frac{\lambda}{n}\right]$.

So we proved that $\sum_{k=\lambda}^{\infty} l_{nk}(f)\,\psi(nx-k)$ and $\sum_{k=-\infty}^{-\lambda} l_{nk}(f)\,\psi(nx-k)$ are continuous on $\mathbb{R}$. Since $\sum_{k=-\lambda+1}^{\lambda-1} l_{nk}(f)\,\psi(nx-k)$ is a finite sum of continuous functions on $\mathbb{R}$, it is also a continuous function on $\mathbb{R}$.

Writing

$$\sum_{k=-\infty}^{\infty} l_{nk}(f)\,\psi(nx-k) = \sum_{k=-\infty}^{-\lambda} l_{nk}(f)\,\psi(nx-k) +$$

$$\sum_{k=-\lambda+1}^{\lambda-1} l_{nk}(f)\,\psi(nx-k) + \sum_{k=\lambda}^{\infty} l_{nk}(f)\,\psi(nx-k) \tag{5.96}$$

we have it as a continuous function on $\mathbb{R}$. Therefore $F_n(f)$, when $N=1$, is a continuous function on $\mathbb{R}$.

When $N=2$ we have

$$F_n(f,x_1,x_2) = \sum_{k_1=-\infty}^{\infty}\sum_{k_2=-\infty}^{\infty} l_{nk}(f)\,\psi(nx_1-k_1)\,\psi(nx_2-k_2) =$$

$$\sum_{k_1=-\infty}^{\infty} \psi(nx_1-k_1)\left(\sum_{k_2=-\infty}^{\infty} l_{nk}(f)\,\psi(nx_2-k_2)\right)$$

(there always exist $\lambda_1, \lambda_2 \in \mathbb{N}$ such that $nx_1 \in [-\lambda_1, \lambda_1]$ and $nx_2 \in [-\lambda_2, \lambda_2]$)

$$= \sum_{k_1=-\infty}^{\infty} \psi(nx_1-k_1)\left[\sum_{k_2=-\infty}^{-\lambda_2} l_{nk}(f)\,\psi(nx_2-k_2) + \right.$$

$$\left.\sum_{k_2=-\lambda_2+1}^{\lambda_2-1} l_{nk}(f)\,\psi(nx_2-k_2) + \sum_{k_2=\lambda_2}^{\infty} l_{nk}(f)\,\psi(nx_2-k_2)\right] =$$

$$= \sum_{k_1=-\infty}^{\infty}\sum_{k_2=-\infty}^{-\lambda_2} l_{nk}(f)\,\psi(nx_1-k_1)\,\psi(nx_2-k_2) +$$

$$\sum_{k_1=-\infty}^{\infty}\sum_{k_2=-\lambda_2+1}^{\lambda_2-1} l_{nk}(f)\,\psi(nx_1-k_1)\,\psi(nx_2-k_2) +$$

$$\sum_{k_1=-\infty}^{\infty}\sum_{k_2=\lambda_2}^{\infty} l_{nk}(f)\,\psi(nx_1-k_1)\,\psi(nx_2-k_2) =: (*)\,.$$

(For convenience call

$$F(k_1,k_2,x_1,x_2) := l_{nk}(f)\,\psi(nx_1-k_1)\,\psi(nx_2-k_2)\,.)$$

Thus

$$(*) = \sum_{k_1=-\infty}^{-\lambda_1}\sum_{k_2=-\infty}^{-\lambda_2} F(k_1,k_2,x_1,x_2) + \sum_{k_1=-\lambda_1+1}^{\lambda_1-1}\sum_{k_2=-\infty}^{-\lambda_2} F(k_1,k_2,x_1,x_2) +$$

$$\sum_{k_1=\lambda_1}^{\infty}\sum_{k_2=-\infty}^{-\lambda_2} F(k_1,k_2,x_1,x_2) + \sum_{k_1=-\infty}^{-\lambda_1}\sum_{k_2=-\lambda_2+1}^{\lambda_2-1} F(k_1,k_2,x_1,x_2) +$$

$$\sum_{k_1=-\lambda_1+1}^{\lambda_1-1}\sum_{k_2=-\lambda_2+1}^{\lambda_2-1} F(k_1,k_2,x_1,x_2) + \sum_{k_1=\lambda_1}^{\infty}\sum_{k_2=-\lambda_2+1}^{\lambda_2-1} F(k_1,k_2,x_1,x_2) +$$

$$\sum_{k_1=-\infty}^{-\lambda_1}\sum_{k_2=\lambda_2}^{\infty} F(k_1,k_2,x_1,x_2) + \sum_{k_1=-\lambda_1+1}^{\lambda_1-1}\sum_{k_2=\lambda_2}^{\infty} F(k_1,k_2,x_1,x_2) + \tag{5.97}$$

$$\sum_{k_1=\lambda_1}^{\infty}\sum_{k_2=\lambda_2}^{\infty} F(k_1,k_2,x_1,x_2)\,.$$

Notice that the finite sum of continuous functions $F(k_1,k_2,x_1,x_2)$, $\sum_{k_1=-\lambda_1+1}^{\lambda_1-1}\sum_{k_2=-\lambda_2+1}^{\lambda_2-1} F(k_1,k_2,x_1,x_2)$ is a continuous function.

The rest of the summands of $F_n(f,x_1,x_2)$ are treated all the same way and similarly to the case of $N=1$. The method is demonstrated as follows.

We will prove that $\sum_{k_1=\lambda_1}^{\infty}\sum_{k_2=-\infty}^{-\lambda_2} l_{nk}(f)\,\psi(nx_1-k_1)\,\psi(nx_2-k_2)$ is continuous in $(x_1,x_2)\in\mathbb{R}^2$.

The continuous function

$$\|l_{nk}(f)\|_\gamma\,\psi(nx_1-k_1)\,\psi(nx_2-k_2) \le \left\|\|f\|_\gamma\right\|_\infty \psi(k_1-\lambda_1)\,\psi(k_2+\lambda_2)\,,$$

and

$$\left\|\|f\|_\gamma\right\|_\infty \sum_{k_1=\lambda_1}^{\infty}\sum_{k_2=-\infty}^{-\lambda_2} \psi(k_1-\lambda_1)\,\psi(k_2+\lambda_2) =$$

$$\left\|\|f\|_\gamma\right\|_\infty \left(\sum_{k_1=\lambda_1}^{\infty}\psi(k_1-\lambda_1)\right)\left(\sum_{k_2=-\infty}^{-\lambda_2}\psi(k_2+\lambda_2)\right) \le$$

$$\left\| \|f\|_{\gamma} \right\|_{\infty} \left(\sum_{k_1'=0}^{\infty} \psi\left(k_1'\right) \right) \left(\sum_{k_2'=-\infty}^{0} \psi\left(k_2'\right) \right) \leq \left\| \|f\|_{\gamma} \right\|_{\infty}.$$

So by the Weierstrass M test we get that $\sum_{k_1=\lambda_1}^{\infty} \sum_{k_2=-\infty}^{-\lambda_2} l_{nk}(f)\, \psi(nx_1 - k_1)\, \psi(nx_2 - k_2)$ is uniformly and absolutely convergent. Therefore it is continuous on $\mathbb{R}^2$.

Next we prove continuity on $\mathbb{R}^2$ of $\sum_{k_1=-\lambda_1+1}^{\lambda_1-1} \sum_{k_2=-\infty}^{-\lambda_2} l_{nk}(f)\, \psi(nx_1 - k_1)\, \psi(nx_2 - k_2)$.

Notice here that

$$\|l_{nk}(f)\|_{\gamma}\, \psi(nx_1 - k_1)\, \psi(nx_2 - k_2) \leq \left\| \|f\|_{\gamma} \right\|_{\infty} \psi(nx_1 - k_1)\, \psi(k_2 + \lambda_2)$$

$$\leq \left\| \|f\|_{\gamma} \right\|_{\infty} \psi(0)\, \psi(k_2 + \lambda_2) = 0.319 \cdot \left\| \|f\|_{\gamma} \right\|_{\infty} \psi(k_2 + \lambda_2),$$

and

$$0.319 \cdot \left\| \|f\|_{\gamma} \right\|_{\infty} \left(\sum_{k_1=-\lambda_1+1}^{\lambda_1-1} 1 \right) \left(\sum_{k_2=-\infty}^{-\lambda_2} \psi(k_2 + \lambda_2) \right) =$$

$$0.319 \cdot \left\| \|f\|_{\gamma} \right\|_{\infty} (2\lambda_1 - 1) \left(\sum_{k_2'=-\infty}^{0} \psi\left(k_2'\right) \right) \leq 18.31 \cdot (2\lambda_1 - 1) \left\| \|f\|_{\gamma} \right\|_{\infty}. \tag{5.98}$$

So the double series under consideration is uniformly convergent and continuous. Clearly $F_n(f, x_1, x_2)$ is proved to be continuous on $\mathbb{R}^2$.

Similarly reasoning one can prove easily now, but with more tedious work, that $F_n(f, x_1, ..., x_N)$ is continuous on $\mathbb{R}^N$, for any $N \geq 1$. We choose to omit this similar extra work. ■

Remark 5.22 By (5.27) it is obvious that $\left\| \|A_n(f)\|_{\gamma} \right\|_{\infty} \leq \left\| \|f\|_{\gamma} \right\|_{\infty} < \infty$, and $A_n(f) \in C\left(\prod_{i=1}^{N} [a_i, b_i], X \right)$, given that $f \in C\left(\prod_{i=1}^{N} [a_i, b_i], X \right)$.

Call L_n any of the operators A_n, B_n, C_n, D_n.

Clearly then

$$\left\| \left\| L_n^2(f) \right\|_{\gamma} \right\|_{\infty} = \left\| \|L_n(L_n(f))\|_{\gamma} \right\|_{\infty} \leq \left\| \|L_n(f)\|_{\gamma} \right\|_{\infty} \leq \left\| \|f\|_{\gamma} \right\|_{\infty}, \tag{5.99}$$

etc.

Therefore we get

$$\left\| \left\| L_n^k(f) \right\|_{\gamma} \right\|_{\infty} \leq \left\| \|f\|_{\gamma} \right\|_{\infty}, \ \forall\, k \in \mathbb{N}, \tag{5.100}$$

the contraction property.

Also we see that

$$\left\| \left\| L_n^k(f) \right\|_\gamma \right\|_\infty \leq \left\| \left\| L_n^{k-1}(f) \right\|_\gamma \right\|_\infty \leq \ldots \leq \left\| \left\| L_n(f) \right\|_\gamma \right\|_\infty \leq \left\| \left\| f \right\|_\gamma \right\|_\infty . \tag{5.101}$$

Here L_n^k are bounded linear operators.

Notation Here $N \in \mathbb{N}$, $0 < \beta < 1$. Denote by

$$c_N := \begin{cases} (4.9737)^N, \text{ if } L_n = A_n, \\ 1, \text{ if } L_n = B_n, C_n, D_n, \end{cases} \tag{5.102}$$

$$\varphi(n) := \begin{cases} \frac{1}{n^\beta}, \text{ if } L_n = A_n, B_n, \\ \frac{1}{n} + \frac{1}{n^\beta}, \text{ if } L_n = C_n, D_n, \end{cases} \tag{5.103}$$

$$\Omega := \begin{cases} C\left(\prod_{i=1}^N [a_i, b_i], X\right), \text{ if } L_n = A_n, \\ C_B\left(\mathbb{R}^N, X\right), \text{ if } L_n = B_n, C_n, D_n, \end{cases} \tag{5.104}$$

and

$$Y := \begin{cases} \prod_{i=1}^N [a_i, b_i], \text{ if } L_n = A_n, \\ \mathbb{R}^N, \text{ if } L_n = B_n, C_n, D_n. \end{cases} \tag{5.105}$$

We give the condensed

Theorem 5.23 *Let $f \in \Omega$, $0 < \beta < 1$, $x \in Y$; $n, N \in \mathbb{N}$ with $n^{1-\beta} > 2$. Then*
(i)

$$\left\| L_n(f, x) - f(x) \right\|_\gamma \leq c_N \left[\omega_1(f, \varphi(n)) + \frac{4 \left\| \left\| f \right\|_\gamma \right\|_\infty}{\pi^2 \left(n^{1-\beta} - 2\right)} \right] =: \tau(n), \tag{5.106}$$

where ω_1 is for $p = \infty$,
and
(ii)

$$\left\| \left\| L_n(f) - f \right\|_\gamma \right\|_\infty \leq \tau(n) \to 0, \text{ as } n \to \infty. \tag{5.107}$$

For f uniformly continuous and in Ω we obtain

$$\lim_{n \to \infty} L_n(f) = f,$$

pointwise and uniformly.

Proof By Theorems 5.8, 5.17, 5.18, 5.19. ∎

Next we do iterated neural network approximation (see also [9]).
We make

Remark 5.24 Let $r \in \mathbb{N}$ and L_n as above. We observe that

$$L_n^r f - f = \left(L_n^r f - L_n^{r-1} f\right) + \left(L_n^{r-1} f - L_n^{r-2} f\right) +$$

$$\left(L_n^{r-2} f - L_n^{r-3} f\right) + ... + \left(L_n^2 f - L_n f\right) + \left(L_n f - f\right).$$

Then

$$\left\| \left\| L_n^r f - f \right\|_\gamma \right\|_\infty \leq \left\| \left\| L_n^r f - L_n^{r-1} f \right\|_\gamma \right\|_\infty + \left\| \left\| L_n^{r-1} f - L_n^{r-2} f \right\|_\gamma \right\|_\infty +$$

$$\left\| \left\| L_n^{r-2} f - L_n^{r-3} f \right\|_\gamma \right\|_\infty + ... + \left\| \left\| L_n^2 f - L_n f \right\|_\gamma \right\|_\infty + \left\| \left\| L_n f - f \right\|_\gamma \right\|_\infty =$$

$$\left\| \left\| L_n^{r-1} \left(L_n f - f\right) \right\|_\gamma \right\|_\infty + \left\| \left\| L_n^{r-2} \left(L_n f - f\right) \right\|_\gamma \right\|_\infty + \left\| \left\| L_n^{r-3} \left(L_n f - f\right) \right\|_\gamma \right\|_\infty$$

$$+ ... + \left\| \left\| L_n \left(L_n f - f\right) \right\|_\gamma \right\|_\infty + \left\| \left\| L_n f - f \right\|_\gamma \right\|_\infty \leq r \left\| \left\| L_n f - f \right\|_\gamma \right\|_\infty. \tag{5.108}$$

That is

$$\left\| \left\| L_n^r f - f \right\|_\gamma \right\|_\infty \leq r \left\| \left\| L_n f - f \right\|_\gamma \right\|_\infty. \tag{5.109}$$

We give

Theorem 5.25 *All here as in Theorem 5.23 and $r \in \mathbb{N}$, $\tau(n)$ as in (5.106). Then*

$$\left\| \left\| L_n^r f - f \right\|_\gamma \right\|_\infty \leq r\tau(n). \tag{5.110}$$

So that the speed of convergence to the unit operator of L_n^r is not worse than of L_n.

Proof By (5.109) and (5.107). ∎

We make

Remark 5.26 Let $m_1, ..., m_r \in \mathbb{N} : m_1 \leq m_2 \leq ... \leq m_r$, $0 < \beta < 1$, $f \in \Omega$. Then $\varphi(m_1) \geq \varphi(m_2) \geq ... \geq \varphi(m_r)$, φ as in (5.103).
Therefore

$$\omega_1\left(f, \varphi(m_1)\right) \geq \omega_1\left(f, \varphi(m_2)\right) \geq ... \geq \omega_1\left(f, \varphi(m_r)\right). \tag{5.111}$$

Assume further that $m_i^{1-\beta} > 2$, $i = 1, ..., r$. Then

$$\frac{2}{\pi^2\left(m_1^{1-\beta}-2\right)} \geq \frac{2}{\pi^2\left(m_2^{1-\beta}-2\right)} \geq \ldots \geq \frac{2}{\pi^2\left(m_r^{1-\beta}-2\right)}. \tag{5.112}$$

Let L_{m_i} as above, $i = 1, \ldots, r$, all of the same kind.

We write

$$L_{m_r}\left(L_{m_{r-1}}\left(\ldots L_{m_2}\left(L_{m_1}f\right)\right)\right) - f =$$

$$L_{m_r}\left(L_{m_{r-1}}\left(\ldots L_{m_2}\left(L_{m_1}f\right)\right)\right) - L_{m_r}\left(L_{m_{r-1}}\left(\ldots L_{m_2}f\right)\right) +$$

$$L_{m_r}\left(L_{m_{r-1}}\left(\ldots L_{m_2}f\right)\right) - L_{m_r}\left(L_{m_{r-1}}\left(\ldots L_{m_3}f\right)\right) +$$

$$L_{m_r}\left(L_{m_{r-1}}\left(\ldots L_{m_3}f\right)\right) - L_{m_r}\left(L_{m_{r-1}}\left(\ldots L_{m_4}f\right)\right) + \ldots + \tag{5.113}$$

$$L_{m_r}\left(L_{m_{r-1}}f\right) - L_{m_r}f + L_{m_r}f - f =$$

$$L_{m_r}\left(L_{m_{r-1}}\left(\ldots L_{m_2}\right)\right)\left(L_{m_1}f - f\right) + L_{m_r}\left(L_{m_{r-1}}\left(\ldots L_{m_3}\right)\right)\left(L_{m_2}f - f\right) +$$

$$L_{m_r}\left(L_{m_{r-1}}\left(\ldots L_{m_4}\right)\right)\left(L_{m_3}f - f\right) + \ldots + L_{m_r}\left(L_{m_{r-1}}f - f\right) + L_{m_r}f - f.$$

Hence by the triangle inequality property of $\left\|\|\cdot\|_\gamma\right\|_\infty$ we get

$$\left\|\left\|L_{m_r}\left(L_{m_{r-1}}\left(\ldots L_{m_2}\left(L_{m_1}f\right)\right)\right) - f\right\|_\gamma\right\|_\infty \leq$$

$$\left\|\left\|L_{m_r}\left(L_{m_{r-1}}\left(\ldots L_{m_2}\right)\right)\left(L_{m_1}f - f\right)\right\|_\gamma\right\|_\infty +$$

$$\left\|\left\|L_{m_r}\left(L_{m_{r-1}}\left(\ldots L_{m_3}\right)\right)\left(L_{m_2}f - f\right)\right\|_\gamma\right\|_\infty +$$

$$\left\|\left\|L_{m_r}\left(L_{m_{r-1}}\left(\ldots L_{m_4}\right)\right)\left(L_{m_3}f - f\right)\right\|_\gamma\right\|_\infty + \ldots +$$

$$\left\|\left\|L_{m_r}\left(L_{m_{r-1}}f - f\right)\right\|_\gamma\right\|_\infty + \left\|\left\|L_{m_r}f - f\right\|_\gamma\right\|_\infty$$

(repeatedly applying (5.99))

$$\leq \left\|\left\|L_{m_1}f - f\right\|_\gamma\right\|_\infty + \left\|\left\|L_{m_2}f - f\right\|_\gamma\right\|_\infty + \left\|\left\|L_{m_3}f - f\right\|_\gamma\right\|_\infty + \ldots +$$

$$\left\|\left\|L_{m_{r-1}}f - f\right\|_\gamma\right\|_\infty + \left\|\left\|L_{m_r}f - f\right\|_\gamma\right\|_\infty = \sum_{i=1}^{r}\left\|\left\|L_{m_i}f - f\right\|_\gamma\right\|_\infty. \tag{5.114}$$

That is, we proved

$$\left\| \left\| L_{m_r}\left(L_{m_{r-1}}\left(...L_{m_2}\left(L_{m_1}f\right)\right)\right) - f \right\|_{\gamma} \right\|_{\infty} \leq \sum_{i=1}^{r} \left\| \left\| L_{m_i}f - f \right\|_{\gamma} \right\|_{\infty}. \quad (5.115)$$

We give

Theorem 5.27 *Let $f \in \Omega$; $N, m_1, m_2, ..., m_r \in \mathbb{N} : m_1 \leq m_2 \leq ... \leq m_r, 0 < \beta < 1$; $m_i^{1-\beta} > 2$, $i = 1, ..., r$, $x \in Y$, and let $\left(L_{m_1}, ..., L_{m_r}\right)$ as $\left(A_{m_1}, ..., A_{m_r}\right)$ or $\left(B_{m_1}, ..., B_{m_r}\right)$ or $\left(C_{m_1}, ..., C_{m_r}\right)$ or $\left(D_{m_1}, ..., D_{m_r}\right)$, $p = \infty$. Then*

$$\left\| L_{m_r}\left(L_{m_{r-1}}\left(...L_{m_2}\left(L_{m_1}f\right)\right)\right)(x) - f(x) \right\|_{\gamma} \leq$$

$$\left\| \left\| L_{m_r}\left(L_{m_{r-1}}\left(...L_{m_2}\left(L_{m_1}f\right)\right)\right) - f \right\|_{\gamma} \right\|_{\infty} \leq$$

$$\sum_{i=1}^{r} \left\| \left\| L_{m_i}f - f \right\|_{\gamma} \right\|_{\infty} \leq$$

$$c_N \sum_{i=1}^{r} \left[\omega_1\left(f, \varphi\left(m_i\right)\right) + \frac{4 \left\| \left\| f \right\|_{\gamma} \right\|_{\infty}}{\pi^2 \left(m_i^{1-\beta} - 2\right)} \right] \leq$$

$$r c_N \left[\omega_1\left(f, \varphi\left(m_1\right)\right) + \frac{4 \left\| \left\| f \right\|_{\gamma} \right\|_{\infty}}{\pi^2 \left(m_1^{1-\beta} - 2\right)} \right]. \quad (5.116)$$

Clearly, we notice that the speed of convergence to the unit operator of the multiply iterated operator is not worse than the speed of L_{m_1}.

Proof Using (5.115), (5.111), (5.112) and (5.106), (5.107). ■

We continue with

Theorem 5.28 *Let all as in Corollary 5.16, and $r \in \mathbb{N}$. Here $\varphi_3(n)$ is as in (5.78). Then*

$$\left\| \left\| A_n^r f - f \right\|_{\gamma} \right\|_{\infty} \leq r \left\| \left\| A_n f - f \right\|_{\gamma} \right\|_{\infty} \leq r \varphi_3(n). \quad (5.117)$$

Proof By (5.109) and (5.78). ■

Application 5.29 *A typical application of all of our results is when $\left(X, \|\cdot\|_{\gamma}\right) = (\mathbb{C}, |\cdot|)$, where $\mathbb{C}$ are the complex numbers.*

References

1. G.A. Anastassiou, *Moments in Probability and Approximation Theory, Pitman Research Notes in Math*, vol. 287 (Longman Sci. & Tech, Harlow, U.K., 1993)
2. G.A. Anastassiou, Rate of convergence of some neural network operators to the unit-univariate case. J. Math. Anal. Appl. **212**, 237–262 (1997)
3. G.A. Anastassiou, *Quantitative Approximations* (Chapman & Hall/CRC, Boca Raton, New York, 2001)
4. G.A. Anastassiou, *Inteligent Systems: Approximation by Artificial Neural Networks, Intelligent Systems Reference Library*, vol. 19 (Springer, Heidelberg, 2011)
5. G.A. Anastassiou, Univariate hyperbolic tangent neural network approximation. Math. Comput. Modell. **53**, 1111–1132 (2011)
6. G.A. Anastassiou, Multivariate hyperbolic tangent neural network approximation. Comput. Math. **61**, 809–821 (2011)
7. G.A. Anastassiou, Multivariate sigmoidal neural network approximation. Neural Netw. **24**, 378–386 (2011)
8. G.A. Anastassiou, Univariate sigmoidal neural network approximation. J. Comput. Anal. Appl. **14**(4), 659–690 (2012)
9. G.A. Anastassiou, Approximation by neural networks iterates, in *Advances in Applied Mathematics and Approximation Theory*, ed. by G. Anastassiou, O. Duman. Springer Proceedings in Mathematics & Statistics (Springer, New York, 2013), pp. 1–20
10. G. Anastassiou, *Intelligent Systems II: Complete Approximation by Neural Network Operators* (Springer, Heidelberg, New York, 2016)
11. G. Anastassiou, *Intelligent Computations: Abstract Fractional Calculus* (Inequalities, Approximations, Springer, Heidelberg, New York, 2018)
12. G.A. Anastassiou, *General multivariate arctangent function activated neural network approximations* (submitted) (2022)
13. H. Cartan, *Differential Calculus* (Hermann, Paris, 1971)
14. Z. Chen, F. Cao, The approximation operators with sigmoidal functions. Comput. Math. Appl. **58**, 758–765 (2009)
15. D. Costarelli, R. Spigler, Approximation results for neural network operators activated by sigmoidal functions. Neural Netw. **44**, 101–106 (2013)
16. D. Costarelli, R. Spigler, Multivariate neural network operators with sigmoidal activation functions. Neural Netw. **48**, 72–77 (2013)
17. S. Haykin, *Neural Networks: A Comprehensive Foundation*, 2nd edn. (Prentice Hall, New York, 1998)
18. W. McCulloch, W. Pitts, A logical calculus of the ideas immanent in nervous activity. Bull. Math. Biophys. **7**, 115–133 (1943)
19. T.M. Mitchell, *Machine Learning* (WCB-McGraw-Hill, New York, 1997)
20. L.B. Rall, *Computational Solution of Nonlinear Operator Equations* (Wiley, New York, 1969)

Chapter 6
Abstract Multivariate Gudermannian Function Induced Neural Network Approximations

Here we present multivariate quantitative approximations of Banach space valued continuous multivariate functions on a box or $\mathbb{R}^N$, $N \in \mathbb{N}$, by the multivariate normalized, quasi-interpolation, Kantorovich type and quadrature type neural network operators. We examine also the case of approximation by iterated operators of the last four types. These approximations are achieved by establishing multidimensional Jackson type inequalities involving the multivariate modulus of continuity of the engaged function or its high order Fréchet derivatives. Our multivariate operators are defined by using a multidimensional density function induced by the Gudermannian sigmoid function. The approximations are pointwise and uniform. The related feed-forward neural network is with one hidden layer. It follows [14].

6.1 Introduction

G. A. Anastassiou in [2, 3], see Chaps. 2–5, was the first to establish neural network approximations to continuous functions with rates by very specifically defined neural network operators of Cardaliaguet-Euvrard and "Squashing" types, by employing the modulus of continuity of the engaged function or its high order derivative, and producing very tight Jackson type inequalities. He treats there both the univariate and multivariate cases. The defining these operators "bell-shaped" and "squashing" functions are assumed to be of compact support. Also in [3] he gives the Nth order asymptotic expansion for the error of weak approximation of these two operators to a special natural class of smooth functions, see Chaps. 4 and 5 there.

Motivations for this chapter are the articles [18] of Z. Chen and F. Cao, and [4–12, 15, 16, 19, 20].

Here we perform multivariate Gudermannian sigmoid function based neural network approximations to continuous functions over boxes or over the whole $\mathbb{R}^N$,

© The Author(s), under exclusive license to Springer Nature Switzerland AG 2023
G. A. Anastassiou, *Banach Space Valued Neural Network*,
Studies in Computational Intelligence 1062,
https://doi.org/10.1007/978-3-031-16400-2_6

$N \in \mathbb{N}$, and also iterated approximations. All convergences here are with rates expressed via the multivariate modulus of continuity of the involved function or its high order Fréchet derivative and given by very tight multidimensional Jackson type inequalities.

We come up with the "right" precisely defined multivariate normalized, quasi-interpolation neural network operators related to boxes or $\mathbb{R}^N$, as well as Kantorovich type and quadrature type related operators on $\mathbb{R}^N$. Our boxes are not necessarily symmetric to the origin. In preparation to prove our results we establish important properties of the basic multivariate density function induced by Gudermannian sigmoid function and defining our operators.

Feed-forward neural networks (FNNs) with one hidden layer, the only type of networks we deal with in this chapter, are mathematically expressed as

$$N_n(x) = \sum_{j=0}^{n} c_j \sigma\left(\langle a_j \cdot x \rangle + b_j\right), \quad x \in \mathbb{R}^s, \quad s \in \mathbb{N},$$

where for $0 \leq j \leq n$, $b_j \in \mathbb{R}$ are the thresholds, $a_j \in \mathbb{R}^s$ are the connection weights, $c_j \in \mathbb{R}$ are the coefficients, $\langle a_j \cdot x \rangle$ is the inner product of a_j and x, and σ is the activation function of the network. In many fundamental network models, the activation function is the Gudermannian sigmoid function. About neural networks see [21–23].

6.2 Background

See also [13, 25].

Here we consider $gd(x)$ the Gudermannian function [25], which is a sigmoid function, as a generator function:

$$\sigma(x) = 2\arctan\left(\tanh\left(\frac{x}{2}\right)\right) = \int_0^x \frac{dt}{\cosh t} =: gd(x), x \in \mathbb{R}. \tag{6.1}$$

Let the normalized generator sigmoid function

$$f(x) := \frac{4}{\pi}\sigma(x) = \frac{4}{\pi}\int_0^x \frac{dt}{\cosh t} = \frac{8}{\pi}\int_0^x \frac{1}{e^t + e^{-t}}dt, \quad x \in \mathbb{R}. \tag{6.2}$$

Here

$$f'(x) = \frac{4}{\pi \cosh x} > 0, \quad \forall x \in \mathbb{R},$$

hence f is strictly increasing on $\mathbb{R}$.

Notice that $\tanh(-x) = -\tanh x$ and $\arctan(-x) = -\arctan x$, $x \in \mathbb{R}$.

So, here the neural network activation function will be:

$$W(x) = \frac{1}{4}[f(x+1) - f(x-1)], x \in \mathbb{R}. \tag{6.3}$$

By [3], we get that

$$W(x) = W(-x), \quad \forall x \in \mathbb{R}, \tag{6.4}$$

i.e. it is even and symmetric with respect to the y-axis. Here we have $f(+\infty) = 1$, $f(-\infty) = -1$ and $f(0) = 0$. Clearly it is

$$f(-x) = -f(x), \; \forall x \in \mathbb{R}, \tag{6.5}$$

an odd function, symmetric with respect to the origin. Since $x + 1 > x - 1$, and $f(x+1) > f(x-1)$, we obtain $W(x) > 0$, $\forall x \in \mathbb{R}$.

By [13], we have that

$$W(0) = \frac{2}{\pi} gd(1) \cong 0.551. \tag{6.6}$$

By [13] W is strictly decreasing on $(0, +\infty)$, and strictly increasing on $(-\infty, 0)$, and $W'(0) = 0$.

Also we have that

$$\lim_{x \to +\infty} W(x) = \lim_{x \to -\infty} W(x) = 0, \tag{6.7}$$

that is the x-axis is the horizontal asymptote for W.

Conclusion, W is a bell shaped symmetric function with maximum $\psi(0) \cong 0.551$.

We need

Theorem 6.1 ([13]) *It holds that*

$$\sum_{i=-\infty}^{\infty} W(x-i) = 1, \; \forall x \in \mathbb{R}. \tag{6.8}$$

Theorem 6.2 ([13]) *We have that*

$$\int_{-\infty}^{\infty} W(x)\, dx = 1. \tag{6.9}$$

So $W(x)$ is a density function.

Theorem 6.3 ([13]) *Let $0 < \alpha < 1$, and $n \in \mathbb{N}$ with $n^{1-\alpha} > 2$. It holds*

$$\sum_{\substack{k=-\infty \\ : |nx-k| \geq n^{1-\alpha}}}^{\infty} W(nx-k) < \frac{4}{\pi e^{(n^{1-\alpha}-2)}} = \frac{4e^2}{\pi e^{n^{1-\alpha}}}. \tag{6.10}$$

Denote by $\lfloor \cdot \rfloor$ the integral part of the number and by $\lceil \cdot \rceil$ the ceiling of the number.

Theorem 6.4 ([13]) *Let* $[a, b] \subset \mathbb{R}$ *and* $n \in \mathbb{N}$, *so that* $\lceil na \rceil \leq \lfloor nb \rfloor$. *It holds*

$$\frac{1}{\sum_{k=\lceil na \rceil}^{\lfloor nb \rfloor} W(nx - k)} < \frac{\pi}{gd(2)} \cong 2.412, \tag{6.11}$$

$\forall\, x \in [a, b]$.

We make

Remark 6.5 ([13]) (i) We have that

$$\lim_{n \to \infty} \sum_{k=\lceil na \rceil}^{\lfloor nb \rfloor} W(nx - k) \neq 1, \tag{6.12}$$

for at least some $x \in [a, b]$.

(ii) Let $[a, b] \subset \mathbb{R}$. For large n we always have $\lceil na \rceil \leq \lfloor nb \rfloor$. Also $a \leq \frac{k}{n} \leq b$, iff $\lceil na \rceil \leq k \leq \lfloor nb \rfloor$.

In general it holds

$$\sum_{k=\lceil na \rceil}^{\lfloor nb \rfloor} W(nx - k) \leq 1. \tag{6.13}$$

We introduce

$$Z(x_1, ..., x_N) := Z(x) := \prod_{i=1}^{N} W(x_i), \quad x = (x_1, ..., x_N) \in \mathbb{R}^N, \ N \in \mathbb{N}. \tag{6.14}$$

It has the properties:

(i) $Z(x) > 0, \ \forall\, x \in \mathbb{R}^N$,

(ii)

$$\sum_{k=-\infty}^{\infty} Z(x - k) := \sum_{k_1=-\infty}^{\infty} \sum_{k_2=-\infty}^{\infty} ... \sum_{k_N=-\infty}^{\infty} Z(x_1 - k_1, ..., x_N - k_N) = 1, \tag{6.15}$$

where $k := (k_1, ..., k_n) \in \mathbb{Z}^N$, $\forall\, x \in \mathbb{R}^N$,

hence

(iii)

$$\sum_{k=-\infty}^{\infty} Z(nx - k) = 1, \tag{6.16}$$

$\forall\, x \in \mathbb{R}^N$; $n \in \mathbb{N}$,

and
(iv)

$$\int_{\mathbb{R}^N} Z(x)\,dx = 1, \tag{6.17}$$

that is Z is a multivariate density function.

Here denote $\|x\|_\infty := \max\{|x_1|, ..., |x_N|\}$, $x \in \mathbb{R}^N$, also set $\infty := (\infty, ..., \infty)$, $-\infty := (-\infty, ..., -\infty)$ upon the multivariate context, and

$$\begin{aligned} \lceil na \rceil &:= (\lceil na_1 \rceil, ..., \lceil na_N \rceil), \\ \lfloor nb \rfloor &:= (\lfloor nb_1 \rfloor, ..., \lfloor nb_N \rfloor), \end{aligned} \tag{6.18}$$

where $a := (a_1, ..., a_N)$, $b := (b_1, ..., b_N)$.

We obviously see that

$$\sum_{k=\lceil na \rceil}^{\lfloor nb \rfloor} Z(nx - k) = \sum_{k=\lceil na \rceil}^{\lfloor nb \rfloor} \left(\prod_{i=1}^{N} W(nx_i - k_i) \right) =$$

$$\sum_{k_1=\lceil na_1 \rceil}^{\lfloor nb_1 \rfloor} \cdots \sum_{k_N=\lceil na_N \rceil}^{\lfloor nb_N \rfloor} \left(\prod_{i=1}^{N} W(nx_i - k_i) \right) = \prod_{i=1}^{N} \left(\sum_{k_i=\lceil na_i \rceil}^{\lfloor nb_i \rfloor} W(nx_i - k_i) \right). \tag{6.19}$$

For $0 < \beta < 1$ and $n \in \mathbb{N}$, a fixed $x \in \mathbb{R}^N$, we have that

$$\sum_{k=\lceil na \rceil}^{\lfloor nb \rfloor} Z(nx - k) =$$

$$\sum_{\begin{cases} k = \lceil na \rceil \\ \left\| \frac{k}{n} - x \right\|_\infty \le \frac{1}{n^\beta} \end{cases}}^{\lfloor nb \rfloor} Z(nx - k) + \sum_{\begin{cases} k = \lceil na \rceil \\ \left\| \frac{k}{n} - x \right\|_\infty > \frac{1}{n^\beta} \end{cases}}^{\lfloor nb \rfloor} Z(nx - k). \tag{6.20}$$

In the last two sums the counting is over disjoint vector sets of k's, because the condition $\left\| \frac{k}{n} - x \right\|_\infty > \frac{1}{n^\beta}$ implies that there exists at least one $\left| \frac{k_r}{n} - x_r \right| > \frac{1}{n^\beta}$, where $r \in \{1, ..., N\}$.

(v) As in [10], pp. 379–380, we derive that

$$\sum_{\begin{cases} k = \lceil na \rceil \\ \left\| \frac{k}{n} - x \right\|_\infty > \frac{1}{n^\beta} \end{cases}}^{\lfloor nb \rfloor} Z(nx - k) \overset{(6.10)}{<} \frac{4e^2}{\pi e^{n^{1-\beta}}}, \quad 0 < \beta < 1, \tag{6.21}$$

with $n \in \mathbb{N} : n^{1-\beta} > 2$, $x \in \prod_{i=1}^{N} [a_i, b_i]$.

(vi) By Theorem 6.4 we get that

$$0 < \frac{1}{\sum_{k=\lceil na \rceil}^{\lfloor nb \rfloor} Z(nx - k)} < \left(\frac{\pi}{gd(2)} \right)^N \cong (2.412)^N, \tag{6.22}$$

$\forall\, x \in \left(\prod_{i=1}^{N} [a_i, b_i] \right)$, $n \in \mathbb{N}$.

It is also clear that

(vii)

$$\sum_{\substack{k=-\infty \\ \left\| \frac{k}{n} - x \right\|_\infty > \frac{1}{n^\beta}}}^{\infty} Z(nx - k) < \frac{4e^2}{\pi e^{n^{1-\beta}}}, \tag{6.23}$$

$0 < \beta < 1$, $n \in \mathbb{N} : n^{1-\beta} > 2$, $x \in \mathbb{R}^N$.

Furthermore it holds

$$\lim_{n \to \infty} \sum_{k=\lceil na \rceil}^{\lfloor nb \rfloor} Z(nx - k) \neq 1, \tag{6.24}$$

for at least some $x \in \left(\prod_{i=1}^{N} [a_i, b_i] \right)$.

Here $(X, \|\cdot\|_\gamma)$ is a Banach space.

Let $f \in C\left(\prod_{i=1}^{N} [a_i, b_i], X \right)$, $x = (x_1, ..., x_N) \in \prod_{i=1}^{N} [a_i, b_i]$, $n \in \mathbb{N}$ such that $\lceil na_i \rceil \leq \lfloor nb_i \rfloor$, $i = 1, ..., N$.

We introduce and define the following multivariate linear normalized neural network operator ($x := (x_1, ..., x_N) \in \left(\prod_{i=1}^{N} [a_i, b_i] \right)$):

$$A_n(f, x_1, ..., x_N) := A_n(f, x) := \frac{\sum_{k=\lceil na \rceil}^{\lfloor nb \rfloor} f\left(\frac{k}{n}\right) Z(nx - k)}{\sum_{k=\lceil na \rceil}^{\lfloor nb \rfloor} Z(nx - k)} =$$

$$\frac{\sum_{k_1=\lceil na_1 \rceil}^{\lfloor nb_1 \rfloor} \sum_{k_2=\lceil na_2 \rceil}^{\lfloor nb_2 \rfloor} \cdots \sum_{k_N=\lceil na_N \rceil}^{\lfloor nb_N \rfloor} f\left(\frac{k_1}{n}, ..., \frac{k_N}{n}\right) \left(\prod_{i=1}^{N} W(nx_i - k_i) \right)}{\prod_{i=1}^{N} \left(\sum_{k_i=\lceil na_i \rceil}^{\lfloor nb_i \rfloor} W(nx_i - k_i) \right)}. \tag{6.25}$$

For large enough $n \in \mathbb{N}$ we always obtain $\lceil na_i \rceil \leq \lfloor nb_i \rfloor$, $i = 1, ..., N$. Also $a_i \leq \frac{k_i}{n} \leq b_i$, iff $\lceil na_i \rceil \leq k_i \leq \lfloor nb_i \rfloor$, $i = 1, ..., N$.

When $g \in C\left(\prod_{i=1}^{N} [a_i, b_i] \right)$ we define the companion operator

$$\widetilde{A}_n(g, x) := \frac{\sum_{k=\lceil na \rceil}^{\lfloor nb \rfloor} g\left(\frac{k}{n}\right) Z(nx - k)}{\sum_{k=\lceil na \rceil}^{\lfloor nb \rfloor} Z(nx - k)}. \tag{6.26}$$

Clearly $\widetilde{A}_n$ is a positive linear operator. We have that

$$\widetilde{A}_n(1,x) = 1, \ \forall\, x \in \left(\prod_{i=1}^{N} [a_i, b_i] \right).$$

Notice that $A_n(f) \in C\left(\prod_{i=1}^{N}[a_i,b_i], X\right)$ and $\widetilde{A}_n(g) \in C\left(\prod_{i=1}^{N}[a_i,b_i]\right)$.

Furthermore it holds

$$\|A_n(f,x)\|_\gamma \le \frac{\sum_{k=\lceil na \rceil}^{\lfloor nb \rfloor} \left\| f\left(\frac{k}{n}\right)\right\|_\gamma Z(nx-k)}{\sum_{k=\lceil na \rceil}^{\lfloor nb \rfloor} Z(nx-k)} = \widetilde{A}_n\left(\|f\|_\gamma, x\right), \tag{6.27}$$

$\forall\, x \in \prod_{i=1}^{N}[a_i,b_i]$.

Clearly $\|f\|_\gamma \in C\left(\prod_{i=1}^{N}[a_i,b_i]\right)$.

So, we have that

$$\|A_n(f,x)\|_\gamma \le \widetilde{A}_n\left(\|f\|_\gamma, x\right), \tag{6.28}$$

$\forall\, x \in \prod_{i=1}^{N}[a_i,b_i]$, $\forall\, n \in \mathbb{N}$, $\forall\, f \in C\left(\prod_{i=1}^{N}[a_i,b_i], X\right)$.

Let $c \in X$ and $g \in C\left(\prod_{i=1}^{N}[a_i,b_i]\right)$, then $cg \in C\left(\prod_{i=1}^{N}[a_i,b_i], X\right)$.

Furthermore it holds

$$A_n(cg,x) = c\widetilde{A}_n(g,x), \ \forall\, x \in \prod_{i=1}^{N}[a_i,b_i]. \tag{6.29}$$

Since $\widetilde{A}_n(1) = 1$, we get that

$$A_n(c) = c, \ \forall\, c \in X. \tag{6.30}$$

We call $\widetilde{A}_n$ the companion operator of A_n.

For convenience we call

$$A_n^*(f,x) := \sum_{k=\lceil na \rceil}^{\lfloor nb \rfloor} f\left(\frac{k}{n}\right) Z(nx-k) =$$

$$\sum_{k_1=\lceil na_1 \rceil}^{\lfloor nb_1 \rfloor} \sum_{k_2=\lceil na_2 \rceil}^{\lfloor nb_2 \rfloor} \cdots \sum_{k_N=\lceil na_N \rceil}^{\lfloor nb_N \rfloor} f\left(\frac{k_1}{n}, \ldots, \frac{k_N}{n}\right) \left(\prod_{i=1}^{N} W(nx_i - k_i) \right), \tag{6.31}$$

$\forall\, x \in \left(\prod_{i=1}^{N}[a_i,b_i]\right)$.

That is

$$A_n(f,x) := \frac{A_n^*(f,x)}{\sum_{k=\lceil na\rceil}^{\lfloor nb\rfloor} Z(nx-k)}, \tag{6.32}$$

$\forall\, x \in \left(\prod_{i=1}^N [a_i, b_i]\right)$, $n \in \mathbb{N}$.

Hence

$$A_n(f,x) - f(x) = \frac{A_n^*(f,x) - f(x)\left(\sum_{k=\lceil na\rceil}^{\lfloor nb\rfloor} Z(nx-k)\right)}{\sum_{k=\lceil na\rceil}^{\lfloor nb\rfloor} Z(nx-k)}. \tag{6.33}$$

Consequently we derive

$$\|A_n(f,x) - f(x)\|_\gamma \overset{(6.22)}{\leq} (2.412)^N \left\| A_n^*(f,x) - f(x) \sum_{k=\lceil na\rceil}^{\lfloor nb\rfloor} Z(nx-k) \right\|_\gamma, \tag{6.34}$$

$\forall\, x \in \left(\prod_{i=1}^N [a_i, b_i]\right)$.

We will estimate the right hand side of (6.34).

For the last and others we need

Definition 6.6 ([11], p. 274) Let M be a convex and compact subset of $\left(\mathbb{R}^N, \|\cdot\|_p\right)$, $p \in [1,\infty]$, and $\left(X, \|\cdot\|_\gamma\right)$ be a Banach space. Let $f \in C(M,X)$. We define the first modulus of continuity of f as

$$\omega_1(f,\delta) := \sup_{\substack{x, y \in M: \\ \|x-y\|_p \leq \delta}} \|f(x) - f(y)\|_\gamma, \quad 0 < \delta \leq diam(M). \tag{6.35}$$

If $\delta > diam(M)$, then

$$\omega_1(f,\delta) = \omega_1(f, diam(M)). \tag{6.36}$$

Notice $\omega_1(f,\delta)$ is increasing in $\delta > 0$. For $f \in C_B(M,X)$ (continuous and bounded functions) $\omega_1(f,\delta)$ is defined similarly.

Lemma 6.7 ([11], p. 274) *We have* $\omega_1(f,\delta) \to 0$ *as* $\delta \downarrow 0$, *iff* $f \in C(M,X)$, *where* M *is a convex compact subset of* $\left(\mathbb{R}^N, \|\cdot\|_p\right)$, $p \in [1,\infty]$.

Clearly we have also: $f \in C_U\left(\mathbb{R}^N, X\right)$ (uniformly continuous functions), iff $\omega_1(f,\delta) \to 0$ as $\delta \downarrow 0$, where ω_1 is defined similarly to (6.35). The space $C_B\left(\mathbb{R}^N, X\right)$ denotes the continuous and bounded functions on $\mathbb{R}^N$.

When $f \in C_B\left(\mathbb{R}^N, X\right)$ we define,

$$B_n(f,x) := B_n(f,x_1,...,x_N) := \sum_{k=-\infty}^{\infty} f\left(\frac{k}{n}\right) Z(nx-k) :=$$

$$\sum_{k_1=-\infty}^{\infty}\sum_{k_2=-\infty}^{\infty}...\sum_{k_N=-\infty}^{\infty} f\left(\frac{k_1}{n},\frac{k_2}{n},...,\frac{k_N}{n}\right)\left(\prod_{i=1}^{N} W(nx_i-k_i)\right), \tag{6.37}$$

$n \in \mathbb{N}, \forall\, x \in \mathbb{R}^N$, $N \in \mathbb{N}$, the multivariate quasi-interpolation neural network operator.

Also for $f \in C_B(\mathbb{R}^N, X)$ we define the multivariate Kantorovich type neural network operator

$$C_n(f,x) := C_n(f,x_1,...,x_N) := \sum_{k=-\infty}^{\infty}\left(n^N \int_{\frac{k}{n}}^{\frac{k+1}{n}} f(t)\,dt\right) Z(nx-k) =$$

$$\sum_{k_1=-\infty}^{\infty}\sum_{k_2=-\infty}^{\infty}...\sum_{k_N=-\infty}^{\infty}\left(n^N \int_{\frac{k_1}{n}}^{\frac{k_1+1}{n}}\int_{\frac{k_2}{n}}^{\frac{k_2+1}{n}}...\int_{\frac{k_N}{n}}^{\frac{k_N+1}{n}} f(t_1,...,t_N)\,dt_1...dt_N\right)$$

$$\cdot\left(\prod_{i=1}^{N} W(nx_i-k_i)\right), \tag{6.38}$$

$n \in \mathbb{N}, \ \forall\, x \in \mathbb{R}^N$.

Again for $f \in C_B(\mathbb{R}^N, X)$, $N \in \mathbb{N}$, we define the multivariate neural network operator of quadrature type $D_n(f,x)$, $n \in \mathbb{N}$, as follows.

Let $\theta = (\theta_1,...,\theta_N) \in \mathbb{N}^N$, $r = (r_1,...,r_N) \in \mathbb{Z}_+^N$, $w_r = w_{r_1,r_2,...r_N} \geq 0$, such that $\sum_{r=0}^{\theta} w_r = \sum_{r_1=0}^{\theta_1}\sum_{r_2=0}^{\theta_2}...\sum_{r_N=0}^{\theta_N} w_{r_1,r_2,...r_N} = 1$; $k \in \mathbb{Z}^N$ and

$$\delta_{nk}(f) := \delta_{n,k_1,k_2,...,k_N}(f) := \sum_{r=0}^{\theta} w_r f\left(\frac{k}{n}+\frac{r}{n\theta}\right) =$$

$$\sum_{r_1=0}^{\theta_1}\sum_{r_2=0}^{\theta_2}...\sum_{r_N=0}^{\theta_N} w_{r_1,r_2,...r_N} f\left(\frac{k_1}{n}+\frac{r_1}{n\theta_1},\frac{k_2}{n}+\frac{r_2}{n\theta_2},...,\frac{k_N}{n}+\frac{r_N}{n\theta_N}\right), \tag{6.39}$$

where $\frac{r}{\theta} := \left(\frac{r_1}{\theta_1},\frac{r_2}{\theta_2},...,\frac{r_N}{\theta_N}\right)$.

We set

$$D_n(f,x) := D_n(f,x_1,...,x_N) := \sum_{k=-\infty}^{\infty} \delta_{nk}(f)\, Z(nx-k) = \tag{6.40}$$

$$\sum_{k_1=-\infty}^{\infty}\sum_{k_2=-\infty}^{\infty}\cdots\sum_{k_N=-\infty}^{\infty}\delta_{n,k_1,k_2,\ldots,k_N}(f)\left(\prod_{i=1}^{N}W(nx_i-k_i)\right),$$

$\forall\, x\in\mathbb{R}^N$.

In this chapter we study the approximation properties of A_n, B_n, C_n, D_n neural network operators and as well of their iterates. That is, the quantitative pointwise and uniform convergence of these operators to the unit operator I.

6.3 Multivariate General Neural Network Approximations

Here we present several vectorial neural network approximations to Banach space valued functions given with rates.

We give

Theorem 6.8 *Let* $f\in C\left(\prod_{i=1}^{N}[a_i,b_i],X\right)$, $0<\beta<1$, $x\in\left(\prod_{i=1}^{N}[a_i,b_i]\right)$, $m,N,n\in\mathbb{N}$ *with* $n^{1-\beta}>2$. *Then*

(1)

$$\|A_n(f,x)-f(x)\|_\gamma\leq(2.412)^N\left[\omega_1\left(f,\frac{1}{n^\beta}\right)+\frac{8e^2\left\|\|f\|_\gamma\right\|_\infty}{\pi e^{n^{1-\beta}}}\right]=:\lambda_1(n), \tag{6.41}$$

and

(2)

$$\left\|\|A_n(f)-f\|_\gamma\right\|_\infty\leq\lambda_1(n). \tag{6.42}$$

We notice that $\lim_{n\to\infty}A_n(f)\overset{\|\cdot\|_\gamma}{=}f$, *pointwise and uniformly.*

Above ω_1 *is with respect to* $p=\infty$.

Proof We observe that

$$\Delta(x):=A_n^*(f,x)-f(x)\sum_{k=\lceil na\rceil}^{\lfloor nb\rfloor}Z(nx-k)=$$

$$\sum_{k=\lceil na\rceil}^{\lfloor nb\rfloor}f\left(\frac{k}{n}\right)Z(nx-k)-\sum_{k=\lceil na\rceil}^{\lfloor nb\rfloor}f(x)Z(nx-k)=$$

$$\sum_{k=\lceil na\rceil}^{\lfloor nb\rfloor}\left(f\left(\frac{k}{n}\right)-f(x)\right)Z(nx-k). \tag{6.43}$$

Thus

$$\|\Delta(x)\|_\gamma \le \sum_{k=\lceil na\rceil}^{\lfloor nb\rfloor} \left\| f\left(\frac{k}{n}\right) - f(x)\right\|_\gamma Z(nx-k) =$$

$$\sum_{\begin{cases} k=\lceil na\rceil \\ \left\|\frac{k}{n}-x\right\|_\infty \le \frac{1}{n^\beta}\end{cases}}^{\lfloor nb\rfloor} \left\| f\left(\frac{k}{n}\right) - f(x)\right\|_\gamma Z(nx-k) +$$

$$\sum_{\begin{cases} k=\lceil na\rceil \\ \left\|\frac{k}{n}-x\right\|_\infty > \frac{1}{n^\beta}\end{cases}}^{\lfloor nb\rfloor} \left\| f\left(\frac{k}{n}\right) - f(x)\right\|_\gamma Z(nx-k) \overset{(6.16)}{\le}$$

$$\omega_1\left(f, \frac{1}{n^\beta}\right) + 2\left\|\|f\|_\gamma\right\|_\infty \sum_{\begin{cases} k=\lceil na\rceil \\ \left\|\frac{k}{n}-x\right\|_\infty > \frac{1}{n^\beta}\end{cases}}^{\lfloor nb\rfloor} Z(nx-k) \overset{(6.21)}{\le}$$

$$\omega_1\left(f, \frac{1}{n^\beta}\right) + \frac{8e^2\left\|\|f\|_\gamma\right\|_\infty}{\pi e^{n^{1-\beta}}}. \tag{6.44}$$

So that

$$\|\Delta(x)\|_\gamma \le \omega_1\left(f, \frac{1}{n^\beta}\right) + \frac{8e^2\left\|\|f\|_\gamma\right\|_\infty}{\pi e^{n^{1-\beta}}}. \tag{6.45}$$

Now using (6.34) we finish the proof. ■

We make

Remark 6.9 ([11], pp. 263–266) Let $\left(\mathbb{R}^N, \|\cdot\|_p\right)$, $N \in \mathbb{N}$; where $\|\cdot\|_p$ is the L_p-norm, $1 \le p \le \infty$. $\mathbb{R}^N$ is a Banach space, and $\left(\mathbb{R}^N\right)^j$ denotes the j-fold product space $\mathbb{R}^N \times ... \times \mathbb{R}^N$ endowed with the max-norm $\|x\|_{\left(\mathbb{R}^N\right)^j} := \max_{1\le\lambda\le j} \|x_\lambda\|_p$, where $x := (x_1, ..., x_j) \in \left(\mathbb{R}^N\right)^j$.

Let $(X, \|\cdot\|_\gamma)$ be a general Banach space. Then the space $L_j := L_j\left(\left(\mathbb{R}^N\right)^j; X\right)$ of all j-multilinear continuous maps $g : \left(\mathbb{R}^N\right)^j \to X$, $j = 1, ..., m$, is a Banach space with norm

$$\|g\| := \|g\|_{L_j} := \sup_{\left(\|x\|_{\left(\mathbb{R}^N\right)^j}=1\right)} \|g(x)\|_\gamma = \sup \frac{\|g(x)\|_\gamma}{\|x_1\|_p \cdots \|x_j\|_p}. \tag{6.46}$$

Let M be a non-empty convex and compact subset of $\mathbb{R}^N$ and $x_0 \in M$ is fixed.

Let O be an open subset of $\mathbb{R}^N : M \subset O$. Let $f : O \to X$ be a continuous function, whose Fréchet derivatives (see [24]) $f^{(j)} : O \to L_j = L_j\left(\left(\mathbb{R}^N\right)^j ; X\right)$ exist and are continuous for $1 \leq j \leq m$, $m \in \mathbb{N}$.

Call $(x - x_0)^j := (x - x_0, ..., x - x_0) \in \left(\mathbb{R}^N\right)^j$, $x \in M$.

We will work with $f|_M$.

Then, by Taylor's formula [17], ([24], p. 124), we get

$$f(x) = \sum_{j=0}^{m} \frac{f^{(j)}(x_0)(x - x_0)^j}{j!} + R_m(x, x_0), \text{ all } x \in M, \tag{6.47}$$

where the remainder is the Riemann integral

$$R_m(x, x_0) := \int_0^1 \frac{(1-u)^{m-1}}{(m-1)!} \left(f^{(m)}(x_0 + u(x - x_0)) - f^{(m)}(x_0)\right)(x - x_0)^m du, \tag{6.48}$$

here we set $f^{(0)}(x_0)(x - x_0)^0 = f(x_0)$.

We consider

$$w := \omega_1\left(f^{(m)}, h\right) := \sup_{\substack{x, y \in M: \\ \|x-y\|_p \leq h}} \left\| f^{(m)}(x) - f^{(m)}(y) \right\|, \tag{6.49}$$

$h > 0$.

We obtain

$$\left\| \left(f^{(m)}(x_0 + u(x - x_0)) - f^{(m)}(x_0)\right)(x - x_0)^m \right\|_\gamma \leq$$

$$\left\| f^{(m)}(x_0 + u(x - x_0)) - f^{(m)}(x_0) \right\| \cdot \|x - x_0\|_p^m \leq$$

$$w \|x - x_0\|_p^m \left\lceil \frac{u \|x - x_0\|_p}{h} \right\rceil, \tag{6.50}$$

by Lemma 7.1.1, [1], p. 208, where $\lceil \cdot \rceil$ is the ceiling.

Therefore for all $x \in M$ (see [1], pp. 121–122):

$$\|R_m(x, x_0)\|_\gamma \leq w \|x - x_0\|_p^m \int_0^1 \left\lceil \frac{u \|x - x_0\|_p}{h} \right\rceil \frac{(1-u)^{m-1}}{(m-1)!} du$$

$$= w \Phi_m\left(\|x - x_0\|_p\right) \tag{6.51}$$

by a change of variable, where

$$\Phi_m(t) := \int_0^{|t|} \left\lceil \frac{s}{h} \right\rceil \frac{(|t|-s)^{m-1}}{(m-1)!} ds = \frac{1}{m!} \left(\sum_{j=0}^{\infty} (|t| - jh)_+^m \right), \ \forall t \in \mathbb{R}, \quad (6.52)$$

is a (polynomial) spline function, see [1], pp. 210–211.

Also from there we get

$$\Phi_m(t) \leq \left(\frac{|t|^{m+1}}{(m+1)!h} + \frac{|t|^m}{2m!} + \frac{h|t|^{m-1}}{8(m-1)!} \right), \ \forall t \in \mathbb{R}, \quad (6.53)$$

with equality true only at $t = 0$.

Therefore it holds

$$\|R_m(x, x_0)\|_\gamma \leq w \left(\frac{\|x - x_0\|_p^{m+1}}{(m+1)!h} + \frac{\|x - x_0\|_p^m}{2m!} + \frac{h\|x - x_0\|_p^{m-1}}{8(m-1)!} \right), \ \forall x \in M. \quad (6.54)$$

We have found that

$$\left\| f(x) - \sum_{j=0}^{m} \frac{f^{(j)}(x_0)(x - x_0)^j}{j!} \right\|_\gamma \leq$$

$$\omega_1\left(f^{(m)}, h\right) \left(\frac{\|x - x_0\|_p^{m+1}}{(m+1)!h} + \frac{\|x - x_0\|_p^m}{2m!} + \frac{h\|x - x_0\|_p^{m-1}}{8(m-1)!} \right) < \infty, \quad (6.55)$$

$\forall x, x_0 \in M$.

Here $0 < \omega_1\left(f^{(m)}, h\right) < \infty$, by M being compact and $f^{(m)}$ being continuous on M.

One can rewrite (6.55) as follows:

$$\left\| f(\cdot) - \sum_{j=0}^{m} \frac{f^{(j)}(x_0)(\cdot - x_0)^j}{j!} \right\|_\gamma \leq$$

$$\omega_1\left(f^{(m)}, h\right) \left(\frac{\|\cdot - x_0\|_p^{m+1}}{(m+1)!h} + \frac{\|\cdot - x_0\|_p^m}{2m!} + \frac{h\|\cdot - x_0\|_p^{m-1}}{8(m-1)!} \right), \ \forall x_0 \in M, \quad (6.56)$$

a pointwise functional inequality on M.

Here $(\cdot - x_0)^j$ maps M into $\left(\mathbb{R}^N\right)^j$ and it is continuous, also $f^{(j)}(x_0)$ maps $\left(\mathbb{R}^N\right)^j$ into X and it is continuous. Hence their composition $f^{(j)}(x_0)(\cdot - x_0)^j$ is continuous from M into X.

Clearly $f(\cdot) - \sum_{j=0}^{m} \frac{f^{(j)}(x_0)(\cdot - x_0)^j}{j!} \in C(M, X)$, hence $\left\| f(\cdot) - \sum_{j=0}^{m} \frac{f^{(j)}(x_0)(\cdot - x_0)^j}{j!} \right\|_\gamma \in C(M)$.

Let $\{\widetilde{L}_N\}_{N\in\mathbb{N}}$ be a sequence of positive linear operators mapping $C(M)$ into $C(M)$.

Therefore we obtain

$$\left(\widetilde{L}_N\left(\left\|f(\cdot)-\sum_{j=0}^{m}\frac{f^{(j)}(x_0)(\cdot-x_0)^j}{j!}\right\|_\gamma\right)\right)(x_0)\leq$$

$$\omega_1\left(f^{(m)},h\right)\left[\frac{\left(\widetilde{L}_N\left(\|\cdot-x_0\|_p^{m+1}\right)\right)(x_0)}{(m+1)!h}+\frac{\left(\widetilde{L}_N\left(\|\cdot-x_0\|_p^{m}\right)\right)(x_0)}{2m!}\right.$$

$$\left.+\frac{h\left(\widetilde{L}_N\left(\|\cdot-x_0\|_p^{m-1}\right)\right)(x_0)}{8(m-1)!}\right], \tag{6.57}$$

$\forall\, N\in\mathbb{N}, \forall\, x_0\in M$.

Clearly (6.57) is valid when $M=\prod_{i=1}^{N}[a_i,b_i]$ and $\widetilde{L}_n=\widetilde{A}_n$, see (6.26).

All the above is preparation for the following theorem, where we assume Fréchet differentiability of functions.

This will be a direct application of Theorem 10.2, [11], pp. 268–270. The operators A_n, $\widetilde{A}_n$ fulfill its assumptions, see (6.25), (6.26), (6.28), (6.29) and (6.30).

We present the following high order approximation results.

Theorem 6.10 *Let O open subset of $\left(\mathbb{R}^N,\|\cdot\|_p\right)$, $p\in[1,\infty]$, such that $\prod_{i=1}^{N}[a_i,b_i]\subset O\subseteq\mathbb{R}^N$, and let $\left(X,\|\cdot\|_\gamma\right)$ be a general Banach space. Let $m\in\mathbb{N}$ and $f\in C^m(O,X)$, the space of $\overline{m}$-times continuously Fréchet differentiable functions from O into X. We study the approximation of $f|_{\prod_{i=1}^{N}[a_i,b_i]}$. Let $x_0\in\left(\prod_{i=1}^{N}[a_i,b_i]\right)$ and $r>0$. Then*

(1)

$$\left\|(A_n(f))(x_0)-\sum_{j=0}^{m}\frac{1}{j!}\left(A_n\left(f^{(j)}(x_0)(\cdot-x_0)^j\right)\right)(x_0)\right\|_\gamma\leq$$

$$\frac{\omega_1\left(f^{(m)},r\left(\left(\widetilde{A}_n\left(\|\cdot-x_0\|_p^{m+1}\right)\right)(x_0)\right)^{\frac{1}{m+1}}\right)}{rm!}\left(\left(\widetilde{A}_n\left(\|\cdot-x_0\|_p^{m+1}\right)\right)(x_0)\right)^{\left(\frac{m}{m+1}\right)}$$

$$\left[\frac{1}{(m+1)}+\frac{r}{2}+\frac{mr^2}{8}\right], \tag{6.58}$$

(2) additionally if $f^{(j)}(x_0)=0$, $j=1,\ldots,\overline{m}$, we have

$$\|(A_n(f))(x_0) - f(x_0)\|_\gamma \leq$$

$$\frac{\omega_1\left(f^{(m)}, r\left(\left(\widetilde{A}_n\left(\|\cdot - x_0\|_p^{m+1}\right)\right)(x_0)\right)^{\frac{1}{m+1}}\right)}{rm!}\left(\left(\widetilde{A}_n\left(\|\cdot - x_0\|_p^{m+1}\right)\right)(x_0)\right)^{\left(\frac{m}{m+1}\right)}$$

$$\left[\frac{1}{(m+1)} + \frac{r}{2} + \frac{mr^2}{8}\right], \tag{6.59}$$

(3)

$$\|(A_n(f))(x_0) - f(x_0)\|_\gamma \leq \sum_{j=1}^{m} \frac{1}{j!}\left\|\left(A_n\left(f^{(j)}(x_0)(\cdot - x_0)^j\right)\right)(x_0)\right\|_\gamma +$$

$$\frac{\omega_1\left(f^{(m)}, r\left(\left(\widetilde{A}_n\left(\|\cdot - x_0\|_p^{m+1}\right)\right)(x_0)\right)^{\frac{1}{m+1}}\right)}{rm!}\left(\left(\widetilde{A}_n\left(\|\cdot - x_0\|_p^{m+1}\right)\right)(x_0)\right)^{\left(\frac{m}{m+1}\right)}$$

$$\left[\frac{1}{(m+1)} + \frac{r}{2} + \frac{mr^2}{8}\right], \tag{6.60}$$

and
(4)

$$\left\|\|A_n(f) - f\|_\gamma\right\|_{\infty, \prod_{i=1}^{N}[a_i, b_i]} \leq$$

$$\sum_{j=1}^{m} \frac{1}{j!}\left\|\left\|\left(A_n\left(f^{(j)}(x_0)(\cdot - x_0)^j\right)\right)(x_0)\right\|_\gamma\right\|_{\infty, x_0 \in \prod_{i=1}^{N}[a_i, b_i]} +$$

$$\frac{\omega_1\left(f^{(m)}, r\left\|\left(\widetilde{A}_n\left(\|\cdot - x_0\|_p^{m+1}\right)\right)(x_0)\right\|_{\infty, x_0 \in \prod_{i=1}^{N}[a_i, b_i]}^{\frac{1}{m+1}}\right)}{rm!}$$

$$\left\|\left(\widetilde{A}_n\left(\|\cdot - x_0\|_p^{m+1}\right)\right)(x_0)\right\|_{\infty, x_0 \in \prod_{i=1}^{N}[a_i, b_i]}^{\left(\frac{m}{m+1}\right)} \tag{6.61}$$

$$\left[\frac{1}{(m+1)} + \frac{r}{2} + \frac{mr^2}{8}\right].$$

We need

Lemma 6.11 *The function* $\left(\widetilde{A}_n\left(\|\cdot - x_0\|_p^m\right)\right)(x_0)$ *is continuous in* $x_0 \in \left(\prod_{i=1}^{N}[a_i, b_i]\right)$, $m \in \mathbb{N}$.

Proof By Lemma 10.3, [11], p. 272. ■

We give

Corollary 6.12 (to Theorem 6.10, case of $m = 1$) *Then*
(1)

$$\|(A_n(f))(x_0) - f(x_0)\|_\gamma \leq \left\|\left(A_n\left(f^{(1)}(x_0)(\cdot - x_0)\right)\right)(x_0)\right\|_\gamma +$$

$$\frac{1}{2r}\omega_1\left(f^{(1)}, r\left(\left(\widetilde{A}_n\left(\|\cdot - x_0\|_p^2\right)\right)(x_0)\right)^{\frac{1}{2}}\right)\left(\left(\widetilde{A}_n\left(\|\cdot - x_0\|_p^2\right)\right)(x_0)\right)^{\frac{1}{2}} \tag{6.62}$$

$$\left[1 + r + \frac{r^2}{4}\right],$$

and
(2)

$$\left\|\|(A_n(f)) - f\|_\gamma\right\|_{\infty, \prod_{i=1}^{N}[a_i, b_i]} \leq$$

$$\left\|\left\|\left(A_n\left(f^{(1)}(x_0)(\cdot - x_0)\right)\right)(x_0)\right\|_\gamma\right\|_{\infty, x_0 \in \prod_{i=1}^{N}[a_i, b_i]} +$$

$$\frac{1}{2r}\omega_1\left(f^{(1)}, r\left\|\left(\widetilde{A}_n\left(\|\cdot - x_0\|_p^2\right)\right)(x_0)\right\|_{\infty, x_0 \in \prod_{i=1}^{N}[a_i, b_i]}^{\frac{1}{2}}\right)$$

$$\left\|\left(\widetilde{A}_n\left(\|\cdot - x_0\|_p^2\right)\right)(x_0)\right\|_{\infty, x_0 \in \prod_{i=1}^{N}[a_i, b_i]}^{\frac{1}{2}}\left[1 + r + \frac{r^2}{4}\right], \tag{6.63}$$

$r > 0$.

We make

Remark 6.13 We estimate $0 < \alpha < 1$, $m, n \in \mathbb{N} : n^{1-\alpha} > 2$,

$$\widetilde{A}_n\left(\|\cdot - x_0\|_\infty^{m+1}\right)(x_0) = \frac{\sum_{k=\lceil na \rceil}^{\lfloor nb \rfloor}\left\|\frac{k}{n} - x_0\right\|_\infty^{m+1} Z(nx_0 - k)}{\sum_{k=\lceil na \rceil}^{\lfloor nb \rfloor} Z(nx_0 - k)} \overset{(6.22)}{<}$$

$$(2.412)^N \sum_{k=\lceil na \rceil}^{\lfloor nb \rfloor} \left\| \frac{k}{n} - x_0 \right\|_\infty^{m+1} Z(nx_0 - k) = \tag{6.64}$$

$$(2.412)^N \left\{ \sum_{\substack{k = \lceil na \rceil \\ : \left\| \frac{k}{n} - x_0 \right\|_\infty \le \frac{1}{n^\alpha}}}^{\lfloor nb \rfloor} \left\| \frac{k}{n} - x_0 \right\|_\infty^{m+1} Z(nx_0 - k) + \right.$$

$$\left. \sum_{\substack{k = \lceil na \rceil \\ : \left\| \frac{k}{n} - x_0 \right\|_\infty > \frac{1}{n^\alpha}}}^{\lfloor nb \rfloor} \left\| \frac{k}{n} - x_0 \right\|_\infty^{m+1} Z(nx_0 - k) \right\} \overset{(6.23)}{\le}$$

$$(2.412)^N \left\{ \frac{1}{n^{\alpha(m+1)}} + \frac{4e^2 \|b - a\|_\infty^{m+1}}{\pi e^{n^{1-\alpha}}} \right\}, \tag{6.65}$$

(where $b - a = (b_1 - a_1, ..., b_N - a_N)$).

We have proved that ($\forall\, x_0 \in \prod_{i=1}^{N} [a_i, b_i]$)

$$\widetilde{A}_n \left(\|\cdot - x_0\|_\infty^{m+1} \right)(x_0) < (2.412)^N \left\{ \frac{1}{n^{\alpha(m+1)}} + \frac{4e^2 \|b - a\|_\infty^{m+1}}{\pi e^{n^{1-\alpha}}} \right\} =: \varphi_1(n) \tag{6.66}$$

($0 < \alpha < 1$, $m, n \in \mathbb{N} : n^{1-\alpha} > 2$).

And, consequently it holds

$$\left\| \widetilde{A}_n \left(\|\cdot - x_0\|_\infty^{m+1} \right)(x_0) \right\|_{\infty, x_0 \in \prod_{i=1}^{N} [a_i, b_i]} <$$

$$(2.412)^N \left\{ \frac{1}{n^{\alpha(m+1)}} + \frac{4e^2 \|b - a\|_\infty^{m+1}}{\pi e^{n^{1-\alpha}}} \right\} = \varphi_1(n) \to 0, \quad \text{as } n \to +\infty. \tag{6.67}$$

So, we have that $\varphi_1(n) \to 0$, as $n \to +\infty$. Thus, when $p \in [1, \infty]$, from Theorem 6.10 we have the convergence to zero in the right hand sides of parts (1), (2).

Next we estimate $\left\| \left(\widetilde{A}_n \left(f^{(j)}(x_0)(\cdot - x_0)^j \right) \right)(x_0) \right\|_\gamma$.

We have that

$$\left(\widetilde{A}_n\left(f^{(j)}(x_0)(\cdot - x_0)^j\right)\right)(x_0) = \frac{\sum_{k=\lceil na \rceil}^{\lfloor nb \rfloor} f^{(j)}(x_0)\left(\frac{k}{n} - x_0\right)^j Z(nx_0 - k)}{\sum_{k=\lceil na \rceil}^{\lfloor nb \rfloor} Z(nx_0 - k)}. \tag{6.68}$$

When $p = \infty$, $j = 1, ..., m$, we obtain

$$\left\| f^{(j)}(x_0)\left(\frac{k}{n} - x_0\right)^j \right\|_\gamma \leq \left\| f^{(j)}(x_0) \right\| \left\| \frac{k}{n} - x_0 \right\|_\infty^j. \tag{6.69}$$

We further have that

$$\left\| \left(\widetilde{A}_n\left(f^{(j)}(x_0)(\cdot - x_0)^j\right)\right)(x_0) \right\|_\gamma \overset{(6.22)}{<}$$

$$(2.412)^N \left(\sum_{k=\lceil na \rceil}^{\lfloor nb \rfloor} \left\| f^{(j)}(x_0)\left(\frac{k}{n} - x_0\right)^j \right\|_\gamma Z(nx_0 - k) \right) \leq$$

$$(2.412)^N \left(\sum_{k=\lceil na \rceil}^{\lfloor nb \rfloor} \left\| f^{(j)}(x_0) \right\| \left\| \frac{k}{n} - x_0 \right\|_\infty^j Z(nx_0 - k) \right) = \tag{6.70}$$

$$(2.412)^N \left\| f^{(j)}(x_0) \right\| \left(\sum_{k=\lceil na \rceil}^{\lfloor nb \rfloor} \left\| \frac{k}{n} - x_0 \right\|_\infty^j Z(nx_0 - k) \right) =$$

$$(2.412)^N \left\| f^{(j)}(x_0) \right\| \left\{ \sum_{\substack{k = \lceil na \rceil \\ : \left\| \frac{k}{n} - x_0 \right\|_\infty \leq \frac{1}{n^\alpha}}}^{\lfloor nb \rfloor} \left\| \frac{k}{n} - x_0 \right\|_\infty^j Z(nx_0 - k) \right.$$

$$\left. + \sum_{\substack{k = \lceil na \rceil \\ : \left\| \frac{k}{n} - x_0 \right\|_\infty > \frac{1}{n^\alpha}}}^{\lfloor nb \rfloor} \left\| \frac{k}{n} - x_0 \right\|_\infty^j Z(nx_0 - k) \right\} \overset{(6.23)}{\leq} \tag{6.71}$$

$$(2.412)^N \left\| f^{(j)}(x_0) \right\| \left\{ \frac{1}{n^{\alpha j}} + \frac{4e^2 \|b - a\|_\infty^j}{\pi e^{n^{1-\alpha}}} \right\} \to 0, \text{ as } n \to \infty.$$

That is

$$\left\| \left(\widetilde{A}_n \left(f^{(j)} (x_0) (\cdot - x_0)^j \right) \right) (x_0) \right\|_\gamma \to 0, \text{ as } n \to \infty.$$

Therefore when $p = \infty$, for $j = 1, ..., m$, we have proved:

$$\left\| \left(\widetilde{A}_n \left(f^{(j)} (x_0) (\cdot - x_0)^j \right) \right) (x_0) \right\|_\gamma <$$

$$(2.412)^N \left\| f^{(j)} (x_0) \right\| \left\{ \frac{1}{n^{\alpha j}} + \frac{4e^2 \|b - a\|_\infty^j}{\pi e^{n^{1-\alpha}}} \right\} \leq \tag{6.72}$$

$$(2.412)^N \left\| f^{(j)} \right\|_\infty \left\{ \frac{1}{n^{\alpha j}} + \frac{4e^2 \|b - a\|_\infty^j}{\pi e^{n^{1-\alpha}}} \right\} =: \varphi_{2j} (n) < \infty,$$

and converges to zero, as $n \to \infty$.

We conclude:

In Theorem 6.10, the right hand sides of (6.60) and (6.61) converge to zero as $n \to \infty$, for any $p \in [1, \infty]$.

Also in Corollary 6.12, the right hand sides of (6.62) and (6.63) converge to zero as $n \to \infty$, for any $p \in [1, \infty]$.

Conclusion 6.14 *We have proved that the left hand sides of (6.58), (6.59), (6.60), (6.61) and (6.62), (6.63) converge to zero as $n \to \infty$, for $p \in [1, \infty]$. Consequently $A_n \to I$ (unit operator) pointwise and uniformly, as $n \to \infty$, where $p \in [1, \infty]$. In the presence of initial conditions we achieve a higher speed of convergence, see (6.59). Higher speed of convergence happens also to the left hand side of (6.58).*

We further give

Corollary 6.15 (to Theorem 6.10) *Let O open subset of $\left(\mathbb{R}^N, \|\cdot\|_\infty\right)$, such that $\prod_{i=1}^N [a_i, b_i] \subset O \subseteq \mathbb{R}^N$, and let $\left(X, \|\cdot\|_\gamma\right)$ be a general Banach space. Let $m \in \mathbb{N}$ and $f \in C^m (O, X)$, the space of $\overline{m}$-times continuously Fréchet differentiable functions from O into X. We study the approximation of $f|_{\prod_{i=1}^N [a_i, b_i]}$. Let $x_0 \in \left(\prod_{i=1}^N [a_i, b_i]\right)$ and $r > 0$. Here $\varphi_1 (n)$ as in (6.67) and $\varphi_{2j} (n)$ as in (6.72), where $n \in \mathbb{N} : n^{1-\alpha} > 2$, $0 < \alpha < 1$, $j = 1, ..., m$. Then*

(1)

$$\left\| (A_n (f)) (x_0) - \sum_{j=0}^m \frac{1}{j!} \left(A_n \left(f^{(j)} (x_0) (\cdot - x_0)^j \right) \right) (x_0) \right\|_\gamma \leq$$

$$\frac{\omega_1 \left(f^{(m)}, r (\varphi_1 (n))^{\frac{1}{m+1}} \right)}{r m!} (\varphi_1 (n))^{\left(\frac{m}{m+1}\right)} \left[\frac{1}{(m+1)} + \frac{r}{2} + \frac{m r^2}{8} \right], \tag{6.73}$$

(2) additionally, if $f^{(j)}(x_0) = 0$, $j = 1, ..., \overline{m}$, *we have*

$$\|(A_n(f))(x_0) - f(x_0)\|_\gamma \le$$

$$\frac{\omega_1\left(f^{(m)}, r(\varphi_1(n))^{\frac{1}{m+1}}\right)}{rm!}(\varphi_1(n))^{\left(\frac{m}{m+1}\right)}\left[\frac{1}{(m+1)} + \frac{r}{2} + \frac{mr^2}{8}\right], \tag{6.74}$$

(3)

$$\left\|\|A_n(f) - f\|_\gamma\right\|_{\infty, \prod_{i=1}^{N}[a_i,b_i]} \le \sum_{j=1}^{m} \frac{\varphi_{2j}(n)}{j!} +$$

$$\frac{\omega_1\left(f^{(m)}, r(\varphi_1(n))^{\frac{1}{m+1}}\right)}{rm!}(\varphi_1(n))^{\left(\frac{m}{m+1}\right)} \tag{6.75}$$

$$\left[\frac{1}{(m+1)} + \frac{r}{2} + \frac{mr^2}{8}\right] =: \varphi_3(n) \to 0, \text{ as } n \to \infty.$$

We continue with

Theorem 6.16 *Let* $f \in C_B(\mathbb{R}^N, X)$, $0 < \beta < 1$, $x \in \mathbb{R}^N$, $m, N, n \in \mathbb{N}$ *with* $n^{1-\beta} > 2$, ω_1 *is for* $p = \infty$. *Then*

(1)

$$\|B_n(f, x) - f(x)\|_\gamma \le \omega_1\left(f, \frac{1}{n^\beta}\right) + \frac{8e^2\left\|\|f\|_\gamma\right\|_\infty}{\pi e^{n^{1-\beta}}} =: \lambda_2(n), \tag{6.76}$$

(2)

$$\left\|\|B_n(f) - f\|_\gamma\right\|_\infty \le \lambda_2(n). \tag{6.77}$$

Given that $f \in (C_U(\mathbb{R}^N, X) \cap C_B(\mathbb{R}^N, X))$, *we obtain* $\lim_{n\to\infty} B_n(f) = f$, *uniformly.*

Proof We have that

$$B_n(f, x) - f(x) \overset{(6.16)}{=} \sum_{k=-\infty}^{\infty} f\left(\frac{k}{n}\right) Z(nx - k) - f(x) \sum_{k=-\infty}^{\infty} Z(nx - k) =$$

$$\sum_{k=-\infty}^{\infty} \left(f\left(\frac{k}{n}\right) - f(x)\right) Z(nx - k). \tag{6.78}$$

Hence

$$\|B_n(f,x) - f(x)\|_\gamma \le \sum_{k=-\infty}^{\infty} \left\| f\left(\frac{k}{n}\right) - f(x)\right\|_\gamma Z(nx-k) =$$

$$\sum_{\begin{cases} k=-\infty \\ \left\|\frac{k}{n}-x\right\|_\infty \le \frac{1}{n^\beta}\end{cases}}^{\infty} \left\| f\left(\frac{k}{n}\right) - f(x)\right\|_\gamma Z(nx-k) +$$

$$\sum_{\begin{cases} k=-\infty \\ \left\|\frac{k}{n}-x\right\|_\infty > \frac{1}{n^\beta}\end{cases}}^{\infty} \left\| f\left(\frac{k}{n}\right) - f(x)\right\|_\gamma Z(nx-k) \overset{(6.16)}{\le}$$

$$\omega_1\left(f, \frac{1}{n^\beta}\right) + 2\left\|\|f\|_\gamma\right\|_\infty \sum_{\begin{cases} k=-\infty \\ \left\|\frac{k}{n}-x\right\|_\infty > \frac{1}{n^\beta}\end{cases}}^{\infty} Z(nx-k) \overset{(6.23)}{\le}$$

$$\omega_1\left(f, \frac{1}{n^\beta}\right) + \frac{8e^2\left\|\|f\|_\gamma\right\|_\infty}{\pi e^{n^{1-\beta}}}, \tag{6.79}$$

proving the claim. ∎

We give

Theorem 6.17 *Let* $f \in C_B(\mathbb{R}^N, X)$, $0 < \beta < 1$, $x \in \mathbb{R}^N$, $m, N, n \in \mathbb{N}$ *with* $n^{1-\beta} > 2$, ω_1 *is for* $p = \infty$. *Then*
(1)

$$\|C_n(f,x) - f(x)\|_\gamma \le \omega_1\left(f, \frac{1}{n} + \frac{1}{n^\beta}\right) + \frac{8e^2\left\|\|f\|_\gamma\right\|_\infty}{\pi e^{n^{1-\beta}}} =: \lambda_3(n), \tag{6.80}$$

(2)

$$\left\|\|C_n(f) - f\|_\gamma\right\|_\infty \le \lambda_3(n). \tag{6.81}$$

Given that $f \in \left(C_U(\mathbb{R}^N, X) \cap C_B(\mathbb{R}^N, X)\right)$, *we obtain* $\lim_{n\to\infty} C_n(f) = f$, *uniformly.*

Proof We notice that

$$\int_{\frac{k}{n}}^{\frac{k+1}{n}} f(t)\,dt = \int_{\frac{k_1}{n}}^{\frac{k_1+1}{n}} \int_{\frac{k_2}{n}}^{\frac{k_2+1}{n}} \ldots \int_{\frac{k_N}{n}}^{\frac{k_N+1}{n}} f(t_1, t_2, \ldots, t_N)\,dt_1 dt_2 \ldots dt_N =$$

$$\int_0^{\frac{1}{n}}\int_0^{\frac{1}{n}}\dots\int_0^{\frac{1}{n}} f\left(t_1+\frac{k_1}{n}, t_2+\frac{k_2}{n}, \dots, t_N+\frac{k_N}{n}\right) dt_1 \dots dt_N = \int_0^{\frac{1}{n}} f\left(t+\frac{k}{n}\right) dt. \tag{6.82}$$

Thus it holds (by (6.38))

$$C_n(f,x) = \sum_{k=-\infty}^{\infty}\left(n^N \int_0^{\frac{1}{n}} f\left(t+\frac{k}{n}\right) dt\right) Z(nx-k). \tag{6.83}$$

We observe that

$$\|C_n(f,x) - f(x)\|_\gamma =$$

$$\left\| \sum_{k=-\infty}^{\infty}\left(n^N \int_0^{\frac{1}{n}} f\left(t+\frac{k}{n}\right) dt\right) Z(nx-k) - \sum_{k=-\infty}^{\infty} f(x) Z(nx-k) \right\|_\gamma =$$

$$\left\| \sum_{k=-\infty}^{\infty}\left(\left(n^N \int_0^{\frac{1}{n}} f\left(t+\frac{k}{n}\right) dt\right) - f(x)\right) Z(nx-k) \right\|_\gamma =$$

$$\left\| \sum_{k=-\infty}^{\infty}\left(n^N \int_0^{\frac{1}{n}} \left(f\left(t+\frac{k}{n}\right) - f(x)\right) dt\right) Z(nx-k) \right\|_\gamma \leq \tag{6.84}$$

$$\sum_{k=-\infty}^{\infty}\left(n^N \int_0^{\frac{1}{n}} \left\| f\left(t+\frac{k}{n}\right) - f(x)\right\|_\gamma dt\right) Z(nx-k) =$$

$$\sum_{\substack{k=-\infty \\ \left\|\frac{k}{n}-x\right\|_\infty \leq \frac{1}{n^\beta}}}^{\infty}\left(n^N \int_0^{\frac{1}{n}} \left\| f\left(t+\frac{k}{n}\right) - f(x)\right\|_\gamma dt\right) Z(nx-k) +$$

$$\sum_{\substack{k=-\infty \\ \left\|\frac{k}{n}-x\right\|_\infty > \frac{1}{n^\beta}}}^{\infty}\left(n^N \int_0^{\frac{1}{n}} \left\| f\left(t+\frac{k}{n}\right) - f(x)\right\|_\gamma dt\right) Z(nx-k) \leq$$

$$\sum_{\substack{k=-\infty \\ \left\|\frac{k}{n}-x\right\|_\infty \leq \frac{1}{n^\beta}}}^{\infty}\left(n^N \int_0^{\frac{1}{n}} \omega_1\left(f, \|t\|_\infty + \left\|\frac{k}{n}-x\right\|_\infty\right) dt\right) Z(nx-k) +$$

$$2\left\|\|f\|_{\gamma}\right\|_{\infty}\left(\sum_{\substack{k=-\infty \\ \left\|\frac{k}{n}-x\right\|_{\infty}>\frac{1}{n^{\beta}}}}^{\infty} Z(|nx-k|)\right) \leq$$

$$\omega_1\left(f, \frac{1}{n}+\frac{1}{n^{\beta}}\right)+\frac{8e^2\left\|\|f\|_{\gamma}\right\|_{\infty}}{\pi e^{n^{1-\beta}}}, \tag{6.85}$$

proving the claim. ■

We also present

Theorem 6.18 *Let $f \in C_B\left(\mathbb{R}^N, X\right)$, $0<\beta<1$, $x \in \mathbb{R}^N$, $m, N, n \in \mathbb{N}$ with $n^{1-\beta}>2$, ω_1 is for $p=\infty$. Then*
(1)

$$\left\|D_n(f, x)-f(x)\right\|_{\gamma} \leq \omega_1\left(f, \frac{1}{n}+\frac{1}{n^{\beta}}\right)+\frac{8\left\|\|f\|_{\gamma}\right\|_{\infty}}{\pi e^{n^{1-\beta}}}=\lambda_4(n), \tag{6.86}$$

(2)

$$\left\|\left\|D_n(f)-f\right\|_{\gamma}\right\|_{\infty} \leq \lambda_4(n). \tag{6.87}$$

Given that $f \in \left(C_U\left(\mathbb{R}^N, X\right) \cap C_B\left(\mathbb{R}^N, X\right)\right)$, we obtain $\lim_{n \rightarrow \infty} D_n(f)=f$, uniformly.

Proof Similar to the proof of Theorem 6.17, as such is omitted. ■

We make

Definition 6.19 Let $f \in C_B\left(\mathbb{R}^N, X\right)$, $N \in \mathbb{N}$, where $\left(X,\|\cdot\|_{\gamma}\right)$ is a Banach space. We define the general neural network operator

$$F_n(f, x):=\sum_{k=-\infty}^{\infty} l_{nk}(f) Z(nx-k)=$$

$$\begin{cases} B_n(f, x), \text{ if } l_{nk}(f)=f\left(\frac{k}{n}\right), \\ C_n(f, x), \text{ if } l_{nk}(f)=n^N \int_{\frac{k}{n}}^{\frac{k+1}{n}} f(t)\, dt, \\ D_n(f, x), \text{ if } l_{nk}(f)=\delta_{nk}(f). \end{cases} \tag{6.88}$$

Clearly $l_{nk}(f)$ is an X-valued bounded linear functional such that $\left\|l_{nk}(f)\right\|_{\gamma} \leq \left\|\|f\|_{\gamma}\right\|_{\infty}$.

Hence $F_n(f)$ is a bounded linear operator with $\left\|\left\|F_n(f)\right\|_{\gamma}\right\|_{\infty} \leq \left\|\|f\|_{\gamma}\right\|_{\infty}$.

We need

Theorem 6.20 *Let $f \in C_B\left(\mathbb{R}^N, X\right)$, $N \geq 1$. Then $F_n(f) \in C_B\left(\mathbb{R}^N, X\right)$.*

Proof Lengthy and similar to the proof of Theorem 21 of [15], as such is omitted. ∎

Remark 6.21 By (6.25) it is obvious that $\left\| \|A_n(f)\|_\gamma \right\|_\infty \leq \left\| \|f\|_\gamma \right\|_\infty < \infty$, and $A_n(f) \in C\left(\prod_{i=1}^N [a_i, b_i], X\right)$, given that $f \in C\left(\prod_{i=1}^N [a_i, b_i], X\right)$.

Call L_n any of the operators A_n, B_n, C_n, D_n.

Clearly then

$$\left\| \left\| L_n^2(f) \right\|_\gamma \right\|_\infty = \left\| \|L_n(L_n(f))\|_\gamma \right\|_\infty \leq \left\| \|L_n(f)\|_\gamma \right\|_\infty \leq \left\| \|f\|_\gamma \right\|_\infty, \tag{6.89}$$

etc.

Therefore we get

$$\left\| \left\| L_n^k(f) \right\|_\gamma \right\|_\infty \leq \left\| \|f\|_\gamma \right\|_\infty, \ \forall\, k \in \mathbb{N}, \tag{6.90}$$

the contraction property.

Also we see that

$$\left\| \left\| L_n^k(f) \right\|_\gamma \right\|_\infty \leq \left\| \left\| L_n^{k-1}(f) \right\|_\gamma \right\|_\infty \leq \ldots \leq \left\| \|L_n(f)\|_\gamma \right\|_\infty \leq \left\| \|f\|_\gamma \right\|_\infty. \tag{6.91}$$

Here L_n^k are bounded linear operators.

Notation 6.22 Here $N \in \mathbb{N}$, $0 < \beta < 1$. Denote by

$$c_N := \begin{cases} (2.412)^N, \text{ if } L_n = A_n, \\ 1, \text{ if } L_n = B_n, C_n, D_n, \end{cases} \tag{6.92}$$

$$\varphi(n) := \begin{cases} \frac{1}{n^\beta}, \text{ if } L_n = A_n, B_n, \\ \frac{1}{n} + \frac{1}{n^\beta}, \text{ if } L_n = C_n, D_n, \end{cases} \tag{6.93}$$

$$\Omega := \begin{cases} C\left(\prod_{i=1}^N [a_i, b_i], X\right), \text{ if } L_n = A_n, \\ C_B\left(\mathbb{R}^N, X\right), \text{ if } L_n = B_n, C_n, D_n, \end{cases} \tag{6.94}$$

and

$$Y := \begin{cases} \prod_{i=1}^N [a_i, b_i], \text{ if } L_n = A_n, \\ \mathbb{R}^N, \text{ if } L_n = B_n, C_n, D_n. \end{cases} \tag{6.95}$$

We give the condensed

Theorem 6.23 *Let $f \in \Omega$, $0 < \beta < 1$, $x \in Y$; $n, m, N \in \mathbb{N}$ with $n^{1-\beta} > 2$. Then*
(i)

$$\|L_n(f, x) - f(x)\|_\gamma \leq c_N \left[\omega_1(f, \varphi(n)) + \frac{8e^2 \left\| \|f\|_\gamma \right\|_\infty}{\pi e^{n^{1-\beta}}} \right] =: \tau(n), \quad (6.96)$$

where ω_1 is for $p = \infty$,
and
(ii)

$$\left\| \|L_n(f) - f\|_\gamma \right\|_\infty \leq \tau(n) \to 0, \text{ as } n \to \infty. \quad (6.97)$$

For f uniformly continuous and in Ω we obtain

$$\lim_{n\to\infty} L_n(f) = f,$$

pointwise and uniformly.

Proof By Theorems 6.8, 6.16, 6.17, 6.18. ■

Next we talk about iterated neural network approximation (see also [9]).
We give

Theorem 6.24 *All here as in Theorem 6.23 and $r \in \mathbb{N}$, $\tau(n)$ as in (6.96). Then*

$$\left\| \left\| L_n^r f - f \right\|_\gamma \right\|_\infty \leq r\tau(n). \quad (6.98)$$

So that the speed of convergence to the unit operator of L_n^r is not worse than of L_n.

Proof As similar to [15] is omitted. ■

We also present

Theorem 6.25 *Let $f \in \Omega$; $m, N, m_1, m_2, ..., m_r \in \mathbb{N} : m_1 \leq m_2 \leq ... \leq m_r$, $0 < \beta < 1$; $m_i^{1-\beta} > 2$, $i = 1, ..., r$, $x \in Y$, and let $(L_{m_1}, ..., L_{m_r})$ as $(A_{m_1}, ..., A_{m_r})$ or $(B_{m_1}, ..., B_{m_r})$ or $(C_{m_1}, ..., C_{m_r})$ or $(D_{m_1}, ..., D_{m_r})$, $p = \infty$. Then*

$$\left\| L_{m_r}\left(L_{m_{r-1}}\left(...L_{m_2}\left(L_{m_1} f\right)\right)\right)(x) - f(x) \right\|_\gamma \leq$$

$$\left\| \left\| L_{m_r}\left(L_{m_{r-1}}\left(...L_{m_2}\left(L_{m_1} f\right)\right)\right) - f \right\|_\gamma \right\|_\infty \leq$$

$$\sum_{i=1}^{r} \left\| \left\| L_{m_i} f - f \right\|_\gamma \right\|_\infty \leq$$

$$c_N \sum_{i=1}^{r} \left[\omega_1 \left(f, \varphi \left(m_i \right) \right) + \frac{8e^2 \left\| \| f \|_\gamma \right\|_\infty}{\pi e^{n^{1-\beta}}} \right] \leq$$

$$r c_N \left[\omega_1 \left(f, \varphi \left(m_1 \right) \right) + \frac{8e^2 \left\| \| f \|_\gamma \right\|_\infty}{\pi e^{n^{1-\beta}}} \right]. \tag{6.99}$$

Clearly, we notice that the speed of convergence to the unit operator of the multiply iterated operator is not worse than the speed of L_{m_1}.

Proof As similar to [15] is omitted. ■

We also give

Theorem 6.26 *Let all as in Corollary 6.15, and* $r \in \mathbb{N}$. *Here* $\varphi_3(n)$ *is as in (6.75). Then*

$$\left\| \left\| A_n^r f - f \right\|_\gamma \right\|_\infty \leq r \left\| \| A_n f - f \|_\gamma \right\|_\infty \leq r \varphi_3(n). \tag{6.100}$$

Proof As similar to [15] is omitted. ■

Application 6.27 *A typical application of all of our results is when* $\left(X, \|\cdot\|_\gamma \right) = (\mathbb{C}, |\cdot|)$, *where* $\mathbb{C}$ *are the complex numbers.*

References

1. G.A. Anastassiou, *Moments in Probability and Approximation Theory, Pitman Research Notes in Math*, vol. 287 (Longman Sci. & Tech, Harlow, U.K., 1993)
2. G.A. Anastassiou, Rate of convergence of some neural network operators to the unit-univariate case. J. Math. Anal. Appl. **212**, 237–262 (1997)
3. G.A. Anastassiou, *Quantitative Approximations* (Chapman & Hall/CRC, Boca Raton, New York, 2001)
4. G.A. Anastassiou, *Inteligent Systems: Approximation by Artificial Neural Networks, Intelligent Systems Reference Library*, vol. 19 (Springer, Heidelberg, 2011)
5. G.A. Anastassiou, Univariate hyperbolic tangent neural network approximation. Math. Comput. Modell. **53**, 1111–1132 (2011)
6. G.A. Anastassiou, Multivariate hyperbolic tangent neural network approximation. Comput. Math. **61**, 809–821 (2011)
7. G.A. Anastassiou, Multivariate sigmoidal neural network approximation. Neural Netw. **24**, 378–386 (2011)
8. G.A. Anastassiou, Univariate sigmoidal neural network approximation. J. Comput. Anal. Appl. **14**(4), 659–690 (2012)
9. G.A. Anastassiou, Approximation by neural networks iterates, in *Advances in Applied Mathematics and Approximation Theory*, ed. by G. Anastassiou, O. Duman. Springer Proceedings in Mathematics & Statistics (Springer, New York, 2013), pp. 1–20
10. G.A. Anastassiou, *Intelligent Systems II: Complete Approximation by Neural Network Operators* (Springer, Heidelberg, New York, 2016)

11. G.A. Anastassiou, *Intelligent Computations: Abstract Fractional Calculus, Inequalities, Approximations* (Springer, Heidelberg, New York, 2018)
12. G.A. Anastassiou, Algebraic function based Banach space valued ordinary and fractional neural network approximations, New Trends in Mathematical Sciences, 10 special issue (1) (2022), pp. 100–125
13. G.A. Anastassiou, Gudermannian function activated Banach space valued ordinary and fractional neural network approximation. Adv. Nonlinear Var. Inequal. **25**(2), 27–64 (2022)
14. G.A. Anastassiou, Abstract multivariate Gudermannian function activated neural network approximations. PanAmerican Math. J. (accepted) (2022)
15. G.A. Anastassiou, General multimariate arctangent function activated neural network approximations (submitted) (2022)
16. G.A. Anastassiou, R. Kozma, Abstract multivariate algebraic function activated neural network approximations (submitted) (2022)
17. H. Cartan, *Differential Calculus* (Hermann, Paris, 1971)
18. Z. Chen, F. Cao, The approximation operators with sigmoidal functions. Comput. Math. Appl. **58**, 758–765 (2009)
19. D. Costarelli, R. Spigler, Approximation results for neural network operators activated by sigmoidal functions. Neural Netw. **44**, 101–106 (2013)
20. D. Costarelli, R. Spigler, Multivariate neural network operators with sigmoidal activation functions. Neural Netw. **48**, 72–77 (2013)
21. S. Haykin, *Neural Networks: A Comprehensive Foundation*, 2nd edn. (Prentice Hall, New York, 1998)
22. W. McCulloch, W. Pitts, A logical calculus of the ideas immanent in nervous activity. Bull. Math. Biophys. **7**, 115–133 (1943)
23. T.M. Mitchell, *Machine Learning* (WCB-McGraw-Hill, New York, 1997)
24. L.B. Rall, *Computational Solution of Nonlinear Operator Equations* (Wiley, New York, 1969)
25. E.W. Weisstein, *Gudermannian*, MathWorld

Chapter 7
Generalized Symmetrical Sigmoid Function Induced Neural Network Multivariate Approximation

Here we exhibit multivariate quantitative approximations of Banach space valued continuous multivariate functions on a box or $\mathbb{R}^N$, $N \in \mathbb{N}$, by the multivariate normalized, quasi-interpolation, Kantorovich type and quadrature type neural network operators. We treat also the case of approximation by iterated operators of the last four types. These approximations are achieved by establishing multidimensional Jackson type inequalities involving the multivariate modulus of continuity of the engaged function or its high order Fréchet derivatives. Our multivariate operators are defined by using a multidimensional density function induced by the generalized symmetrical sigmoid function. The approximations are pointwise and uniform. The related feed-forward neural network is with one hidden layer. It follows [16].

7.1 Introduction

Anastassiou in [2, 3], see Chap. 2–5, was the first to establish neural network approximations to continuous functions with rates by very specifically defined neural network operators of Cardaliaguet-Euvrard and "Squashing" types, by employing the modulus of continuity of the engaged function or its high order derivative, and producing very tight Jackson type inequalities. He treats there both the univariate and multivariate cases. The defining these operators "bell-shaped" and "squashing" functions are assumed to be of compact support. Also in [3] he gives the Nth order asymptotic expansion for the error of weak approximation of these two operators to a special natural class of smooth functions, see Chaps. 4–5 there.

Motivations for this chapter are the articles [19] of Chen and Cao, and [4–15, 17, 20, 21].

Here we perform multivariate generalized symmetrical sigmoid function based neural network approximations to continuous functions over boxes or over the whole

© The Author(s), under exclusive license to Springer Nature Switzerland AG 2023

G. A. Anastassiou, *Banach Space Valued Neural Network*,
Studies in Computational Intelligence 1062,
https://doi.org/10.1007/978-3-031-16400-2_7

$\mathbb{R}^N$, $N \in \mathbb{N}$, and also iterated approximations. All convergences here are with rates expressed via the multivariate modulus of continuity of the involved function or its high order Fréchet derivative and given by very tight multidimensional Jackson type inequalities.

We come up with the "right" precisely defined multivariate normalized, quasi-interpolation neural network operators related to boxes or $\mathbb{R}^N$, as well as Kantorovich type and quadrature type related operators on $\mathbb{R}^N$. Our boxes are not necessarily symmetric to the origin. In preparation to prove our results we establish important properties of the basic multivariate density function induced by generalized symmetrical sigmoid function and defining our operators.

Feed-forward neural networks (FNNs) with one hidden layer, the only type of networks we deal with in this chapter, are mathematically expressed as

$$N_n(x) = \sum_{j=0}^{n} c_j \sigma\left(\left\langle a_j \cdot x\right\rangle + b_j\right), \quad x \in \mathbb{R}^s, \ s \in \mathbb{N},$$

where for $0 \le j \le n$, $b_j \in \mathbb{R}$ are the thresholds, $a_j \in \mathbb{R}^s$ are the connection weights, $c_j \in \mathbb{R}$ are the coefficients, $\left\langle a_j \cdot x\right\rangle$ is the inner product of a_j and x, and σ is the activation function of the network. In many fundamental network models, the activation function is the generalized symmetrical sigmoid function. About neural networks see [23–25].

7.2 Auxiliary Results (See Also [14])

Here we consider the generalized symmetrical sigmoid function [22]

$$f_1(x) = \frac{x}{\left(1 + |x|^{\mu}\right)^{\frac{1}{\mu}}}, \ \mu > 0, x \in \mathbb{R}. \tag{7.1}$$

This has applications in immunology and protection from disease together with probability theory. It is also called a symmetrical protection curve.

The parameter μ is a shape parameter controling how fast the curve approaches the asymptotes for a given slope at the inflection point. When $\mu = 1$ f_1 is the absolute sigmoid function, and when $\mu = 2$, f_1 is the square root sigmoid function. When $\mu = 1.5$ the function approximates the arctangent function, when $\mu = 2.9$ it approximates the logistic function, and when $\mu = 3.4$ it approximates the error function. Parameter μ is estimated in the likelihood maximization [22]. For more see [22].

Next we study the particular generator sigmoid function

$$f_2(x) = \frac{x}{\left(1 + |x|^{\lambda}\right)^{\frac{1}{\lambda}}}, \ \lambda \text{ is an odd number, } x \in \mathbb{R}. \tag{7.2}$$

We have that $f_2(0) = 0$, and

$$f_2(-x) = -f_2(x), \tag{7.3}$$

so f_2 is symmetric with respect to zero.

When $x \geq 0$, we get that [14]

$$f_2'(x) = \frac{1}{\left(1 + x^{\lambda}\right)^{\frac{\lambda+1}{\lambda}}} > 0, \tag{7.4}$$

that is f_2 is strictly increasing on $[0, +\infty)$ and f_2 is strictly increasing on $(-\infty, 0]$. Hence f_2 is strictly increasing on $\mathbb{R}$.

We also have $f_2(+\infty) = f_2(-\infty) = 1$.

Let us consider the activation function [14]:

$$\chi(x) = \frac{1}{4}\left[f_2(x+1) - f_2(x-1)\right] =$$

$$\frac{1}{4}\left[\frac{(x+1)}{\left(1 + |x+1|^{\lambda}\right)^{\frac{1}{\lambda}}} - \frac{(x-1)}{\left(1 + |x-1|^{\lambda}\right)^{\frac{1}{\lambda}}}\right]. \tag{7.5}$$

Clearly it holds [14]

$$\chi(x) = \chi(-x), \ \forall x \in \mathbb{R}. \tag{7.6}$$

and

$$\chi(0) = \frac{1}{2\sqrt[\lambda]{2}}, \tag{7.7}$$

and $\chi(x) > 0$, $\forall\, x \in \mathbb{R}$.

Following [14], we have that χ is strictly decreasing over $[0, +\infty)$, and χ is strictly increasing on $(-\infty, 0]$, by χ-symmetry with respect to y-axis, and $\chi'(0) = 0$.

Clearly it is

$$\lim_{x \to +\infty} \chi(x) = \lim_{x \to -\infty} \chi(x) = 0, \tag{7.8}$$

therefore the x-axis is the horizontal asymptote of $\chi(x)$.

The value

$$\chi(0) = \frac{1}{2\sqrt[\lambda]{2}}, \ \lambda \text{ is an odd number}, \tag{7.9}$$

is the maximum of χ, which is a bell shaped function.

We need

Theorem 7.1 ([14]) *It holds*

$$\sum_{i=-\infty}^{\infty} \chi(x-i) = 1, \forall x \in \mathbb{R}. \tag{7.10}$$

Theorem 7.2 ([14]) *We have that*

$$\int_{-\infty}^{\infty} \chi(x)\, dx = 1. \tag{7.11}$$

So that $\chi(x)$ is a density function on $\mathbb{R}$.
We need

Theorem 7.3 ([14]) *Let* $0 < \alpha < 1$, *and* $n \in \mathbb{N}$ *with* $n^{1-\alpha} > 2$. *It holds*

$$\sum_{\substack{j=-\infty \\ : |nx-j| \geq n^{1-\alpha}}}^{\infty} \chi(nx-j) < \frac{1}{2\lambda\left(n^{1-\alpha}-2\right)^{\lambda}}, \tag{7.12}$$

where $\lambda \in \mathbb{N}$ *is an odd number.*

Denote by $\lfloor \cdot \rfloor$ the integral part of the number and by $\lceil \cdot \rceil$ the ceiling of the number.
We also need

Theorem 7.4 ([14]) *Let* $[a,b] \subset \mathbb{R}$ *and* $n \in \mathbb{N}$ *so that* $\lceil na \rceil \leq \lfloor nb \rfloor$. *Then*

$$\frac{1}{\sum_{k=\lceil na \rceil}^{\lfloor nb \rfloor} \chi(|nx-k|)} < 2\sqrt[\lambda]{1+2^{\lambda}}, \tag{7.13}$$

where λ *is an odd number,* $\forall\, x \in [a,b]$.

We make

Remark 7.5 ([14]) (1) We have that

$$\lim_{n\to\infty} \sum_{k=\lceil na \rceil}^{\lfloor nb \rfloor} \chi(nx-k) \neq 1, \text{ for at least some } x \in [a,b]. \tag{7.14}$$

(2) Let $[a,b] \subset \mathbb{R}$. For large enough n we always obtain $\lceil na \rceil \leq \lfloor nb \rfloor$. Also $a \leq \frac{k}{n} \leq b$, iff $\lceil na \rceil \leq k \leq \lfloor nb \rfloor$.
In general it holds that

$$\sum_{k=\lceil na \rceil}^{\lfloor nb \rfloor} \chi(nx-k) \leq 1. \tag{7.15}$$

We introduce

$$Z(x_1, ..., x_N) := Z(x) := \prod_{i=1}^{N} \chi(x_i), \quad x = (x_1, ..., x_N) \in \mathbb{R}^N, \ N \in \mathbb{N}. \tag{7.16}$$

It has the properties:

(i) $Z(x) > 0, \ \forall\, x \in \mathbb{R}^N$,

(ii)

$$\sum_{k=-\infty}^{\infty} Z(x-k) := \sum_{k_1=-\infty}^{\infty} \sum_{k_2=-\infty}^{\infty} \ldots \sum_{k_N=-\infty}^{\infty} Z(x_1 - k_1, ..., x_N - k_N) = 1, \tag{7.17}$$

where $k := (k_1, ..., k_n) \in \mathbb{Z}^N, \forall\, x \in \mathbb{R}^N$,

hence

(iii)

$$\sum_{k=-\infty}^{\infty} Z(nx-k) = 1, \tag{7.18}$$

$\forall\, x \in \mathbb{R}^N; \ n \in \mathbb{N}$,

and

(iv)

$$\int_{\mathbb{R}^N} Z(x)\, dx = 1, \tag{7.19}$$

that is Z is a multivariate density function.

Here denote $\|x\|_\infty := \max\{|x_1|, ..., |x_N|\}$, $x \in \mathbb{R}^N$, also set $\infty := (\infty, ..., \infty)$, $-\infty := (-\infty, ..., -\infty)$ upon the multivariate context, and

$$\begin{aligned} \lceil na \rceil &:= (\lceil na_1 \rceil, ..., \lceil na_N \rceil), \\ \lfloor nb \rfloor &:= (\lfloor nb_1 \rfloor, ..., \lfloor nb_N \rfloor), \end{aligned} \tag{7.20}$$

where $a := (a_1, ..., a_N)$, $b := (b_1, ..., b_N)$.

We obviously see that

$$\sum_{k=\lceil na \rceil}^{\lfloor nb \rfloor} Z(nx-k) = \sum_{k=\lceil na \rceil}^{\lfloor nb \rfloor} \left(\prod_{i=1}^{N} \chi(nx_i - k_i) \right) =$$

$$\sum_{k_1=\lceil na_1 \rceil}^{\lfloor nb_1 \rfloor} \ldots \sum_{k_N=\lceil na_N \rceil}^{\lfloor nb_N \rfloor} \left(\prod_{i=1}^{N} \chi(nx_i - k_i) \right) = \prod_{i=1}^{N} \left(\sum_{k_i=\lceil na_i \rceil}^{\lfloor nb_i \rfloor} \chi(nx_i - k_i) \right). \tag{7.21}$$

For $0 < \beta < 1$ and $n \in \mathbb{N}$, a fixed $x \in \mathbb{R}^N$, we have that

$$\sum_{k=\lceil na \rceil}^{\lfloor nb \rfloor} Z(nx-k) =$$

$$\sum_{\begin{cases} k=\lceil na \rceil \\ \left\| \frac{k}{n} - x \right\|_{\infty} \le \frac{1}{n^{\beta}} \end{cases}}^{\lfloor nb \rfloor} Z(nx-k) + \sum_{\begin{cases} k=\lceil na \rceil \\ \left\| \frac{k}{n} - x \right\|_{\infty} > \frac{1}{n^{\beta}} \end{cases}}^{\lfloor nb \rfloor} Z(nx-k). \tag{7.22}$$

In the last two sums the counting is over disjoint vector sets of k's, because the condition $\left\| \frac{k}{n} - x \right\|_{\infty} > \frac{1}{n^{\beta}}$ implies that there exists at least one $\left| \frac{k_r}{n} - x_r \right| > \frac{1}{n^{\beta}}$, where $r \in \{1, ..., N\}$.

(v) As in [10], pp. 379–380, we derive that

$$\sum_{\begin{cases} k=\lceil na \rceil \\ \left\| \frac{k}{n} - x \right\|_{\infty} > \frac{1}{n^{\beta}} \end{cases}}^{\lfloor nb \rfloor} Z(nx-k) \overset{(7.12)}{<} \frac{1}{2\lambda \left(n^{1-\beta}-2\right)^{\lambda}}, \ 0<\beta<1, \ \lambda \text{ is odd,} \tag{7.23}$$

with $n \in \mathbb{N} : n^{1-\beta} > 2$, $x \in \prod_{i=1}^{N} [a_i, b_i]$.

(vi) By Theorem 7.4 we get that

$$0 < \frac{1}{\sum_{k=\lceil na \rceil}^{\lfloor nb \rfloor} Z(nx-k)} < \left(2\sqrt[\lambda]{1+2^{\lambda}}\right)^N, \tag{7.24}$$

$\forall\, x \in \left(\prod_{i=1}^{N} [a_i, b_i]\right)$, $n \in \mathbb{N}$, λ is odd.

It is also clear that

(vii)

$$\sum_{\begin{cases} k=-\infty \\ \left\| \frac{k}{n} - x \right\|_{\infty} > \frac{1}{n^{\beta}} \end{cases}}^{\infty} Z(nx-k) < \frac{1}{2\lambda \left(n^{1-\beta}-2\right)^{\lambda}}, \tag{7.25}$$

$0<\beta<1$, $n \in \mathbb{N} : n^{1-\beta} > 2$, $x \in \mathbb{R}^N$, λ is odd.

Furthermore it holds

$$\lim_{n \to \infty} \sum_{k=\lceil na \rceil}^{\lfloor nb \rfloor} Z(nx-k) \neq 1, \tag{7.26}$$

for at least some $x \in \left(\prod_{i=1}^{N} [a_i, b_i]\right)$.

Here $\left(X, \|\cdot\|_{\gamma}\right)$ is a Banach space.

Let $f \in C\left(\prod_{i=1}^{N}[a_i, b_i], X\right)$, $x = (x_1, ..., x_N) \in \prod_{i=1}^{N}[a_i, b_i]$, $n \in \mathbb{N}$ such that $\lceil na_i \rceil \leq \lfloor nb_i \rfloor$, $i = 1, ..., N$.

We introduce and define the following multivariate linear normalized neural network operator ($x := (x_1, ..., x_N) \in \left(\prod_{i=1}^{N}[a_i, b_i]\right)$):

$$A_n(f, x_1, ..., x_N) := A_n(f, x) := \frac{\sum_{k=\lceil na \rceil}^{\lfloor nb \rfloor} f\left(\frac{k}{n}\right) Z(nx - k)}{\sum_{k=\lceil na \rceil}^{\lfloor nb \rfloor} Z(nx - k)} =$$

$$\frac{\sum_{k_1=\lceil na_1 \rceil}^{\lfloor nb_1 \rfloor} \sum_{k_2=\lceil na_2 \rceil}^{\lfloor nb_2 \rfloor} \cdots \sum_{k_N=\lceil na_N \rceil}^{\lfloor nb_N \rfloor} f\left(\frac{k_1}{n}, ..., \frac{k_N}{n}\right)\left(\prod_{i=1}^{N} \chi(nx_i - k_i)\right)}{\prod_{i=1}^{N}\left(\sum_{k_i=\lceil na_i \rceil}^{\lfloor nb_i \rfloor} \chi(nx_i - k_i)\right)}. \quad (7.27)$$

For large enough $n \in \mathbb{N}$ we always obtain $\lceil na_i \rceil \leq \lfloor nb_i \rfloor$, $i = 1, ..., N$. Also $a_i \leq \frac{k_i}{n} \leq b_i$, iff $\lceil na_i \rceil \leq k_i \leq \lfloor nb_i \rfloor$, $i = 1, ..., N$.

When $g \in C\left(\prod_{i=1}^{N}[a_i, b_i]\right)$ we define the companion operator

$$\widetilde{A}_n(g, x) := \frac{\sum_{k=\lceil na \rceil}^{\lfloor nb \rfloor} g\left(\frac{k}{n}\right) Z(nx - k)}{\sum_{k=\lceil na \rceil}^{\lfloor nb \rfloor} Z(nx - k)}. \quad (7.28)$$

Clearly $\widetilde{A}_n$ is a positive linear operator. We have that

$$\widetilde{A}_n(1, x) = 1, \forall x \in \left(\prod_{i=1}^{N}[a_i, b_i]\right).$$

Notice that $A_n(f) \in C\left(\prod_{i=1}^{N}[a_i, b_i], X\right)$ and $\widetilde{A}_n(g) \in C\left(\prod_{i=1}^{N}[a_i, b_i]\right)$.

Furthermore it holds

$$\|A_n(f, x)\|_\gamma \leq \frac{\sum_{k=\lceil na \rceil}^{\lfloor nb \rfloor} \left\|f\left(\frac{k}{n}\right)\right\|_\gamma Z(nx - k)}{\sum_{k=\lceil na \rceil}^{\lfloor nb \rfloor} Z(nx - k)} = \widetilde{A}_n\left(\|f\|_\gamma, x\right), \quad (7.29)$$

$\forall x \in \prod_{i=1}^{N}[a_i, b_i]$.

Clearly $\|f\|_\gamma \in C\left(\prod_{i=1}^{N}[a_i, b_i]\right)$.

So, we have that

$$\|A_n(f, x)\|_\gamma \leq \widetilde{A}_n\left(\|f\|_\gamma, x\right), \quad (7.30)$$

$\forall x \in \prod_{i=1}^{N}[a_i, b_i]$, $\forall n \in \mathbb{N}$, $\forall f \in C\left(\prod_{i=1}^{N}[a_i, b_i], X\right)$.

Let $c \in X$ and $g \in C\left(\prod_{i=1}^{N}[a_i, b_i]\right)$, then $cg \in C\left(\prod_{i=1}^{N}[a_i, b_i], X\right)$.

Furthermore it holds

$$A_n(cg, x) = c\widetilde{A}_n(g, x), \forall x \in \prod_{i=1}^{N} [a_i, b_i]. \tag{7.31}$$

Since $\widetilde{A}_n(1) = 1$, we get that

$$A_n(c) = c, \forall c \in X. \tag{7.32}$$

We call $\widetilde{A}_n$ the companion operator of A_n.

For convinience we call

$$A_n^*(f, x) := \sum_{k=\lceil na \rceil}^{\lfloor nb \rfloor} f\left(\frac{k}{n}\right) Z(nx - k) =$$

$$\sum_{k_1=\lceil na_1 \rceil}^{\lfloor nb_1 \rfloor} \sum_{k_2=\lceil na_2 \rceil}^{\lfloor nb_2 \rfloor} \dots \sum_{k_N=\lceil na_N \rceil}^{\lfloor nb_N \rfloor} f\left(\frac{k_1}{n}, \dots, \frac{k_N}{n}\right) \left(\prod_{i=1}^{N} \chi(nx_i - k_i)\right), \tag{7.33}$$

$\forall x \in \left(\prod_{i=1}^{N} [a_i, b_i]\right)$.

That is

$$A_n(f, x) := \frac{A_n^*(f, x)}{\sum_{k=\lceil na \rceil}^{\lfloor nb \rfloor} Z(nx - k)}, \tag{7.34}$$

$\forall x \in \left(\prod_{i=1}^{N} [a_i, b_i]\right), n \in \mathbb{N}$.

Hence

$$A_n(f, x) - f(x) = \frac{A_n^*(f, x) - f(x)\left(\sum_{k=\lceil na \rceil}^{\lfloor nb \rfloor} Z(nx - k)\right)}{\sum_{k=\lceil na \rceil}^{\lfloor nb \rfloor} Z(nx - k)}. \tag{7.35}$$

Consequently we derive

$$\|A_n(f, x) - f(x)\|_\gamma \overset{(7.24.)}{\leq} \left(2\sqrt[\lambda]{1 + 2^\lambda}\right)^N \left\| A_n^*(f, x) - f(x) \sum_{k=\lceil na \rceil}^{\lfloor nb \rfloor} Z(nx - k) \right\|_\gamma, \tag{7.36}$$

$\forall x \in \left(\prod_{i=1}^{N} [a_i, b_i]\right)$.

We will estimate the right hand side of (7.36).

For the last and others we need

Definition 7.6 ([11], p. 274) Let M be a convex and compact subset of $\left(\mathbb{R}^N, \|\cdot\|_p\right)$, $p \in [1, \infty]$, and $\left(X, \|\cdot\|_\gamma\right)$ be a Banach space. Let $f \in C(M, X)$. We define the first modulus of continuity of f as

$$\omega_1(f,\delta) := \sup_{\substack{x, y \in M : \\ \|x-y\|_p \le \delta}} \|f(x) - f(y)\|_\gamma, \; 0 < \delta \le diam(M). \tag{7.37}$$

If $\delta > diam(M)$, then

$$\omega_1(f,\delta) = \omega_1(f, diam(M)). \tag{7.38}$$

Notice $\omega_1(f,\delta)$ is increasing in $\delta > 0$. For $f \in C_B(M, X)$ (continuous and bounded functions) $\omega_1(f,\delta)$ is defined similarly.

Lemma 7.7 ([11], p. 274) *We have* $\omega_1(f,\delta) \to 0$ *as* $\delta \downarrow 0$, *iff* $f \in C(M, X)$, *where* M *is a convex compact subset of* $\left(\mathbb{R}^N, \|\cdot\|_p\right)$, $p \in [1, \infty]$.

Clearly we have also: $f \in C_U\left(\mathbb{R}^N, X\right)$ (uniformly continuous functions), iff $\omega_1(f,\delta) \to 0$ as $\delta \downarrow 0$, where ω_1 is defined similarly to (7.37). The space $C_B\left(\mathbb{R}^N, X\right)$ denotes the continuous and bounded functions on $\mathbb{R}^N$.

When $f \in C_B\left(\mathbb{R}^N, X\right)$ we define,

$$B_n(f,x) := B_n(f, x_1, ..., x_N) := \sum_{k=-\infty}^{\infty} f\left(\frac{k}{n}\right) Z(nx-k) :=$$

$$\sum_{k_1=-\infty}^{\infty} \sum_{k_2=-\infty}^{\infty} ... \sum_{k_N=-\infty}^{\infty} f\left(\frac{k_1}{n}, \frac{k_2}{n}, ..., \frac{k_N}{n}\right) \left(\prod_{i=1}^{N} \chi(nx_i - k_i)\right), \tag{7.39}$$

$n \in \mathbb{N}$, $\forall\, x \in \mathbb{R}^N$, $N \in \mathbb{N}$, the multivariate quasi-interpolation neural network operator.

Also for $f \in C_B\left(\mathbb{R}^N, X\right)$ we define the multivariate Kantorovich type neural network operator

$$C_n(f,x) := C_n(f, x_1, ..., x_N) := \sum_{k=-\infty}^{\infty} \left(n^N \int_{\frac{k}{n}}^{\frac{k+1}{n}} f(t)\,dt\right) Z(nx-k) =$$

$$\sum_{k_1=-\infty}^{\infty} \sum_{k_2=-\infty}^{\infty} ... \sum_{k_N=-\infty}^{\infty} \left(n^N \int_{\frac{k_1}{n}}^{\frac{k_1+1}{n}} \int_{\frac{k_2}{n}}^{\frac{k_2+1}{n}} ... \int_{\frac{k_N}{n}}^{\frac{k_N+1}{n}} f(t_1, ..., t_N)\,dt_1 ... dt_N\right)$$

$$\cdot \left(\prod_{i=1}^{N} \chi(nx_i - k_i)\right), \tag{7.40}$$

$n \in \mathbb{N}$, $\forall\, x \in \mathbb{R}^N$.

Again for $f \in C_B\left(\mathbb{R}^N, X\right)$, $N \in \mathbb{N}$, we define the multivariate neural network operator of quadrature type $D_n(f,x)$, $n \in \mathbb{N}$, as follows.

Let $\theta = (\theta_1, ..., \theta_N) \in \mathbb{N}^N$, $r = (r_1, ..., r_N) \in \mathbb{Z}_+^N$, $w_r = w_{r_1,r_2,...r_N} \geq 0$, such that $\sum_{r=0}^{\theta} w_r = \sum_{r_1=0}^{\theta_1} \sum_{r_2=0}^{\theta_2} ... \sum_{r_N=0}^{\theta_N} w_{r_1,r_2,...r_N} = 1$; $k \in \mathbb{Z}^N$ and

$$\delta_{nk}(f) := \delta_{n,k_1,k_2,...,k_N}(f) := \sum_{r=0}^{\theta} w_r f\left(\frac{k}{n} + \frac{r}{n\theta}\right) =$$

$$\sum_{r_1=0}^{\theta_1} \sum_{r_2=0}^{\theta_2} \cdots \sum_{r_N=0}^{\theta_N} w_{r_1,r_2,...r_N} f\left(\frac{k_1}{n} + \frac{r_1}{n\theta_1}, \frac{k_2}{n} + \frac{r_2}{n\theta_2}, ..., \frac{k_N}{n} + \frac{r_N}{n\theta_N}\right), \tag{7.41}$$

where $\frac{r}{\theta} := \left(\frac{r_1}{\theta_1}, \frac{r_2}{\theta_2}, ..., \frac{r_N}{\theta_N}\right)$.

We set

$$D_n(f, x) := D_n(f, x_1, ..., x_N) := \sum_{k=-\infty}^{\infty} \delta_{nk}(f) Z(nx - k) = \tag{7.42}$$

$$\sum_{k_1=-\infty}^{\infty} \sum_{k_2=-\infty}^{\infty} \cdots \sum_{k_N=-\infty}^{\infty} \delta_{n,k_1,k_2,...,k_N}(f) \left(\prod_{i=1}^{N} \chi(nx_i - k_i)\right),$$

$\forall\, x \in \mathbb{R}^N$.

In this chapter we study the approximation properties of A_n, B_n, C_n, D_n neural network operators and as well of their iterates. That is, the quantitative pointwise and uniform convergence of these operators to the unit operator I.

7.3 Multivariate General Neural Network Approximations

Here we present several vectorial neural network approximations to Banach space valued functions given with rates.

We give

Theorem 7.8 *Let* $f \in C\left(\prod_{i=1}^N [a_i, b_i], X\right)$, $0 < \beta < 1$, $x \in \left(\prod_{i=1}^N [a_i, b_i]\right)$, λ *is odd,* $N, n \in \mathbb{N}$ *with* $n^{1-\beta} > 2$. *Then*

(1)

$$\|A_n(f, x) - f(x)\|_\gamma \leq \left(2\sqrt[\lambda]{1 + 2^\lambda}\right)^N \left[\omega_1\left(f, \frac{1}{n^\beta}\right) + \frac{\| \|f\|_\gamma \|_\infty}{\lambda\left(n^{1-\beta} - 2\right)^\lambda}\right] =: \lambda_1(n), \tag{7.43}$$

and

(2)

$$\left\| \|A_n(f) - f\|_\gamma \right\|_\infty \leq \lambda_1(n). \tag{7.44}$$

We notice that $\lim_{n\to\infty} A_n(f) \overset{\|\cdot\|_\gamma}{=} f$, *pointwise and uniformly.*

Above ω_1 *is with respect to* $p = \infty$.

Proof We observe that

$$\Delta(x) := A_n^*(f, x) - f(x) \sum_{k=\lceil na \rceil}^{\lfloor nb \rfloor} Z(nx - k) =$$

$$\sum_{k=\lceil na \rceil}^{\lfloor nb \rfloor} f\left(\frac{k}{n}\right) Z(nx - k) - \sum_{k=\lceil na \rceil}^{\lfloor nb \rfloor} f(x) Z(nx - k) =$$

$$\sum_{k=\lceil na \rceil}^{\lfloor nb \rfloor} \left(f\left(\frac{k}{n}\right) - f(x) \right) Z(nx - k). \tag{7.45}$$

Thus

$$\|\Delta(x)\|_\gamma \leq \sum_{k=\lceil na \rceil}^{\lfloor nb \rfloor} \left\| f\left(\frac{k}{n}\right) - f(x) \right\|_\gamma Z(nx - k) =$$

$$\sum_{\substack{k=\lceil na \rceil \\ \left\| \frac{k}{n} - x \right\|_\infty \leq \frac{1}{n^\beta}}}^{\lfloor nb \rfloor} \left\| f\left(\frac{k}{n}\right) - f(x) \right\|_\gamma Z(nx - k) +$$

$$\sum_{\substack{k=\lceil na \rceil \\ \left\| \frac{k}{n} - x \right\|_\infty > \frac{1}{n^\beta}}}^{\lfloor nb \rfloor} \left\| f\left(\frac{k}{n}\right) - f(x) \right\|_\gamma Z(nx - k) \overset{(7.18.)}{\leq}$$

$$\omega_1\left(f, \frac{1}{n^\beta}\right) + 2 \left\| \|f\|_\gamma \right\|_\infty \sum_{\substack{k=\lceil na \rceil \\ \left\| \frac{k}{n} - x \right\|_\infty > \frac{1}{n^\beta}}}^{\lfloor nb \rfloor} Z(nx - k) \overset{(7.23.)}{\leq}$$

$$\omega_1\left(f, \frac{1}{n^\beta}\right) + \frac{\left\| \|f\|_\gamma \right\|_\infty}{\lambda \left(n^{1-\beta} - 2\right)^\lambda}. \tag{7.46}$$

So that

$$\left\| \Delta (x) \right\|_{\gamma} \leq \omega_1 \left(f, \frac{1}{n^{\beta}} \right) + \frac{\left\| \left\| f \right\|_{\gamma} \right\|_{\infty}}{\lambda \left(n^{1-\beta} - 2 \right)^{\lambda}}. \tag{7.47}$$

Now using (7.36) we finish the proof. ■

We make

Remark 7.9 ([11], pp. 263–266) Let $\left(\mathbb{R}^N, \left\| \cdot \right\|_p \right)$, $N \in \mathbb{N}$; where $\left\| \cdot \right\|_p$ is the L_p-norm, $1 \leq p \leq \infty$. $\mathbb{R}^N$ is a Banach space, and $\left(\mathbb{R}^N \right)^j$ denotes the j-fold product space $\mathbb{R}^N \times ... \times \mathbb{R}^N$ endowed with the max-norm $\left\| x \right\|_{\left(\mathbb{R}^N \right)^j} := \max_{1 \leq \lambda \leq j} \left\| x_{\lambda} \right\|_p$, where $x := \left(x_1, ..., x_j \right) \in \left(\mathbb{R}^N \right)^j$.

Let $\left(X, \left\| \cdot \right\|_{\gamma} \right)$ be a general Banach space. Then the space $L_j := L_j \left(\left(\mathbb{R}^N \right)^j ; X \right)$ of all j-multilinear continuous maps $g : \left(\mathbb{R}^N \right)^j \to X$, $j = 1, ..., m$, is a Banach space with norm

$$\left\| g \right\| := \left\| g \right\|_{L_j} := \sup_{\left(\left\| x \right\|_{\left(\mathbb{R}^N \right)^j} = 1 \right)} \left\| g (x) \right\|_{\gamma} = \sup \frac{\left\| g (x) \right\|_{\gamma}}{\left\| x_1 \right\|_p ... \left\| x_j \right\|_p}. \tag{7.48}$$

Let M be a non-empty convex and compact subset of $\mathbb{R}^N$ and $x_0 \in M$ is fixed.

Let O be an open subset of $\mathbb{R}^N : M \subset O$. Let $f : O \to X$ be a continuous function, whose Fréchet derivatives (see [26]) $f^{(j)} : O \to L_j = L_j \left(\left(\mathbb{R}^N \right)^j ; X \right)$ exist and are continuous for $1 \leq j \leq m$, $m \in \mathbb{N}$.

Call $(x - x_0)^j := (x - x_0, ..., x - x_0) \in \left(\mathbb{R}^N \right)^j$, $x \in M$.

We will work with $f|_M$.

Then, by Taylor's formula [18], ([26], p. 124), we get

$$f (x) = \sum_{j=0}^{m} \frac{f^{(j)} (x_0) (x - x_0)^j}{j!} + R_m (x, x_0), \text{ all } x \in M, \tag{7.49}$$

where the remainder is the Riemann integral

$$R_m (x, x_0) := \int_0^1 \frac{(1 - u)^{m-1}}{(m - 1)!} \left(f^{(m)} (x_0 + u (x - x_0)) - f^{(m)} (x_0) \right) (x - x_0)^m du, \tag{7.50}$$

here we set $f^{(0)} (x_0) (x - x_0)^0 = f (x_0)$.

We consider

$$w := \omega_1 \left(f^{(m)}, h \right) := \sup_{\substack{x, y \in M: \\ \left\| x - y \right\|_p \leq h}} \left\| f^{(m)} (x) - f^{(m)} (y) \right\|, \tag{7.51}$$

$h > 0$.

We obtain

$$\left\| \left(f^{(m)} \left(x_0 + u \left(x - x_0 \right) \right) - f^{(m)} \left(x_0 \right) \right) \left(x - x_0 \right)^m \right\|_\gamma \leq$$

$$\left\| f^{(m)} \left(x_0 + u \left(x - x_0 \right) \right) - f^{(m)} \left(x_0 \right) \right\| \cdot \left\| x - x_0 \right\|_p^m \leq$$

$$w \left\| x - x_0 \right\|_p^m \left\lceil \frac{u \left\| x - x_0 \right\|_p}{h} \right\rceil, \tag{7.52}$$

by Lemma 7.1.1, [1], p. 208, where $\lceil \cdot \rceil$ is the ceiling.

Therefore for all $x \in M$ (see [1], pp. 121–122):

$$\left\| R_m \left(x, x_0 \right) \right\|_\gamma \leq w \left\| x - x_0 \right\|_p^m \int_0^1 \left\lceil \frac{u \left\| x - x_0 \right\|_p}{h} \right\rceil \frac{(1-u)^{m-1}}{(m-1)!} du$$

$$= w \Phi_m \left(\left\| x - x_0 \right\|_p \right) \tag{7.53}$$

by a change of variable, where

$$\Phi_m(t) := \int_0^{|t|} \left\lceil \frac{s}{h} \right\rceil \frac{(|t| - s)^{m-1}}{(m-1)!} ds = \frac{1}{m!} \left(\sum_{j=0}^{\infty} (|t| - jh)_+^m \right), \ \forall t \in \mathbb{R}, \tag{7.54}$$

is a (polynomial) spline function, see [1], pp. 210–211.

Also from there we get

$$\Phi_m(t) \leq \left(\frac{|t|^{m+1}}{(m+1)!h} + \frac{|t|^m}{2m!} + \frac{h |t|^{m-1}}{8(m-1)!} \right), \ \forall t \in \mathbb{R}, \tag{7.55}$$

with equality true only at $t = 0$.

Therefore it holds

$$\left\| R_m \left(x, x_0 \right) \right\|_\gamma \leq w \left(\frac{\left\| x - x_0 \right\|_p^{m+1}}{(m+1)!h} + \frac{\left\| x - x_0 \right\|_p^m}{2m!} + \frac{h \left\| x - x_0 \right\|_p^{m-1}}{8(m-1)!} \right), \ \forall x \in M. \tag{7.56}$$

We have found that

$$\left\| f(x) - \sum_{j=0}^{m} \frac{f^{(j)} \left(x_0 \right) \left(x - x_0 \right)^j}{j!} \right\|_\gamma \leq$$

$$\omega_1 \left(f^{(m)}, h \right) \left(\frac{\left\| x - x_0 \right\|_p^{m+1}}{(m+1)!h} + \frac{\left\| x - x_0 \right\|_p^m}{2m!} + \frac{h \left\| x - x_0 \right\|_p^{m-1}}{8(m-1)!} \right) < \infty, \tag{7.57}$$

$\forall\, x, x_0 \in M$.

Here $0 < \omega_1 \left(f^{(m)}, h\right) < \infty$, by M being compact and $f^{(m)}$ being continuous on M.

One can rewrite (7.57) as follows:

$$\left\| f(\cdot) - \sum_{j=0}^{m} \frac{f^{(j)}(x_0)(\cdot - x_0)^j}{j!} \right\|_{\gamma} \leq$$

$$\omega_1 \left(f^{(m)}, h\right) \left(\frac{\|\cdot - x_0\|_p^{m+1}}{(m+1)!h} + \frac{\|\cdot - x_0\|_p^{m}}{2m!} + \frac{h \|\cdot - x_0\|_p^{m-1}}{8(m-1)!} \right), \ \forall\, x_0 \in M, \quad (7.58)$$

a pointwise functional inequality on M.

Here $(\cdot - x_0)^j$ maps M into $\left(\mathbb{R}^N\right)^j$ and it is continuous, also $f^{(j)}(x_0)$ maps $\left(\mathbb{R}^N\right)^j$ into X and it is continuous. Hence their composition $f^{(j)}(x_0)(\cdot - x_0)^j$ is continuous from M into X.

Clearly $f(\cdot) - \sum_{j=0}^{m} \frac{f^{(j)}(x_0)(\cdot - x_0)^j}{j!} \in C(M, X)$, hence $\left\| f(\cdot) - \sum_{j=0}^{m} \frac{f^{(j)}(x_0)(\cdot - x_0)^j}{j!} \right\|_{\gamma} \in C(M)$.

Let $\left\{\widetilde{L}_N\right\}_{N \in \mathbb{N}}$ be a sequence of positive linear operators mapping $C(M)$ into $C(M)$.

Therefore we obtain

$$\left(\widetilde{L}_N \left(\left\| f(\cdot) - \sum_{j=0}^{m} \frac{f^{(j)}(x_0)(\cdot - x_0)^j}{j!} \right\|_{\gamma} \right) \right) (x_0) \leq$$

$$\omega_1 \left(f^{(m)}, h\right) \left[\frac{\left(\widetilde{L}_N \left(\|\cdot - x_0\|_p^{m+1}\right)\right)(x_0)}{(m+1)!h} + \frac{\left(\widetilde{L}_N \left(\|\cdot - x_0\|_p^{m}\right)\right)(x_0)}{2m!} \right.$$

$$\left. + \frac{h \left(\widetilde{L}_N \left(\|\cdot - x_0\|_p^{m-1}\right)\right)(x_0)}{8(m-1)!} \right], \quad (7.59)$$

$\forall\, N \in \mathbb{N}, \forall\, x_0 \in M$.

Clearly (7.59) is valid when $M = \prod_{i=1}^{N} [a_i, b_i]$ and $\widetilde{L}_n = \widetilde{A}_n$, see (7.28).

All the above is preparation for the following theorem, where we assume Fréchet differentiability of functions.

This will be a direct application of Theorem 10.2, [11], pp. 268–270. The operators A_n, $\widetilde{A}_n$ fulfill its assumptions, see (7.27), (7.28), (7.30), (7.31) and (7.32).

We present the following high order approximation results.

Theorem 7.10 *Let O open subset of $\left(\mathbb{R}^N, \|\cdot\|_p\right)$, $p \in [1, \infty]$, such that $\prod_{i=1}^{N} [a_i, b_i] \subset O \subseteq \mathbb{R}^N$, and let $\left(X, \|\cdot\|_\gamma\right)$ be a general Banach space. Let $m \in \mathbb{N}$ and $f \in C^m(O, X)$, the space of m-times continuously Fréchet differentiable functions from O into X. We study the approximation of $f|_{\prod_{i=1}^{N}[a_i,b_i]}$. Let $x_0 \in \left(\prod_{i=1}^{N} [a_i, b_i]\right)$ and $r > 0$. Then*

(1)

$$\left\| (A_n(f))(x_0) - \sum_{j=0}^{m} \frac{1}{j!} \left(A_n\left(f^{(j)}(x_0)(\cdot - x_0)^j\right)\right)(x_0) \right\|_\gamma \leq$$

$$\frac{\omega_1\left(f^{(m)}, r\left(\left(\widetilde{A}_n\left(\|\cdot - x_0\|_p^{m+1}\right)\right)(x_0)\right)^{\frac{1}{m+1}}\right)}{rm!} \left(\left(\widetilde{A}_n\left(\|\cdot - x_0\|_p^{m+1}\right)\right)(x_0)\right)^{\left(\frac{m}{m+1}\right)}$$

$$\left[\frac{1}{(m+1)} + \frac{r}{2} + \frac{mr^2}{8}\right], \tag{7.60}$$

(2) additionally if $f^{(j)}(x_0) = 0$, $j = 1, ..., m$, we have

$$\|(A_n(f))(x_0) - f(x_0)\|_\gamma \leq$$

$$\frac{\omega_1\left(f^{(m)}, r\left(\left(\widetilde{A}_n\left(\|\cdot - x_0\|_p^{m+1}\right)\right)(x_0)\right)^{\frac{1}{m+1}}\right)}{rm!} \left(\left(\widetilde{A}_n\left(\|\cdot - x_0\|_p^{m+1}\right)\right)(x_0)\right)^{\left(\frac{m}{m+1}\right)}$$

$$\left[\frac{1}{(m+1)} + \frac{r}{2} + \frac{mr^2}{8}\right], \tag{7.61}$$

(3)

$$\|(A_n(f))(x_0) - f(x_0)\|_\gamma \leq \sum_{j=1}^{m} \frac{1}{j!} \left\| \left(A_n\left(f^{(j)}(x_0)(\cdot - x_0)^j\right)\right)(x_0) \right\|_\gamma +$$

$$\frac{\omega_1\left(f^{(m)}, r\left(\left(\widetilde{A}_n\left(\|\cdot - x_0\|_p^{m+1}\right)\right)(x_0)\right)^{\frac{1}{m+1}}\right)}{rm!} \left(\left(\widetilde{A}_n\left(\|\cdot - x_0\|_p^{m+1}\right)\right)(x_0)\right)^{\left(\frac{m}{m+1}\right)}$$

$$\left[\frac{1}{(m+1)} + \frac{r}{2} + \frac{mr^2}{8}\right], \tag{7.62}$$

and
(4)

$$\left\| \left\| A_n(f) - f \right\|_{\gamma} \right\|_{\infty, \prod_{i=1}^{N}[a_i,b_i]} \leq$$

$$\sum_{j=1}^{m} \frac{1}{j!} \left\| \left\| \left(A_n \left(f^{(j)}(x_0)(\cdot - x_0)^j \right) \right)(x_0) \right\|_{\gamma} \right\|_{\infty, x_0 \in \prod_{i=1}^{N}[a_i,b_i]} +$$

$$\frac{\omega_1 \left(f^{(m)}, r \left\| \left(\widetilde{A}_n \left(\|\cdot - x_0\|_p^{m+1} \right) \right)(x_0) \right\|_{\infty, x_0 \in \prod_{i=1}^{N}[a_i,b_i]}^{\frac{1}{m+1}} \right)}{rm!}$$

$$\left\| \left(\widetilde{A}_n \left(\|\cdot - x_0\|_p^{m+1} \right) \right)(x_0) \right\|_{\infty, x_0 \in \prod_{i=1}^{N}[a_i,b_i]}^{\left(\frac{m}{m+1}\right)} \tag{7.63}$$

$$\left[\frac{1}{(m+1)} + \frac{r}{2} + \frac{mr^2}{8} \right].$$

We need

Lemma 7.11 *The function* $\left(\widetilde{A}_n \left(\|\cdot - x_0\|_p^m \right) \right)(x_0)$ *is continuous in* $x_0 \in \left(\prod_{i=1}^{N} [a_i, b_i] \right)$, $m \in \mathbb{N}$.

Proof By Lemma 10.3, [11], p. 272. ■

We give

Corollary 7.12 (to Theorem 7.10, case of $m = 1$) *Then*
(1)

$$\| (A_n(f))(x_0) - f(x_0) \|_{\gamma} \leq \left\| \left(A_n \left(f^{(1)}(x_0)(\cdot - x_0) \right) \right)(x_0) \right\|_{\gamma} +$$

$$\frac{1}{2r} \omega_1 \left(f^{(1)}, r \left(\left(\widetilde{A}_n \left(\|\cdot - x_0\|_p^2 \right) \right)(x_0) \right)^{\frac{1}{2}} \right) \left(\left(\widetilde{A}_n \left(\|\cdot - x_0\|_p^2 \right) \right)(x_0) \right)^{\frac{1}{2}} \tag{7.64}$$

$$\left[1 + r + \frac{r^2}{4} \right],$$

and
(2)

$$\left\| \| (A_n(f)) - f \|_\gamma \right\|_{\infty, \prod_{i=1}^{N} [a_i, b_i]} \leq$$

$$\left\| \left\| \left(A_n \left(f^{(1)} (x_0) (\cdot - x_0) \right) \right) (x_0) \right\|_\gamma \right\|_{\infty, x_0 \in \prod_{i=1}^{N} [a_i, b_i]} +$$

$$\frac{1}{2r} \omega_1 \left(f^{(1)}, r \left\| \left(\widetilde{A}_n \left(\| \cdot - x_0 \|_p^2 \right) \right) (x_0) \right\|_{\infty, x_0 \in \prod_{i=1}^{N} [a_i, b_i]}^{\frac{1}{2}} \right)$$

$$\left\| \left(\widetilde{A}_n \left(\| \cdot - x_0 \|_p^2 \right) \right) (x_0) \right\|_{\infty, x_0 \in \prod_{i=1}^{N} [a_i, b_i]}^{\frac{1}{2}} \left[1 + r + \frac{r^2}{4} \right], \tag{7.65}$$

$r > 0$.

We make

Remark 7.13 We estimate $0 < \alpha < 1$, $m, n \in \mathbb{N} : n^{1-\alpha} > 2$,

$$\widetilde{A}_n \left(\| \cdot - x_0 \|_\infty^{m+1} \right) (x_0) = \frac{\sum_{k=\lceil na \rceil}^{\lfloor nb \rfloor} \left\| \frac{k}{n} - x_0 \right\|_\infty^{m+1} Z(nx_0 - k)}{\sum_{k=\lceil na \rceil}^{\lfloor nb \rfloor} Z(nx_0 - k)} \overset{(7.24.)}{<}$$

$$\left(2 \sqrt[\lambda]{1 + 2^\lambda} \right)^N \sum_{k=\lceil na \rceil}^{\lfloor nb \rfloor} \left\| \frac{k}{n} - x_0 \right\|_\infty^{m+1} Z(nx_0 - k) = \tag{7.66}$$

$$\left(2 \sqrt[\lambda]{1 + 2^\lambda} \right)^N \left\{ \sum_{\substack{k = \lceil na \rceil \\ : \left\| \frac{k}{n} - x_0 \right\|_\infty \leq \frac{1}{n^\alpha}}}^{\lfloor nb \rfloor} \left\| \frac{k}{n} - x_0 \right\|_\infty^{m+1} Z(nx_0 - k) + \right.$$

$$\left. \sum_{\substack{k = \lceil na \rceil \\ : \left\| \frac{k}{n} - x_0 \right\|_\infty > \frac{1}{n^\alpha}}}^{\lfloor nb \rfloor} \left\| \frac{k}{n} - x_0 \right\|_\infty^{m+1} Z(nx_0 - k) \right\} \overset{(7.25.)}{\leq}$$

$$\left(2 \sqrt[\lambda]{1 + 2^\lambda} \right)^N \left\{ \frac{1}{n^{\alpha(m+1)}} + \frac{\| b - a \|_\infty^{m+1}}{2\lambda \left(n^{1-\alpha} - 2 \right)^\lambda} \right\}, \tag{7.67}$$

(where $b - a = (b_1 - a_1, ..., b_N - a_N)$).

We have proved that ($\forall\, x_0 \in \prod_{i=1}^{N} [a_i, b_i]$)

$$\widetilde{A}_n \left(\|\cdot - x_0\|_\infty^{m+1}\right)(x_0) < \left(2\sqrt[\lambda]{1+2^\lambda}\right)^N \left\{\frac{1}{n^{\alpha(m+1)}} + \frac{\|b-a\|_\infty^{m+1}}{2\lambda\left(n^{1-\alpha}-2\right)^\lambda}\right\} =: \varphi_1(n) \tag{7.68}$$

($0 < \alpha < 1, m, n \in \mathbb{N} : n^{1-\alpha} > 2$).

And, consequently it holds

$$\left\|\widetilde{A}_n \left(\|\cdot - x_0\|_\infty^{m+1}\right)(x_0)\right\|_{\infty, x_0 \in \prod_{i=1}^{N}[a_i, b_i]} <$$

$$\left(2\sqrt[\lambda]{1+2^\lambda}\right)^N \left\{\frac{1}{n^{\alpha(m+1)}} + \frac{\|b-a\|_\infty^{m+1}}{2\lambda\left(n^{1-\alpha}-2\right)^\lambda}\right\} = \varphi_1(n) \to 0, \text{ as } n \to +\infty. \tag{7.69}$$

So, we have that $\varphi_1(n) \to 0$, as $n \to +\infty$. Thus, when $p \in [1, \infty]$, from Theorem 7.10 we have the convergence to zero in the right hand sides of parts (1), (2).

Next we estimate $\left\|\left(\widetilde{A}_n \left(f^{(j)}(x_0)(\cdot - x_0)^j\right)\right)(x_0)\right\|_\gamma$.

We have that

$$\left(\widetilde{A}_n \left(f^{(j)}(x_0)(\cdot - x_0)^j\right)\right)(x_0) = \frac{\sum_{k=\lceil na\rceil}^{\lfloor nb\rfloor} f^{(j)}(x_0)\left(\frac{k}{n} - x_0\right)^j Z(nx_0 - k)}{\sum_{k=\lceil na\rceil}^{\lfloor nb\rfloor} Z(nx_0 - k)}. \tag{7.70}$$

When $p = \infty$, $j = 1, ..., m$, we obtain

$$\left\|f^{(j)}(x_0)\left(\frac{k}{n} - x_0\right)^j\right\|_\gamma \leq \left\|f^{(j)}(x_0)\right\| \left\|\frac{k}{n} - x_0\right\|_\infty^j. \tag{7.71}$$

We further have that

$$\left\|\left(\widetilde{A}_n \left(f^{(j)}(x_0)(\cdot - x_0)^j\right)\right)(x_0)\right\|_\gamma \overset{(7.24.)}{<}$$

$$\left(2\sqrt[\lambda]{1+2^\lambda}\right)^N \left(\sum_{k=\lceil na\rceil}^{\lfloor nb\rfloor} \left\|f^{(j)}(x_0)\left(\frac{k}{n} - x_0\right)^j\right\|_\gamma Z(nx_0 - k)\right) \leq$$

$$\left(2\sqrt[\lambda]{1+2^\lambda}\right)^N \left(\sum_{k=\lceil na\rceil}^{\lfloor nb\rfloor} \left\|f^{(j)}(x_0)\right\| \left\|\frac{k}{n} - x_0\right\|_\infty^j Z(nx_0 - k)\right) = \tag{7.72}$$

$$\left(2\sqrt[\lambda]{1+2^{\lambda}}\right)^{N}\left\|f^{(j)}(x_0)\right\|\left(\sum_{k=\lceil na\rceil}^{\lfloor nb\rfloor}\left\|\frac{k}{n}-x_0\right\|_{\infty}^{j}Z(nx_0-k)\right)=$$

$$\left(2\sqrt[\lambda]{1+2^{\lambda}}\right)^{N}\left\|f^{(j)}(x_0)\right\|\left\{\sum_{\substack{k=\lceil na\rceil \\ \left\{:\left\|\frac{k}{n}-x_0\right\|_{\infty}\leq\frac{1}{n^{\alpha}}\right.}}^{\lfloor nb\rfloor}\left\|\frac{k}{n}-x_0\right\|_{\infty}^{j}Z(nx_0-k)\right.$$

$$\left.+\sum_{\substack{k=\lceil na\rceil \\ \left\{:\left\|\frac{k}{n}-x_0\right\|_{\infty}>\frac{1}{n^{\alpha}}\right.}}^{\lfloor nb\rfloor}\left\|\frac{k}{n}-x_0\right\|_{\infty}^{j}Z(nx_0-k)\right\}\overset{(7.25.)}{\leq} \tag{7.73}$$

$$\left(2\sqrt[\lambda]{1+2^{\lambda}}\right)^{N}\left\|f^{(j)}(x_0)\right\|\left\{\frac{1}{n^{\alpha j}}+\frac{\|b-a\|_{\infty}^{j}}{2\lambda\left(n^{1-\alpha}-2\right)^{\lambda}}\right\}\to 0,\text{ as } n\to\infty.$$

That is

$$\left\|\left(\widetilde{A}_n\left(f^{(j)}(x_0)(\cdot-x_0)^{j}\right)\right)(x_0)\right\|_{\gamma}\to 0,\text{ as } n\to\infty.$$

Therefore when $p=\infty$, for $j=1,\ldots,m$, we have proved:

$$\left\|\left(\widetilde{A}_n\left(f^{(j)}(x_0)(\cdot-x_0)^{j}\right)\right)(x_0)\right\|_{\gamma}<$$

$$\left(2\sqrt[\lambda]{1+2^{\lambda}}\right)^{N}\left\|f^{(j)}(x_0)\right\|\left\{\frac{1}{n^{\alpha j}}+\frac{\|b-a\|_{\infty}^{j}}{2\lambda\left(n^{1-\alpha}-2\right)^{\lambda}}\right\}\leq \tag{7.74}$$

$$\left(2\sqrt[\lambda]{1+2^{\lambda}}\right)^{N}\left\|f^{(j)}\right\|_{\infty}\left\{\frac{1}{n^{\alpha j}}+\frac{\|b-a\|_{\infty}^{j}}{2\lambda\left(n^{1-\alpha}-2\right)^{\lambda}}\right\}=:\varphi_{2j}(n)<\infty,$$

and converges to zero, as $n\to\infty$.

We conclude:

In Theorem 7.10, the right hand sides of (7.62) and (7.63) converge to zero as $n\to\infty$, for any $p\in[1,\infty]$.

Also in Corollary 7.12, the right hand sides of (7.64) and (7.65) converge to zero as $n\to\infty$, for any $p\in[1,\infty]$.

Conclusion 7.14 *We have proved that the left hand sides of (7.60), (7.61), (7.62), (7.63) and (7.64), (7.65) converge to zero as $n \to \infty$, for $p \in [1, \infty]$. Consequently $A_n \to I$ (unit operator) pointwise and uniformly, as $n \to \infty$, where $p \in [1, \infty]$. In the presence of initial conditions we achieve a higher speed of convergence, see (7.61). Higher speed of convergence happens also to the left hand side of (7.60).*

We further give

Corollary 7.15 (to Theorem 7.10) *Let O open subset of $\left(\mathbb{R}^N, \|\cdot\|_\infty\right)$, such that $\prod_{i=1}^{N} [a_i, b_i] \subset O \subseteq \mathbb{R}^N$, and let $\left(X, \|\cdot\|_\gamma\right)$ be a general Banach space. Let $m \in \mathbb{N}$ and $f \in C^m(O, X)$, the space of $m-$times continuously Fréchet differentiable functions from O into X. We study the approximation of $f|_{\prod_{i=1}^{N}[a_i,b_i]}$. Let $x_0 \in \left(\prod_{i=1}^{N} [a_i, b_i]\right)$ and $r > 0$. Here $\varphi_1(n)$ as in (7.69) and $\varphi_{2j}(n)$ as in (7.74), where $n \in \mathbb{N} : n^{1-\alpha} > 2$, $0 < \alpha < 1$, $j = 1, ..., m$. Then*

(1)

$$\left\| (A_n(f))(x_0) - \sum_{j=0}^{m} \frac{1}{j!} \left(A_n\left(f^{(j)}(x_0)(\cdot - x_0)^j\right)\right)(x_0) \right\|_\gamma \leq$$

$$\frac{\omega_1\left(f^{(m)}, r(\varphi_1(n))^{\frac{1}{m+1}}\right)}{rm!} (\varphi_1(n))^{\left(\frac{m}{m+1}\right)} \left[\frac{1}{(m+1)} + \frac{r}{2} + \frac{mr^2}{8}\right], \tag{7.75}$$

(2) additionally, if $f^{(j)}(x_0) = 0$, $j = 1, ..., m$, we have

$$\|(A_n(f))(x_0) - f(x_0)\|_\gamma \leq$$

$$\frac{\omega_1\left(f^{(m)}, r(\varphi_1(n))^{\frac{1}{m+1}}\right)}{rm!} (\varphi_1(n))^{\left(\frac{m}{m+1}\right)} \left[\frac{1}{(m+1)} + \frac{r}{2} + \frac{mr^2}{8}\right], \tag{7.76}$$

(3)

$$\left\| \|A_n(f) - f\|_\gamma \right\|_{\infty, \prod_{i=1}^{N}[a_i,b_i]} \leq \sum_{j=1}^{m} \frac{\varphi_{2j}(n)}{j!} +$$

$$\frac{\omega_1\left(f^{(m)}, r(\varphi_1(n))^{\frac{1}{m+1}}\right)}{rm!} (\varphi_1(n))^{\left(\frac{m}{m+1}\right)} \tag{7.77}$$

$$\left[\frac{1}{(m+1)} + \frac{r}{2} + \frac{mr^2}{8}\right] =: \varphi_3(n) \to 0, \text{ as } n \to \infty.$$

We continue with

Theorem 7.16 *Let $f \in C_B\left(\mathbb{R}^N, X\right)$, $0 < \beta < 1$, $x \in \mathbb{R}^N$, λ is odd, $N, n \in \mathbb{N}$ with $n^{1-\beta} > 2$, ω_1 is for $p = \infty$. Then*

(1)

$$\left\| B_n(f, x) - f(x) \right\|_\gamma \leq \omega_1\left(f, \frac{1}{n^\beta}\right) + \frac{\left\| \|f\|_\gamma \right\|_\infty}{\lambda\left(n^{1-\beta} - 2\right)^\lambda} =: \lambda_2(n), \tag{7.78}$$

(2)

$$\left\| \left\| B_n(f) - f \right\|_\gamma \right\|_\infty \leq \lambda_2(n). \tag{7.79}$$

Given that $f \in \left(C_U\left(\mathbb{R}^N, X\right) \cap C_B\left(\mathbb{R}^N, X\right)\right)$, we obtain $\lim_{n \to \infty} B_n(f) = f$, uniformly.

Proof We have that

$$B_n(f, x) - f(x) \overset{(7.18.)}{=} \sum_{k=-\infty}^{\infty} f\left(\frac{k}{n}\right) Z(nx - k) - f(x) \sum_{k=-\infty}^{\infty} Z(nx - k) =$$
$$\sum_{k=-\infty}^{\infty} \left(f\left(\frac{k}{n}\right) - f(x)\right) Z(nx - k). \tag{7.80}$$

Hence

$$\left\| B_n(f, x) - f(x) \right\|_\gamma \leq \sum_{k=-\infty}^{\infty} \left\| f\left(\frac{k}{n}\right) - f(x) \right\|_\gamma Z(nx - k) =$$

$$\sum_{\begin{cases} k = -\infty \\ \left\| \frac{k}{n} - x \right\|_\infty \leq \frac{1}{n^\beta} \end{cases}}^{\infty} \left\| f\left(\frac{k}{n}\right) - f(x) \right\|_\gamma Z(nx - k) +$$

$$\sum_{\begin{cases} k = -\infty \\ \left\| \frac{k}{n} - x \right\|_\infty > \frac{1}{n^\beta} \end{cases}}^{\infty} \left\| f\left(\frac{k}{n}\right) - f(x) \right\|_\gamma Z(nx - k) \overset{(7.18.)}{\leq}$$

$$\omega_1\left(f, \frac{1}{n^\beta}\right) + 2\left\| \|f\|_\gamma \right\|_\infty \sum_{\begin{cases} k = -\infty \\ \left\| \frac{k}{n} - x \right\|_\infty > \frac{1}{n^\beta} \end{cases}}^{\infty} Z(nx - k) \overset{(7.25.)}{\leq}$$

$$\omega_1\left(f, \frac{1}{n^\beta}\right) + \frac{\left\| \|f\|_\gamma \right\|_\infty}{\lambda\left(n^{1-\beta} - 2\right)^\lambda}, \tag{7.81}$$

proving the claim. ■

We give

Theorem 7.17 *Let $f \in C_B(\mathbb{R}^N, X)$, $0 < \beta < 1$, $x \in \mathbb{R}^N$, λ is odd, $N, n \in \mathbb{N}$ with $n^{1-\beta} > 2$, ω_1 is for $p = \infty$. Then*

(1)

$$\|C_n(f,x) - f(x)\|_\gamma \leq \omega_1\left(f, \frac{1}{n} + \frac{1}{n^\beta}\right) + \frac{\| \|f\|_\gamma \|_\infty}{\lambda\left(n^{1-\beta} - 2\right)^\lambda} =: \lambda_3(n), \quad (7.82)$$

(2)

$$\| \|C_n(f) - f\|_\gamma \|_\infty \leq \lambda_3(n). \quad (7.83)$$

Given that $f \in \left(C_U(\mathbb{R}^N, X) \cap C_B(\mathbb{R}^N, X)\right)$, we obtain $\lim_{n\to\infty} C_n(f) = f$, uniformly.

Proof We notice that

$$\int_{\frac{k}{n}}^{\frac{k+1}{n}} f(t)\,dt = \int_{\frac{k_1}{n}}^{\frac{k_1+1}{n}} \int_{\frac{k_2}{n}}^{\frac{k_2+1}{n}} \dots \int_{\frac{k_N}{n}}^{\frac{k_N+1}{n}} f(t_1, t_2, \dots, t_N)\,dt_1 dt_2 \dots dt_N =$$

$$\int_0^{\frac{1}{n}} \int_0^{\frac{1}{n}} \dots \int_0^{\frac{1}{n}} f\left(t_1 + \frac{k_1}{n}, t_2 + \frac{k_2}{n}, \dots, t_N + \frac{k_N}{n}\right) dt_1 \dots dt_N = \int_0^{\frac{1}{n}} f\left(t + \frac{k}{n}\right) dt. \quad (7.84)$$

Thus it holds (by (7.40))

$$C_n(f,x) = \sum_{k=-\infty}^{\infty} \left(n^N \int_0^{\frac{1}{n}} f\left(t + \frac{k}{n}\right) dt\right) Z(nx - k). \quad (7.85)$$

We observe that

$$\|C_n(f,x) - f(x)\|_\gamma =$$

$$\left\| \sum_{k=-\infty}^{\infty} \left(n^N \int_0^{\frac{1}{n}} f\left(t + \frac{k}{n}\right) dt\right) Z(nx-k) - \sum_{k=-\infty}^{\infty} f(x) Z(nx-k) \right\|_\gamma =$$

$$\left\| \sum_{k=-\infty}^{\infty} \left(\left(n^N \int_0^{\frac{1}{n}} f\left(t + \frac{k}{n}\right) dt\right) - f(x)\right) Z(nx-k) \right\|_\gamma =$$

$$\left\| \sum_{k=-\infty}^{\infty} \left(n^N \int_0^{\frac{1}{n}} \left(f\left(t + \frac{k}{n}\right) - f(x)\right) dt\right) Z(nx-k) \right\|_\gamma \leq \quad (7.86)$$

$$\sum_{k=-\infty}^{\infty}\left(n^N\int_0^{\frac{1}{n}}\left\|f\left(t+\frac{k}{n}\right)-f(x)\right\|_\gamma dt\right)Z(nx-k)=$$

$$\sum_{\substack{k=-\infty \\ \left\|\frac{k}{n}-x\right\|_\infty\le\frac{1}{n^\beta}}}^{\infty}\left(n^N\int_0^{\frac{1}{n}}\left\|f\left(t+\frac{k}{n}\right)-f(x)\right\|_\gamma dt\right)Z(nx-k)+$$

$$\sum_{\substack{k=-\infty \\ \left\|\frac{k}{n}-x\right\|_\infty>\frac{1}{n^\beta}}}^{\infty}\left(n^N\int_0^{\frac{1}{n}}\left\|f\left(t+\frac{k}{n}\right)-f(x)\right\|_\gamma dt\right)Z(nx-k)\le$$

$$\sum_{\substack{k=-\infty \\ \left\|\frac{k}{n}-x\right\|_\infty\le\frac{1}{n^\beta}}}^{\infty}\left(n^N\int_0^{\frac{1}{n}}\omega_1\left(f,\|t\|_\infty+\left\|\frac{k}{n}-x\right\|_\infty\right)dt\right)Z(nx-k)+$$

$$2\left\|\|f\|_\gamma\right\|_\infty\left(\sum_{\substack{k=-\infty \\ \left\|\frac{k}{n}-x\right\|_\infty>\frac{1}{n^\beta}}}^{\infty}Z(|nx-k|)\right)\le$$

$$\omega_1\left(f,\frac{1}{n}+\frac{1}{n^\beta}\right)+\frac{\left\|\|f\|_\gamma\right\|_\infty}{\lambda\left(n^{1-\beta}-2\right)^\lambda}, \tag{7.87}$$

proving the claim. ■

We also present

Theorem 7.18 *Let $f\in C_B\left(\mathbb{R}^N,X\right)$, $0<\beta<1$, $x\in\mathbb{R}^N$, λ is odd, $N,n\in\mathbb{N}$ with $n^{1-\beta}>2$, ω_1 is for $p=\infty$. Then*

(1)

$$\|D_n(f,x)-f(x)\|_\gamma\le\omega_1\left(f,\frac{1}{n}+\frac{1}{n^\beta}\right)+\frac{\left\|\|f\|_\gamma\right\|_\infty}{\lambda\left(n^{1-\beta}-2\right)^\lambda}=\lambda_3(n), \tag{7.88}$$

(2)

$$\left\|\|D_n(f)-f\|_\gamma\right\|_\infty\le\lambda_3(n). \tag{7.89}$$

Given that $f \in \left(C_U\left(\mathbb{R}^N, X\right) \cap C_B\left(\mathbb{R}^N, X\right)\right)$, *we obtain* $\lim_{n\to\infty} D_n(f) = f$, *uniformly.*

Proof Similar to the proof of Theorem 7.17, as such is omitted. ■

We make

Definition 7.19 Let $f \in C_B\left(\mathbb{R}^N, X\right)$, $N \in \mathbb{N}$, where $\left(X, \|\cdot\|_\gamma\right)$ is a Banach space. We define the general neural network operator

$$F_n(f, x) := \sum_{k=-\infty}^{\infty} l_{nk}(f) Z(nx - k) =$$

$$\begin{cases} B_n(f, x), & \text{if } l_{nk}(f) = f\left(\frac{k}{n}\right), \\ C_n(f, x), & \text{if } l_{nk}(f) = n^N \int_{\frac{k}{n}}^{\frac{k+1}{n}} f(t)\,dt, \\ D_n(f, x), & \text{if } l_{nk}(f) = \delta_{nk}(f). \end{cases} \tag{7.90}$$

Clearly $l_{nk}(f)$ is an X-valued bounded linear functional such that $\|l_{nk}(f)\|_\gamma \leq \left\|\|f\|_\gamma\right\|_\infty$.

Hence $F_n(f)$ is a bounded linear operator with $\left\|\|F_n(f)\|_\gamma\right\|_\infty \leq \left\|\|f\|_\gamma\right\|_\infty$.

We need

Theorem 7.20 *Let* $f \in C_B\left(\mathbb{R}^N, X\right)$, $N \geq 1$. *Then* $F_n(f) \in C_B\left(\mathbb{R}^N, X\right)$.

Proof Lengthy and similar to the proof of Theorem 21 of [15], as such is omitted. ■

Remark 7.21 By (7.27) it is obvious that $\left\|\|A_n(f)\|_\gamma\right\|_\infty \leq \left\|\|f\|_\gamma\right\|_\infty < \infty$, and $A_n(f) \in C\left(\prod_{i=1}^N [a_i, b_i], X\right)$, given that $f \in C\left(\prod_{i=1}^N [a_i, b_i], X\right)$.

Call L_n any of the operators A_n, B_n, C_n, D_n.

Clearly then

$$\left\|\left\|L_n^2(f)\right\|_\gamma\right\|_\infty = \left\|\|L_n(L_n(f))\|_\gamma\right\|_\infty \leq \left\|\|L_n(f)\|_\gamma\right\|_\infty \leq \left\|\|f\|_\gamma\right\|_\infty, \tag{7.91}$$

etc.

Therefore we get

$$\left\|\left\|L_n^k(f)\right\|_\gamma\right\|_\infty \leq \left\|\|f\|_\gamma\right\|_\infty, \ \forall k \in \mathbb{N}, \tag{7.92}$$

the contraction property.

Also we see that

$$\left\|\left\|L_n^k(f)\right\|_\gamma\right\|_\infty \leq \left\|\left\|L_n^{k-1}(f)\right\|_\gamma\right\|_\infty \leq \dots \leq \left\|\|L_n(f)\|_\gamma\right\|_\infty \leq \left\|\|f\|_\gamma\right\|_\infty. \tag{7.93}$$

Here L_n^k are bounded linear operators.

Notation 7.22 *Here $N \in \mathbb{N}$, $0 < \beta < 1$. Denote by*

$$c_N := \begin{cases} \left(2\sqrt[\lambda]{1+2^{\lambda}}\right)^N, \text{ if } L_n = A_n, \\ 1, \text{ if } L_n = B_n, C_n, D_n, \end{cases} \tag{7.94}$$

$$\varphi(n) := \begin{cases} \frac{1}{n^{\beta}}, \text{ if } L_n = A_n, B_n, \\ \frac{1}{n} + \frac{1}{n^{\beta}}, \text{ if } L_n = C_n, D_n, \end{cases} \tag{7.95}$$

$$\Omega := \begin{cases} C\left(\prod_{i=1}^{N}[a_i, b_i], X\right), \text{ if } L_n = A_n, \\ C_B\left(\mathbb{R}^N, X\right), \text{ if } L_n = B_n, C_n, D_n, \end{cases} \tag{7.96}$$

and

$$Y := \begin{cases} \prod_{i=1}^{N}[a_i, b_i], \text{ if } L_n = A_n, \\ \mathbb{R}^N, \text{ if } L_n = B_n, C_n, D_n. \end{cases} \tag{7.97}$$

We give the condensed

Theorem 7.23 *Let $f \in \Omega$, $0 < \beta < 1$, $x \in Y$; n, $N \in \mathbb{N}$ with $n^{1-\beta} > 2$, λ is odd. Then*

(i)

$$\|L_n(f,x) - f(x)\|_{\gamma} \leq c_N \left[\omega_1(f, \varphi(n)) + \frac{\| \|f\|_{\gamma} \|_{\infty}}{\lambda\left(n^{1-\beta}-2\right)^{\lambda}}\right] =: \tau(n), \tag{7.98}$$

where ω_1 is for $p = \infty$,

and

(ii)

$$\| \|L_n(f) - f\|_{\gamma} \|_{\infty} \leq \tau(n) \to 0, \text{ as } n \to \infty. \tag{7.99}$$

For f uniformly continuous and in Ω we obtain

$$\lim_{n\to\infty} L_n(f) = f,$$

pointwise and uniformly.

Proof By Theorems 7.8, 7.16, 7.17 and 7.18. ■

Next we talk about iterated neural network approximation (see also [9]).
We give

Theorem 7.24 *All here as in Theorem 7.23 and $r \in \mathbb{N}$, $\tau(n)$ as in (7.98). Then*

$$\left\| \left\| L_n^r f - f \right\|_\gamma \right\|_\infty \leq r\tau(n). \tag{7.100}$$

So that the speed of convergence to the unit operator of L_n^r is not worse than of L_n.

Proof As similar to [15] is omitted. ∎

We also present

Theorem 7.25 *Let $f \in \Omega$; λ is odd, $N, m_1, m_2, ..., m_r \in \mathbb{N} : m_1 \leq m_2 \leq ... \leq m_r$, $0 < \beta < 1$; $m_i^{1-\beta} > 2$, $i = 1, ..., r$, $x \in Y$, and let $\left(L_{m_1}, ..., L_{m_r}\right)$ as $\left(A_{m_1}, ..., A_{m_r}\right)$ or $\left(B_{m_1}, ..., B_{m_r}\right)$ or $\left(C_{m_1}, ..., C_{m_r}\right)$ or $\left(D_{m_1}, ..., D_{m_r}\right)$, $p = \infty$. Then*

$$\left\| L_{m_r}\left(L_{m_{r-1}}\left(...L_{m_2}\left(L_{m_1} f\right)\right)\right)(x) - f(x) \right\|_\gamma \leq$$

$$\left\| \left\| L_{m_r}\left(L_{m_{r-1}}\left(...L_{m_2}\left(L_{m_1} f\right)\right)\right) - f \right\|_\gamma \right\|_\infty \leq$$

$$\sum_{i=1}^r \left\| \left\| L_{m_i} f - f \right\|_\gamma \right\|_\infty \leq$$

$$c_N \sum_{i=1}^r \left[\omega_1\left(f, \varphi\left(m_i\right)\right) + \frac{\left\| \left\| f \right\|_\gamma \right\|_\infty}{\lambda\left(n^{1-\beta} - 2\right)^\lambda} \right] \leq$$

$$r c_N \left[\omega_1\left(f, \varphi\left(m_1\right)\right) + \frac{\left\| \left\| f \right\|_\gamma \right\|_\infty}{\lambda\left(n^{1-\beta} - 2\right)^\lambda} \right]. \tag{7.101}$$

Clearly, we notice that the speed of convergence to the unit operator of the multiply iterated operator is not worse than the speed of L_{m_1}.

Proof As similar to [15] is omitted. ∎

We also give

Theorem 7.26 *Let all as in Corollary 7.15, and $r \in \mathbb{N}$. Here $\varphi_3(n)$ is as in (7.77). Then*

$$\left\| \left\| A_n^r f - f \right\|_\gamma \right\|_\infty \leq r \left\| \left\| A_n f - f \right\|_\gamma \right\|_\infty \leq r\varphi_3(n). \tag{7.102}$$

Proof As similar to [15] is omitted. ∎

Application 7.27 *A typical application of all of our results is when $\left(X, \|\cdot\|_\gamma\right) = (\mathbb{C}, |\cdot|)$, where $\mathbb{C}$ are the complex numbers.*

References

1. G.A. Anastassiou, *Moments in Probability and Approximation Theory, Pitman Research Notes in Mathematics*, vol. 287 (Longman Scientific & Technical, Harlow, UK, 1993)
2. G.A. Anastassiou, Rate of convergence of some neural network operators to the unit-univariate case. J. Math. Anal. Appli. **212**, 237–262 (1997)
3. G.A. Anastassiou, *Quantitative Approximations* (Chapman & Hall/CRC, Boca Raton, New York, 2001)
4. G.A. Anastassiou, *Inteligent Systems: Approximation by Artificial Neural Networks, Intelligent Systems Reference Library*, vol. 19 (Springer, Heidelberg, 2011)
5. G.A. Anastassiou, Univariate hyperbolic tangent neural network approximation. Math. Comput. Model. **53**, 1111–1132 (2011)
6. G.A. Anastassiou, Multivariate hyperbolic tangent neural network approximation. Comput. Math. **61**, 809–821 (2011)
7. G.A. Anastassiou, Multivariate sigmoidal neural network approximation. Neural Netw. **24**, 378–386 (2011)
8. G.A. Anastassiou, Univariate sigmoidal neural network approximation. J. Comput. Anal. Appl. **14**(4), 659–690 (2012)
9. G.A. Anastassiou, Approximation by neural networks iterates, in *Advances in Applied Mathematics and Approximation Theory*. Springer Proceedings in Mathematics & Statistics, ed. by G. Anastassiou, O. Duman (Springer, New York, 2013), pp. 1–20
10. G.A. Anastassiou, *Intelligent Systems II: Complete Approximation by Neural Network Operators* (Springer, Heidelberg, New York, 2016)
11. G.A. Anastassiou, *Intelligent Computations: Abstract Fractional Calculus* Inequalities Approximations. (Springer, Heidelberg, New York, 2018)
12. G.A. Anastassiou, Algebraic function based Banach space valued ordinary and fractional neural network approximations, *New Trends in Mathematical Sciences*, vol. 10 special issue (1), 100–125 (2022)
13. G.A. Anastassiou, Gudermannian function activated Banach space valued ordinary and fractional neural network approximation. Advances in Nonlinear Variational Inequalities **25**(2), 27–64 (2022)
14. G.A. Anastassiou, *Generalized Symmetrical Sigmoid Function Activated Banach Space Valued Ordinary and Fractional Neural Network Approximation* (Fasc. Matematica, Accepted for Publication, Analele Universităţii Oradea, Oradea, 2022)
15. G.A. Anastassiou, *General Multivariate Arctangent Function Activated Neural Network Approximations*. Submitted (2022)
16. G.A. Anastassiou, *Generalized Symmetrical Sigmoid Function Activated Neural Network Multivariate Approximation*. Submitted (2022)
17. G.A. Anastassiou, R. Kozma, *Abstract Multivariate Algebraic Function Activated Neural Network Approximations*. Submitted (2022)
18. H. Cartan, *Differential Calculus* (Hermann, Paris, 1971)
19. Z. Chen, F. Cao, The approximation operators with sigmoidal functions. Comput. Math. Appl. **58**, 758–765 (2009)
20. D. Costarelli, R. Spigler, Approximation results for neural network operators activated by sigmoidal functions. Neural Netw. **44**, 101–106 (2013)
21. D. Costarelli, R. Spigler, Multivariate neural network operators with sigmoidal activation functions. Neural Netw. **48**, 72–77 (2013)
22. A.J. Dunning, J. Kensler, L. Goudeville, F. Bailleux, Some extensions in continuous methods for immunological correlates of protection. BMC Med. Res. Methodol. **15**(107) (2015). https://doi.org/10.1186/s12874-015-0096-9
23. S. Haykin, *Neural Networks: A Comprehensive Foundation*, 2nd edn. (Prentice Hall, New York, 1998)

24. W. McCulloch, W. Pitts, A logical calculus of the ideas immanent in nervous activity. Bull. Math. Biophys. **7**, 115–133 (1943)
25. T.M. Mitchell, *Machine Learning* (WCB-McGraw-Hill, New York, 1997)
26. L.B. Rall, *Computational Solution of Nonlinear Operator Equations* (Wiley, New York, 1969)

Chapter 8
Quantitative Approximation by Kantorovich-Choquet Quasi-Interpolation Neural Network Operators Revisited

In this chapter we exhibit univariate and multivariate quantitative approximation by Kantorovich-Choquet type quasi-interpolation neural network operators with respect to supremum norm. This is done with rates using the first univariate and multivariate moduli of continuity. We approximate continuous and bounded functions on $\mathbb{R}^N$, $N \in \mathbb{N}$. When they are also uniformly continuous we have pointwise and uniform convergences. Our activation functions are induced by the arctangent, algebraic, Gudermannian and generalized symmetrical sigmoid functions. It follows [19].

8.1 Introduction

The author in [1, 2], see Chaps. 2–5, was the first to establish neural network approximations to continuous functions with rates by very specifically defined neural network operators of Cardaliaguet-Euvrard and "Squashing" types, by employing the modulus of continuity of the engaged function or its high order derivative, and producing very tight Jackson type inequalities. He treats there both the univariate and multivariate cases. The defining these operators "bell-shaped" and "squashing" functions are assumed to be compact support. Also in [2] he gives the Nth order asymptotic expansion for the error of weak approximation of these two operators to a special natural class of smooth functions, see Chaps. 4 and 5 there.

The author inspired by [21], continued his studies on neural networks approximation by introducing and using the proper quasi-interpolation operators of sigmoidal and hyperbolic tangent type which resulted into [3–7], by treating both the univariate and multivariate cases. He did also the corresponding fractional case [8]. For recent works see [9–18, 20].

The author here performs univariate and multivariate arctangent-algebraic-Gudermannian-generalized symmetrical sigmoid activation functions based neural

© The Author(s), under exclusive license to Springer Nature Switzerland AG 2023
G. A. Anastassiou, *Banach Space Valued Neural Network*,
Studies in Computational Intelligence 1062,
https://doi.org/10.1007/978-3-031-16400-2_8

network approximations to continuous functions over the whole $\mathbb{R}^N$, $N \in \mathbb{N}$, then he extends his results to complex valued functions. All convergences here are with rates expressed via the modulus of continuity of the involved function and given by very tight Jackson type inequalities. This is a continuation of [12], Chap. 1.

The author comes up with the "right" precisely defined flexible quasi-interpolation, Kantorovich-Choquet type integral coefficient neural networks operators associated with: arctangent-algebraic-Gudermannian-generalized symmetrical sigmoid activation functions. In preparation to prove our results we establish important properties of the basic density functions defining our operators.

Feed-forward neural networks (FNNs) with one hidden layer, the only type of networks we deal with in this chapter, are mathematically expressed as

$$N_n(x) = \sum_{j=0}^{n} c_j \sigma\left(\langle a_j \cdot x \rangle + b_j\right), \; x \in \mathbb{R}^s, \; s \in \mathbb{N},$$

where for $0 \leq j \leq n$, $b_j \in \mathbb{R}$ are the thresholds, $a_j \in \mathbb{R}^s$ are the connection weights, $c_j \in \mathbb{R}$ are the coefficients, $\langle a_j \cdot x \rangle$ is the inner product of a_j and x, and σ is the activation function of the network. In many fundamental neural network models, the activation functions are the arctangent-algebraic-Gudermannian-generalized symmetrical sigmoid activation functions. About neural networks in general read [26–28].

8.2 Background

Next we present briefly about the Choquet integral.

We make

Definition 8.1 Consider $\Omega \neq \varnothing$ and let $\mathcal{C}$ be a σ-algebra of subsets in Ω.

(i) (see, e.g., [29], p. 63) The set function $\mu : \mathcal{C} \to [0, +\infty]$ is called a monotone set function (or capacity) if $\mu(\varnothing) = 0$ and $\mu(A) \leq \mu(B)$ for all $A, B \in \mathcal{C}$, with $A \subset B$. Also, μ is called submodular if

$$\mu(A \cup B) + \mu(A \cap B) \leq \mu(A) + \mu(B), \text{ for all } A, B \in \mathcal{C}.$$

μ is called bounded if $\mu(\Omega) < +\infty$ and normalized if $\mu(\Omega) = 1$.

(ii) (see, e.g., [29], p. 233, or [22]) If μ is a monotone set function on $\mathcal{C}$ and if $f : \Omega \to \mathbb{R}$ is $\mathcal{C}$-measurable (that is, for any Borel subset $B \subset \mathbb{R}$ it follows $f^{-1}(B) \in \mathcal{C}$), the for any $A \in \mathcal{C}$, the Choquet integral is defined by

$$(C)\int_A f d\mu = \int_0^{+\infty} \mu\left(F_\beta(f) \cap A\right) d\beta + \int_{-\infty}^{0} \left[\mu\left(F_\beta(f) \cap A\right) - \mu(A)\right] d\beta,$$

where we used the notation $F_\beta(f) = \{\omega \in \Omega : f(\omega) \geq \beta\}$. Notice that if $f \geq 0$ on A, then in the above formula we get $\int_{-\infty}^{0} = 0$.

The integrals on the right-hand side are the usual Riemann integral.

The function f will be called Choquet integrable on A if $(C)\int_A f d\mu \in \mathbb{R}$.

Next we list some well known properties of the Choquet integral.

Remark 8.2 If $\mu : \mathcal{C} \to [0, +\infty]$ is a monotone set function, then the following properties hold:

(i) For all $a \geq 0$ we have $(C)\int_A af d\mu = a \cdot (C)\int_A f d\mu$ (if $f \geq 0$ then see, e.g., [29], Theorem 11.2, (5), p. 228 and if f is arbitrary sign, then see, e.g., [23], p. 64, Proposition 5.1, (ii)).

(ii) For all $c \in \mathbb{R}$ and f of arbitrary sign, we have (see, e.g., [29], pp. 232–233, or [23], p. 65) $(C)\int_A (f + c)\, d\mu = (C)\int_A f d\mu + c \cdot \mu(A)$.

If μ is submodular too, then for all f, g of arbitrary sign and lower bounded, we have (see, e.g., [23], p. 75, Theorem 6.3)

$$(C)\int_A (f + g)\, d\mu \leq (C)\int_A f d\mu + (C)\int_A g d\mu.$$

(iii) If $f \leq g$ on A then $(C)\int_A f d\mu \leq (C)\int_A g d\mu$ (see, e.g., [29], p. 228, Theorem 11.2, (3) if $f, g \geq 0$ and p. 232 if f, g are of arbitrary sign).

(iv) Let $f \geq 0$. If $A \subset B$ then $(C)\int_A f d\mu \leq (C)\int_B f d\mu$. In addition, if μ is finitely aubadditive, then

$$(C)\int_{A \cup B} f d\mu \leq (C)\int_A f d\mu + (C)\int_B f d\mu.$$

(v) It is immediate that $(C)\int_A 1 \cdot d\mu(t) = \mu(A)$.

(vi) The formula $\mu(A) = \gamma(M(A))$, where $\gamma : [0, 1] \to [0, 1]$ is an increasing and concave function, with $\gamma(0) = 0$, $\gamma(1) = 1$ and M is a probability measure (or only finitely additive) on a σ-algebra on Ω (that is, $M(\emptyset) = 0$, $M(\Omega) = 1$ and M is countably additive), gives simple examples of normalized, monotone and submodular set functions (see, e.g., [23], pp. 16–17, Example 2.1). Such of set functions μ are also called distorsions of countably normalized, additive measures (or distorted measures). For a simple example, we can take $\gamma(t) = \frac{2t}{1+t}$, $\gamma(t) = \sqrt{t}$.

If the above γ function is increasing, concave and satisfies only $\gamma(0) = 0$, then for any bounded Borel measure m, $\mu(A) = \gamma(m(A))$ gives a simple example of bounded, monotone and submodular set function.

(vii) If μ is a countably additive bounded measure, then the Choquet integral $(C)\int_A f d\mu$ reduces to the usual Lebesgue type integral (see, e.g., [23], p. 62, or [29], p. 226).

(viii) If $f \geq 0$, then $(C)\int_A f d\mu \geq 0$.

(ix) Let $\mu = \sqrt{M}$, where M is the Lebesgue measure on $[0, +\infty)$, then μ is a monotone and submodular set function, furthermore μ is strictly positive, see [25].

(x) If $\Omega = \mathbb{R}^N$, $N \in \mathbb{N}$, we call μ strictly positive if $\mu(A) > 0$, for any open subset $A \subseteq \mathbb{R}^N$.

8.2.1 About the Arctangent Activation Function

We consider the

$$\arctan x = \int_0^x \frac{dz}{1+z^2}, \; x \in \mathbb{R}. \tag{8.1}$$

We will be using

$$h(x) := \frac{2}{\pi} \arctan\left(\frac{\pi}{2}x\right) = \frac{2}{\pi} \int_0^{\frac{\pi x}{2}} \frac{dz}{1+z^2}, \; x \in \mathbb{R}, \tag{8.2}$$

which is a sigmoid type function and it is strictly increasing. We have that

$$h(0) = 0, \; h(-x) = -h(x), \; h(+\infty) = 1, \; h(-\infty) = -1,$$

and

$$h'(x) = \frac{4}{4+\pi^2 x^2} > 0, \; \text{all } x \in \mathbb{R}. \tag{8.3}$$

We consider the activation function

$$\psi_1(x) := \frac{1}{4}(h(x+1) - h(x-1)), \; x \in \mathbb{R}, \tag{8.4}$$

and we notice that

$$\psi_1(-x) = \psi_1(x), \tag{8.5}$$

it is an even function.

Since $x+1 > x-1$, then $h(x+1) > h(x-1)$, and $\psi_1(x) > 0$, all $x \in \mathbb{R}$.

We see that

$$\psi_1(0) = \frac{1}{\pi} \arctan \frac{\pi}{2} \cong 0.319. \tag{8.6}$$

Let $x > 0$, we have that

$$\psi_1'(x) = \frac{1}{4}\left(h'(x+1) - h'(x-1)\right) =$$

$$\frac{-4\pi^2 x}{\left(4+\pi^2(x+1)^2\right)\left(4+\pi^2(x-1)^2\right)} < 0. \tag{8.7}$$

That is

$$\psi_1'(x) < 0, \text{ for } x > 0. \tag{8.8}$$

That is ψ_1 is strictly decreasing on $[0, \infty)$ and clearly is strictly increasing on $(-\infty, 0]$, and $\psi_1'(0) = 0$.

Observe that

$$\begin{gathered} \lim_{x \to +\infty} \psi_1(x) = \tfrac{1}{4}(h(+\infty) - h(+\infty)) = 0, \\ \text{and} \\ \lim_{x \to -\infty} \psi_1(x) = \tfrac{1}{4}(h(-\infty) - h(-\infty)) = 0. \end{gathered} \tag{8.9}$$

That is the x-axis is the horizontal asymptote on ψ_1.

All in all, ψ_1 is a bell symmetric function with maximum $\psi_1(0) \cong 0.319$.

We need

Theorem 8.3 ([11], p. 286) *We have that*

$$\sum_{i=-\infty}^{\infty} \psi_1(x - i) = 1, \ \forall x \in \mathbb{R}. \tag{8.10}$$

Theorem 8.4 ([11], p. 287) *It holds*

$$\int_{-\infty}^{\infty} \psi_1(x)\, dx = 1. \tag{8.11}$$

So that $\psi_1(x)$ is a density function on $\mathbb{R}$.

We mention

Theorem 8.5 ([11], p. 288) *Let* $0 < \alpha < 1$, *and* $n \in \mathbb{N}$ *with* $n^{1-\alpha} > 2$. *It holds*

$$\sum_{\substack{k = -\infty \\ : |nx - k| \geq n^{1-\alpha}}}^{\infty} \psi_1(nx - k) < \frac{2}{\pi^2 \left(n^{1-\alpha} - 2\right)}. \tag{8.12}$$

We introduce (see [17])

$$Z_1(x_1, ..., x_N) := Z_1(x) := \prod_{i=1}^{N} \psi_1(x_i), \ x = (x_1, ..., x_N) \in \mathbb{R}^N, \ N \in \mathbb{N}. \tag{8.13}$$

It has the properties:

(i) $Z_1(x) > 0, \ \forall x \in \mathbb{R}^N$,

(ii)

$$\sum_{k=-\infty}^{\infty} Z_1(x-k) := \sum_{k_1=-\infty}^{\infty} \sum_{k_2=-\infty}^{\infty} \dots \sum_{k_N=-\infty}^{\infty} Z_1(x_1-k_1, \dots, x_N-k_N) = 1, \tag{8.14}$$

where $k := (k_1, \dots, k_n) \in \mathbb{Z}^N$, $\forall\, x \in \mathbb{R}^N$,

hence

(iii)

$$\sum_{k=-\infty}^{\infty} Z_1(nx-k) = 1, \tag{8.15}$$

$\forall\, x \in \mathbb{R}^N$; $n \in \mathbb{N}$,

and

(iv)

$$\int_{\mathbb{R}^N} Z_1(x)\,dx = 1, \tag{8.16}$$

that is Z_1 is a multivariate density function.

(v) It is clear that

$$\sum_{\begin{cases} k = -\infty \\ \left\| \frac{k}{n} - x \right\|_{\infty} > \frac{1}{n^{\beta}} \end{cases}}^{\infty} Z_1(nx-k) < \frac{2}{\pi^2\left(n^{1-\beta}-2\right)} =: c_1(\beta, n), \tag{8.17}$$

$0 < \beta < 1$, $n \in \mathbb{N} : n^{1-\beta} > 2$, $x \in \mathbb{R}^N$.

Above it is $\|x\|_{\infty} := \max\{|x_1|, \dots, |x_N|\}$, $x \in \mathbb{R}^N$, also set $\infty := (\infty, \dots, \infty)$, $-\infty = (-\infty, \dots -\infty)$ upon the multivariate context.

8.2.2 *About the Algebraic Activation Function*

Here see also [17].

We consider the generator algebraic function

$$\varphi(x) = \frac{x}{\sqrt[2m]{1+x^{2m}}}, \quad m \in \mathbb{N}, x \in \mathbb{R}, \tag{8.18}$$

which is a sigmoidal type of function and is a strictly increasing function.

We see that $\varphi(-x) = -\varphi(x)$ with $\varphi(0) = 0$. We get that

$$\varphi'(x) = \frac{1}{\left(1+x^{2m}\right)^{\frac{2m+1}{2m}}} > 0, \ \forall\, x \in \mathbb{R}, \tag{8.19}$$

proving φ as strictly increasing over $\mathbb{R}$, $\varphi'(x) = \varphi'(-x)$. We easily find that $\lim_{x\to+\infty} \varphi(x) = 1$, $\varphi(+\infty) = 1$, and $\lim_{x\to-\infty} \varphi(x) = -1$, $\varphi(-\infty) = -1$.

We consider the activation function

$$\psi_2(x) = \frac{1}{4}[\varphi(x+1) - \varphi(x-1)]. \tag{8.20}$$

Clearly it is $\psi_2(x) = \psi_2(-x)$, $\forall x \in \mathbb{R}$, so that ψ_2 is an even function and symmetric with respect to the y-axis. Clealry $\psi_2(x) > 0$, $\forall x \in \mathbb{R}$.

Also it is

$$\psi_2(0) = \frac{1}{2\sqrt[2m]{2}}. \tag{8.21}$$

By [13], we have that $\psi_2'(x) < 0$ for $x > 0$. That is ψ_2 is strictly decreasing over $(0, +\infty)$.

Clearly, ψ_2 is strictly increasing over $(-\infty, 0)$ and $\psi_2'(0) = 0$.

Furthermore we obtain that

$$\lim_{x\to+\infty} \psi_2(x) = \frac{1}{4}[\varphi(+\infty) - \varphi(+\infty)] = 0, \tag{8.22}$$

and

$$\lim_{x\to-\infty} \psi_2(x) = \frac{1}{4}[\varphi(-\infty) - \varphi(-\infty)] = 0. \tag{8.23}$$

That is the x-axis is the horizontal asymptote of ψ_2.

Conclusion, ψ_2 is a bell shape symmetric function with maximum

$$\psi_2(0) = \frac{1}{2\sqrt[2m]{2}}, \quad m \in \mathbb{N}. \tag{8.24}$$

We need

Theorem 8.6 ([13]) *We have that*

$$\sum_{i=-\infty}^{\infty} \psi_2(x-i) = 1, \ \forall x \in \mathbb{R}. \tag{8.25}$$

Theorem 8.7 ([13]) *It holds*

$$\int_{-\infty}^{\infty} \psi_2(x)\,dx = 1. \tag{8.26}$$

Theorem 8.8 ([13]) *Let* $0 < \alpha < 1$, *and* $n \in \mathbb{N}$ *with* $n^{1-\alpha} > 2$. *It holds*

$$\sum_{\begin{cases} k = -\infty \\ : |nx - k| \geq n^{1-\alpha} \end{cases}}^{\infty} \psi_2 (nx - k) < \frac{1}{4m \left(n^{1-\alpha} - 2\right)^{2m}}, \ m \in \mathbb{N}. \tag{8.27}$$

We introduce (see also [20])

$$Z_2 (x_1, ..., x_N) := Z_2 (x) := \prod_{i=1}^{N} \psi_2 (x_i), \ x = (x_1, ..., x_N) \in \mathbb{R}^N, \ N \in \mathbb{N}. \tag{8.28}$$

It has the properties:

(i) $Z_2 (x) > 0, \ \forall x \in \mathbb{R}^N$,

(ii)

$$\sum_{k=-\infty}^{\infty} Z_2 (x - k) := \sum_{k_1=-\infty}^{\infty} \sum_{k_2=-\infty}^{\infty} ... \sum_{k_N=-\infty}^{\infty} Z_2 (x_1 - k_1, ..., x_N - k_N) = 1, \tag{8.29}$$

where $k := (k_1, ..., k_n) \in \mathbb{Z}^N, \forall x \in \mathbb{R}^N$,

hence

(iii)

$$\sum_{k=-\infty}^{\infty} Z_2 (nx - k) = 1, \tag{8.30}$$

$\forall x \in \mathbb{R}^N; n \in \mathbb{N}$,

and

(iv)

$$\int_{\mathbb{R}^N} Z_2 (x) \, dx = 1, \tag{8.31}$$

that is Z_2 is a multivariate density function.

(v) It is clear that

$$\sum_{\begin{cases} k = -\infty \\ \left\| \frac{k}{n} - x \right\|_{\infty} > \frac{1}{n^{\beta}} \end{cases}}^{\infty} Z_2 (nx - k) < \frac{1}{4m \left(n^{1-\beta} - 2\right)^{2m}} =: c_2 (\beta, n), \tag{8.32}$$

$0 < \beta < 1, n \in \mathbb{N} : n^{1-\beta} > 2, x \in \mathbb{R}^N, m \in \mathbb{N}$.

8.2.3 About the Gudermannian Activation Function

See also [14, 30].

Here we consider $gd(x)$ the Gudermannian function [30], which is a sigmoid function, as a generator function:

$$\sigma(x) = 2\arctan\left(\tanh\left(\frac{x}{2}\right)\right) = \int_0^x \frac{dt}{\cosh t} =: gd(x), \ x \in \mathbb{R}. \tag{8.33}$$

Let the normalized generator sigmoid function

$$f(x) := \frac{4}{\pi}\sigma(x) = \frac{4}{\pi}\int_0^x \frac{dt}{\cosh t} = \frac{8}{\pi}\int_0^x \frac{1}{e^t + e^{-t}}dt, \ \ x \in \mathbb{R}. \tag{8.34}$$

Here

$$f'(x) = \frac{4}{\pi \cosh x} > 0, \ \forall x \in \mathbb{R},$$

hence f is strictly increasing on $\mathbb{R}$.

Notice that $\tanh(-x) = -\tanh x$ and $\arctan(-x) = -\arctan x$, $x \in \mathbb{R}$.

So, here the neural network activation function will be:

$$\psi_3(x) = \frac{1}{4}\left[f(x+1) - f(x-1)\right], x \in \mathbb{R}. \tag{8.35}$$

By [14], we get that

$$\psi_3(x) = \psi_3(-x), \ \forall x \in \mathbb{R}, \tag{8.36}$$

i.e. it is even and symmetric with respect to the y-axis. Here we have $f(+\infty) = 1$, $f(-\infty) = -1$ and $f(0) = 0$. Clearly it is

$$f(-x) = -f(x), \ \forall x \in \mathbb{R}, \tag{8.37}$$

an odd function, symmetric with respect to the origin. Since $x + 1 > x - 1$, and $f(x+1) > f(x-1)$, we obtain $\psi_3(x) > 0$, $\forall\, x \in \mathbb{R}$.

By [14], we have that

$$\psi_3(0) = \frac{2}{\pi} gd(1) \cong 0.551. \tag{8.38}$$

By [?] ψ_3 is strictly decreasing on $(0, +\infty)$, and strictly increasing on $(-\infty, 0)$, and $\psi_3'(0) = 0$.

Also we have that

$$\lim_{x\to+\infty}\psi_3(x) = \lim_{x\to-\infty}\psi_3(x) = 0, \tag{8.39}$$

that is the x-axis is the horizontal asymptote for ψ_3.

Conclusion, ψ_3 is a bell shaped symmetric function with maximum $\psi_3(0) \cong 0.551$.

We need

Theorem 8.9 ([14]) *It holds that*

$$\sum_{i=-\infty}^{\infty} \psi_3(x-i) = 1, \forall x \in \mathbb{R}. \tag{8.40}$$

Theorem 8.10 ([14]) *We have that*

$$\int_{-\infty}^{\infty} \psi_3(x)\, dx = 1. \tag{8.41}$$

So $\psi_3(x)$ is a density function.

Theorem 8.11 ([14]) *Let $0 < \alpha < 1$, and $n \in \mathbb{N}$ with $n^{1-\alpha} > 2$. It holds*

$$\sum_{\begin{cases} k = -\infty \\ : |nx-k| \geq n^{1-\alpha} \end{cases}}^{\infty} \psi_3(nx-k) < \frac{4}{\pi e^{(n^{1-\alpha}-2)}} = \frac{4e^2}{\pi e^{n^{1-\alpha}}}. \tag{8.42}$$

We introduce (see also [16])

$$Z_3(x_1, ..., x_N) := Z_3(x) := \prod_{i=1}^{N} \psi_3(x_i), \ \ x = (x_1, ..., x_N) \in \mathbb{R}^N, \ N \in \mathbb{N}. \tag{8.43}$$

It has the properties:

(i) $Z_3(x) > 0, \ \forall x \in \mathbb{R}^N$,

(ii)

$$\sum_{k=-\infty}^{\infty} Z_3(x-k) := \sum_{k_1=-\infty}^{\infty} \sum_{k_2=-\infty}^{\infty} ... \sum_{k_N=-\infty}^{\infty} Z_3(x_1-k_1, ..., x_N-k_N) = 1, \tag{8.44}$$

where $k := (k_1, ..., k_n) \in \mathbb{Z}^N, \forall x \in \mathbb{R}^N$,

hence

(iii)

$$\sum_{k=-\infty}^{\infty} Z_3(nx-k) = 1, \tag{8.45}$$

$\forall x \in \mathbb{R}^N; n \in \mathbb{N}$,

and

(iv)

$$\int_{\mathbb{R}^N} Z_3(x)\,dx = 1, \tag{8.46}$$

that is Z_3 is a multivariate density function.

(v) It is also clear that

$$\sum_{\substack{k=-\infty \\ \left\| \frac{k}{n} - x \right\|_\infty > \frac{1}{n^\beta}}}^{\infty} Z_3(nx-k) < \frac{4e^2}{\pi e^{n^{1-\beta}}} = c_3(\beta, n), \tag{8.47}$$

$0 < \beta < 1, n \in \mathbb{N} : n^{1-\beta} > 2, x \in \mathbb{R}^N, m \in \mathbb{N}.$

8.2.4 About the Generalized Symmetrical Activation Function

Here we consider the generalized symmetrical sigmoid function [15, 24]

$$f_1(x) = \frac{x}{(1+|x|^\mu)^{\frac{1}{\mu}}}, \ \mu > 0, x \in \mathbb{R}. \tag{8.48}$$

This has applications in immunology and protection from disease together with probability theory. It is also called a symmetrical protection curve.

The parameter μ is a shape parameter controling how fast the curve approaches the asymptotes for a given slope at the inflection point. When $\mu = 1$ f_1 is the absolute sigmoid function, and when $\mu = 2$, f_1 is the square root sigmoid function. When $\mu = 1.5$ the function approximates the arctangent function, when $\mu = 2.9$ it approximates the logistic function, and when $\mu = 3.4$ it approximates the error function. Parameter μ is estimated in the likelihood maximization [24]. For more see [24].

Next we study the particular generator sigmoid function

$$f_2(x) = \frac{x}{\left(1+|x|^\lambda\right)^{\frac{1}{\lambda}}}, \ \lambda \text{ is an odd number, } x \in \mathbb{R}. \tag{8.49}$$

We have that $f_2(0) = 0$, and

$$f_2(-x) = -f_2(x), \tag{8.50}$$

so f_2 is symmetric with respect to zero.

When $x \geq 0$, we get that [15]

$$f_2'(x) = \frac{1}{\left(1 + x^{\lambda}\right)^{\frac{\lambda+1}{\lambda}}} > 0, \tag{8.51}$$

that is f_2 is strictly increasing on $[0, +\infty)$ and f_2 is strictly increasing on $(-\infty, 0]$. Hence f_2 is strictly increasing on $\mathbb{R}$.

We also have $f_2(+\infty) = f_2(-\infty) = 1$.

Let us consider the activation function [15]:

$$\psi_4(x) = \frac{1}{4}\left[f_2(x+1) - f_2(x-1)\right] =$$

$$\frac{1}{4}\left[\frac{(x+1)}{\left(1 + |x+1|^{\lambda}\right)^{\frac{1}{\lambda}}} - \frac{(x-1)}{\left(1 + |x-1|^{\lambda}\right)^{\frac{1}{\lambda}}}\right]. \tag{8.52}$$

Clearly it holds [15]

$$\psi_4(x) = \psi_4(-x), \ \forall x \in \mathbb{R}. \tag{8.53}$$

and

$$\psi_4(0) = \frac{1}{2\sqrt[\lambda]{2}}, \tag{8.54}$$

and $\psi_4(x) > 0$, $\forall x \in \mathbb{R}$.

Following [15], we have that ψ_4 is strictly decreasing over $[0, +\infty)$, and ψ_4 is strictly increasing on $(-\infty, 0]$, by ψ_4-symmetry with respect to y-axis, and $\psi_4'(0) = 0$.

Clearly it is

$$\lim_{x \to +\infty} \psi_4(x) = \lim_{x \to -\infty} \psi_4(x) = 0, \tag{8.55}$$

therefore the x-axis is the horizontal asymptote of $\psi_4(x)$.

The value

$$\psi_4(0) = \frac{1}{2\sqrt[\lambda]{2}}, \ \lambda \text{ is an odd number}, \tag{8.56}$$

is the maximum of ψ_4, which is a bell shaped function.

We need

Theorem 8.12 ([15]) *It holds*

$$\sum_{i=-\infty}^{\infty} \psi_4(x - i) = 1, \ \forall x \in \mathbb{R}. \tag{8.57}$$

Theorem 8.13 ([15]) *We have that*

$$\int_{-\infty}^{\infty} \psi_4(x)\, dx = 1. \tag{8.58}$$

So that $\psi_4(x)$ is a density function on $\mathbb{R}$.

We need

Theorem 8.14 ([15]) *Let $0 < \alpha < 1$, and $n \in \mathbb{N}$ with $n^{1-\alpha} > 2$. It holds*

$$\sum_{\substack{j=-\infty \\ : |nx-j| \geq n^{1-\alpha}}}^{\infty} \psi_4(nx-j) < \frac{1}{2\lambda\left(n^{1-\alpha}-2\right)^{\lambda}}, \tag{8.59}$$

where $\lambda \in \mathbb{N}$ is an odd number.

We introduce (see also [18])

$$Z_4(x_1, ..., x_N) := Z_4(x) := \prod_{i=1}^{N} \psi_4(x_i), \ \ x = (x_1, ..., x_N) \in \mathbb{R}^N, \ N \in \mathbb{N}. \tag{8.60}$$

It has the properties:

(i) $Z_4(x) > 0, \ \forall\, x \in \mathbb{R}^N$,

(ii)

$$\sum_{k=-\infty}^{\infty} Z_4(x-k) := \sum_{k_1=-\infty}^{\infty} \sum_{k_2=-\infty}^{\infty} ... \sum_{k_N=-\infty}^{\infty} Z_4(x_1-k_1, ..., x_N-k_N) = 1, \tag{8.61}$$

where $k := (k_1, ..., k_n) \in \mathbb{Z}^N, \forall\, x \in \mathbb{R}^N$,

hence

(iii)

$$\sum_{k=-\infty}^{\infty} Z_4(nx-k) = 1, \tag{8.62}$$

$\forall\, x \in \mathbb{R}^N; n \in \mathbb{N}$,

and

(iv)

$$\int_{\mathbb{R}^N} Z_4(x)\, dx = 1, \tag{8.63}$$

that is Z_4 is a multivariate density function.

(v) It is clear that

$$\sum_{\substack{k=-\infty \\ \left\| \frac{k}{n}-x \right\|_{\infty} > \frac{1}{n^{\beta}}}}^{\infty} Z_4 (nx-k) < \frac{1}{2\lambda \left(n^{1-\beta}-2\right)^{\lambda}} =: c_4 (\beta, n), \tag{8.64}$$

$0 < \beta < 1, n \in \mathbb{N} : n^{1-\beta} > 2, x \in \mathbb{R}^N$, λ is odd.

For $f \in C_B^+ \left(\mathbb{R}^N\right)$ (continuous and bounded functions from $\mathbb{R}^N$ into $\mathbb{R}_+$), we define the first modulus of continuity

$$\omega_1 (f, \delta) := \sup_{\substack{x, y \in \mathbb{R}^N \\ \|x-y\|_{\infty} \leq h}} |f(x) - f(y)|, h > 0. \tag{8.65}$$

Given that $f \in C_U^+ \left(\mathbb{R}^N\right)$ (uniformly continuous from $\mathbb{R}^N$ into $\mathbb{R}_+$, same definition for ω_1), we have that

$$\lim_{h \to 0} \omega_1 (f, h) = 0. \tag{8.66}$$

When $N = 1$, ω_1 is defined as in (8.65) with $\|\cdot\|_{\infty}$ collapsing to $|\cdot|$ and has the property (8.66).

8.3 Main Results

We need

Definition 8.15 Let $\mathcal{L}$ be the Lebesgue σ-algebra on $\mathbb{R}^N$, $N \in \mathbb{N}$, and the set function $\mu : \mathcal{L} \to [0, +\infty)$, which is assumed to be monotone, submodular and strictly positive. For $f \in C_B^+ \left(\mathbb{R}^N\right)$, we define the general multivariate Kantorovich-Choquet type neural network operators for any $x \in \mathbb{R}^N$ ($j = 1, 2, 3, 4$):

$${}_jK_n^{\mu} (f, x) = {}_jK_n^{\mu} (f, x_1, ..., x_N) := \tag{8.67}$$

$$\sum_{k=-\infty}^{\infty} \left(\frac{(C) \int_{\left[0, \frac{1}{n}\right]^N} f\left(t + \frac{k}{n}\right) d\mu (t)}{\mu \left(\left[0, \frac{1}{n}\right]^N\right)} \right) Z_j (nx-k) =$$

$$\sum_{k_1=-\infty}^{\infty} \sum_{k_2=-\infty}^{\infty} \cdots \sum_{k_N=-\infty}^{\infty} \left(\frac{(C) \int_0^{\frac{1}{n}} \cdots \int_0^{\frac{1}{n}} f\left(t_1 + \frac{k_1}{n}, t_2 + \frac{k_2}{n}, ..., t_N + \frac{k_N}{n}\right) d\mu (t_1, ..., t_N)}{\mu \left(\left[0, \frac{1}{n}\right]^N\right)} \right)$$

$$\left(\prod_{i=1}^{N} \psi_j (nx_i - k_i) \right),$$

where $x = (x_1, ..., x_N) \in \mathbb{R}^N$, $k = (k_1, ..., k_N)$, $t = (t_1, ..., t_N)$, $n \in \mathbb{N}$.

Clearly here $\mu\left(\left[0, \frac{1}{n}\right]^N\right) > 0$, $\forall\, n \in \mathbb{N}$.

Above we notice that

$$\left\| {}_jK_n^{\mu}(f)\right\|_{\infty} \leq \|f\|_{\infty}, \tag{8.68}$$

so that ${}_jK_n^{\mu}(f, x)$ is well-defined, $j = 1, 2, 3, 4$.

We make

Remark 8.16 Let $f \in C_B^+\left(\mathbb{R}^N\right)$, $t \in \left[0, \frac{1}{n}\right]^N$ and $x \in \mathbb{R}^N$, then

$$f\left(t + \frac{k}{n}\right) = f\left(t + \frac{k}{n}\right) - f(x) + f(x) \leq \left|f\left(t + \frac{k}{n}\right) - f(x)\right| + f(x),$$

hence

$$(C)\int_{\left[0,\frac{1}{n}\right]^N} f\left(t + \frac{k}{n}\right) d\mu(t) \leq$$

$$(C)\int_{\left[0,\frac{1}{n}\right]^N} \left|f\left(t + \frac{k}{n}\right) - f(x)\right| d\mu(t) + (C)\int_{\left[0,\frac{1}{n}\right]^N} f(x)\, d\mu(t) = \tag{8.69}$$

$$(C)\int_{\left[0,\frac{1}{n}\right]^N} \left|f\left(t + \frac{k}{n}\right) - f(x)\right| d\mu(t) + f(x)\,\mu\left(\left[0, \frac{1}{n}\right]^N\right).$$

That is

$$(C)\int_{\left[0,\frac{1}{n}\right]^N} f\left(t + \frac{k}{n}\right) d\mu(t) - f(x)\,\mu\left(\left[0, \frac{1}{n}\right]^N\right) \leq$$

$$(C)\int_{\left[0,\frac{1}{n}\right]^N} \left|f\left(t + \frac{k}{n}\right) - f(x)\right| d\mu(t). \tag{8.70}$$

Similarly, we have that

$$f(x) = f(x) - f\left(t + \frac{k}{n}\right) + f\left(t + \frac{k}{n}\right) \leq \left|f\left(t + \frac{k}{n}\right) - f(x)\right| + f\left(t + \frac{k}{n}\right).$$

Hence

$$(C)\int_{\left[0,\frac{1}{n}\right]^N} f(x)\,\mu(dt) \leq$$

$$(C)\int_{\left[0,\frac{1}{n}\right]^N} \left|f\left(t + \frac{k}{n}\right) - f(x)\right| d\mu(t) + (C)\int_{\left[0,\frac{1}{n}\right]^N} f\left(t + \frac{k}{n}\right) \mu(dt),$$

and

$$f(x)\,\mu\left(\left[0,\frac{1}{n}\right]^N\right)-(C)\int_{\left[0,\frac{1}{n}\right]^N} f\left(t+\frac{k}{n}\right)\mu(dt)\leq$$

$$(C)\int_{\left[0,\frac{1}{n}\right]^N}\left|f\left(t+\frac{k}{n}\right)-f(x)\right|d\mu(t). \tag{8.71}$$

By (8.70) and (8.71) we derive that

$$\left|(C)\int_{\left[0,\frac{1}{n}\right]^N} f\left(t+\frac{k}{n}\right)\mu(dt)-f(x)\,\mu\left(\left[0,\frac{1}{n}\right]^N\right)\right|\leq$$

$$(C)\int_{\left[0,\frac{1}{n}\right]^N}\left|f\left(t+\frac{k}{n}\right)-f(x)\right|d\mu(t). \tag{8.72}$$

In particular, it holds

$$\left|\left(\frac{(C)\int_{\left[0,\frac{1}{n}\right]^N} f\left(t+\frac{k}{n}\right)\mu(dt)}{\mu\left(\left[0,\frac{1}{n}\right]^N\right)}\right)-f(x)\right|\leq\frac{(C)\int_{\left[0,\frac{1}{n}\right]^N}\left|f\left(t+\frac{k}{n}\right)-f(x)\right|d\mu(t)}{\mu\left(\left[0,\frac{1}{n}\right]^N\right)}. \tag{8.73}$$

We present the following approximation result.

Theorem 8.17 *Let $f\in C_B^+\left(\mathbb{R}^N\right)$, $0<\beta<1$, $x\in\mathbb{R}^N$, $N,n\in\mathbb{N}$ with $n^{1-\beta}>2$; $j=1,2,3,4$. Then*
(i)

$$\sup_{\mu}\left|{}_jK_n^{\mu}(f,x)-f(x)\right|\leq\omega_1\left(f,\frac{1}{n}+\frac{1}{n^{\beta}}\right)+2\|f\|_{\infty}c_j(\beta,n)=:\rho_{jn}, \tag{8.74}$$

and
(ii)

$$\sup_{\mu}\left\|{}_jK_n^{\mu}(f)-f\right\|_{\infty}\leq\rho_{jn}. \tag{8.75}$$

Given that $f\in\left(C_U^+\left(\mathbb{R}^N\right)\cap C_B^+\left(\mathbb{R}^N\right)\right)$, we obtain $\lim_{n\to\infty}{}_jK_n^k(f)=f$, uniformly.
Above $c_j(\beta,n)$ are as in (8.17), (8.32), (8.47) and (8.64), respectively.

Proof We observe that

$$\left|{}_jK_n^{\mu}(f,x)-f(x)\right|=$$

$$\left|\sum_{k=-\infty}^{\infty}\left(\frac{(C)\int_{\left[0,\frac{1}{n}\right]^N} f\left(t+\frac{k}{n}\right)d\mu(t)}{\mu\left(\left[0,\frac{1}{n}\right]^N\right)}\right)Z_j(nx-k)-\sum_{k=-\infty}^{\infty}f(x)Z_j(nx-k)\right|$$

$$= \left| \sum_{k=-\infty}^{\infty} \left(\left(\frac{(C) \int_{[0,\frac{1}{n}]^N} f\left(t+\frac{k}{n}\right) d\mu(t)}{\mu\left(\left[0,\frac{1}{n}\right]^N\right)} \right) - f(x) \right) Z_j(nx-k) \right| \leq \quad (8.76)$$

$$\sum_{k=-\infty}^{\infty} \left| \left(\frac{(C) \int_{[0,\frac{1}{n}]^N} f\left(t+\frac{k}{n}\right) d\mu(t)}{\mu\left(\left[0,\frac{1}{n}\right]^N\right)} \right) - f(x) \right| Z_j(nx-k) \overset{(8.73.)}{\leq}$$

$$\sum_{k=-\infty}^{\infty} \left(\frac{(C) \int_{[0,\frac{1}{n}]^N} \left| f\left(t+\frac{k}{n}\right) - f(x) \right| d\mu(t)}{\mu\left(\left[0,\frac{1}{n}\right]^N\right)} \right) Z_j(nx-k) =$$

$$\sum_{\begin{cases} k=-\infty \\ \left\| \frac{k}{n} - x \right\|_\infty \leq \frac{1}{n^\beta} \end{cases}}^{\infty} \left(\frac{(C) \int_{[0,\frac{1}{n}]^N} \left| f\left(t+\frac{k}{n}\right) - f(x) \right| d\mu(t)}{\mu\left(\left[0,\frac{1}{n}\right]^N\right)} \right) Z_j(nx-k) +$$

$$(8.77)$$

$$\sum_{\begin{cases} k=-\infty \\ \left\| \frac{k}{n} - x \right\|_\infty > \frac{1}{n^\beta} \end{cases}}^{\infty} \left(\frac{(C) \int_{[0,\frac{1}{n}]^N} \left| f\left(t+\frac{k}{n}\right) - f(x) \right| d\mu(t)}{\mu\left(\left[0,\frac{1}{n}\right]^N\right)} \right) Z_j(nx-k) \leq$$

$$\sum_{\begin{cases} k=-\infty \\ \left\| \frac{k}{n} - x \right\|_\infty \leq \frac{1}{n^\beta} \end{cases}}^{\infty} \left(\frac{(C) \int_{[0,\frac{1}{n}]^N} \omega_1\left(f, \|t\|_\infty + \left\| \frac{k}{n} - x \right\|_\infty\right) d\mu(t)}{\mu\left(\left[0,\frac{1}{n}\right]^N\right)} \right) Z_j(nx-k) +$$

$$2\|f\|_\infty \left(\sum_{\begin{cases} k=-\infty \\ \left\| \frac{k}{n} - x \right\|_\infty > \frac{1}{n^\beta} \end{cases}}^{\infty} Z_j(|nx-k|) \right) \text{ (by (8.17), (8.32), (8.47), (8.64))}$$

$$\leq \omega_1\left(f, \frac{1}{n} + \frac{1}{n^\beta}\right) + 2\|f\|_\infty c_j(\beta, n), \quad (8.78)$$

proving the claim. ∎

Additionally we give

Definition 8.18 Denote $C_B^+(\mathbb{R}^N, \mathbb{C}) = \{f : \mathbb{R}^N \to \mathbb{C} | f = f_1 + if_2$, where $f_1, f_2 \in C_B^+(\mathbb{R}^N)\}$. We set for $f \in C_B^+(\mathbb{R}^N, \mathbb{C})$ that

$$ {}_jK_n^{\mu}(f, x) := {}_jK_n^{\mu}(f_1, x) + i\, {}_jK_n^{\mu}(f_2, x), \tag{8.79}$$

$\forall\, n \in \mathbb{N}, x \in \mathbb{R}^N;\ j = 1, 2, 3, 4;\ i = \sqrt{-1}$.

We give

Theorem 8.19 *Let $f \in C_B^+(\mathbb{R}^N, \mathbb{C})$, $f = f_1 + if_2$, $N \in \mathbb{N}$, $0 < \beta < 1$, $x \in \mathbb{R}^N$, $n \in \mathbb{N}$ with $n^{1-\beta} > 2$; $j = 1, 2, 3, 4$. Then*
(i)

$$\sup_{\mu} \left| {}_jK_n^{\mu}(f, x) - f(x) \right| \leq \left(\omega_1\left(f_1, \frac{1}{n} + \frac{1}{n^{\beta}}\right) + \omega_1\left(f_2, \frac{1}{n} + \frac{1}{n^{\beta}}\right)\right)$$

$$+ 2\left(\|f_1\|_{\infty} + \|f_2\|_{\infty}\right) c_j(\beta, n) =: \gamma_{jn}, \tag{8.80}$$

and
(ii)

$$\sup_{\mu} \left\| {}_jK_n^{\mu}(f) - f \right\|_{\infty} \leq \gamma_{jn}.$$

Proof We have that

$$\left| {}_jK_n^{\mu}(f, x) - f(x) \right| = \left| {}_jK_n^{\mu}(f_1, x) + i\, {}_jK_n^{\mu}(f_2, x) - f_1(x) - if_2(x) \right| =$$

$$\left| \left({}_jK_n^{\mu}(f_1, x) - f_1(x)\right) + i\left({}_jK_n^{\mu}(f_2, x) - f_2(x)\right) \right| \leq$$

$$\left| {}_jK_n^{\mu}(f_1, x) - f_1(x) \right| + \left| {}_jK_n^{\mu}(f_2, x) - f_2(x) \right| \overset{(8.74.)}{\leq}$$

$$\left(\omega_1\left(f_1, \frac{1}{n} + \frac{1}{n^{\beta}}\right) + 2\|f_1\|_{\infty} c_j(\beta, n) \right) + \tag{8.81}$$

$$\left(\omega_1\left(f_2, \frac{1}{n} + \frac{1}{n^{\beta}}\right) + 2\|f_2\|_{\infty} c_j(\beta, n) \right),$$

proving the claim. ∎

We need

Definition 8.20 Let $\mathcal{L}^*$ be the Lebesgue σ-algebra on $\mathbb{R}$, and the set function $\mu^* : \mathcal{L}^* \to [0, +\infty]$, which is assumed to be monotone, submodular and strictly positive.

For $f \in C_B^+(\mathbb{R})$, we define the general univariate Kantorovich-Choquet type neural network operator for any $x \in \mathbb{R}$ $(j = 1, 2, 3, 4)$:

$$ {}_jM_n^{\mu^*}(f, x) = \sum_{k=-\infty}^{\infty} \left(\frac{(C)\int_0^{\frac{1}{n}} f\left(t + \frac{k}{n}\right) d\mu^*(t)}{\mu^*\left(\left[0, \frac{1}{n}\right]\right)} \right) \psi_j(nx - k). \tag{8.82} $$

Clearly here $\mu^*\left(\left[0, \frac{1}{n}\right]\right) > 0$, $\forall\, n \in \mathbb{N}$.

Above we notice that

$$ \left\| {}_jM_n^{\mu^*}(f) \right\|_\infty \le \|f\|_\infty, \tag{8.83} $$

so that ${}_jM_n^{\mu^*}(f, x)$ is well-defined, $j = 1, 2, 3, 4$.

Notice that ${}_jK_n^{\mu}$, when $N = 1$, collapses to ${}_jM_n^{\mu^*}$, $j = 1, 2, 3, 4$.

It follows another appropiate result.

Corollary 8.21 (to Theorem 8.17 when $N = 1$) *Let $f \in C_B^+(\mathbb{R})$, $0 < \beta < 1$, $x \in \mathbb{R}$; $n \in \mathbb{N}$ with $n^{1-\beta} > 2$; $j = 1, 2, 3, 4$. Then*

(i)

$$ \sup_{\mu^*} \left| {}_jM_n^{\mu^*}(f, x) - f(x) \right| \le \omega_1\left(f, \frac{1}{n} + \frac{1}{n^\beta}\right) + 2\|f\|_\infty c_j(\beta, n) =: \varepsilon_{jn}, \tag{8.84} $$

and

(ii)

$$ \sup_{\mu} \left\| {}_jM_n^{\mu^*}(f) - f \right\|_\infty \le \varepsilon_{jn}. \tag{8.85} $$

Given that $f \in \left(C_U^+(\mathbb{R}) \cap C_B^+(\mathbb{R})\right)$, we obtain $\lim_{n\to\infty} {}_jM_n^{\mu^*}(f) = f$, uniformly.

Above $c_j(\beta, n)$ are as in (8.17), (8.32), (8.47) and (8.64), respectively.

Proof As similar to Theorem 8.17 is omitted. ∎

We need

Definition 8.22 Let $f \in C_B^+(\mathbb{R}, \mathbb{C})$ where $f = f_1 + if_2$ with $f_1, f_2 \in C_B^+(\mathbb{R})$. We set

$$ {}_jM_n^{\mu^*}(f, x) := {}_jM_n^{\mu^*}(f_1, x) + i\, {}_jM_n^{\mu^*}(f_2, x), \tag{8.86} $$

$\forall\, n \in \mathbb{N}$, $x \in \mathbb{R}$; $j = 1, 2, 3, 4$.

We finish with

Corollary 8.23 (to Theorem 8.19 when $N = 1$) *Let $f \in C_B^+(\mathbb{R}, \mathbb{C})$, $f = f_1 + if_2$, $0 < \beta < 1$, $x \in \mathbb{R}$, $n \in \mathbb{N}$ with $n^{1-\beta} > 2$; $j = 1, 2, 3, 4$. Then*

(i)

$$\sup_{\mu^*}\left|{}_jM_n^{\mu^*}(f,x)-f(x)\right|\leq\left(\omega_1\left(f_1,\frac{1}{n}+\frac{1}{n^{\beta}}\right)+\omega_1\left(f_2,\frac{1}{n}+\frac{1}{n^{\beta}}\right)\right)+$$

$$2\left(\|f_1\|_{\infty}+\|f_2\|_{\infty}\right)c_j(\beta,n)=:\delta_{jn}, \tag{8.87}$$

and

(ii)

$$\sup_{\mu^*}\left\|{}_jM_n^{\mu^*}(f)-f\right\|_{\infty}\leq\delta_{jn}.$$

Proof As similar to Theorem 8.19 is omitted. ■

References

1. G.A. Anastassiou, Rate of convergence of some neural network operators to the unit-univariate case. J. Math. Anal. Appl. **212**, 237–262 (1997)
2. G.A. Anastassiou, *Quantitative Approximations* (Chapman & Hall/CRC, Boca Raton, New York, 2001)
3. G.A. Anastassiou, *Intelligent Systems: Approximation by Artificial Neural Networks, Intelligent Systems Reference Library*, vol. 19 (Springer, Heidelberg, 2011)
4. G.A. Anastassiou, Univariate hyperbolic tangent neural network approximation. Math. Comput. Model. **53**, 1111–1132 (2011)
5. G.A. Anastassiou, Multivariate hyperbolic tangent neural network approximation. Comput. Math. **61**, 809–821 (2011)
6. G.A. Anastassiou, Multivariate sigmoidal neural network approximation. Neural Netw. **24**, 378–386 (2011)
7. G.A. Anastassiou, Univariate sigmoidal neural network approximation. J. Comput. Anal. Appl. **14**(4), 659–690 (2012)
8. G.A. Anastassiou, Fractional neural network approximation. Comput. Math. Appl. **64**, 1655–1676 (2012)
9. G.A. Anastassiou, Univariate error function based neural network approximation. Indian J. Math. **57**(2), 243–291 (2015)
10. G.A. Anastassiou, *Intelligent Systems II: Complete Approximation by Neural Network Operators* (Springer, Heidelberg, New York, 2016)
11. G.A. Anastassiou, *Intelligent Computations: Abstract Fractional Calculus* Inequalities Approximations. (Springer, Heidelberg, New York, 2018)
12. G.A. Anastassiou, *Ordinary and Fractional Approximation by Non-additive Integrals: Choquet* Shilkret and Sugeno Integral Approximators. (Springer, Heidelberg, New York, 2019)
13. G.A. Anastassiou, Algebraic function based Banach space valued ordinary and fractional neural network approximations. New Trends Math. Sci. **10**(1), 100–125 (2022)
14. G.A. Anastassiou, Gudermannian function activated Banach space valued ordinary and fractional neural network approximation. Adv. Nonlinear Var. Inequal. **25**(2), 27–64 (2022)
15. G.A. Anastassiou, *Generalized Symmetrical Sigmoid Function Activated Banach Space Valued Ordinary and Fractional Neural Network Approximation* (Fasc. Matematica, Accepted for Publication, Analele Universităţii Oradea, 2022)

16. G.A. Anastassiou, Abstract multivariate Gudermannian function activated neural network approximations. Panamerican Math. J. (2022)
17. G.A. Anastassiou, *General Multivariate Arctangent Function Activated Neural Network Approximations*. Submitted (2022)
18. G.A. Anastassiou, *Generalized Symmetrical Sigmoid Function Activated Neural Network Multivariate Approximation*. Submitted (2022)
19. G.A. Anastassiou, *Degree of Approximation by Kantorovich-Choquet Quasi-Interpolation Neural Network Operators Revisited*. Submitted (2022)
20. G.A. Anastassiou, R. Kozma, *Abstract Multivariate Algebraic Function Activated Neural Network Approximations*. Submitted (2022)
21. Z. Chen, F. Cao, The approximation operators with sigmoidal functions. Comput. Math. Appl. **58**, 758–765 (2009)
22. G. Choquet, Theory of capacities. Ann. Inst. Fourier (Grenoble) **5**, 131–295 (1954)
23. D. Denneberg, *Non-additive Measure and Integral* (Kluwer, Dordrecht, 1994)
24. A.J. Dunning, J. Kensler, L. Goudeville, F. Bailleux, Some extensions in continuous methods for immunological correlates of protection. BMC Med. Res. Methodol. **15**(107) (2015). https://doi.org/10.1186/s12874-015-0096-9
25. S. Gal, Uniform and Pointwise Quantitative Approximation by Kantorovich-Choquet type integral Operators with respect to monotone and submodular set functions. Mediterr. J. Math. **14**(5), Art. 205, 12 pp (2017)
26. I.S. Haykin, *Neural Networks: A Comprehensive Foundation*, 2nd edn. (Prentice Hall, New York, 1998)
27. W. McCulloch, W. Pitts, A logical calculus of the ideas immanent in nervous activity. Bull. Math. Biophys. **7**, 115–133 (1943)
28. T.M. Mitchell, *Machine Learning* (WCB-McGraw-Hill, New York, 1997)
29. Z. Wang, G.J. Klir, *Generalized Measure Theory* (Springer, New York, 2009)
30. E.W. Weisstein, *Gudermannian*. MathWorld

Chapter 9
Quantitative Approximation by Kantorovich-Shilkret Quasi-interpolation Neural Network Operators Revisited

In this chapter we exhibit multivariate basic approximation by a Kantorovich-Shilkret type quasi-interpolation neural network operators with respect to supremum norm. This is done with rates using the multivariate modulus of continuity. We approximate continuous and bounded functions on $\mathbb{R}^N$, $N \in \mathbb{N}$. When they are additionally uniformly continuous we derive pointwise and uniform convergences. We include also the related Complex approximation. Our activation functions are induced by the arctangent, algebraic, Gudermannian and generalized symmetrical sigmoid functions. It follows [8].

9.1 Introduction

Here we are motivated by [1].

The author here performs multivariate arctangent-algebraic-Gudermannian-generalized symmetrical activation functions based neural network approximation to continuous functions over $\mathbb{R}^N$, $N \in \mathbb{N}$, and then he extends his results to complex valued functions. The convergences here are with rates expressed via the multivariate modulus of continuity of the involved function and given by very tight Jackson type inequalities.

The author comes up with the "right" precisely defined flexible quasi-interpolation Kantorovich-Shilkret type integral coefficient neural network operators associated to the arctangent-algebraic-Gudermannian-generalized symmetrical activation functions. This is a continuation of [3], Chap. 11.

Feed-forward neural network (FNNs) with one hidden layer with deal with, are expressed mathematically as

© The Author(s), under exclusive license to Springer Nature Switzerland AG 2023

G. A. Anastassiou, *Banach Space Valued Neural Network*,
Studies in Computational Intelligence 1062,
https://doi.org/10.1007/978-3-031-16400-2_9

$$N_n(x) = \sum_{j=0}^{n} c_j \sigma\left(\left\langle a_j \cdot x\right\rangle + b_j\right), \quad x \in \mathbb{R}^s, \, s \in \mathbb{N},$$

where for $0 \leq j \leq n$, $b_j \in \mathbb{R}$ are the thresholds, $a_j \in \mathbb{R}^s$ are the connection weights, $c_j \in \mathbb{R}$ are the coefficients, $\left\langle a_j \cdot x\right\rangle$ is the inner product of a_j and x, and σ is the activation function of the network. In many fundamental neural network models the activation functions are the arctangent- algebraic- Gudermannian- 9.

Generalized symmetrical activation functions.

About neural networks in general you may read [13–15]. In recent years non-additive integrals, like the N. Shilkret one [16], have become fashionable and more useful in Economic theory, etc.

9.2 Background

Here we follow [16].

Let $\mathcal{F}$ be a σ-field of subsets of an arbitrary set Ω. An extended non-negative real valued function μ on $\mathcal{F}$ is called maxitive if $\mu(\varnothing) = 0$ and

$$\mu\left(\cup_{i \in I} E_i\right) = \sup_{i \in I} \mu\left(E_i\right), \tag{9.1}$$

where the set I is of cardinality at most countable. We also call μ a maxitive measure. Here f stands for a non-negative measurable function on Ω. In [16], Niel Shilkret developed his non-additive integral defined as follows:

$$\left(N^*\right) \int_D f d\mu := \sup_{y \in Y} \left\{y \cdot \mu\left(D \cap \{f \geq y\}\right)\right\}, \tag{9.2}$$

where $Y = [0, m]$ or $Y = [0, m)$ with $0 < m \leq \infty$, and $D \in \mathcal{F}$. Here we take $Y = [0, \infty)$.

It is easily proved that

$$\left(N^*\right) \int_D f d\mu = \sup_{y > 0} \left\{y \cdot \mu\left(D \cap \{f > y\}\right)\right\}. \tag{9.3}$$

The Shilkret integral takes values in $[0, \infty]$.

The Shilkret integral [16] has the following properties:

$$\left(N^*\right) \int_\Omega \chi_E d\mu = \mu(E), \tag{9.4}$$

where χ_E is the indicator function on $E \in \mathcal{F}$,

$$(N*) \int_D cf d\mu = c\left(N^*\right) \int_D f d\mu,\ c \geq 0, \tag{9.5}$$

$$\left(N^*\right) \int_D \sup_{n \in \mathbb{N}} f_n d\mu = \sup_{n \in \mathbb{N}} \left(N^*\right) \int_D f_n d\mu, \tag{9.6}$$

where f_n, $n \in \mathbb{N}$, is an increasing sequence of elementary (countably valued) functions converging uniformly to f. Furthermore we have

$$\left(N^*\right) \int_D f d\mu \geq 0, \tag{9.7}$$

$$f \geq g \text{ implies } \left(N^*\right) \int_D f d\mu \geq \left(N^*\right) \int_D g d\mu, \tag{9.8}$$

where $f, g : \Omega \to [0, \infty]$ are measurable.

Let $a \leq f(\omega) \leq b$ for almost every $\omega \in E$, then

$$a\mu(E) \leq \left(N^*\right) \int_E f d\mu \leq b\mu(E);$$

$$\left(N^*\right) \int_E 1 d\mu = \mu(E);$$

$f > 0$ almost everywhere and $(N^*) \int_E f d\mu = 0$ imply $\mu(E) = 0$;

$(N^*) \int_\Omega f d\mu = 0$ if and only $f = 0$ almost everywhere;

$(N^*) \int_\Omega f d\mu < \infty$ implies that

$$\overline{N}(f) := \{\omega \in \Omega | f(\omega) \neq 0\} \text{ has } \sigma\text{-finite measure}; \tag{9.9}$$

$$\left(N^*\right) \int_D (f+g) d\mu \leq \left(N^*\right) \int_D f d\mu + \left(N^*\right) \int_D g d\mu;$$

and

$$\left| \left(N^*\right) \int_D f d\mu - \left(N^*\right) \int_D g d\mu \right| \leq \left(N^*\right) \int_D |f-g| d\mu. \tag{9.10}$$

From now on in this chapter we assume that $\mu : \mathcal{F} \to [0, +\infty)$.

9.2.1 About the Arctangent Activation Function

We consider the

$$\arctan x = \int_0^x \frac{dz}{1+z^2}, \ x \in \mathbb{R}. \tag{9.11}$$

We will be using

$$h(x) := \frac{2}{\pi} \arctan\left(\frac{\pi}{2}x\right) = \frac{2}{\pi} \int_0^{\frac{\pi x}{2}} \frac{dz}{1+z^2}, \ x \in \mathbb{R}, \tag{9.12}$$

which is a sigmoid type function and it is strictly increasing. We have that

$$h(0) = 0, \ h(-x) = -h(x), \ h(+\infty) = 1, \ h(-\infty) = -1,$$

and

$$h'(x) = \frac{4}{4+\pi^2 x^2} > 0, \ \text{all } x \in \mathbb{R}. \tag{9.13}$$

We consider the activation function

$$\psi_1(x) := \frac{1}{4}(h(x+1) - h(x-1)), \ x \in \mathbb{R}, \tag{9.14}$$

and we notice that

$$\psi_1(-x) = \psi_1(x), \tag{9.15}$$

it is an even function.

Since $x+1 > x-1$, then $h(x+1) > h(x-1)$, and $\psi_1(x) > 0$, all $x \in \mathbb{R}$.

We see that

$$\psi_1(0) = \frac{1}{\pi} \arctan\frac{\pi}{2} \cong 0.319. \tag{9.16}$$

Let $x > 0$, we have that

$$\psi_1'(x) = \frac{1}{4}\left(h'(x+1) - h'(x-1)\right) =$$

$$\frac{-4\pi^2 x}{\left(4+\pi^2(x+1)^2\right)\left(4+\pi^2(x-1)^2\right)} < 0. \tag{9.17}$$

That is

$$\psi_1'(x) < 0, \text{ for } x > 0. \tag{9.18}$$

That is ψ_1 is strictly decreasing on $[0, \infty)$ and clearly is strictly increasing on $(-\infty, 0]$, and $\psi_1'(0) = 0$.

Observe that

$$\begin{aligned}&\lim_{x\to+\infty}\psi_1(x)=\tfrac{1}{4}(h(+\infty)-h(+\infty))=0,\\&\text{and}\\&\lim_{x\to-\infty}\psi_1(x)=\tfrac{1}{4}(h(-\infty)-h(-\infty))=0.\end{aligned} \tag{9.19}$$

That is the x-axis is the horizontal asymptote on ψ_1.

All in all, ψ_1 is a bell symmetric function with maximum $\psi_1(0)\cong 0.319$.

We need

Theorem 9.1 ([2], p. 286) *We have that*

$$\sum_{i=-\infty}^{\infty}\psi_1(x-i)=1,\ \forall\, x\in\mathbb{R}. \tag{9.20}$$

Theorem 9.2 ([2], p. 287) *It holds*

$$\int_{-\infty}^{\infty}\psi_1(x)\,dx=1. \tag{9.21}$$

So that $\psi_1(x)$ is a density function on $\mathbb{R}$.

We mention

Theorem 9.3 ([2], p. 288) *Let* $0<\alpha<1$, *and* $n\in\mathbb{N}$ *with* $n^{1-\alpha}>2$. *It holds*

$$\sum_{\begin{cases}k=-\infty\\ :|nx-k|\geq n^{1-\alpha}\end{cases}}^{\infty}\psi_1(nx-k)<\frac{2}{\pi^2\left(n^{1-\alpha}-2\right)}. \tag{9.22}$$

We introduce (see [9])

$$Z_1(x_1,\ldots,x_N):=Z_1(x):=\prod_{i=1}^{N}\psi_1(x_i),\ \ x=(x_1,\ldots,x_N)\in\mathbb{R}^N,\ \ N\in\mathbb{N}. \tag{9.23}$$

It has the properties:

(i) $Z_1(x)>0,\ \forall\, x\in\mathbb{R}^N$,

(ii)

$$\sum_{k=-\infty}^{\infty}Z_1(x-k):=\sum_{k_1=-\infty}^{\infty}\sum_{k_2=-\infty}^{\infty}\ldots\sum_{k_N=-\infty}^{\infty}Z_1(x_1-k_1,\ldots,x_N-k_N)=1, \tag{9.24}$$

where $k := (k_1, ..., k_n) \in \mathbb{Z}^N, \forall\, x \in \mathbb{R}^N$,

hence

(iii)

$$\sum_{k=-\infty}^{\infty} Z_1 (nx - k) = 1, \tag{9.25}$$

$\forall\, x \in \mathbb{R}^N; n \in \mathbb{N}$,

and

(iv)

$$\int_{\mathbb{R}^N} Z_1 (x)\, dx = 1, \tag{9.26}$$

that is Z_1 is a multivariate density function.

(v) It is clear that

$$\sum_{\begin{cases} k = -\infty \\ \left\| \frac{k}{n} - x \right\|_{\infty} > \frac{1}{n^{\beta}} \end{cases}}^{\infty} Z_1 (nx - k) < \frac{2}{\pi^2 \left(n^{1-\beta} - 2\right)} =: c_1 (\beta, n), \tag{9.27}$$

$0 < \beta < 1, n \in \mathbb{N} : n^{1-\beta} > 2, x \in \mathbb{R}^N$.

Above it is $\|x\|_{\infty} := \max \{|x_1|, ..., |x_N|\}$, $x \in \mathbb{R}^N$, also set $\infty := (\infty, ..., \infty)$, $-\infty = (-\infty, ... - \infty)$ upon the multivariate context.

9.2.2 About the Algebraic Activation Function

Here see also [4].

We consider the generator algebraic function

$$\varphi (x) = \frac{x}{\sqrt[2m]{1 + x^{2m}}}, \quad m \in \mathbb{N}, x \in \mathbb{R}, \tag{9.28}$$

which is a sigmoidal type of function and is a strictly increasing function.

We see that $\varphi (-x) = -\varphi (x)$ with $\varphi (0) = 0$. We get that

$$\varphi' (x) = \frac{1}{\left(1 + x^{2m}\right)^{\frac{2m+1}{2m}}} > 0, \forall\, x \in \mathbb{R}, \tag{9.29}$$

proving φ as strictly increasing over $\mathbb{R}$, $\varphi' (x) = \varphi' (-x)$. We easily find that $\lim_{x \to +\infty} \varphi (x) = 1$, $\varphi (+\infty) = 1$, and $\lim_{x \to -\infty} \varphi (x) = -1$, $\varphi (-\infty) = -1$.

We consider the activation function

$$\psi_2(x) = \frac{1}{4}[\varphi(x+1) - \varphi(x-1)]. \tag{9.30}$$

Clearly it is $\psi_2(x) = \psi_2(-x)$, $\forall\, x \in \mathbb{R}$, so that ψ_2 is an even function and symmetric with respect to the y-axis. Clearly $\psi_2(x) > 0$, $\forall\, x \in \mathbb{R}$.

Also it is

$$\psi_2(0) = \frac{1}{2\sqrt[2m]{2}}. \tag{9.31}$$

By [4], we have that $\psi_2'(x) < 0$ for $x > 0$. That is ψ_2 is strictly decreasing over $(0, +\infty)$.

Clearly, ψ_2 is strictly increasing over $(-\infty, 0)$ and $\psi_2'(0) = 0$.

Furthermore we obtain that

$$\lim_{x\to+\infty} \psi_2(x) = \frac{1}{4}[\varphi(+\infty) - \varphi(+\infty)] = 0, \tag{9.32}$$

and

$$\lim_{x\to-\infty} \psi_2(x) = \frac{1}{4}[\varphi(-\infty) - \varphi(-\infty)] = 0. \tag{9.33}$$

That is the x-axis is the horizontal asymptote of ψ_2.

Conclusion, ψ_2 is a bell shape symmetric function with maximum

$$\psi_2(0) = \frac{1}{2\sqrt[2m]{2}}, \quad m \in \mathbb{N}. \tag{9.34}$$

We need

Theorem 9.4 ([4]) *We have that*

$$\sum_{i=-\infty}^{\infty} \psi_2(x-i) = 1, \ \forall\, x \in \mathbb{R}. \tag{9.35}$$

Theorem 9.5 ([4]) *It holds*

$$\int_{-\infty}^{\infty} \psi_2(x)\,dx = 1. \tag{9.36}$$

Theorem 9.6 ([4]) *Let* $0 < \alpha < 1$, *and* $n \in \mathbb{N}$ *with* $n^{1-\alpha} > 2$. *It holds*

$$\sum_{\substack{k=-\infty \\ :|nx-k| \geq n^{1-\alpha}}}^{\infty} \psi_2(nx-k) < \frac{1}{4m\left(n^{1-\alpha}-2\right)^{2m}}, \quad m \in \mathbb{N}. \tag{9.37}$$

We introduce (see also [11])

$$Z_2(x_1, ..., x_N) := Z_2(x) := \prod_{i=1}^{N} \psi_2(x_i), \quad x = (x_1, ..., x_N) \in \mathbb{R}^N, \quad N \in \mathbb{N}. \tag{9.38}$$

It has the properties:

(i) $Z_2(x) > 0, \ \forall\, x \in \mathbb{R}^N$,

(ii)

$$\sum_{k=-\infty}^{\infty} Z_2(x-k) := \sum_{k_1=-\infty}^{\infty} \sum_{k_2=-\infty}^{\infty} \dots \sum_{k_N=-\infty}^{\infty} Z_2(x_1-k_1, ..., x_N-k_N) = 1, \tag{9.39}$$

where $k := (k_1, ..., k_n) \in \mathbb{Z}^N$, $\forall\, x \in \mathbb{R}^N$,

hence

(iii)

$$\sum_{k=-\infty}^{\infty} Z_2(nx-k) = 1, \tag{9.40}$$

$\forall\, x \in \mathbb{R}^N$; $n \in \mathbb{N}$,

and

(iv)

$$\int_{\mathbb{R}^N} Z_2(x)\, dx = 1, \tag{9.41}$$

that is Z_2 is a multivariate density function.

(v) It is clear that

$$\sum_{\substack{k=-\infty \\ \left\| \frac{k}{n} - x \right\|_\infty > \frac{1}{n^\beta}}}^{\infty} Z_2(nx-k) < \frac{1}{4m\left(n^{1-\beta}-2\right)^{2m}} =: c_2(\beta, n), \tag{9.42}$$

$0 < \beta < 1$, $n \in \mathbb{N} : n^{1-\beta} > 2$, $x \in \mathbb{R}^N$, $m \in \mathbb{N}$.

9.2.3 About the Gudermannian Activation Function

See also [5, 17].

Here we consider $gd(x)$ the Gudermannian function [17], which is a sigmoid function, as a generator function:

$$\sigma(x) = 2\arctan\left(\tanh\left(\frac{x}{2}\right)\right) = \int_0^x \frac{dt}{\cosh t} =: gd(x), \ x \in \mathbb{R}. \tag{9.43}$$

Let the normalized generator sigmoid function

$$f(x) := \frac{4}{\pi}\sigma(x) = \frac{4}{\pi}\int_0^x \frac{dt}{\cosh t} = \frac{8}{\pi}\int_0^x \frac{1}{e^t + e^{-t}}dt, \ x \in \mathbb{R}. \tag{9.44}$$

Here

$$f'(x) = \frac{4}{\pi \cosh x} > 0, \ \forall\, x \in \mathbb{R},$$

hence f is strictly increasing on $\mathbb{R}$.

Notice that $\tanh(-x) = -\tanh x$ and $\arctan(-x) = -\arctan x$, $x \in \mathbb{R}$.

So, here the neural network activation function will be:

$$\psi_3(x) = \frac{1}{4}[f(x+1) - f(x-1)], \ x \in \mathbb{R}. \tag{9.45}$$

By [5], we get that

$$\psi_3(x) = \psi_3(-x), \ \forall\, x \in \mathbb{R}, \tag{9.46}$$

i.e. it is even and symmetric with respect to the y-axis. Here we have $f(+\infty) = 1$, $f(-\infty) = -1$ and $f(0) = 0$. Clearly it is

$$f(-x) = -f(x), \ \forall\, x \in \mathbb{R}, \tag{9.47}$$

an odd function, symmetric with respect to the origin. Since $x + 1 > x - 1$, and $f(x+1) > f(x-1)$, we obtain $\psi_3(x) > 0$, $\forall\, x \in \mathbb{R}$.

By [5], we have that

$$\psi_3(0) = \frac{2}{\pi}gd(1) \cong 0.551. \tag{9.48}$$

By [5] ψ_3 is strictly decreasing on $(0, +\infty)$, and strictly increasing on $(-\infty, 0)$, and $\psi_3'(0) = 0$.

Also we have that

$$\lim_{x\to+\infty} \psi_3(x) = \lim_{x\to-\infty} \psi_3(x) = 0, \tag{9.49}$$

that is the x-axis is the horizontal asymptote for ψ_3.

Conclusion, ψ_3 is a bell shaped symmetric function with maximum $\psi_3(0) \cong 0.551$.

We need

Theorem 9.7 ([5]) *It holds that*

$$\sum_{i=-\infty}^{\infty} \psi_3 (x - i) = 1, \forall x \in \mathbb{R}. \tag{9.50}$$

Theorem 9.8 ([5]) *We have that*

$$\int_{-\infty}^{\infty} \psi_3 (x) \, dx = 1. \tag{9.51}$$

So $\psi_3 (x)$ is a density function.

Theorem 9.9 ([5]) *Let $0 < \alpha < 1$, and $n \in \mathbb{N}$ with $n^{1-\alpha} > 2$. It holds*

$$\sum_{\begin{cases} k = -\infty \\ : |nx - k| \geq n^{1-\alpha} \end{cases}}^{\infty} \psi_3 (nx - k) < \frac{4}{\pi e^{(n^{1-\alpha}-2)}} = \frac{4e^2}{\pi e^{n^{1-\alpha}}}. \tag{9.52}$$

We introduce (see also [7])

$$Z_3 (x_1, ..., x_N) := Z_3 (x) := \prod_{i=1}^{N} \psi_3 (x_i), \ \ x = (x_1, ..., x_N) \in \mathbb{R}^N, \ \ N \in \mathbb{N}. \tag{9.53}$$

It has the properties:

(i) $Z_3 (x) > 0, \ \forall x \in \mathbb{R}^N$,

(ii)

$$\sum_{k=-\infty}^{\infty} Z_3 (x - k) := \sum_{k_1=-\infty}^{\infty} \sum_{k_2=-\infty}^{\infty} ... \sum_{k_N=-\infty}^{\infty} Z_3 (x_1 - k_1, ..., x_N - k_N) = 1, \tag{9.54}$$

where $k := (k_1, ..., k_n) \in \mathbb{Z}^N, \forall x \in \mathbb{R}^N$,

hence

(iii)

$$\sum_{k=-\infty}^{\infty} Z_3 (nx - k) = 1, \tag{9.55}$$

$\forall x \in \mathbb{R}^N; n \in \mathbb{N}$,

and

(iv)

$$\int_{\mathbb{R}^N} Z_3 (x) \, dx = 1, \tag{9.56}$$

that is Z_3 is a multivariate density function.

(v) It is also clear that

$$\sum_{\substack{k=-\infty \\ \left\| \frac{k}{n} - x \right\|_{\infty} > \frac{1}{n^{\beta}}}}^{\infty} Z_3(nx - k) < \frac{4e^2}{\pi e^{n^{1-\beta}}} = c_3(\beta, n), \tag{9.57}$$

$0 < \beta < 1, n \in \mathbb{N} : n^{1-\beta} > 2, x \in \mathbb{R}^N, m \in \mathbb{N}$.

9.2.4 About the Generalized Symmetrical Activation Function

Here we consider the generalized symmetrical sigmoid function [6, 12]

$$f_1(x) = \frac{x}{(1 + |x|^{\mu})^{\frac{1}{\mu}}}, \ \mu > 0, \ x \in \mathbb{R}. \tag{9.58}$$

This has applications in immunology and protection from disease together with probability theory. It is also called a symmetrical protection curve.

The parameter μ is a shape parameter controlling how fast the curve approaches the asymptotes for a given slope at the inflection point. When $\mu = 1$ f_1 is the absolute sigmoid function, and when $\mu = 2$, f_1 is the square root sigmoid function. When $\mu = 1.5$ the function approximates the arctangent function, when $\mu = 2.9$ it approximates the logistic function, and when $\mu = 3.4$ it approximates the error function. Parameter μ is estimated in the likelihood maximization [12]. For more see [12].

Next we study the particular generator sigmoid function

$$f_2(x) = \frac{x}{\left(1 + |x|^{\lambda}\right)^{\frac{1}{\lambda}}}, \ \lambda \text{ is an odd number, } x \in \mathbb{R}. \tag{9.59}$$

We have that $f_2(0) = 0$, and

$$f_2(-x) = -f_2(x), \tag{9.60}$$

so f_2 is symmetric with respect to zero.

When $x \geq 0$, we get that [6]

$$f_2'(x) = \frac{1}{\left(1 + x^{\lambda}\right)^{\frac{\lambda+1}{\lambda}}} > 0, \tag{9.61}$$

that is f_2 is strictly increasing on $[0, +\infty)$ and f_2 is strictly increasing on $(-\infty, 0]$. Hence f_2 is strictly increasing on $\mathbb{R}$.

We also have $f_2(+\infty) = f_2(-\infty) = 1$.

Let us consider the activation function [6]:

$$\psi_4(x) = \frac{1}{4}\left[f_2(x+1) - f_2(x-1)\right] =$$

$$\frac{1}{4}\left[\frac{(x+1)}{\left(1+|x+1|^{\lambda}\right)^{\frac{1}{\lambda}}} - \frac{(x-1)}{\left(1+|x-1|^{\lambda}\right)^{\frac{1}{\lambda}}}\right]. \tag{9.62}$$

Clearly it holds [6]

$$\psi_4(x) = \psi_4(-x), \ \forall\, x \in \mathbb{R}. \tag{9.63}$$

and

$$\psi_4(0) = \frac{1}{2\sqrt[\lambda]{2}}, \tag{9.64}$$

and $\psi_4(x) > 0$, $\forall\, x \in \mathbb{R}$.

Following [6], we have that ψ_4 is strictly decreasing over $[0, +\infty)$, and ψ_4 is strictly increasing on $(-\infty, 0]$, by ψ_4-symmetry with respect to y-axis, and $\psi_4'(0) = 0$.

Clearly it is

$$\lim_{x\to+\infty} \psi_4(x) = \lim_{x\to-\infty} \psi_4(x) = 0, \tag{9.65}$$

therefore the x-axis is the horizontal asymptote of $\psi_4(x)$.

The value

$$\psi_4(0) = \frac{1}{2\sqrt[\lambda]{2}}, \ \lambda \text{ is an odd number}, \tag{9.66}$$

is the maximum of ψ_4, which is a bell shaped function.

We need

Theorem 9.10 ([6]) *It holds*

$$\sum_{i=-\infty}^{\infty} \psi_4(x-i) = 1, \ \forall\, x \in \mathbb{R}. \tag{9.67}$$

Theorem 9.11 ([6]) *We have that*

$$\int_{-\infty}^{\infty} \psi_4(x)\, dx = 1. \tag{9.68}$$

So that $\psi_4(x)$ is a density function on $\mathbb{R}$.

We need

Theorem 9.12 ([6]) *Let* $0 < \alpha < 1$, *and* $n \in \mathbb{N}$ *with* $n^{1-\alpha} > 2$. *It holds*

$$\sum_{\substack{j=-\infty \\ : |nx-j| \geq n^{1-\alpha}}}^{\infty} \psi_4 (nx - j) < \frac{1}{2\lambda \left(n^{1-\alpha} - 2\right)^{\lambda}}, \tag{9.69}$$

where $\lambda \in \mathbb{N}$ is an odd number.

We introduce (see also [10])

$$Z_4 (x_1, ..., x_N) := Z_4 (x) := \prod_{i=1}^{N} \psi_4 (x_i), \quad x = (x_1, ..., x_N) \in \mathbb{R}^N, \quad N \in \mathbb{N}. \tag{9.70}$$

It has the properties:
(i) $Z_4 (x) > 0, \ \forall \, x \in \mathbb{R}^N$,
(ii)

$$\sum_{k=-\infty}^{\infty} Z_4 (x - k) := \sum_{k_1=-\infty}^{\infty} \sum_{k_2=-\infty}^{\infty} \cdots \sum_{k_N=-\infty}^{\infty} Z_4 (x_1 - k_1, ..., x_N - k_N) = 1, \tag{9.71}$$

where $k := (k_1, ..., k_n) \in \mathbb{Z}^N, \forall \, x \in \mathbb{R}^N$,
hence
(iii)

$$\sum_{k=-\infty}^{\infty} Z_4 (nx - k) = 1, \tag{9.72}$$

$\forall \, x \in \mathbb{R}^N; \, n \in \mathbb{N}$,
and
(iv)

$$\int_{\mathbb{R}^N} Z_4 (x) \, dx = 1, \tag{9.73}$$

that is Z_4 is a multivariate density function.
(v) It is clear that

$$\sum_{\substack{k=-\infty \\ \left\| \frac{k}{n} - x \right\|_{\infty} > \frac{1}{n^{\beta}}}}^{\infty} Z_4 (nx - k) < \frac{1}{2\lambda \left(n^{1-\beta} - 2\right)^{\lambda}} =: c_4 (\beta, n), \tag{9.74}$$

$0 < \beta < 1, n \in \mathbb{N} : n^{1-\beta} > 2, x \in \mathbb{R}^N$, λ is odd.

For $f \in C_B^+ \left(\mathbb{R}^N\right)$ (continuous and bounded functions from $\mathbb{R}^N$ into $\mathbb{R}_+$), we define the first modulus of continuity

$$\omega_1 (f, \delta) := \sup_{\substack{x, y \in \mathbb{R}^N \\ \|x - y\|_\infty \leq h}} |f (x) - f (y)| , h > 0. \tag{9.75}$$

Given that $f \in C_U^+ (\mathbb{R}^N)$ (uniformly continuous from $\mathbb{R}^N$ into $\mathbb{R}_+$, same definition for ω_1), we have that

$$\lim_{h \to 0} \omega_1 (f, h) = 0. \tag{9.76}$$

When $N = 1$, ω_1 is defined as in (9.75) with $\|\cdot\|_\infty$ collapsing to $|\cdot|$ and has the property (9.76).

9.3 Main Results

We need

Definition 9.13 Let $\mathcal{L}$ be the Lebesgue σ-algebra on $\mathbb{R}^N$, $N \in \mathbb{N}$, and the maxitive measure $\mu : \mathcal{L} \to [0, +\infty)$, such that for any $A \in \mathcal{L}$ with $A \neq \varnothing$, we get $\mu (A) > 0$.

For $f \in C_B^+ (\mathbb{R}^N)$, we define the multivariate Kantorovich-Shilkret type neural network operators for any $x \in \mathbb{R}^N$:

$$_jT_n^\mu (f, x) = {}_jT_n^\mu (f, x_1, ..., x_N) :=$$

$$\sum_{k=-\infty}^{\infty} \left(\frac{(N^*) \int_{[0, \frac{1}{N}]^N} f \left(t + \frac{k}{n}\right) d\mu (t)}{\mu \left(\left[0, \frac{1}{n}\right]^N\right)} \right) Z_j (nx - k) =$$

$$\sum_{k_1=-\infty}^{\infty} \sum_{k_2=-\infty}^{\infty} \cdots \sum_{k_N=-\infty}^{\infty} \left(\frac{(N^*) \int_0^{\frac{1}{n}} ... \int_0^{\frac{1}{n}} f \left(t_1 + \frac{k_1}{n}, t_2 + \frac{k_2}{n}, ..., t_N + \frac{k_N}{n}\right) d\mu (t_1, ..., t_N)}{\mu \left(\left[0, \frac{1}{n}\right]^N\right)} \right) \tag{9.77}$$

$$\cdot \left(\prod_{i=1}^{N} \psi_j (nx_i - k_i) \right),$$

where $x = (x_1, ..., x_N) \in \mathbb{R}^N$, $k = (k_1, ..., k_N)$, $t = (t_1, ..., t_N)$, $n \in \mathbb{N}$; $j = 1, 2, 3, 4$.

Clearly here $\mu \left(\left[0, \frac{1}{n}\right]^N\right) > 0$, $\forall\, n \in \mathbb{N}$.

Above we notice that

$$\left\| {}_jT_n^\mu (f) \right\|_\infty \leq \|f\|_\infty , \tag{9.78}$$

so that $_jT_n^\mu (f, x)$ is well-defined, $j = 1, 2, 3, 4$.

We make

Remark 9.14 Let $t \in \left[0, \frac{1}{n}\right]^N$ and $x \in \mathbb{R}^N$, then

$$f\left(t+\frac{k}{n}\right) = f\left(t+\frac{k}{n}\right) - f(x) + f(x) \leq \left|f\left(t+\frac{k}{n}\right) - f(x)\right| + f(x), \tag{9.79}$$

hence

$$(N^*)\int_{[0,\frac{1}{n}]^N} f\left(t+\frac{k}{n}\right) d\mu(t) \leq$$

$$(N^*)\int_{[0,\frac{1}{n}]^N} \left|f\left(t+\frac{k}{n}\right) - f(x)\right| d\mu(t) + f(x)\,\mu\left(\left[0,\frac{1}{n}\right]^N\right). \tag{9.80}$$

That is

$$(N^*)\int_{[0,\frac{1}{n}]^N} f\left(t+\frac{k}{n}\right) d\mu(t) - f(x)\,\mu\left(\left[0,\frac{1}{n}\right]^N\right) \leq \tag{9.81}$$

$$(N^*)\int_{[0,\frac{1}{n}]^N} \left|f\left(t+\frac{k}{n}\right) - f(x)\right| d\mu(t).$$

Similarly, we have

$$f(x) = f(x) - f\left(t+\frac{k}{n}\right) + f\left(t+\frac{k}{n}\right) \leq \left|f\left(t+\frac{k}{n}\right) - f(x)\right| + f\left(t+\frac{k}{n}\right),$$

hence

$$(N^*)\int_{[0,\frac{1}{n}]^N} f(x)\,d\mu(t) \leq (N^*)\int_{[0,\frac{1}{n}]^N} \left|f\left(t+\frac{k}{n}\right) - f(x)\right| d\mu(t)$$

$$+ (N^*)\int_{[0,\frac{1}{n}]^N} f\left(t+\frac{k}{n}\right) d\mu(t).$$

That is

$$f(x)\,\mu\left(\left[0,\frac{1}{n}\right]^N\right) - (N^*)\int_{[0,\frac{1}{n}]^N} f\left(t+\frac{k}{n}\right) d\mu(t) \leq \tag{9.82}$$

$$(N^*)\int_{[0,\frac{1}{n}]^N} \left|f\left(t+\frac{k}{n}\right) - f(x)\right| d\mu(t).$$

By (9.81) and (9.82) we derive

$$\left| (N^*) \int_{\left[0, \frac{1}{n}\right]^N} f\left(t + \frac{k}{n}\right) d\mu(t) - f(x)\, \mu\left(\left[0, \frac{1}{n}\right]^N\right) \right| \leq$$

$$(N^*) \int_{\left[0, \frac{1}{n}\right]^N} \left| f\left(t + \frac{k}{n}\right) - f(x) \right| d\mu(t). \tag{9.83}$$

In particular it holds

$$\left| \frac{(N^*) \int_{\left[0, \frac{1}{n}\right]^N} f\left(t + \frac{k}{n}\right) d\mu(t)}{\mu\left(\left[0, \frac{1}{n}\right]^N\right)} - f(x) \right| \leq$$

$$\frac{(N^*) \int_{\left[0, \frac{1}{n}\right]^N} \left| f\left(t + \frac{k}{n}\right) - f(x) \right| d\mu(t)}{\mu\left(\left[0, \frac{1}{n}\right]^N\right)}. \tag{9.84}$$

We present the following approximation result.

Theorem 9.15 *Let* $f \in C_B^+\left(\mathbb{R}^N\right)$, $0 < \beta < 1$, $x \in \mathbb{R}^N$; $N, n \in \mathbb{N}$ *with* $n^{1-\beta} > 2$; $j = 1, 2, 3, 4$. *Then*

(i)

$$\sup_{\mu} \left| {}_jT_n^{\mu}(f, x) - f(x) \right| \leq \omega_1\left(f, \frac{1}{n} + \frac{1}{n^{\beta}}\right) + 2\|f\|_{\infty} c_j(\beta, n) =: \lambda_{jn}, \tag{9.85}$$

and

(ii)

$$\sup_{\mu} \left\| {}_jT_n^{\mu}(f) - f \right\|_{\infty} \leq \lambda_{jn}. \tag{9.86}$$

Given that $f \in \left(C_U^+\left(\mathbb{R}^N\right) \cap C_B^+\left(\mathbb{R}^N\right)\right)$, *we obtain* $\lim_{n \to \infty} {}_jT_n^{\mu}(f) = f$, *uniformly. Above* $c_j(\beta, n)$ *are as in (9.27), (9.42), (9.57) and (9.74), respectively.*

Proof We observe that

$$\left| {}_jT_n^{\mu}(f, x) - f(x) \right| =$$

$$\left| \sum_{k=-\infty}^{\infty} \left(\frac{(N^*) \int_{\left[0, \frac{1}{n}\right]^N} f\left(t + \frac{k}{n}\right) d\mu(t)}{\mu\left(\left[0, \frac{1}{n}\right]^N\right)} \right) Z_j(nx - k) - \sum_{k=-\infty}^{\infty} f(x) Z_j(nx - k) \right| =$$

$$\left| \sum_{k=-\infty}^{\infty} \left(\left(\frac{(N^*) \int_{\left[0, \frac{1}{n}\right]^N} f\left(t + \frac{k}{n}\right) d\mu(t)}{\mu\left(\left[0, \frac{1}{n}\right]^N\right)} \right) - f(x) \right) Z_j(nx - k) \right| \leq$$

$$\sum_{k=-\infty}^{\infty}\left|\left(\frac{(N^*)\int_{[0,\frac{1}{n}]^N} f\left(t+\frac{k}{n}\right)d\mu(t)}{\mu\left(\left[0,\frac{1}{n}\right]^N\right)}\right)-f(x)\right| Z_j(nx-k) \overset{(9.84)}{\leq}$$

$$\sum_{k=-\infty}^{\infty}\left(\frac{(N^*)\int_{[0,\frac{1}{n}]^N}\left|f\left(t+\frac{k}{n}\right)-f(x)\right|d\mu(t)}{\mu\left(\left[0,\frac{1}{n}\right]^N\right)}\right) Z_j(nx-k) =$$

$$\sum_{\begin{cases} k=-\infty \\ :\left\|\frac{k}{n}-x\right\|_\infty \leq \frac{1}{n^\beta}\end{cases}}^{\infty}\left(\frac{(N^*)\int_{[0,\frac{1}{n}]^N}\left|f\left(t+\frac{k}{n}\right)-f(x)\right|d\mu(t)}{\mu\left(\left[0,\frac{1}{n}\right]^N\right)}\right) Z_j(nx-k)+$$

$$\sum_{\begin{cases} k=-\infty \\ :\left\|\frac{k}{n}-x\right\|_\infty > \frac{1}{n^\beta}\end{cases}}^{\infty}\left(\frac{(N^*)\int_{[0,\frac{1}{n}]^N}\left|f\left(t+\frac{k}{n}\right)-f(x)\right|d\mu(t)}{\mu\left(\left[0,\frac{1}{n}\right]^N\right)}\right) Z_j(nx-k) \leq \tag{9.87}$$

$$\sum_{\begin{cases} k=-\infty \\ :\left\|\frac{k}{n}-x\right\|_\infty \leq \frac{1}{n^\beta}\end{cases}}^{\infty}\left(\frac{(N^*)\int_{[0,\frac{1}{n}]^N}\omega_1\left(f,\|t\|_\infty+\left\|\frac{k}{n}-x\right\|_\infty\right)d\mu(t)}{\mu\left(\left[0,\frac{1}{n}\right]^N\right)}\right) Z_j(nx-k)$$

$$+2\|f\|_\infty\left(\sum_{\begin{cases} k=-\infty \\ :\left\|\frac{k}{n}-x\right\|_\infty > \frac{1}{n^\beta}\end{cases}}^{\infty} Z_j(nx-k)\right) \text{ (by (9.27), (9.42), (9.57), (9.74))}$$

$$\leq \omega_1\left(f,\frac{1}{n}+\frac{1}{n^\beta}\right)+2\|f\|_\infty c_j(\beta,n), \tag{9.88}$$

proving the claim. ∎

Additionally we give

Definition 9.16 Denote by $C_B^+\left(\mathbb{R}^N,\mathbb{C}\right)=\{f:\mathbb{R}^N\to\mathbb{C}\,|\,f=f_1+if_2,$ where $f_1,f_2\in C_B^+\left(\mathbb{R}^N\right), N\in\mathbb{N}\}$. We set for $f\in C_B^+\left(\mathbb{R}^N,\mathbb{C}\right)$ that

$${}_jT_n^\mu(f,x) := {}_jT_n^\mu(f_1,x)+i\ {}_jT_n^\mu(f_2,x),\ j=1,2,3,4, \tag{9.89}$$

$\forall\, n\in\mathbb{N}, x\in\mathbb{R}^N, i=\sqrt{-1}$.

Theorem 9.17 *Let $f \in C_B^+(\mathbb{R}^N, \mathbb{C})$, $f = f_1 + if_2$, $N \in \mathbb{N}$, $0 < \beta < 1$, $x \in \mathbb{R}^N$; $n \in \mathbb{N}$ with $n^{1-\beta} > 2$; $j = 1, 2, 3, 4$. Then*

(i)

$$\sup_{\mu} \left| {}_jT_n^{\mu}(f, x) - f(x) \right| \leq \left[\omega_1 \left(f_1, \frac{1}{n} + \frac{1}{n^{\beta}} \right) + \omega_1 \left(f_2, \frac{1}{n} + \frac{1}{n^{\beta}} \right) \right]$$

$$+ 2 \left(\|f_1\|_{\infty} + \|f_2\|_{\infty} \right) c_j(\beta, n) =: l_{jn}, \tag{9.90}$$

and

(ii)

$$\sup_{\mu} \left\| {}_jT_n^{\mu}(f) - f \right\| \leq l_{jn}. \tag{9.91}$$

Proof

$$\left| {}_jT_n^{\mu}(f, x) - f(x) \right| = \left| {}_jT_n^{\mu}(f_1, x) + i \; {}_jT_n^{\mu}(f_2, x) - f_1(x) - if_2(x) \right| =$$

$$\left| \left({}_jT_n^{\mu}(f_1, x) - f_1(x) \right) + i \left({}_jT_n^{\mu}(f_2, x) - f_2(x) \right) \right| \leq$$

$$\left| {}_jT_n^{\mu}(f_1, x) - f_1(x) \right| + \left| {}_jT_n^{\mu}(f_2, x) - f_2(x) \right| \overset{(9.85)}{\leq} \tag{9.92}$$

$$\left(\omega_1 \left(f_1, \frac{1}{n} + \frac{1}{n^{\beta}} \right) + 2 \|f_1\|_{\infty} c_j(\beta, n) \right) +$$

$$\left(\omega_1 \left(f_2, \frac{1}{n} + \frac{1}{n^{\beta}} \right) + 2 \|f_2\|_{\infty} c_j(\beta, n) \right) =$$

$$\left[\omega_1 \left(f_1, \frac{1}{n} + \frac{1}{n^{\beta}} \right) + \omega_1 \left(f_2, \frac{1}{n} + \frac{1}{n^{\beta}} \right) \right] +$$

$$2 \left(\|f_1\|_{\infty} + \|f_2\|_{\infty} \right) c_j(\beta, n), \tag{9.93}$$

proving the claim. ∎

References

1. G.A. Anastassiou, *Intelligent Systems II: Complete Approximation by Neural Network Operators* (Springer, Heidelberg, New York, 2016)
2. G.A. Anastassiou, *Intelligent Computations: Abstract Fractional Calculus, Inequalities, Approximations* (Springer, Heidelberg, New York, 2018)
3. G.A. Anastassiou, *Ordinary and Fractional Approximation by Non-additive Integrals: Choquet, Shilkret and Sugeno Integral Approximators* (Springer, Heidelberg, New York, 2019)
4. G.A. Anastassiou, Algebraic function based Banach space valued ordinary and fractional neural network approximations. New Trends Math. Sci. 10 special issues (1), 100–125 (2022)

5. G.A. Anastassiou, Gudermannian function activated Banach space valued ordinary and fractional neural network approximation. Adv. Nonlinear Var. Inequal. **25**(2), 27–64 (2022)
6. G.A. Anastassiou, *Generalized symmetrical sigmoid function activated Banach space valued ordinary and fractional neural network approximation*. Analele Universităţii Oradea, Fasc. Matematica, accepted for publication (2022)
7. G.A. Anastassiou, Abstract multivariate Gudermannian function activated neural network approximations Panam. Math. J. accepted (2022)
8. G.A. Anastassiou, *Degree of Approximation by Kantorovich-Shilkret quasi-interpolation neural network operators revisited*. Analele Universitatii Oradea, Fasc. Matematica, accepted (2022)
9. G.A. Anastassiou, *General multivariate arctangent function activated neural network approximations*, submitted (2022)
10. G.A. Anastassiou, *Generalized symmetrical sigmoid function activated neural network multivariate approximation*, submitted (2022)
11. G.A. Anastassiou, R. Kozma, *Abstract multivariate algebraic function activated neural network approximations*, submitted (2022)
12. A.J. Dunning, J. Kensler, L. Goudeville, F. Bailleux, Some extensions in continuous methods for immunological correlates of protection. BMC Med. Res. Methodol. **15**(107) (2015). https://doi.org/10.1186/s12874-015-0096-9
13. I.S. Haykin, *Neural Networks: A Comprehensive Foundation*, 2nd edn. (Prentice Hall, New York, 1998)
14. W. McCulloch, W. Pitts, A logical calculus of the ideas immanent in nervous activity. Bull. Math. Biophys. **7**, 115–133 (1943)
15. T.M. Mitchell, *Machine Learning* (WCB-McGraw-Hill, New York, 1997)
16. N. Shilkret, Maxitive measure and integration. Indagationes Mathematicae **33**, 109–116 (1971)
17. E.W. Weisstein, *Gudermannian*, MathWorld

Chapter 10
Voronouskaya Univariate and Multivariate Asymptotic Expansions for Sigmoid Functions Induced Quasi-interpolation Neural Network Operators Revisited

Here we study further the quasi-interpolation arctangent-algebraic-Gudermannian-generalized symmetrical activation functions relied neural network operators of one hidden layer. Based on fractional calculus theory we derive fractional Voronovskaya type asymptotic expansions for the approximation of these operators to the unit operator, as we are studying the univariate case. We treat also analogously the multivariate case by using Fréchet derivatives. The functions under approximation are Banach space valued. It follows [17].

10.1 Background

This is a continuation and generalization of [8].

10.1.1 About the Arctangent Activation Function

We consider the

$$\arctan x = \int_0^x \frac{dz}{1+z^2}, \ x \in \mathbb{R}. \tag{10.1}$$

We will be using

$$h(x) := \frac{2}{\pi} \arctan\left(\frac{\pi}{2}x\right) = \frac{2}{\pi} \int_0^{\frac{\pi x}{2}} \frac{dz}{1+z^2}, \ x \in \mathbb{R}, \tag{10.2}$$

© The Author(s), under exclusive license to Springer Nature Switzerland AG 2023
G. A. Anastassiou, *Banach Space Valued Neural Network*,
Studies in Computational Intelligence 1062,
https://doi.org/10.1007/978-3-031-16400-2_10

which is a sigmoid type function and it is strictly increasing. We have that

$$h(0) = 0,\ h(-x) = -h(x), h(+\infty) = 1,\ h(-\infty) = -1,$$

and

$$h'(x) = \frac{4}{4 + \pi^2 x^2} > 0, \text{ all } x \in \mathbb{R}. \tag{10.3}$$

We consider the activation function

$$\psi_1(x) := \frac{1}{4}(h(x+1) - h(x-1)),\ x \in \mathbb{R}, \tag{10.4}$$

and we notice that

$$\psi_1(-x) = \psi_1(x), \tag{10.5}$$

it is an even function.

Since $x + 1 > x - 1$, then $h(x+1) > h(x-1)$, and $\psi_1(x) > 0$, all $x \in \mathbb{R}$.

We see that

$$\psi_1(0) = \frac{1}{\pi} \arctan \frac{\pi}{2} \cong 0.319. \tag{10.6}$$

Let $x > 0$, we have that

$$\psi_1'(x) = \frac{1}{4}\left(h'(x+1) - h'(x-1)\right) =$$

$$\frac{-4\pi^2 x}{\left(4 + \pi^2 (x+1)^2\right)\left(4 + \pi^2 (x-1)^2\right)} < 0. \tag{10.7}$$

That is

$$\psi_1'(x) < 0, \text{ for } x > 0. \tag{10.8}$$

That is ψ_1 is strictly decreasing on $[0, \infty)$ and clearly is strictly increasing on $(-\infty, 0]$, and $\psi_1'(0) = 0$.

Observe that

$$\begin{gathered}\lim_{x \to +\infty} \psi_1(x) = \tfrac{1}{4}(h(+\infty) - h(+\infty)) = 0, \\ \text{and} \\ \lim_{x \to -\infty} \psi_1(x) = \tfrac{1}{4}(h(-\infty) - h(-\infty)) = 0.\end{gathered} \tag{10.9}$$

That is the x-axis is the horizontal asymptote on ψ_1.

All in all, ψ_1 is a bell symmetric function with maximum $\psi_1(0) \cong 0.319$.

We need

Theorem 10.1 ([10], p. 286) *We have that*

$$\sum_{i=-\infty}^{\infty} \psi_1 (x - i) = 1, \ \forall x \in \mathbb{R}. \tag{10.10}$$

Theorem 10.2 ([10], p. 287) *It holds*

$$\int_{-\infty}^{\infty} \psi_1 (x) \, dx = 1. \tag{10.11}$$

So that $\psi_1 (x)$ is a density function on $\mathbb{R}$.
We mention

Theorem 10.3 ([10], p. 288) *Let* $0 < \alpha < 1$, *and* $n \in \mathbb{N}$ *with* $n^{1-\alpha} > 2$. *It holds*

$$\sum_{\substack{k=-\infty \\ : |nx-k| \geq n^{1-\alpha}}}^{\infty} \psi_1 (nx - k) < \frac{2}{\pi^2 \left(n^{1-\alpha} - 2\right)} =: c_1 (\alpha, n) . \tag{10.12}$$

Denote by $\lfloor \cdot \rfloor$ the integral part of the number and by $\lceil \cdot \rceil$ the ceiling of the number.
We need

Theorem 10.4 ([10], p. 289) *Let* $x \in [a, b] \subset \mathbb{R}$ *and* $n \in \mathbb{N}$ *so that* $\lceil na \rceil \leq \lfloor nb \rfloor$. *It holds*

$$\frac{1}{\sum_{k=\lceil na \rceil}^{\lfloor nb \rfloor} \psi_1 (nx - k)} < \frac{1}{\psi_1 (1)} \cong 4.9737 =: \alpha_1, \ \forall x \in [a, b] . \tag{10.13}$$

Note 10.5 ([10], pp. 290–291)
(i) We have that

$$\lim_{n \to \infty} \sum_{k=\lceil na \rceil}^{\lfloor nb \rfloor} \psi_1 (nx - k) \neq 1, \tag{10.14}$$

for at least some $x \in [a, b]$.
(ii) For large enough $n \in \mathbb{N}$ we always obtain $\lceil na \rceil \leq \lfloor nb \rfloor$. Also $a \leq \frac{k}{n} \leq b$, iff $\lceil na \rceil \leq k \leq \lfloor nb \rfloor$.
In general, by Theorem 10.1, it holds

$$\sum_{k=\lceil na \rceil}^{\lfloor nb \rfloor} \psi_1 (nx - k) \leq 1. \tag{10.15}$$

We introduce (see [15])

$$Z_1(x_1, ..., x_N) := Z_1(x) := \prod_{i=1}^{N} \psi_1(x_i), \quad x = (x_1, ..., x_N) \in \mathbb{R}^N, \quad N \in \mathbb{N}. \tag{10.16}$$

Denote by $a = (a_1, ..., a_N)$ and $b = (b_1, ..., b_N)$.

It has the properties:

(i) $Z_1(x) > 0, \ \forall\, x \in \mathbb{R}^N$,

(ii)

$$\sum_{k=-\infty}^{\infty} Z_1(x-k) := \sum_{k_1=-\infty}^{\infty} \sum_{k_2=-\infty}^{\infty} ... \sum_{k_N=-\infty}^{\infty} Z_1(x_1-k_1, ..., x_N-k_N) = 1, \tag{10.17}$$

where $k := (k_1, ..., k_n) \in \mathbb{Z}^N, \forall\, x \in \mathbb{R}^N$,

hence

(iii)

$$\sum_{k=-\infty}^{\infty} Z_1(nx-k) = 1, \tag{10.18}$$

$\forall\, x \in \mathbb{R}^N; n \in \mathbb{N}$,

and

(iv)

$$\int_{\mathbb{R}^N} Z_1(x)\,dx = 1, \tag{10.19}$$

that is Z_1 is a multivariate density function.

(v) It is clear that

$$\sum_{\begin{cases} k=-\infty \\ \left\| \frac{k}{n} - x \right\|_\infty > \frac{1}{n^\beta} \end{cases}}^{\infty} Z_1(nx-k) < \frac{2}{\pi^2 \left(n^{1-\beta}-2\right)} = c_1(\beta, n), \tag{10.20}$$

$0 < \beta < 1, n \in \mathbb{N} : n^{1-\beta} > 2, x \in \mathbb{R}^N$.

(vi) By Theorem 10.4 we get that

$$0 < \frac{1}{\sum_{k=\lceil na \rceil}^{\lfloor nb \rfloor} Z_1(nx-k)} < \frac{1}{(\psi_1(1))^N} \cong (4.9737)^N =: \gamma_1(N), \tag{10.21}$$

$\forall\, x \in \left(\prod_{i=1}^{N} [a_i, b_i]\right), \ n \in \mathbb{N}$.

Furthermore it holds

$$\lim_{n \to \infty} \sum_{k=\lceil na \rceil}^{\lfloor nb \rfloor} Z_1 (nx - k) \neq 1, \tag{10.22}$$

for at least some $x \in \left(\prod_{i=1}^{N} [a_i, b_i] \right)$.

Above it is $\|x\|_\infty := \max \{|x_1|, ..., |x_N|\}$, $x \in \mathbb{R}^N$, also set $\infty := (\infty, ..., \infty)$, $-\infty = (-\infty, ... - \infty)$ upon the multivariate context.

10.1.2 About the Algebraic Activation Function

Here see also [11].

We consider the generator algebraic function

$$\varphi (x) = \frac{x}{\sqrt[2m]{1 + x^{2m}}}, \quad m \in \mathbb{N}, x \in \mathbb{R}, \tag{10.23}$$

which is a sigmoidal type of function and is a strictly increasing function.

We see that $\varphi (-x) = -\varphi (x)$ with $\varphi (0) = 0$. We get that

$$\varphi' (x) = \frac{1}{\left(1 + x^{2m}\right)^{\frac{2m+1}{2m}}} > 0, \ \forall \, x \in \mathbb{R}, \tag{10.24}$$

proving φ as strictly increasing over $\mathbb{R}$, $\varphi' (x) = \varphi' (-x)$. We easily find that $\lim_{x \to +\infty} \varphi (x) = 1, \varphi (+\infty) = 1$, and $\lim_{x \to -\infty} \varphi (x) = -1, \varphi (-\infty) = -1$.

We consider the activation function

$$\psi_2 (x) = \frac{1}{4} \left[\varphi (x + 1) - \varphi (x - 1) \right]. \tag{10.25}$$

Clearly it is $\psi_2 (x) = \psi_2 (-x)$, $\forall \, x \in \mathbb{R}$, so that ψ_2 is an even function and symmetric with respect to the y-axis. Clearly $\psi_2 (x) > 0$, $\forall \, x \in \mathbb{R}$.

Also it is

$$\psi_2 (0) = \frac{1}{2 \sqrt[2m]{2}}. \tag{10.26}$$

By [11], we have that $\psi_2' (x) < 0$ for $x > 0$. That is ψ_2 is strictly decreasing over $(0, +\infty)$.

Clearly, ψ_2 is strictly increasing over $(-\infty, 0)$ and $\psi_2' (0) = 0$.

Furthermore we obtain that

$$\lim_{x \to +\infty} \psi_2 (x) = \frac{1}{4} \left[\varphi (+\infty) - \varphi (+\infty) \right] = 0, \tag{10.27}$$

and

$$\lim_{x \to -\infty} \psi_2(x) = \frac{1}{4}[\varphi(-\infty) - \varphi(-\infty)] = 0. \tag{10.28}$$

That is the x-axis is the horizontal asymptote of ψ_2.

Conclusion, ψ_2 is a bell shape symmetric function with maximum

$$\psi_2(0) = \frac{1}{2\sqrt[2m]{2}}, \quad m \in \mathbb{N}. \tag{10.29}$$

We need

Theorem 10.6 ([11]) *We have that*

$$\sum_{i=-\infty}^{\infty} \psi_2(x-i) = 1, \ \forall \, x \in \mathbb{R}. \tag{10.30}$$

Theorem 10.7 ([11]) *It holds*

$$\int_{-\infty}^{\infty} \psi_2(x)\, dx = 1. \tag{10.31}$$

Theorem 10.8 ([11]) *Let* $0 < \alpha < 1$, *and* $n \in \mathbb{N}$ *with* $n^{1-\alpha} > 2$. *It holds*

$$\sum_{\substack{k=-\infty \\ : |nx-k| \geq n^{1-\alpha}}}^{\infty} \psi_2(nx-k) < \frac{1}{4m\left(n^{1-\alpha}-2\right)^{2m}} =: c_2(\alpha, n), \quad m \in \mathbb{N}. \tag{10.32}$$

We need

Theorem 10.9 ([11]) *Let* $[a, b] \subset \mathbb{R}$ *and* $n \in \mathbb{N}$ *so that* $\lceil na \rceil \leq \lfloor nb \rfloor$. *It holds*

$$\frac{1}{\sum_{k=\lceil na \rceil}^{\lfloor nb \rfloor} \psi_2(nx-k)} < 2\left(\sqrt[2m]{1+4^m}\right) =: \alpha_2, \tag{10.33}$$

$\forall \, x \in [a, b]$, $m \in \mathbb{N}$.

Note 10.10 (1) By [11] we have that

$$\lim_{n \to \infty} \sum_{k=\lceil na \rceil}^{\lfloor nb \rfloor} \psi_2(nx-k) \neq 1, \tag{10.34}$$

for at least some $x \in [a, b]$.

(2) Let $[a, b] \subset \mathbb{R}$. For large $n \in \mathbb{N}$ we always have $\lceil na \rceil \leq \lfloor nb \rfloor$. Also $a \leq \frac{k}{n} \leq b$, iff $\lceil na \rceil \leq k \leq \lfloor nb \rfloor$.

In general it holds that

$$\sum_{k=\lceil na \rceil}^{\lfloor nb \rfloor} \psi_2 (nx - k) \leq 1. \tag{10.35}$$

We introduce (see also [18])

$$Z_2 (x_1, ..., x_N) := Z_2 (x) := \prod_{i=1}^{N} \psi_2 (x_i), \quad x = (x_1, ..., x_N) \in \mathbb{R}^N, \ N \in \mathbb{N}. \tag{10.36}$$

It has the properties:

(i) $Z_2 (x) > 0, \ \forall \, x \in \mathbb{R}^N$,

(ii)

$$\sum_{k=-\infty}^{\infty} Z_2 (x - k) := \sum_{k_1=-\infty}^{\infty} \sum_{k_2=-\infty}^{\infty} ... \sum_{k_N=-\infty}^{\infty} Z_2 (x_1 - k_1, ..., x_N - k_N) = 1, \tag{10.37}$$

where $k := (k_1, ..., k_n) \in \mathbb{Z}^N, \forall \, x \in \mathbb{R}^N$,

hence

(iii)

$$\sum_{k=-\infty}^{\infty} Z_2 (nx - k) = 1, \tag{10.38}$$

$\forall \, x \in \mathbb{R}^N; n \in \mathbb{N}$,

and

(iv)

$$\int_{\mathbb{R}^N} Z_2 (x) \, dx = 1, \tag{10.39}$$

that is Z_2 is a multivariate density function.

(v) It is clear that

$$\sum_{\substack{k = -\infty \\ \left\| \frac{k}{n} - x \right\|_\infty > \frac{1}{n^\beta}}}^{\infty} Z_2 (nx - k) < \frac{1}{4m \left(n^{1-\beta} - 2\right)^{2m}} = c_2 (\beta, n), \tag{10.40}$$

$0 < \beta < 1, n \in \mathbb{N} : n^{1-\beta} > 2, x \in \mathbb{R}^N, m \in \mathbb{N}$.

(vi) By Theorem 10.9 we get that

$$0 < \frac{1}{\sum_{k=\lceil na \rceil}^{\lfloor nb \rfloor} Z_2(nx-k)} < \frac{1}{(\psi_2(1))^N} \cong \left[2\left(\sqrt[2m]{1+4^m} \right) \right]^N := \gamma_2(N), \tag{10.41}$$

$\forall\, x \in \left(\prod_{i=1}^N [a_i, b_i] \right)$, $n \in \mathbb{N}$.

Furthermore it holds

$$\lim_{n \to \infty} \sum_{k=\lceil na \rceil}^{\lfloor nb \rfloor} Z_2(nx-k) \neq 1, \tag{10.42}$$

for at least some $x \in \left(\prod_{i=1}^N [a_i, b_i] \right)$.

10.1.3 About the Gudermannian Activation Function

See also [12, 30].

Here we consider $gd(x)$ the Gudermannian function [30], which is a sigmoid function, as a generator function:

$$\sigma(x) = 2\arctan\left(\tanh\left(\frac{x}{2} \right) \right) = \int_0^x \frac{dt}{\cosh t} =: gd(x), \; x \in \mathbb{R}. \tag{10.43}$$

Let the normalized generator sigmoid function

$$f(x) := \frac{4}{\pi}\sigma(x) = \frac{4}{\pi}\int_0^x \frac{dt}{\cosh t} = \frac{8}{\pi}\int_0^x \frac{1}{e^t + e^{-t}} dt, \;\; x \in \mathbb{R}. \tag{10.44}$$

Here

$$f'(x) = \frac{4}{\pi \cosh x} > 0, \;\; \forall\, x \in \mathbb{R},$$

hence f is strictly increasing on $\mathbb{R}$.

Notice that $\tanh(-x) = -\tanh x$ and $\arctan(-x) = -\arctan x$, $x \in \mathbb{R}$.

So, here the neural network activation function will be:

$$\psi_3(x) = \frac{1}{4}\left[f(x+1) - f(x-1) \right], x \in \mathbb{R}. \tag{10.45}$$

By [12], we get that

$$\psi_3(x) = \psi_3(-x), \;\; \forall\, x \in \mathbb{R}, \tag{10.46}$$

i.e. it is even and symmetric with respect to the y-axis. Here we have $f(+\infty) = 1$, $f(-\infty) = -1$ and $f(0) = 0$. Clearly it is

$$f(-x) = -f(x), \ \forall x \in \mathbb{R}, \tag{10.47}$$

an odd function, symmetric with respect to the origin. Since $x + 1 > x - 1$, and $f(x+1) > f(x-1)$, we obtain $\psi_3(x) > 0$, $\forall x \in \mathbb{R}$.

By [12], we have that

$$\psi_3(0) = \frac{2}{\pi} gd(1) \cong 0.551. \tag{10.48}$$

By [12] ψ_3 is strictly decreasing on $(0, +\infty)$, and strictly increasing on $(-\infty, 0)$, and $\psi_3'(0) = 0$.

Also we have that

$$\lim_{x \to +\infty} \psi_3(x) = \lim_{x \to -\infty} \psi_3(x) = 0, \tag{10.49}$$

that is the x-axis is the horizontal asymptote for ψ_3.

Conclusion, ψ_3 is a bell shaped symmetric function with maximum $\psi_3(0) \cong 0.551$.

We need

Theorem 10.11 ([12]) *It holds that*

$$\sum_{i=-\infty}^{\infty} \psi_3(x - i) = 1, \ \forall x \in \mathbb{R}. \tag{10.50}$$

Theorem 10.12 ([12]) *We have that*

$$\int_{-\infty}^{\infty} \psi_3(x)\, dx = 1. \tag{10.51}$$

So $\psi_3(x)$ is a density function.

Theorem 10.13 ([12]) *Let $0 < \alpha < 1$, and $n \in \mathbb{N}$ with $n^{1-\alpha} > 2$. It holds*

$$\sum_{\begin{cases} k = -\infty \\ : |nx - k| \geq n^{1-\alpha} \end{cases}}^{\infty} \psi_3(nx - k) < \frac{4}{\pi e^{(n^{1-\alpha} - 2)}} = \frac{4e^2}{\pi e^{n^{1-\alpha}}} =: c_3(\alpha, n). \tag{10.52}$$

Theorem 10.14 ([12]) *Let $[a, b] \subset \mathbb{R}$ and $n \in \mathbb{N}$, so that $\lceil na \rceil \leq \lfloor nb \rfloor$. It holds*

$$\frac{1}{\sum_{k=\lceil na \rceil}^{\lfloor nb \rfloor} \psi_3(nx - k)} < \frac{\pi}{gd(2)} \cong 2.412 =: \alpha_3, \tag{10.53}$$

$\forall x \in [a, b]$.

We make

Remark 10.15 ([12])
(i) We have that

$$\lim_{n\to\infty} \sum_{k=\lceil na \rceil}^{\lfloor nb \rfloor} \psi_3 (nx - k) \neq 1, \tag{10.54}$$

for at least some $x \in [a, b]$.
(ii) Let $[a, b] \subset \mathbb{R}$. For large n we always have $\lceil na \rceil \leq \lfloor nb \rfloor$. Also $a \leq \frac{k}{n} \leq b$, iff $\lceil na \rceil \leq k \leq \lfloor nb \rfloor$.
In general it holds

$$\sum_{k=\lceil na \rceil}^{\lfloor nb \rfloor} \psi_3 (nx - k) \leq 1. \tag{10.55}$$

We introduce (see also [14])

$$Z_3 (x_1, ..., x_N) := Z_3 (x) := \prod_{i=1}^{N} \psi_3 (x_i), \quad x = (x_1, ..., x_N) \in \mathbb{R}^N, \ N \in \mathbb{N}. \tag{10.56}$$

It has the properties:
(i) $Z_3 (x) > 0, \ \forall \, x \in \mathbb{R}^N$,
(ii)

$$\sum_{k=-\infty}^{\infty} Z_3 (x - k) := \sum_{k_1=-\infty}^{\infty} \sum_{k_2=-\infty}^{\infty} ... \sum_{k_N=-\infty}^{\infty} Z_3 (x_1 - k_1, ..., x_N - k_N) = 1, \tag{10.57}$$

where $k := (k_1, ..., k_n) \in \mathbb{Z}^N, \forall \, x \in \mathbb{R}^N$,
hence
(iii)

$$\sum_{k=-\infty}^{\infty} Z_3 (nx - k) = 1, \tag{10.58}$$

$\forall \, x \in \mathbb{R}^N; \, n \in \mathbb{N}$,
and
(iv)

$$\int_{\mathbb{R}^N} Z_3 (x) \, dx = 1, \tag{10.59}$$

that is Z_3 is a multivariate density function.

(v) It is also clear that

$$\sum_{\substack{k=-\infty \\ \left\| \frac{k}{n} - x \right\|_{\infty} > \frac{1}{n^{\beta}}}}^{\infty} Z_3(nx - k) < \frac{4e^2}{\pi e^{n^{1-\beta}}} = c_3(\beta, n), \tag{10.60}$$

$0 < \beta < 1, n \in \mathbb{N} : n^{1-\beta} > 2, x \in \mathbb{R}^N, m \in \mathbb{N}$.

(vi) By Theorem 10.14 we get that

$$0 < \frac{1}{\sum_{k=\lceil na \rceil}^{\lfloor nb \rfloor} Z_3(nx - k)} < \left(\frac{\pi}{gd(2)} \right)^N \cong (2.412)^N =: \gamma_3(N), \tag{10.61}$$

$\forall\, x \in \left(\prod_{i=1}^{N} [a_i, b_i] \right)$, $n \in \mathbb{N}$.

Furthermore it holds

$$\lim_{n \to \infty} \sum_{k=\lceil na \rceil}^{\lfloor nb \rfloor} Z_3(nx - k) \neq 1, \tag{10.62}$$

for at least some $x \in \left(\prod_{i=1}^{N} [a_i, b_i] \right)$.

10.1.4 About the Generalized Symmetrical Activation Function

Here we consider the generalized symmetrical sigmoid function [13, 23]

$$f_1(x) = \frac{x}{\left(1 + |x|^{\mu}\right)^{\frac{1}{\mu}}}, \quad \mu > 0, x \in \mathbb{R}. \tag{10.63}$$

This has applications in immunology and protection from disease together with probability theory. It is also called a symmetrical protection curve.

The parameter μ is a shape parameter controlling how fast the curve approaches the asymptotes for a given slope at the inflection point. When $\mu = 1$ f_1 is the absolute sigmoid function, and when $\mu = 2$, f_1 is the square root sigmoid function. When $\mu = 1.5$ the function approximates the arctangent function, when $\mu = 2.9$ it approximates the logistic function, and when $\mu = 3.4$ it approximates the error function. Parameter μ is estimated in the likelihood maximization [23]. For more see [23].

Next we study the particular generator sigmoid function

$$f_2(x) = \frac{x}{\left(1 + |x|^{\lambda}\right)^{\frac{1}{\lambda}}}, \ \lambda \text{ is an odd number, } x \in \mathbb{R}. \tag{10.64}$$

We have that $f_2(0) = 0$, and

$$f_2(-x) = -f_2(x), \tag{10.65}$$

so f_2 is symmetric with respect to zero.

When $x \geq 0$, we get that [13]

$$f_2'(x) = \frac{1}{\left(1 + x^{\lambda}\right)^{\frac{\lambda+1}{\lambda}}} > 0, \tag{10.66}$$

that is f_2 is strictly increasing on $[0, +\infty)$ and f_2 is strictly increasing on $(-\infty, 0]$. Hence f_2 is strictly increasing on $\mathbb{R}$.

We also have $f_2(+\infty) = f_2(-\infty) = 1$.

Let us consider the activation function [13]:

$$\psi_4(x) = \frac{1}{4}\left[f_2(x+1) - f_2(x-1)\right] =$$

$$\frac{1}{4}\left[\frac{(x+1)}{\left(1 + |x+1|^{\lambda}\right)^{\frac{1}{\lambda}}} - \frac{(x-1)}{\left(1 + |x-1|^{\lambda}\right)^{\frac{1}{\lambda}}}\right]. \tag{10.67}$$

Clearly it holds [13]

$$\psi_4(x) = \psi_4(-x), \ \forall \, x \in \mathbb{R}. \tag{10.68}$$

and

$$\psi_4(0) = \frac{1}{2\sqrt[\lambda]{2}}, \tag{10.69}$$

and $\psi_4(x) > 0$, $\forall \, x \in \mathbb{R}$.

Following [13], we have that ψ_4 is strictly decreasing over $[0, +\infty)$, and ψ_4 is strictly increasing on $(-\infty, 0]$, by ψ_4-symmetry with respect to y-axis, and $\psi_4'(0) = 0$.

Clearly it is

$$\lim_{x \to +\infty} \psi_4(x) = \lim_{x \to -\infty} \psi_4(x) = 0, \tag{10.70}$$

therefore the x-axis is the horizontal asymptote of $\psi_4(x)$.

The value

$$\psi_4(0) = \frac{1}{2\sqrt[\lambda]{2}}, \ \lambda \text{ is an odd number,} \tag{10.71}$$

is the maximum of ψ_4, which is a bell shaped function.

We need

Theorem 10.16 ([13]) *It holds*

$$\sum_{i=-\infty}^{\infty} \psi_4(x-i) = 1, \ \forall x \in \mathbb{R}. \tag{10.72}$$

Theorem 10.17 ([13]) *We have that*

$$\int_{-\infty}^{\infty} \psi_4(x)\, dx = 1. \tag{10.73}$$

So that $\psi_4(x)$ is a density function on $\mathbb{R}$.

We need

Theorem 10.18 ([13]) *Let* $0 < \alpha < 1$, *and* $n \in \mathbb{N}$ *with* $n^{1-\alpha} > 2$. *It holds*

$$\sum_{\substack{j=-\infty \\ : |nx-j| \geq n^{1-\alpha}}}^{\infty} \psi_4(nx-j) < \frac{1}{2\lambda\left(n^{1-\alpha}-2\right)^{\lambda}} =: c_4(\alpha, n), \tag{10.74}$$

where $\lambda \in \mathbb{N}$ *is an odd number.*

We also need

Theorem 10.19 ([13]) *Let* $[a, b] \subset \mathbb{R}$ *and* $n \in \mathbb{N}$ *so that* $\lceil na \rceil \leq \lfloor nb \rfloor$. *Then*

$$\frac{1}{\sum_{k=\lceil na \rceil}^{\lfloor nb \rfloor} \psi_4(|nx-k|)} < 2\sqrt[\lambda]{1+2^{\lambda}} =: \alpha_4, \tag{10.75}$$

where λ *is an odd number,* $\forall x \in [a, b]$.

We make

Remark 10.20 ([13]) (1) We have that

$$\lim_{n\to\infty} \sum_{k=\lceil na \rceil}^{\lfloor nb \rfloor} \psi_4(nx-k) \neq 1, \ \text{for at least some } x \in [a, b]. \tag{10.76}$$

(2) Let $[a, b] \subset \mathbb{R}$. For large enough n we always obtain $\lceil na \rceil \leq \lfloor nb \rfloor$. Also $a \leq \frac{k}{n} \leq b$, iff $\lceil na \rceil \leq k \leq \lfloor nb \rfloor$.

In general it holds that

$$\sum_{k=\lceil na \rceil}^{\lfloor nb \rfloor} \psi_4 (nx - k) \leq 1. \tag{10.77}$$

We introduce (see also [16])

$$Z_4 (x_1, ..., x_N) := Z_4 (x) := \prod_{i=1}^{N} \psi_4 (x_i), \quad x = (x_1, ..., x_N) \in \mathbb{R}^N, \ N \in \mathbb{N}. \tag{10.78}$$

It has the properties:

(i) $Z_4 (x) > 0, \ \forall \, x \in \mathbb{R}^N$,

(ii)

$$\sum_{k=-\infty}^{\infty} Z_4 (x - k) := \sum_{k_1=-\infty}^{\infty} \sum_{k_2=-\infty}^{\infty} ... \sum_{k_N=-\infty}^{\infty} Z_4 (x_1 - k_1, ..., x_N - k_N) = 1, \tag{10.79}$$

where $k := (k_1, ..., k_n) \in \mathbb{Z}^N, \forall \, x \in \mathbb{R}^N$,

hence

(iii)

$$\sum_{k=-\infty}^{\infty} Z_4 (nx - k) = 1, \tag{10.80}$$

$\forall \, x \in \mathbb{R}^N; \, n \in \mathbb{N}$,

and

(iv)

$$\int_{\mathbb{R}^N} Z_4 (x) \, dx = 1, \tag{10.81}$$

that is Z_4 is a multivariate density function.

(v) It is clear that

$$\sum_{\substack{k = -\infty \\ \left\| \frac{k}{n} - x \right\|_{\infty} > \frac{1}{n^{\beta}}}}^{\infty} Z_4 (nx - k) < \frac{1}{2\lambda \left(n^{1-\beta} - 2\right)^{\lambda}} = c_4 (\beta, n), \tag{10.82}$$

$0 < \beta < 1, n \in \mathbb{N} : n^{1-\beta} > 2, x \in \mathbb{R}^N$, λ is odd.

(vi) By Theorem 10.19 we get that

$$0 < \frac{1}{\sum_{k=\lceil na \rceil}^{\lfloor nb \rfloor} Z_4(nx-k)} < \left(2\sqrt[\lambda]{1+2^{\lambda}}\right)^N =: \gamma_4(N), \tag{10.83}$$

$\forall\, x \in \left(\prod_{i=1}^N [a_i, b_i]\right)$, $n \in \mathbb{N}$, λ is odd.

Furthermore it holds

$$\lim_{n\to\infty} \sum_{k=\lceil na \rceil}^{\lfloor nb \rfloor} Z_4(nx-k) \neq 1, \tag{10.84}$$

for at least some $x \in \left(\prod_{i=1}^N [a_i, b_i]\right)$.

The next integrals are of Bochner type [26].

We need

Definition 10.21 ([10]) Let $[a, b] \subset \mathbb{R}$, X be a Banach space, $\alpha > 0$; $m = \lceil \alpha \rceil \in \mathbb{N}$, ($\lceil \cdot \rceil$ is the ceiling of the number), $f : [a, b] \to X$. We assume that $f^{(m)} \in L_1([a, b], X)$. We call the Caputo-Bochner left fractional derivative of order α:

$$\left(D_{*a}^{\alpha} f\right)(x) := \frac{1}{\Gamma(m-\alpha)} \int_a^x (x-t)^{m-\alpha-1} f^{(m)}(t)\, dt,\ \forall\, x \in [a, b]. \tag{10.85}$$

If $\alpha \in \mathbb{N}$, we set $D_{*a}^{\alpha} f := f^{(m)}$ the ordinary X-valued derivative (defined similar to numerical one [29]), and also set $D_{*a}^0 f := f$.

By [10], $\left(D_{*a}^{\alpha} f\right)(x)$ exists almost everywhere in $x \in [a, b]$ and $D_{*a}^{\alpha} f \in L_1([a, b], X)$.

If $\left\| f^{(m)} \right\|_{L_\infty([a,b],X)} < \infty$, then by [10], $D_{*a}^{\alpha} f \in C([a, b], X)$, hence $\left\| D_{*a}^{\alpha} f \right\| \in C([a, b])$.

We mention

Lemma 10.22 ([10]) *Let $\alpha > 0$, $\alpha \notin \mathbb{N}$, $m = \lceil \alpha \rceil$, $f \in C^{m-1}([a, b], X)$ and $f^{(m)} \in L_\infty([a, b], X)$. Then $D_{*a}^{\alpha} f(a) = 0$.*

We mention

Definition 10.23 ([10]) Let $[a, b] \subset \mathbb{R}$, X be a Banach space, $\alpha > 0$, $m := \lceil \alpha \rceil$. We assume that $f^{(m)} \in L_1([a, b], X)$, where $f : [a, b] \to X$. We call the Caputo-Bochner right fractional derivative of order α:

$$\left(D_{b-}^{\alpha} f\right)(x) := \frac{(-1)^m}{\Gamma(m-\alpha)} \int_x^b (z-x)^{m-\alpha-1} f^{(m)}(z)\, dz,\ \forall\, x \in [a, b]. \tag{10.86}$$

We observe that $\left(D_{b-}^{m} f\right)(x) = (-1)^m f^{(m)}(x)$, for $m \in \mathbb{N}$, and $\left(D_{b-}^{0} f\right)(x) = f(x)$.

By [10], $\left(D_{b-}^{\alpha} f\right)(x)$ exists almost everywhere on $[a,b]$ and $\left(D_{b-}^{\alpha} f\right) \in L_1([a,b], X)$.

If $\left\| f^{(m)} \right\|_{L_\infty([a,b],X)} < \infty$, and $\alpha \notin \mathbb{N}$, by [10], $D_{b-}^{\alpha} f \in C([a,b], X)$, hence $\left\| D_{b-}^{\alpha} f \right\| \in C([a,b])$.

We need

Lemma 10.24 ([10]) *Let $f \in C^{m-1}([a,b], X)$, $f^{(m)} \in L_\infty([a,b], X)$, $m = \lceil \alpha \rceil$, $\alpha > 0$, $\alpha \notin \mathbb{N}$. Then $D_{b-}^{\alpha} f(b) = 0$.*

We mention the left fractional vector Taylor formula

Theorem 10.25 ([10]) *Let $m \in \mathbb{N}$ and $f \in C^m([a,b], X)$, where $[a,b] \subset \mathbb{R}$ and X is a Banach space, and let $\alpha > 0 : m = \lceil \alpha \rceil$. Then*

$$f(x) = \sum_{i=0}^{m-1} \frac{(x-a)^i}{i!} f^{(i)}(a) + \frac{1}{\Gamma(\alpha)} \int_a^x (x-z)^{\alpha-1} \left(D_{*a}^{\alpha} f\right)(z)\, dz, \quad (10.87)$$

$\forall\, x \in [a,b]$.

We also mention the right fractional vector Taylor formula

Theorem 10.26 ([10]) *Let $[a,b] \subset \mathbb{R}$, X be a Banach space, $\alpha > 0$, $m = \lceil \alpha \rceil$, $f \in C^m([a,b], X)$. Then*

$$f(x) = \sum_{i=0}^{m-1} \frac{(x-b)^i}{i!} f^{(i)}(b) + \frac{1}{\Gamma(\alpha)} \int_x^b (z-x)^{\alpha-1} \left(D_{b-}^{\alpha} f\right)(z)\, dz, \quad (10.88)$$

$\forall\, x \in [a,b]$.

Convention 10.27 *We assume that*

$$D_{*x_0}^{\alpha} f(x) = 0, \text{ for } x < x_0, \quad (10.89)$$

and

$$D_{x_0-}^{\alpha} f(x) = 0, \text{ for } x > x_0, \quad (10.90)$$

for all $x, x_0 \in [a,b]$.

We mention

Proposition 10.28 ([10]) *Let $f \in C^n([a,b], X)$, $n = \lceil \nu \rceil$, $\nu > 0$. Then $D_{*a}^{\nu} f(x)$ is continuous in $x \in [a,b]$.*

Proposition 10.29 ([10]) *Let $f \in C^m([a,b], X)$, $m = \lceil \alpha \rceil$, $\alpha > 0$. Then $D_{b-}^{\nu} f(x)$ is continuous in $x \in [a,b]$.*

We also mention

Proposition 10.30 ([10]) *Let $f \in C^{m-1}([a,b], X)$, $f^{(m)} \in L_\infty([a,b], X)$, $m = \lceil \alpha \rceil$, $\alpha > 0$ and*

$$D_{*x_0}^{\alpha} f(x) = \frac{1}{\Gamma(m-\alpha)} \int_{x_0}^{x} (x-t)^{m-\alpha-1} f^{(m)}(t)\, dt, \tag{10.91}$$

for all $x, x_0 \in [a,b] : x \geq x_0$.

Then $D_{*x_0}^{\alpha} f(x)$ is continuous in x_0.

Proposition 10.31 ([10]) *Let $f \in C^{m-1}([a,b], X)$, $f^{(m)} \in L_\infty([a,b], X)$, $m = \lceil \alpha \rceil$, $\alpha > 0$ and*

$$D_{x_0-}^{\alpha} f(x) = \frac{(-1)^m}{\Gamma(m-\alpha)} \int_{x}^{x_0} (\zeta - x)^{m-\alpha-1} f^{(m)}(\zeta)\, d\zeta, \tag{10.92}$$

for all $x, x_0 \in [a,b] : x_0 \geq x$.

Then $D_{x_0-}^{\alpha} f(x)$ is continuous in x_0.

Corollary 10.32 ([10]) *Let $f \in C^m([a,b], X)$, $m = \lceil \alpha \rceil$, $\alpha > 0$, $x, x_0 \in [a,b]$. Then $D_{*x_0}^{a} f(x)$, $D_{x_0-}^{a} f(x)$ are jointly continuous functions in (x, x_0) from $[a,b]^2$ into X, X is a Banach space.*

We make

Remark 10.33 ([10], pp. 263–266) Let $\left(\mathbb{R}^N, \|\cdot\|_p\right)$, $N \in \mathbb{N}$; where $\|\cdot\|_p$ is the L_p-norm, $1 \leq p \leq \infty$. $\mathbb{R}^N$ is a Banach space, and $\left(\mathbb{R}^N\right)^{j_*}$ denotes the j_*-fold product space $\mathbb{R}^N \times \ldots \times \mathbb{R}^N$ endowed with the max-norm $\|x\|_{\left(\mathbb{R}^N\right)^{j_*}} := \max_{1 \leq \lambda \leq j_*} \|x_\lambda\|_p$, where $x := \left(x_1, \ldots, x_{j_*}\right) \in \left(\mathbb{R}^N\right)^{j_*}$.

Let $\left(X, \|\cdot\|_\gamma\right)$ be a general Banach space. Then the space $L_{j_*} := L_{j_*}\left(\left(\mathbb{R}^N\right)^{j_*}, X\right)$ of all j_*-multilinear continuous maps $g : \left(\mathbb{R}^N\right)^{j_*} \to X$, $j_* = 1, \ldots, m$, is a Banach space with norm

$$\|g\| := \|g\|_{L_{j_*}} := \sup_{\left(\|x\|_{\left(\mathbb{R}^N\right)^{j_*}} = 1\right)} \|g(x)\|_\gamma = \sup \frac{\|g(x)\|_\gamma}{\|x_1\|_p \ldots \left\|x_{j_*}\right\|_p}. \tag{10.93}$$

Let M be a non-empty convex and compact subset of $\mathbb{R}^N$ and $x_0 \in M$ is fixed.

Let O be an open subset of $\mathbb{R}^N : M \subset O$. Let $f : O \to X$ be a continuous function, whose Fréchet derivatives (see [28]) $f^{(j_*)} : O \to L_{j_*} = L_{j_*}\left(\left(\mathbb{R}^N\right)^{j_*}; X\right)$ exist and are continuous for $1 \leq j_* \leq \overline{m}$, $\overline{m} \in \mathbb{N}$.

Call $(x - x_0)^{j_*} := (x - x_0, \ldots, x - x_0) \in \left(R^N\right)^{j_*}$, $x \in M$.

We will work with $f|_M$.

Then, by Taylor's formula ([19], [28], p. 124), we get

$$f(x) = \sum_{j_*=0}^{\overline{m}-1} \frac{f^{(j_*)}(x_0)(x-x_0)^{j_*}}{j_*!} + R_{\overline{m}}(x, x_0), \text{ all } x \in M, \tag{10.94}$$

where the remainder is the Riemann integral

$$R_{\overline{m}}(x, x_0) := \int_0^1 \frac{(1-u)^{\overline{m}-1}}{(\overline{m}-1)!} f^{(\overline{m})}(x_0 + u(x-x_0))(x-x_0)^{\overline{m}} du, \tag{10.95}$$

here we set $f^{(0)}(x_0)(x-x_0)^0 = f(x_0)$.

We obtain

$$\left\| f^{(\overline{m})}(x_0 + u(x-x_0))(x-x_0)^{\overline{m}} \right\|_\gamma \leq$$

$$\left\| f^{(\overline{m})}(x_0 + u(x-x_0)) \right\| \|x-x_0\|_p^{\overline{m}} \leq \left\| \left\| f^{(\overline{m})} \right\| \right\|_\infty \|x-x_0\|_p^{\overline{m}}, \tag{10.96}$$

and

$$\|R_{\overline{m}}(x, x_0)\|_\gamma \leq \frac{\left\| \left\| f^{(\overline{m})} \right\| \right\|_\infty}{m!} \|x-x_0\|_p^{\overline{m}}. \tag{10.97}$$

Let $(X, \|\cdot\|_\gamma)$ be a general Banach space.

We will study the following neural network operators.

Definition 10.34 Let $f \in C([a,b], X)$, $n \in \mathbb{N}$. We set

$${}_jA_n(f, x) := \frac{\sum_{k=\lceil na \rceil}^{\lfloor nb \rfloor} f\left(\frac{k}{n}\right) \psi_j(nx-k)}{\sum_{k=\lceil na \rceil}^{\lfloor nb \rfloor} \psi_j(nx-k)}, \ \forall\, x \in [a.b], \ j = 1, 2, 3, 4. \tag{10.98}$$

These are univariate neural network operators.

Definition 10.35 Let $f \in C\left(\prod_{i=1}^N [a_i, b_i], X\right)$ and $n \in \mathbb{N}$ such that $\lceil na_i \rceil \leq \lfloor nb_i \rfloor$, $i = 1, ..., N$. We will study the following multivariate linear neural network operators $(x := (x_1, ..., x_N) \in \left(\prod_{i=1}^N [a_i, b_i]\right))$

$${}_jH_n(f, x) := {}_jH_n(f, x_1, ..., x_N) := \frac{\sum_{k=\lceil na \rceil}^{\lfloor nb \rfloor} f\left(\frac{k}{n}\right) Z_j(nx-k)}{\sum_{k=\lceil na \rceil}^{\lfloor nb \rfloor} Z_j(nx-k)} = \tag{10.99}$$

$$\frac{\sum_{k_1=\lceil na_1 \rceil}^{\lfloor nb_1 \rfloor} \sum_{k_2=\lceil na_2 \rceil}^{\lfloor nb_2 \rfloor} \cdots \sum_{k_N=\lceil na_N \rceil}^{\lfloor nb_N \rfloor} f\left(\frac{k_1}{n}, ..., \frac{k_N}{n}\right) \left(\prod_{i=1}^N \psi_j(nx_i - k_i)\right)}{\prod_{i=1}^N \left(\sum_{k_i=\lceil na_i \rceil}^{\lfloor nb_i \rfloor} \psi_j(nx_i - k_i)\right)},$$

for $j = 1, 2, 3, 4$.

For large enough n we always obtain $\lceil na_i \rceil \leq \lfloor nb_i \rfloor, i = 1, ..., N$. Also $a_i \leq \frac{k_i}{n} \leq b_i$, iff $\lceil na_i \rceil \leq k_i \leq \lfloor nb_i \rfloor, i = 1, ..., N$.

In this chapter first we find fractional Voronskaya type asymptotic expansions for ${}_jA_n(f, x)$, $x \in [a, b]$, then we find multivariate Voronskaya type asymptotic expansions for ${}_jH_n(f, x)$, $x \in \left(\prod_{i=1}^{N} [a_i, b_i]\right)$; $n \in \mathbb{N}$; $j = 1, 2, 3, 4$.

Our considered neural networks here are of one hidden layer.

For earlier motivational neural networks related work, see [1–10]. For neural networks in general, read [24, 25, 27].

10.2 Main Results

We present our first univariate main result, as Voronovskaya type asymptotic expansion.

Theorem 10.36 *Let $(X, \|\cdot\|_\gamma)$ be a Banach space, $0 < \beta < \frac{1}{2}$ and $0 < \alpha \leq \frac{1-\beta}{\beta}$, $N \in \mathbb{N} : N = \lceil \alpha \rceil$, $f \in C^N([a, b], X)$, $x \in [a, b]$, $n \in \mathbb{N}$ large enough : $n^{1-\beta} > 2$. Assume that $\left\| \left\| D_{x-}^{\alpha} f \right\|_\gamma \right\|_{\infty,[a,x]}, \left\| \left\| D_{*x}^{\alpha} f \right\|_\gamma \right\|_{\infty,[x,b]} \leq M$, $M > 0$. Then*

$$ {}_jA_n(f, x) - f(x) = \sum_{j_*=1}^{N-1} \frac{f^{(j_*)}(x)}{j_*!} \, {}_jA_n\left((\cdot - x)^{j_*}\right)(x) + o\left(\frac{1}{n^{\beta(\alpha-\varepsilon)}}\right), \quad (10.100) $$

where $0 < \varepsilon \leq \alpha$.

If $N = 1$, the sum in (10.100) collapses.

The last (10.100) implies that

$$ n^{\beta(\alpha-\varepsilon)} \left[{}_jA_n(f, x) - f(x) - \sum_{j_*=1}^{N-1} \frac{f^{(j_*)}(x)}{j_*!} \, {}_jA_n\left((\cdot - x)^{j_*}\right)(x) \right] \to 0, \quad (10.101) $$

as $n \to \infty$, $0 < \varepsilon \leq \alpha$.

When $N = 1$, or $f^{(j_*)}(x) = 0$, $j_* = 1, ..., N - 1$, then we derive that

$$ n^{\beta(\alpha-\varepsilon)} \left[{}_jA_n(f, x) - f(x) \right] \to 0 $$

as $n \to \infty$, $0 < \varepsilon \leq \alpha$. Of great interest is the case of $\alpha = \frac{1}{2}$.

Above it is $j = 1, 2, 3, 4$.

Proof From Theorem 10.25 (10.87), we get by the left Caputo fractional vector Taylor formula that

$$f\left(\frac{k}{n}\right)=\sum_{j_*=0}^{N-1}\frac{f^{(j_*)}(x)}{j_*!}\left(\frac{k}{n}-x\right)^{j_*}+\frac{1}{\Gamma(\alpha)}\int_x^{\frac{k}{n}}\left(\frac{k}{n}-J\right)^{\alpha-1}D_{*x}^{\alpha}f(J)\,dJ, \tag{10.102}$$

for all $x\leq\frac{k}{n}\leq b$.

Also from Theorem 10.26 (10.88), using the right Caputo fractional vector Taylor formula we get

$$f\left(\frac{k}{n}\right)=\sum_{j_*=0}^{N-1}\frac{f^{(j_*)}(x)}{j_*!}\left(\frac{k}{n}-x\right)^{j_*}+\frac{1}{\Gamma(\alpha)}\int_{\frac{k}{n}}^{x}\left(J-\frac{k}{n}\right)^{\alpha-1}D_{x-}^{\alpha}f(J)\,dJ, \tag{10.103}$$

for all $a\leq\frac{k}{n}\leq x$.

We call

$$W_j(x):=\sum_{k=\lceil na\rceil}^{\lfloor nb\rfloor}\psi_j(nx-k),\ \ j=1,2,3,4. \tag{10.104}$$

Hence we have

$$\frac{f\left(\frac{k}{n}\right)\psi_j(nx-k)}{W_j(x)}=\sum_{j_*=0}^{N-1}\frac{f^{(j_*)}(x)}{j_*!}\frac{\psi_j(nx-k)}{W_j(x)}\left(\frac{k}{n}-x\right)^{j_*}+ \tag{10.105}$$

$$\frac{\psi_j(nx-k)}{W_j(x)\Gamma(\alpha)}\int_x^{\frac{k}{n}}\left(\frac{k}{n}-J\right)^{\alpha-1}D_{*x}^{\alpha}f(J)\,dJ,$$

all $x\leq\frac{k}{n}\leq b$, iff $\lceil nx\rceil\leq k\leq\lfloor nb\rfloor$, and

$$\frac{f\left(\frac{k}{n}\right)\psi_j(nx-k)}{W_j(x)}=\sum_{j_*=0}^{N-1}\frac{f^{(j_*)}(x)}{j_*!}\frac{\psi_j(nx-k)}{W_j(x)}\left(\frac{k}{n}-x\right)^{j_*}+ \tag{10.106}$$

$$\frac{\psi_j(nx-k)}{W_j(x)\Gamma(\alpha)}\int_{\frac{k}{n}}^{x}\left(J-\frac{k}{n}\right)^{\alpha-1}D_{x-}^{\alpha}f(J)\,dJ,$$

for all $a\leq\frac{k}{n}\leq x$, iff $\lceil na\rceil\leq k\leq\lfloor nx\rfloor$.

We have that $\lceil nx\rceil\leq\lfloor nx\rfloor+1$.

Therefore it holds

$$\sum_{k=\lfloor nx\rfloor+1}^{\lfloor nb\rfloor} \frac{f\left(\frac{k}{n}\right)\psi_j(nx-k)}{W_j(x)} = \sum_{j_*=0}^{N-1} \frac{f^{(j_*)}(x)}{j_*!} \sum_{k=\lfloor nx\rfloor+1}^{\lfloor nb\rfloor} \frac{\psi_j(nx-k)\left(\frac{k}{n}-x\right)^{j_*}}{W_j(x)} +$$

$$\frac{1}{\Gamma(\alpha)}\left(\frac{\sum_{k=\lfloor nx\rfloor+1}^{\lfloor nb\rfloor}\psi_j(nx-k)}{W_j(x)}\int_x^{\frac{k}{n}}\left(\frac{k}{n}-J\right)^{\alpha-1} D_{*x}^{\alpha} f(J)\,dJ\right), \tag{10.107}$$

and

$$\sum_{k=\lceil na\rceil}^{\lfloor nx\rfloor} f\left(\frac{k}{n}\right)\frac{\psi_j(nx-k)}{W_j(x)} = \sum_{j_*=0}^{N-1} \frac{f^{(j_*)}(x)}{j_*!} \sum_{k=\lceil na\rceil}^{\lfloor nx\rfloor} \frac{\psi_j(nx-k)}{W_j(x)}\left(\frac{k}{n}-x\right)^{j_*} +$$

$$\frac{1}{\Gamma(\alpha)}\left(\sum_{k=\lceil na\rceil}^{\lfloor nx\rfloor}\frac{\psi_j(nx-k)}{W_j(x)}\int_{\frac{k}{n}}^{x}\left(J-\frac{k}{n}\right)^{\alpha-1} D_{x-}^{\alpha} f(J)\,dJ\right). \tag{10.108}$$

Adding the last two equalities (10.107) and (10.108) we obtain

$${}_jA_n(f,x) = \sum_{k=\lceil na\rceil}^{\lfloor nb\rfloor} f\left(\frac{k}{n}\right)\frac{\psi_j(nx-k)}{W_j(x)} = \tag{10.109}$$

$$\sum_{j_*=0}^{N-1} \frac{f^{(j_*)}(x)}{j_*!} \sum_{k=\lceil na\rceil}^{\lfloor nb\rfloor} \frac{\psi_j(nx-k)}{W_j(x)}\left(\frac{k}{n}-x\right)^{j_*} +$$

$$\frac{1}{\Gamma(\alpha)W_j(x)}\left\{\sum_{k=\lceil na\rceil}^{\lfloor nx\rfloor}\psi_j(nx-k)\int_{\frac{k}{n}}^{x}\left(J-\frac{k}{n}\right)^{\alpha-1} D_{x-}^{\alpha} f(J)\,dJ +\right.$$

$$\left.\sum_{k=\lfloor nx\rfloor+1}^{\lfloor nb\rfloor}\psi_j(nx-k)\int_x^{\frac{k}{n}}\left(\frac{k}{n}-J\right)^{\alpha-1}\left(D_{*x}^{\alpha} f(J)\right)dJ\right\}.$$

So we have derived ($j = 1, 2, 3, 4$)

$$\theta_j(x) := {}_jA_n(f,x) - f(x) - \sum_{j_*=1}^{N-1}\frac{f^{(j_*)}(x)}{j_*!}\,{}_jA_n\left((\cdot - x)^{j_*}\right)(x) = \theta_{jn}^*(x), \tag{10.110}$$

where

$$\theta_{jn}^*(x) := \frac{1}{\Gamma(\alpha)W_j(x)}\left\{\sum_{k=\lceil na\rceil}^{\lfloor nx\rfloor}\psi_j(nx-k)\int_{\frac{k}{n}}^{x}\left(J-\frac{k}{n}\right)^{\alpha-1} D_{x-}^{\alpha} f(J)\,dJ\right.$$

$$+\sum_{k=\lfloor nx\rfloor+1}^{\lfloor nb\rfloor}\psi_j(nx-k)\int_x^{\frac{k}{n}}\left(\frac{k}{n}-J\right)^{\alpha-1}D_{*x}^{\alpha}f(J)\,dJ\Bigg\}. \quad (10.111)$$

We set

$${}_j\theta_{1n}^{*}(x):=\frac{1}{\Gamma(\alpha)}\left(\frac{\sum_{k=\lceil na\rceil}^{\lfloor nx\rfloor}\psi_j(nx-k)}{W_j(x)}\int_{\frac{k}{n}}^{x}\left(J-\frac{k}{n}\right)^{\alpha-1}D_{x-}^{\alpha}f(J)\,dJ\right), \quad (10.112)$$

and

$${}_j\theta_{2n}^{*}:=\frac{1}{\Gamma(\alpha)}\left(\frac{\sum_{k=\lfloor nx\rfloor+1}^{\lfloor nb\rfloor}\psi_j(nx-k)}{W_j(x)}\int_x^{\frac{k}{n}}\left(\frac{k}{n}-J\right)^{\alpha-1}D_{*x}^{\alpha}f(J)\,dJ\right), \quad (10.113)$$

i.e.

$$\theta_{jn}^{*}(x)={}_j\theta_{1n}^{*}(x)+{}_j\theta_{2n}^{*}(x),\ j=1,2,3,4. \quad (10.114)$$

We assume $b-a>\frac{1}{n^{\beta}}$, $0<\beta<1$, which is always the case for large enough $n\in\mathbb{N}$, that is when $n>\left\lceil(b-a)^{-\frac{1}{\beta}}\right\rceil$. It is always true that either $\left|\frac{k}{n}-x\right|\leq\frac{1}{n^{\beta}}$ or $\left|\frac{k}{n}-x\right|>\frac{1}{n^{\beta}}$.

For $k=\lceil na\rceil,\ldots,\lfloor nx\rfloor$, we consider

$$\gamma_{1k}:=\left\|\int_{\frac{k}{n}}^{x}\left(J-\frac{k}{n}\right)^{\alpha-1}D_{x-}^{\alpha}f(J)\,dJ\right\|_{\gamma}\leq \quad (10.115)$$

$$\int_{\frac{k}{n}}^{x}\left(J-\frac{k}{n}\right)^{\alpha-1}\left\|D_{x-}^{\alpha}f(J)\right\|_{\gamma}dJ$$

$$\leq\left\|\left\|D_{x-}^{\alpha}f\right\|_{\gamma}\right\|_{\infty,[a,x]}\frac{\left(x-\frac{k}{n}\right)^{\alpha}}{\alpha}\leq\left\|\left\|D_{x-}^{\alpha}f\right\|_{\gamma}\right\|_{\infty,[a,x]}\frac{(x-a)^{\alpha}}{\alpha}. \quad (10.116)$$

That is

$$\gamma_{1k}\leq\left\|\left\|D_{x-}^{\alpha}f\right\|_{\gamma}\right\|_{\infty,[a,x]}\frac{(x-a)^{\alpha}}{\alpha}, \quad (10.117)$$

for $k=\lceil na\rceil,\ldots,\lfloor nx\rfloor$.

Also we have in case of $\left|\frac{k}{n}-x\right| \leq \frac{1}{n^\beta}$ that

$$\gamma_{1k} \leq \int_{\frac{k}{n}}^{x} \left(J-\frac{k}{n}\right)^{\alpha-1} \left\| D_{x-}^{\alpha} f(J) \right\|_{\gamma} dJ \tag{10.118}$$

$$\leq \left\| \left\| D_{x-}^{\alpha} f \right\|_{\gamma} \right\|_{\infty,[a,x]} \frac{\left(x-\frac{\kappa}{n}\right)^{\alpha}}{\alpha} \leq \left\| \left\| D_{x-}^{\alpha} f \right\|_{\gamma} \right\|_{\infty,[a,x]} \frac{1}{n^{\alpha\beta}\alpha}.$$

So that, when $\left|x-\frac{k}{n}\right| \leq \frac{1}{n^\beta}$, we get

$$\gamma_{1k} \leq \left\| \left\| D_{x-}^{\alpha} f \right\|_{\gamma} \right\|_{\infty,[a,x]} \frac{1}{\alpha n^{a\beta}}. \tag{10.119}$$

Therefore

$$\left\| {}_j\theta_{1n}^{*}(x) \right\|_{\gamma} \leq \frac{1}{\Gamma(\alpha)} \left(\frac{\sum_{k=\lceil na \rceil}^{\lfloor nx \rfloor} \psi_j(nx-k)}{W_j(x)} \gamma_{1k} \right) = \frac{1}{\Gamma(\alpha)} \cdot$$

$$\left\{ \frac{\sum_{\substack{k=\lceil na \rceil \\ : \left|\frac{k}{n}-x\right| \leq \frac{1}{n^\beta}}}^{\lfloor nx \rfloor} \psi_j(nx-k)}{W_j(x)} \gamma_{1k} + \frac{\sum_{\substack{k=\lceil na \rceil \\ : \left|\frac{k}{n}-x\right| > \frac{1}{n^\beta}}}^{\lfloor nx \rfloor} \psi_j(nx-k)}{W_j(x)} \gamma_{1k} \right\}$$

$$\leq \frac{1}{\Gamma(\alpha)} \left\{ \left(\frac{\sum_{\substack{k=\lceil na \rceil \\ : \left|\frac{k}{n}-x\right| \leq \frac{1}{n^\beta}}}^{\lfloor nx \rfloor} \psi_j(nx-k)}{W_j(x)} \right) \left\| \left\| D_{x-}^{\alpha} f \right\|_{\gamma} \right\|_{\infty,[a,x]} \frac{1}{\alpha n^{\alpha\beta}} + \right.$$

$$\left. \frac{1}{W_j(x)} \left(\sum_{\substack{k=\lceil na \rceil \\ : \left|\frac{k}{n}-x\right| > \frac{1}{n^\beta}}}^{\lfloor nx \rfloor} \psi_j(nx-k) \right) \left\| \left\| D_{x-}^{\alpha} f \right\|_{\gamma} \right\|_{\infty,[a,x]} \frac{(x-a)^{\alpha}}{\alpha} \right\} \tag{10.120}$$

(by (10.12), (10.13); (10.32), (10.33); (10.52), (10.53); (10.74), (10.75))

$$\leq \frac{\left\| \left\| D_{x-}^{\alpha} f \right\|_{\gamma} \right\|_{\infty,[a,x]}}{\Gamma(\alpha+1)} \left\{ \frac{1}{n^{\alpha\beta}} + \alpha_j c_j(\beta,n)(x-a)^{\alpha} \right\}.$$

Therefore we proved

$$\left\|{}_j\theta^*_{1n}(x)\right\|_\gamma \le \frac{\left\|\left\|D^\alpha_{x-}f\right\|_\gamma\right\|_{\infty,[a,x]}}{\Gamma(\alpha+1)}\left\{\frac{1}{n^{\alpha\beta}}+\alpha_j c_j(\beta,n)(x-a)^\alpha\right\}. \tag{10.121}$$

But for large enough $n \in \mathbb{N}$ we get

$$\left\|{}_j\theta^*_{1n}(x)\right\|_\gamma \le \frac{2\left\|\left\|D^\alpha_{x-}f\right\|_\gamma\right\|_{\infty,[a,x]}}{\Gamma(\alpha+1)\,n^{\alpha\beta}}. \tag{10.122}$$

Similarly, we have that

$$\gamma_{2k} := \left\|\int_x^{\frac{k}{n}}\left(\frac{k}{n}-J\right)^{\alpha-1} D^\alpha_{*x}f(J)\,dJ\right\|_\gamma \le$$

$$\int_x^{\frac{k}{n}}\left(\frac{k}{n}-J\right)^{\alpha-1}\left\|D^\alpha_{*x}f(J)\right\|_\gamma dJ \le$$

$$\left\|\left\|D^\alpha_{*x}f\right\|_\gamma\right\|_{\infty,[x,b]}\frac{\left(\frac{k}{n}-x\right)^\alpha}{\alpha} \le \left\|\left\|D^\alpha_{*x}f\right\|_\gamma\right\|_{\infty,[x,b]}\frac{(b-x)^\alpha}{\alpha}. \tag{10.123}$$

That is

$$\gamma_{2k} \le \left\|\left\|D^\alpha_{*x}f\right\|_\gamma\right\|_{\infty,[x,b]}\frac{(b-x)^\alpha}{\alpha}, \tag{10.124}$$

for $k = \lfloor nx\rfloor + 1, \ldots, \lfloor nb\rfloor$.

Also we have in case of $\left|\frac{k}{n}-x\right| \le \frac{1}{n^\beta}$ that

$$\gamma_{2k} \le \frac{\left\|\left\|D^\alpha_{*x}f\right\|_\gamma\right\|_{\infty,[x,b]}}{\alpha n^{\alpha\beta}}. \tag{10.125}$$

Consequently it holds

$$\left\|{}_j\theta^*_{2n}(x)\right\|_\gamma \le \frac{1}{\Gamma(\alpha)}\left(\frac{\sum_{k=\lfloor nx\rfloor+1}^{\lfloor nb\rfloor}\psi_j(nx-k)}{W_j(x)}\gamma_{2k}\right) =$$

$$\frac{1}{\Gamma(\alpha)}\left\{\left(\frac{\sum_{\begin{cases}k=\lfloor nx\rfloor+1\\ :\left|\frac{k}{n}-x\right|\le\frac{1}{n^\beta}\end{cases}}^{\lfloor nb\rfloor}\psi_j(nx-k)}{W_j(x)}\right)\frac{\left\|\left\|D^\alpha_{*x}f\right\|_\gamma\right\|_{\infty,[x,b]}}{\alpha n^{\alpha\beta}}+\right.$$

$$\frac{1}{W_j(x)}\left(\sum_{\left\{\begin{array}{l} k=\lfloor nx\rfloor+1 \\ :\left|\frac{k}{n}-x\right|>\frac{1}{n^{\beta}}\end{array}\right.}^{\lfloor nb\rfloor}\psi_j(nx-k)\right)\left\|\left\|D_{*x}^{\alpha}f\right\|_{\gamma}\right\|_{\infty,[x,b]}\frac{(b-x)^{\alpha}}{\alpha}\right\}\leq$$

$$\frac{\left\|\left\|D_{*x}^{\alpha}f\right\|_{\gamma}\right\|_{\infty,[x,b]}}{\Gamma(\alpha+1)}\left\{\frac{1}{n^{\alpha\beta}}+\alpha_j c_j(\beta,n)(b-x)^{\alpha}\right\}. \tag{10.126}$$

That is

$$\left\|{}_j\theta_{2n}^{*}(x)\right\|_{\gamma}\leq\frac{\left\|\left\|D_{*x}^{\alpha}f\right\|_{\gamma}\right\|_{\infty,[x,b]}}{\Gamma(\alpha+1)}\left\{\frac{1}{n^{\alpha\beta}}+\alpha_j c_j(\beta,n)(b-x)^{\alpha}\right\}. \tag{10.127}$$

But for large enough $n\in\mathbb{N}$ we get

$$\left\|{}_j\theta_{2n}^{*}(x)\right\|_{\gamma}\leq\frac{2\left\|\left\|D_{*x}^{\alpha}f\right\|_{\gamma}\right\|_{\infty,[x,b]}}{\Gamma(\alpha+1)n^{\alpha\beta}}. \tag{10.128}$$

Since $\left\|\left\|D_{x-}^{\alpha}f\right\|_{\gamma}\right\|_{\infty,[a,x]}, \left\|\left\|D_{*x}^{\alpha}f\right\|_{\gamma}\right\|_{\infty,[x,b]}\leq M$, $M>0$, we derive

$$\left\|\theta_{jn}^{*}(x)\right\|_{\gamma}\leq\left\|{}_j\theta_{1n}^{*}(x)\right\|_{\gamma}+\left\|{}_j\theta_{2n}^{*}(x)\right\|_{\gamma}\overset{\text{(by (10.122), 10.128))}}{\leq}\frac{4M}{\Gamma(\alpha+1)n^{\alpha\beta}}. \tag{10.129}$$

That is for large enough $n\in\mathbb{N}$ we get

$$\left\|\theta_j(x)\right\|_{\gamma}=\left\|\theta_{jn}^{*}(x)\right\|_{\gamma}\leq\left(\frac{4M}{\Gamma(\alpha+1)}\right)\left(\frac{1}{n^{\alpha\beta}}\right), \tag{10.130}$$

resulting to

$$\|\theta(x)\|_{\gamma}=O\left(\frac{1}{n^{\alpha\beta}}\right), \tag{10.131}$$

and

$$\|\theta(x)\|_{\gamma}=o(1). \tag{10.132}$$

And, letting $0<\varepsilon\leq\alpha$, we derive

$$\frac{\|\theta(x)\|_{\gamma}}{\left(\frac{1}{n^{\beta(\alpha-\varepsilon)}}\right)}\leq\left(\frac{4M}{\Gamma(\alpha+1)}\right)\left(\frac{1}{n^{\beta\varepsilon}}\right)\to 0, \tag{10.133}$$

as $n\to\infty$.

I.e.

$$\|\theta(x)\|_{\gamma} = o\left(\frac{1}{n^{\beta(\alpha-\varepsilon)}}\right), \tag{10.134}$$

proving the claim. ■

It follows a multivariate Voronovskaya type asymptotic expansion.

Theorem 10.37 *Let $\left(X, \|\cdot\|_{\gamma}\right)$ be a Banach space, $\overline{m} \in \mathbb{N}$ such that $\overline{m} \leq \frac{1-\beta}{\beta}$, where $0 < \beta < \frac{1}{2}$. Let $f \in C^{\overline{m}}\left(\prod_{i=1}^{N}[a_i, b_i], X\right)$ ($\overline{m}$-times continuously Fréchet differentiable functions), $x \in \prod_{i=1}^{N}[a_i, b_i]$, and $n \in \mathbb{N} : n^{1-\beta} > 2$; $j = 1, 2, 3, 4$. Then*

$$_{j}H_n(f, x) - f(x) =$$

$$\sum_{j_*=1}^{\overline{m}-1} \frac{1}{j_*!}\, {}_{j}H_n\left(f^{(j_*)}(x)(\cdot - x)^{j_*}, x\right) + o\left(\frac{1}{n^{\beta(\overline{m}-\varepsilon)}}\right), \tag{10.135}$$

where $0 < \varepsilon \leq \overline{m}$.

If $\overline{m} = 1$, the sum in (10.135) collapses.

The last (10.135) implies that

$$n^{\beta(\overline{m}-\varepsilon)}\left[{}_{j}H_n(f, x) - f(x) - \sum_{j_*=1}^{\overline{m}-1} \frac{1}{j_*!}\, {}_{j}H_n\left(f^{(j_*)}(x)(\cdot - x)^{j_*}, x\right)\right] \to 0, \tag{10.136}$$

as $n \to \infty$, $0 < \varepsilon \leq \overline{m}$.

When $\overline{m} = 1$, or $f^{(j_*)}(x) = 0$, $j_* = 1, ..., \overline{m} - 1$, then we derive that

$$n^{\beta(\overline{m}-\varepsilon)}\left[{}_{j}H_n(f, x) - f(x)\right] \to 0, \tag{10.137}$$

as $n \to \infty$, $0 < \varepsilon \leq \overline{m}$.

Above it is $j = 1, 2, 3, 4$.

Proof We have that

$$f\left(\frac{k}{n}\right) - f(x) = \sum_{j_*=1}^{\overline{m}-1} \frac{f^{(j_*)}(x)}{j_*!}\left(\frac{k}{n} - x\right)^{j_*} + R_{\overline{m}}\left(\frac{k}{n}, x\right), \tag{10.138}$$

where

$$R_{\overline{m}}\left(\frac{k}{n}, x\right) := \int_0^1 \frac{(1-u)^{\overline{m}-1}}{(\overline{m}-1)!} f^{(\overline{m})}\left(x + u\left(\frac{k}{n} - x\right)\right)\left(\frac{k}{n} - x\right)^{\overline{m}} du, \tag{10.139}$$

here we set $f^{(0)}(x)\left(\frac{k}{n} - x\right)^0 = f(x)$.

By (10.97) we get that

$$\left\| R_{\overline{m}} \left(\frac{k}{n}, x \right) \right\|_{\gamma} \leq \frac{\left\| \left\| f^{(\overline{m})} \right\| \right\|_{\infty}}{\overline{m}!} \left\| \frac{k}{n} - x \right\|_{\infty}^{\overline{m}} \tag{10.140}$$

$$\leq \frac{\left\| \left\| f^{(\overline{m})} \right\| \right\|_{\infty}}{\overline{m}!} \left\| b - a \right\|_{\infty}^{\overline{m}}.$$

Call

$$V_j(x) := \sum_{k=\lceil na \rceil}^{\lfloor nb \rfloor} Z_j(nx - k), \tag{10.141}$$

for $j = 1, 2, 3, 4$.

Hence, we have

$${}_jU_n(x) := \frac{\sum_{k=\lceil na \rceil}^{\lfloor nb \rfloor} Z_j(nx-k) R_{\overline{m}}\left(\frac{k}{n}, x\right)}{V_j(x)} =$$

$$\frac{\sum_{\begin{cases} k = \lceil na \rceil \\ : \left\| \frac{k}{n} - x \right\|_{\infty} \leq \frac{1}{n^{\beta}} \end{cases}}^{\lfloor nb \rfloor} Z_j(nx-k) R_{\overline{m}}\left(\frac{k}{n}, x\right)}{V_j(x)} + \tag{10.142}$$

$$\frac{\sum_{\begin{cases} k = \lceil na \rceil \\ : \left\| \frac{k}{n} - x \right\|_{\infty} > \frac{1}{n^{\beta}} \end{cases}}^{\lfloor nb \rfloor} Z_j(nx-k) R_{\overline{m}}\left(\frac{k}{n}, x\right)}{V_j(x)}.$$

Therefore, we obtain ($j = 1, 2, 3, 4$)

$$\left\| {}_jU_n(x) \right\|_{\gamma} \leq \left(\sum_{\begin{cases} k = \lceil na \rceil \\ : \left\| \frac{k}{n} - x \right\|_{\infty} \leq \frac{1}{n^{\beta}} \end{cases}}^{\lfloor nb \rfloor} \frac{Z_j(nx-k)}{V_j(x)} \right) \frac{\left\| \left\| f^{(\overline{m})} \right\| \right\|_{\infty}}{\overline{m}!} \frac{1}{n^{\overline{m}\beta}} +$$

$$\left(\sum_{\begin{cases} k = \lceil na \rceil \\ : \left\| \frac{k}{n} - x \right\|_{\infty} > \frac{1}{n^{\beta}} \end{cases}}^{\lfloor nb \rfloor} \frac{Z_j(nx-k)}{V_j(x)} \right) \frac{\left\| \left\| f^{(\overline{m})} \right\| \right\|_{\infty}}{\overline{m}!} \left\| b - a \right\|_{\infty}^{\overline{m}} \tag{10.143}$$

(by (10.20), (10.21); (10.40), (10.41); (10.60), (10.61); (10.82), (10.83))

$$\leq \frac{\left\| \left\| f^{(\overline{m})} \right\| \right\|_{\infty}}{\overline{m}!} \left[\frac{1}{n^{\beta \overline{m}}} + \gamma_j(N) c_j(\beta, n) \|b - a\|_{\infty}^{\overline{m}} \right].$$

Consequently, we get that

$$\left\| {}_jU_n(x) \right\|_{\gamma} \leq \frac{\left\| \left\| f^{(\overline{m})} \right\| \right\|_{\infty}}{\overline{m}!} \left[\frac{1}{n^{\beta \overline{m}}} + \gamma_j(N) c_j(\beta, n) \|b - a\|_{\infty}^{\overline{m}} \right]. \tag{10.144}$$

For large enough $n \in \mathbb{N}$, we get

$$\left\| {}_jU_n(x) \right\|_{\gamma} \leq \frac{2 \left\| \left\| f^{(\overline{m})} \right\| \right\|_{\infty}}{\overline{m}!} \left(\frac{1}{n^{\beta \overline{m}}} \right). \tag{10.145}$$

That is

$$\left\| {}_jU_n(x) \right\|_{\gamma} = O\left(\frac{1}{n^{\beta \overline{m}}} \right), \tag{10.146}$$

and

$$\left\| {}_jU_n(x) \right\|_{\gamma} = o(1). \tag{10.147}$$

And, letting $0 < \varepsilon \leq \overline{m}$, we derive

$$\frac{\left\| {}_jU_n(x) \right\|_{\gamma}}{\left(\frac{1}{n^{\beta(\overline{m}-\varepsilon)}} \right)} \leq \left(\frac{2 \left\| \left\| f^{(\overline{m})} \right\| \right\|_{\infty}}{\overline{m}!} \right) \frac{1}{n^{\beta\varepsilon}} \to 0, \tag{10.148}$$

as $n \to \infty$.

I.e.

$$\left\| {}_jU_n(x) \right\|_{\gamma} = o\left(\frac{1}{n^{\beta(\overline{m}-\varepsilon)}} \right). \tag{10.149}$$

By (10.138) we observe that

$$\frac{\sum_{k=\lceil na \rceil}^{\lfloor nb \rfloor} f\left(\frac{k}{n}\right) Z_j(nx - k)}{V_j(x)} - f(x) = \tag{10.150}$$

$$\sum_{j_*=1}^{\overline{m}-1} \left(\frac{\sum_{k=\lceil na \rceil}^{\lfloor nb \rfloor} \left(f^{(j_*)}(x) \left(\frac{k}{n} - x\right)^{j_*} \right) Z_j(nx - k)}{j_*! V_j(x)} \right) +$$

$$\frac{\sum_{k=\lceil na \rceil}^{\lfloor nb \rfloor} Z_j(nx - k) R_{\overline{m}}\left(\frac{k}{n}, x\right)}{V_j(x)}.$$

The last says ($j = 1, 2, 3, 4$)

$$
{}_jH_n(f, x) - f(x) - \sum_{j_*=1}^{\overline{m}-1} \frac{1}{j_*!} \, {}_jH_n\left(f^{(j_*)}(x)(\cdot - x)^{j_*}, x\right) = {}_jU_n(x). \quad (10.151)
$$

The proof of the theorem is complete. ■

References

1. G.A. Anastassiou, Rate of convergence of some neural network operators to the unit-univariate case. J. Math. Anal. Appl. **212**, 237–262 (1997)
2. G.A. Anastassiou, *Inteligent Systems: Approximation by Artificial Neural Networks, Intelligent Systems Reference Library*, vol. 19 (Springer, Heidelberg, 2011)
3. G.A. Anastassiou, Univariate hyperbolic tangent neural network approximation. Math. Comput. Model. **53**, 1111–1132 (2011)
4. G.A. Anastassiou, Multivariate hyperbolic tangent neural network approximation. Comput. Math. **61**, 809–821 (2011)
5. G.A. Anastassiou, Multivariate sigmoidal neural network approximation. Neural Netw. **24**, 378–386 (2011)
6. G.A. Anastassiou, Univariate sigmoidal neural network approximation. J. Comput. Anal. Appl. **14**(4), 659–690 (2012)
7. G.A. Anastassiou, Fractional neural network approximation. Comput. Math. Appl. **64**, 1655–1676 (2012)
8. G.A. Anastassiou, Voronovskaya type asymptotic expansions for error function based quasi-interpolation neural network operators. Revista Colombiana De Matematicas **49**(1), 171–192 (2015)
9. G.A. Anastassiou, *Intelligent Systems II: Complete Approximation by Neural Network Operators* (Springer, Heidelberg, New York, 2016)
10. G.A. Anastassiou, *Intelligent Computations: Abstract Fractional Calculus, Inequalities, Approximations* (Springer, Heidelberg, New York, 2018)
11. G.A. Anastassiou, Algebraic function based Banach space valued ordinary and fractional neural network approximations. New Trends Math. Sci. 10 special issues (1), 100–125 (2022)
12. G.A. Anastassiou, Gudermannian function activated Banach space valued ordinary and fractional neural network approximation. Adv. Nonlinear Var. Inequal. **25**(2), 27–64 (2022)
13. G.A. Anastassiou, *Generalized symmetrical sigmoid function activated Banach space valued ordinary and fractional neural network approximation.* Analele Universităţii Oradea, Fasc. Matematica, accepted for publication (2022)
14. G.A. Anastassiou, *Abstract multivariate Gudermannian function activated neural network approximations*. Panam. Math. J., accepted (2022)
15. G.A. Anastassiou, *General multivariate arctangent function activated neural network approximations*, submitted (2022)
16. G.A. Anastassiou, *Generalized symmetrical sigmoid function activated neural network multivariate approximation*, submitted (2022)
17. G.A. Anastassiou, *Vector Voronsovkaya type asymptotic expansions for sigmoid functions induced quasi-interpolation neural network operators revisited*, submitted (2022)
18. G.A. Anastassiou, R. Kozma, *Abstract multivariate algebraic function activated neural network approximations*, submitted (2022)
19. H. Cartan, *Differential Calculus* (Hermann, Paris, 1971)

20. Z. Chen, F. Cao, The approximation operators with sigmoidal functions. Comput. Math. Appl. **58**, 758–765 (2009)
21. D. Costarelli, R. Spigler, Approximation results for neural network operators activated by sigmoidal functions. Neural Netw. **44**, 101–106 (2013)
22. D. Costarelli, R. Spigler, Multivariate neural network operators with sigmoidal activation functions. Neural Netw. **48**, 72–77 (2013)
23. A.J. Dunning, J. Kensler, L. Goudeville, F. Bailleux, Some extensions in continuous methods for immunological correlates of protection. BMC Med. Res. Methodol. **15** (107) (2015). https://doi.org/10.1186/s12874-015-0096-9
24. I.S. Haykin, *Neural Networks: A Comprehensive Foundation*, 2nd edn. (Prentice Hall, New York, 1998)
25. W. McCulloch, W. Pitts, A logical calculus of the ideas immanent in nervous activity. Bull. Math. Biophys. **7**, 115–133 (1943)
26. J. Mikusinski, *The Bochner Integral* (Academic Press, New York, 1978)
27. T.M. Mitchell, *Machine Learning* (WCB-McGraw-Hill, New York, 1997)
28. L.B. Rall, *Computational Solution of Nonlinear Operator Equations* (Wiley, New York, 1969)
29. G.E. Shilov, *Elementary Functional Analysis* (Dover, New York, 1996)
30. E.W. Weisstein, *Gudermannian*, MathWorld

Chapter 11
Univariate Fuzzy Fractional Various Sigmoid Function Activated Neural Network Approximations Revisited

Here we study the univariate fuzzy fractional quantitative approximation of fuzzy real valued functions on a compact interval by quasi-interpolation arctangent-algebraic-Gudermannian-generalized symmetrical activation function relied fuzzy neural network operators. These approximations are derived by establishing fuzzy Jackson type inequalities involving the fuzzy moduli of continuity of the right and left Caputo fuzzy fractional derivatives of the involved function. The approximations are fuzzy pointwise and fuzzy uniform. The related feed-forward fuzzy neural networks are with one hidden layer. We study also the fuzzy integer derivative and just fuzzy continuous cases. Our fuzzy fractional approximation result using higher order fuzzy differentiation converges better than in the fuzzy just continuous case. It follows [27].

11.1 Introduction

The author in [1, 2], see Chaps. 2–5, was the first to derive quantitative neural network approximations to continuous functions with rates by very specifically defined neural network operators of Cardaliaguet-Euvrard and "Squashing" types, by employing the modulus of continuity of the engaged function or its high order derivative, and producing very tight Jackson type inequalities. He studied there both the univariate and multivariate cases. The defining these operators "bell-shaped" and "squashing" function are assumed to be of compact support.

The author inspired by [29], continued his studies on neural networks approximation by introducing and using the proper quasi-interpolation operators of sigmoidal and hyperbolic tangent type which resulted into [10, 13–20], by treating both the univariate and multivariate cases.

© The Author(s), under exclusive license to Springer Nature Switzerland AG 2023

G. A. Anastassiou, *Banach Space Valued Neural Network*,
Studies in Computational Intelligence 1062,
https://doi.org/10.1007/978-3-031-16400-2_11

Continuation of the author's works ([17, 18] and [19], Chap. 20) is this chapter where fuzzy neural network approximation based on arctangent-algebraic-Gudermannian-generalized symmetrical activation functions is taken at the fractional and ordinary levels resulting into higher rates of approximation. We involve the fuzzy ordinary derivatives and the right and left Caputo fuzzy fractional derivatives of the fuzzy function under approximation and we establish tight fuzzy Jackson type inequalities. An extensive background is given on fuzzyness, fractional calculus and neural networks, all needed to present our work.

Our fuzzy feed-forward neural networks (FFNNs) are with one hidden layer. About neural networks in general study [36, 39, 40].

11.2 Fuzzy Fractional Mathematical Analysis Basics

(See also [19], pp. 432–444)

We need the following basic background

Definition 11.1 (see [44]) Let $\mu : \mathbb{R} \to [0, 1]$ with the following properties:

(i) is normal, i.e., $\exists\, x_0 \in \mathbb{R};\ \mu(x_0) = 1$.

(ii) $\mu(\lambda x + (1-\lambda) y) \geq \min\{\mu(x), \mu(y)\}, \forall\, x, y \in \mathbb{R}, \forall\, \lambda \in [0, 1]$ (μ is called a convex fuzzy subset).

(iii) μ is upper semicontinuous on $\mathbb{R}$, i.e. $\forall\, x_0 \in \mathbb{R}$ and $\forall\, \varepsilon > 0$, $\exists$ neighborhood $V(x_0) : \mu(x) \leq \mu(x_0) + \varepsilon, \forall\, x \in V(x_0)$.

(iv) The set $\overline{\operatorname{supp}(\mu)}$ is compact in $\mathbb{R}$ (where $\operatorname{supp}(\mu) := \{x \in \mathbb{R} : \mu(x) > 0\}$).

We call μ a fuzzy real number. Denote the set of all μ with $\mathbb{R}_{\mathcal{F}}$.

E.g. $\chi_{\{x_0\}} \in \mathbb{R}_{\mathcal{F}}$, for any $x_0 \in \mathbb{R}$, where $\chi_{\{x_0\}}$ is the characteristic function at x_0.

For $0 < r \leq 1$ and $\mu \in \mathbb{R}_{\mathcal{F}}$ define

$$[\mu]^r := \{x \in \mathbb{R} : \mu(x) \geq r\}$$

and

$$[\mu]^0 := \overline{\{x \in \mathbb{R} : \mu(x) \geq 0\}}.$$

Then it is well known that for each $r \in [0, 1]$, $[\mu]^r$ is a closed and bounded interval on $\mathbb{R}$ [35].

For $u, v \in \mathbb{R}_{\mathcal{F}}$ and $\lambda \in \mathbb{R}$, we define uniquely the sum $u \oplus v$ and the product $\lambda \odot u$ by

$$[u \oplus v]^r = [u]^r + [v]^r, \quad [\lambda \odot u]^r = \lambda [u]^r, \forall\, r \in [0, 1],$$

where

$[u]^r + [v]^r$ means the usual addition of two intervals (as substes of $\mathbb{R}$) and $\lambda [u]^r$ means the usual product between a scalar and a subset of $\mathbb{R}$ (see, e.g. [44]).

Notice $1 \odot u = u$ and it holds

$$u \oplus v = v \oplus u, \ \lambda \odot u = u \odot \lambda.$$

If $0 \leq r_1 \leq r_2 \leq 1$ then

$$[u]^{r_2} \subseteq [u]^{r_1}.$$

Actually $[u]^r = \left[u_-^{(r)}, u_+^{(r)}\right]$, where $u_-^{(r)} \leq u_+^{(r)}, u_-^{(r)}, u_+^{(r)} \in \mathbb{R}, \forall r \in [0, 1]$.

For $\lambda > 0$ one has $\lambda u_\pm^{(r)} = (\lambda \odot u)_\pm^{(r)}$, respectively.

Define $D : \mathbb{R}_\mathcal{F} \times \mathbb{R}_\mathcal{F} \to \mathbb{R}_\mathcal{F}$ by

$$D(u, v) := \sup_{r \in [0,1]} \max \left\{ \left| u_-^{(r)} - v_-^{(r)} \right|, \left| u_+^{(r)} - v_+^{(r)} \right| \right\},$$

where

$$[v]^r = \left[v_-^{(r)}, v_+^{(r)}\right]; \ u, v \in \mathbb{R}_\mathcal{F}.$$

We have that D is a metric on $\mathbb{R}_\mathcal{F}$.

Then $(\mathbb{R}_\mathcal{F}, D)$ is a complete metric space, see [44, 45].

Here $\overset{*}{\sum}$ stands for fuzzy summation and $\widetilde{o} := \chi_{\{0\}} \in \mathbb{R}_\mathcal{F}$ is the neural element with respect to $\oplus$, i.e.,

$$u \oplus \widetilde{0} = \widetilde{0} \oplus u = u, \ \forall u \in \mathbb{R}_\mathcal{F}.$$

Denote

$$D^*(f, g) = \sup_{x \in X \subseteq \mathbb{R}} D(f, g),$$

where $f, g : X \to \mathbb{R}_\mathcal{F}$.

We mention

Definition 11.2 Let $f : X \subseteq \mathbb{R} \to \mathbb{R}_\mathcal{F}$, X interval, we define the (first) fuzzy modulus of continuity of f by

$$\omega_1^{(\mathcal{F})}(f, \delta)_X = \sup_{x, y \in X, \ |x - y| \leq \delta} D(f(x), f(y)), \ \delta > 0.$$

When $g : X \subseteq \mathbb{R} \to \mathbb{R}$, we define

$$\omega_1(g, \delta) = \omega_1(g, \delta)_X = \sup_{x, y \in X, \ |x - y| \leq \delta} |g(x) - g(y)|.$$

We define by $C_\mathcal{F}^U(\mathbb{R})$ the space of fuzzy uniformly continuous functions from $\mathbb{R} \to \mathbb{R}_\mathcal{F}$, also $C_\mathcal{F}(\mathbb{R})$ is the space of fuzzy continuous functions on $\mathbb{R}$, and $C_b(\mathbb{R}, \mathbb{R}_\mathcal{F})$ is the fuzzy continuous and bounded functions.

We mention

Proposition 11.3 ([5]) *Let* $f \in C_{\mathcal{F}}^{U}(X)$. *Then* $\omega_1^{(\mathcal{F})}(f, \delta)_X < \infty$, *for any* $\delta > 0$.

By [9], p. 129 we have that $C_{\mathcal{F}}^{U}([a, b]) = C_{\mathcal{F}}([a, b])$, fuzzy continuous functions on $[a, b] \subset \mathbb{R}$.

Proposition 11.4 ([5]) *It holds*

$$\lim_{\delta \to 0} \omega_1^{(\mathcal{F})}(f, \delta)_X = \omega_1^{(\mathcal{F})}(f, 0)_X = 0,$$

iff $f \in C_{\mathcal{F}}^{U}(X)$, *where* X *is a compact interval.*

Proposition 11.5 ([5]) *Here* $[f]^r = \left[f_-^{(r)}, f_+^{(r)}\right]$, $r \in [0, 1]$. *Let* $f \in C_{\mathcal{F}}(\mathbb{R})$. *Then* $f_{\pm}^{(r)}$ *are equicontinuous with respect to* $r \in [0, 1]$ *over* $\mathbb{R}$, *respectively in* $\pm$.

Note 11.6 It is clear by Propositions 11.4, 11.5, that if $f \in C_{\mathcal{F}}^{U}(\mathbb{R})$, then $f_{\pm}^{(r)} \in C_U(\mathbb{R})$ (uniformly continuous on $\mathbb{R}$). Also if $f \in C_b(\mathbb{R}, \mathbb{R}_{\mathcal{F}})$ implies $f_{\pm}^{(r)} \in C_b(\mathbb{R})$ (continuous and bounded functions on $\mathbb{R}$).

Proposition 11.7 *Let* $f : \mathbb{R} \to \mathbb{R}_{\mathcal{F}}$. *Assume that* $\omega_1^{\mathcal{F}}(f, \delta)_X$, $\omega_1\left(f_-^{(r)}, \delta\right)_X$, $\omega_1\left(f_+^{(r)}, \delta\right)_X$ *are finite for any* $\delta > 0$, $r \in [0, 1]$, *where* X *any interval of* $\mathbb{R}$.
Then

$$\omega_1^{(\mathcal{F})}(f, \delta)_X = \sup_{r \in [0,1]} \max\left\{\omega_1\left(f_-^{(r)}, \delta\right)_X, \omega_1\left(f_+^{(r)}, \delta\right)_X\right\}.$$

Proof Similar to Proposition 14.15, p. 246 of [9]. ■

We need

Remark 11.8 ([3]) Here $r \in [0, 1]$, $x_i^{(r)}, y_i^{(r)} \in \mathbb{R}$, $i = 1, ..., m \in \mathbb{N}$. Suppose that

$$\sup_{r \in [0,1]} \max\left(x_i^{(r)}, y_i^{(r)}\right) \in \mathbb{R}, \text{ for } i = 1, ..., m.$$

Then one sees easily that

$$\sup_{r \in [0,1]} \max\left(\sum_{i=1}^{m} x_i^{(r)}, \sum_{i=1}^{m} y_i^{(r)}\right) \leq \sum_{i=1}^{m} \sup_{r \in [0,1]} \max\left(x_i^{(r)}, y_i^{(r)}\right). \tag{11.1}$$

We need

Definition 11.9 Let $x, y \in \mathbb{R}_{\mathcal{F}}$. If there exists $z \in \mathbb{R}_{\mathcal{F}} : x = y \oplus z$, then we call z the H-difference on x and y, denoted $x - y$.

Definition 11.10 ([43]) Let $T := [x_0, x_0 + \beta] \subset \mathbb{R}$, with $\beta > 0$. A function $f : T \to \mathbb{R}_{\mathcal{F}}$ is H-differentiable at $x \in T$ if there exists an $f'(x) \in \mathbb{R}_{\mathcal{F}}$ such that the limits (with respect to D)

$$\lim_{h \to 0+} \frac{f(x+h) - f(x)}{h}, \lim_{h \to 0+} \frac{f(x) - f(x-h)}{h} \tag{11.2}$$

exist and are equal to $f'(x)$.

We call f' the H-derivative or fuzzy derivative of f at x.

Above is assumed that the H-differences $f(x+h) - f(x)$, $f(x) - f(x-h)$ exists in $\mathbb{R}_{\mathcal{F}}$ in a neighborhood of x.

Higher order H-fuzzy derivatives are defined the obvious way, like in the real case.

We denote by $C_{\mathcal{F}}^N(\mathbb{R})$, $N \geq 1$, the space of all N-times continuously H-fuzzy differentiable functions from $\mathbb{R}$ into $\mathbb{R}_{\mathcal{F}}$, similarly is defined $C_{\mathcal{F}}^N([a,b])$, $[a,b] \subset \mathbb{R}$.

We mention

Theorem 11.11 ([37]) *Let $f : \mathbb{R} \to \mathbb{R}_{\mathcal{F}}$ be H-fuzzy differentiable. Let $t \in \mathbb{R}$, $0 \leq r \leq 1$. Clearly*

$$[f(t)]^r = \left[f(t)_-^{(r)}, f(t)_+^{(r)}\right] \subseteq \mathbb{R}.$$

Then $(f(t))_{\pm}^{(r)}$ are differentiable and

$$[f'(t)]^r = \left[\left(f(t)_-^{(r)}\right)', \left(f(t)_+^{(r)}\right)'\right].$$

I.e.

$$(f')_{\pm}^{(r)} = \left(f_{\pm}^{(r)}\right)', \quad \forall\, r \in [0,1].$$

Remark 11.12 ([4]) Let $f \in C_{\mathcal{F}}^N(\mathbb{R})$, $N \geq 1$. Then by Theorem 11.11 we obtain

$$\left[f^{(i)}(t)\right]^r = \left[\left(f(t)_-^{(r)}\right)^{(i)}, \left(f(t)_+^{(r)}\right)^{(i)}\right],$$

for $i = 0, 1, 2, ..., N$, and in particular we have that

$$\left(f^{(i)}\right)_{\pm}^{(r)} = \left(f_{\pm}^{(r)}\right)^{(i)},$$

for any $r \in [0,1]$, all $i = 0, 1, 2, ..., N$.

Note 11.13 ([4]) Let $f \in C_{\mathcal{F}}^N(\mathbb{R})$, $N \geq 1$. Then by Theorem 11.11 we have $f_{\pm}^{(r)} \in C^N(\mathbb{R})$, for any $r \in [0,1]$.

Items 11–13 are valid also on $[a,b]$.

By [9], p. 131, if $f \in C_{\mathcal{F}}([a,b])$, then f is a fuzzy bounded function.

We need also a particular case of the Fuzzy Henstock integral ($\delta(x) = \frac{\delta}{2}$), see [44].

Definition 11.14 ([34], p. 644) Let $f : [a,b] \to \mathbb{R}_{\mathcal{F}}$. We say that f is Fuzzy-Riemann integrable to $I \in \mathbb{R}_{\mathcal{F}}$ if for any $\varepsilon > 0$, there exists $\delta > 0$ such that for any division $P = \{[u,v]; \xi\}$ of $[a,b]$ with the norms $\Delta(P) < \delta$, we have

$$D\left(\sum_{P}^{*}(v-u) \odot f(\xi), I\right) < \varepsilon.$$

We write

$$I := (FR)\int_a^b f(x)\,dx. \tag{11.3}$$

We mention

Theorem 11.15 ([35]) *Let $f : [a,b] \to \mathbb{R}_{\mathcal{F}}$ be fuzzy continuous. Then*

$$(FR)\int_a^b f(x)\,dx$$

exists and belongs to $\mathbb{R}_{\mathcal{F}}$, furthermore it holds

$$\left[(FR)\int_a^b f(x)\,dx\right]^r = \left[\int_a^b (f)_-^{(r)}(x)\,dx, \int_a^b (f)_+^{(r)}(x)\,dx\right],$$

$\forall\, r \in [0,1]$.

For the definition of general fuzzy integral we follow [38] next.

Definition 11.16 Let (Ω, Σ, μ) be a complete σ-finite measure space. We call $F : \Omega \to R_{\mathcal{F}}$ measurable iff $\forall$ closed $B \subseteq \mathbb{R}$ the function $F^{-1}(B) : \Omega \to [0,1]$ defined by

$$F^{-1}(B)(w) := \sup_{x \in B} F(w)(x), \text{ all } w \in \Omega$$

is measurable, see [38].

Theorem 11.17 ([38]) *For $F : \Omega \to \mathbb{R}_{\mathcal{F}}$,*

$$F(w) = \left\{\left(F_-^{(r)}(w), F_+^{(r)}(w)\right) | 0 \le r \le 1\right\},$$

the following are equivalent

(1) F is measurable,

(2) $\forall\, r \in [0,1]$, $F_-^{(r)}$, $F_+^{(r)}$ are measurable.

Following [38], given that for each $r \in [0,1]$, $F_-^{(r)}$, $F_+^{(r)}$ are integrable we have that the parametrized representation

$$\left\{\left(\int_A F_-^{(r)} d\mu, \int_A F_+^{(r)} d\mu\right) | 0 \le r \le 1\right\} \tag{11.4}$$

is a fuzzy real number for each $A \in \Sigma$.

The last fact leads to

Definition 11.18 ([38]) A measurable function $F : \Omega \to \mathbb{R}_{\mathcal{F}}$,

$$F(w) = \left\{\left(F_-^{(r)}(w), F_+^{(r)}(w)\right) | 0 \le r \le 1\right\}$$

is integrable if for each $r \in [0,1]$, $F_\pm^{(r)}$ are integrable, or equivalently, if $F_\pm^{(0)}$ are integrable.

In this case, the fuzzy integral of F over $A \in \Sigma$ is defined by

$$\int_A F d\mu := \left\{\left(\int_A F_-^{(r)} d\mu, \int_A F_+^{(r)} d\mu\right) | 0 \le r \le 1\right\}.$$

By [38], F is integrable iff $w \to \|F(w)\|_{\mathcal{F}}$ is real-valued integrable.

Here denote

$$\|u\|_{\mathcal{F}} := D\left(u, \widetilde{0}\right), \ \forall\, u \in \mathbb{R}_{\mathcal{F}}.$$

We need also

Theorem 11.19 ([38]) *Let $F, G : \Omega \to \mathbb{R}_{\mathcal{F}}$ be integrable. Then*

(1) Let $a, b \in \mathbb{R}$, then $aF + bG$ is integrable and for each $A \in \Sigma$,

$$\int_A (aF + bG)\, d\mu = a \int_A F d\mu + b \int_A G d\mu;$$

(2) $D(F, G)$ is a real- valued integrable function and for each $A \in \Sigma$,

$$D\left(\int_A F d\mu, \int_A G d\mu\right) \le \int_A D(F, G)\, d\mu.$$

In particular,

$$\left\|\int_A F d\mu\right\|_{\mathcal{F}} \le \int_A \|F\|_{\mathcal{F}}\, d\mu.$$

Above μ could be the Lebesgue measure, with all the basic properties valid here too.

Basically here we have

$$\left[\int_A F d\mu\right]^r = \left[\int_A F_-^{(r)} d\mu, \int_A F_+^{(r)} d\mu\right], \tag{11.5}$$

i.e.

$$\left(\int_A F d\mu\right)_{\pm}^{(r)} = \int_A F_{\pm}^{(r)} d\mu, \ \forall r \in [0, 1].$$

We need

Definition 11.20 Let $\nu \geq 0$, $n = \lceil \nu \rceil$ ($\lceil \cdot \rceil$ is the ceiling of the number), $f \in AC^n$ $([a, b])$ (space of functions f with $f^{(n-1)} \in AC([a, b])$, absolutely continuous functions). We call left Caputo fractional derivative (see [30], pp. 49–52, [33, 41]) the function

$$D_{*a}^{\nu} f(x) = \frac{1}{\Gamma(n-\nu)} \int_a^x (x-t)^{n-\nu-1} f^{(n)}(t) \, dt, \tag{11.6}$$

$\forall x \in [a, b]$, where Γ is the gamma function $\Gamma(\nu) = \int_0^{\infty} e^{-t} t^{\nu-1} dt$, $\nu > 0$.

Notice $D_{*a}^{\nu} f \in L_1([a, b])$ and $D_{*a}^{\nu} f$ exists a.e. on $[a, b]$.

We set $D_{*a}^{0} f(x) = f(x)$, $\forall x \in [a, b]$.

Lemma 11.21 ([8]) *Let* $\nu > 0$, $\nu \notin \mathbb{N}$, $n = \lceil \nu \rceil$, $f \in C^{n-1}([a, b])$ *and* $f^{(n)} \in L_{\infty}$ $([a, b])$. *Then* $D_{*a}^{\nu} f(a) = 0$.

Definition 11.22 (see also [6, 32, 33]) Let $f \in AC^m([a, b])$, $m = \lceil \beta \rceil$, $\beta > 0$. The right Caputo fractional derivative of order $\beta > 0$ is given by

$$D_{b-}^{\beta} f(x) = \frac{(-1)^m}{\Gamma(m-\beta)} \int_x^b (\zeta - x)^{m-\beta-1} f^{(m)}(\zeta) \, d\zeta, \tag{11.7}$$

$\forall x \in [a, b]$. We set $D_{b-}^{0} f(x) = f(x)$. Notice that $D_{b-}^{\beta} f \in L_1([a, b])$ and $D_{b-}^{\beta} f$ exists a.e. on $[a, b]$.

Lemma 11.23 ([8]) *Let* $f \in C^{m-1}([a, b])$, $f^{(m)} \in L_{\infty}([a, b])$, $m = \lceil \beta \rceil$, $\beta > 0$, $\beta \notin \mathbb{N}$. *Then* $D_{b-}^{\beta} f(b) = 0$.

Convention 11.24 *We assume that*

$$D_{*x_0}^{\beta} f(x) = 0, \text{ for } x < x_0, \tag{11.8}$$

and

$$D_{x_0-}^{\beta} f(x) = 0, \text{ for } x > x_0, \tag{11.9}$$

for all $x, x_0 \in [a, b]$.

We mention

Proposition 11.25 ([8]) *Let* $f \in C^n([a, b])$, $n = \lceil \nu \rceil$, $\nu > 0$. *Then* $D_{*a}^{\nu} f(x)$ *is continuous in* $x \in [a, b]$.

Also we have

Proposition 11.26 ([8]) *Let $f \in C^m([a,b])$, $m = \lceil \beta \rceil$, $\beta > 0$. Then $D_{b-}^{\beta} f(x)$ is continuous in $x \in [a,b]$.*

We further mention

Proposition 11.27 ([8]) *Let $f \in C^{m-1}([a,b])$, $f^{(m)} \in L_\infty([a,b])$, $m = \lceil \beta \rceil$, $\beta > 0$ and*

$$D_{*x_0}^{\beta} f(x) = \frac{1}{\Gamma(m-\beta)} \int_{x_0}^{x} (x-t)^{m-\beta-1} f^{(m)}(t)\,dt, \tag{11.10}$$

for all $x, x_0 \in [a,b] : x \geq x_0$.

Then $D_{*x_0}^{\beta} f(x)$ is continuous in x_0.

Proposition 11.28 ([8]) *Let $f \in C^{m-1}([a,b])$, $f^{(m)} \in L_\infty([a,b])$, $m = \lceil \beta \rceil$, $\beta > 0$ and*

$$D_{x_0-}^{\beta} f(x) = \frac{(-1)^m}{\Gamma(m-\beta)} \int_{x}^{x_0} (\zeta-x)^{m-\beta-1} f^{(m)}(\zeta)\,d\zeta, \tag{11.11}$$

for all $x, x_0 \in [a,b] : x \leq x_0$.

Then $D_{x_0-}^{\beta} f(x)$ is continuous in x_0.

We need

Proposition 11.29 ([8]) *Let $g \in C([a,b])$, $0 < c < 1$, $x, x_0 \in [a,b]$. Define*

$$L(x,x_0) = \int_{x_0}^{x} (x-t)^{c-1} g(t)\,dt, \text{ for } x \geq x_0, \tag{11.12}$$

and $L(x,x_0) = 0$, for $x < x_0$.

Then L is jointly continuous in (x,x_0) on $[a,b]^2$.

We mention

Proposition 11.30 ([8]) *Let $g \in C([a,b])$, $0 < c < 1$, $x, x_0 \in [a,b]$. Define*

$$K(x,x_0) = \int_{x_0}^{x} (\zeta-x)^{c-1} g(\zeta)\,d\zeta, \text{ for } x \leq x_0, \tag{11.13}$$

and $K(x,x_0) = 0$, for $x > x_0$.

Then $K(x,x_0)$ is jointly continuous from $[a,b]^2$ into $\mathbb{R}$.

Based on Propositions 11.29, 11.30 we derive

Corollary 11.31 ([8]) *Let $f \in C^m([a,b])$, $m = \lceil \beta \rceil$, $\beta > 0$, $\beta \notin \mathbb{N}$, $x, x_0 \in [a,b]$. Then $D_{*x_0}^{\beta} f(x)$, $D_{x_0-}^{\beta} f(x)$ are jointly continuous functions in (x,x_0) from $[a,b]^2$ into $\mathbb{R}$.*

We need

Theorem 11.32 ([8]) *Let $f : [a,b]^2 \to \mathbb{R}$ be jointly continuous. Consider*

$$G(x) = \omega_1 (f(\cdot, x), \delta)_{[x,b]}, \tag{11.14}$$

$\delta > 0$, $x \in [a,b]$.

Then G is continuous in $x \in [a,b]$.

Also it holds

Theorem 11.33 ([8]) *Let $f : [a,b]^2 \to \mathbb{R}$ be jointly continous. Then*

$$H(x) = \omega_1 (f(\cdot, x), \delta)_{[a,x]}, \tag{11.15}$$

$x \in [a,b]$, is continuous in $x \in [a,b]$, $\delta > 0$.

So that for $f \in C^m([a,b])$, $m = \lceil \beta \rceil$, $\beta > 0$, $\beta \notin \mathbb{N}$, $x, x_0 \in [a,b]$, we have that $\omega_1 \left(D_{*x}^{\beta} f, h \right)_{[x,b]}$, $\omega_1 \left(D_{x-}^{\beta} f, h \right)_{[a,x]}$ are continuous functions in $x \in [a,b]$, $h > 0$ is fixed.

We make

Remark 11.34 ([8]) Let $f \in C^{n-1}([a,b])$, $f^{(n)} \in L_\infty([a,b])$, $n = \lceil \nu \rceil$, $\nu > 0$, $\nu \notin \mathbb{N}$. Then we have

$$\left| D_{*a}^{\nu} f(x) \right| \leq \frac{\left\| f^{(n)} \right\|_\infty}{\Gamma(n-\nu+1)} (x-a)^{n-\nu}, \quad \forall\, x \in [a,b]. \tag{11.16}$$

Thus we observe

$$\omega_1 \left(D_{*a}^{\nu} f, \delta \right) = \sup_{\substack{x,y \in [a,b] \\ |x-y| \leq \delta}} \left| D_{*a}^{\nu} f(x) - D_{*a}^{\nu} f(y) \right| \tag{11.17}$$

$$\leq \sup_{\substack{x,y \in [a,b] \\ |x-y| \leq \delta}} \left(\frac{\left\| f^{(n)} \right\|_\infty}{\Gamma(n-\nu+1)} (x-a)^{n-\nu} + \frac{\left\| f^{(n)} \right\|_\infty}{\Gamma(n-\nu+1)} (y-a)^{n-\nu} \right)$$

$$\leq \frac{2 \left\| f^{(n)} \right\|_\infty}{\Gamma(n-\nu+1)} (b-a)^{n-\nu}. \tag{11.18}$$

Consequently

$$\omega_1 \left(D_{*a}^{\nu} f, \delta \right) \leq \frac{2 \left\| f^{(n)} \right\|_\infty}{\Gamma(n-\nu+1)} (b-a)^{n-\nu}. \tag{11.19}$$

Similarly, let $f \in C^{m-1}([a,b])$, $f^{(m)} \in L_\infty([a,b])$, $m = \lceil \beta \rceil$, $\beta > 0$, $\beta \notin \mathbb{N}$, then

$$\omega_1 \left(D^{\beta}_{b-} f, \delta \right) \leq \frac{2 \left\| f^{(m)} \right\|_{\infty}}{\Gamma (m - \beta + 1)} (b - a)^{m-\beta} . \tag{11.20}$$

So for $f \in C^{m-1} ([a, b])$, $f^{(m)} \in L_{\infty} ([a, b])$, $m = \lceil \beta \rceil$, $\beta > 0$, $\beta \notin \mathbb{N}$, we find

$$\sup_{x_0 \in [a,b]} \omega_1 \left(D^{\beta}_{*x_0} f, \delta \right)_{[x_0, b]} \leq \frac{2 \left\| f^{(m)} \right\|_{\infty}}{\Gamma (m - \beta + 1)} (b - a)^{m-\beta} , \tag{11.21}$$

and

$$\sup_{x_0 \in [a,b]} \omega_1 \left(D^{\beta}_{x_0-} f, \delta \right)_{[a, x_0]} \leq \frac{2 \left\| f^{(m)} \right\|_{\infty}}{\Gamma (m - \beta + 1)} (b - a)^{m-\beta} . \tag{11.22}$$

By Proposition 15.114, p. 388 of [7], we get here that $D^{\beta}_{*x_0} f \in C ([x_0, b])$, and by [12] we obtain that $D^{\beta}_{x_0-} f \in C ([a, x_0])$.

We need

Definition 11.35 ([11]) Let $f \in C_{\mathcal{F}} ([a, b])$ (fuzzy continuous on $[a, b] \subset \mathbb{R}$), $\nu > 0$.

We define the Fuzzy Fractional left Riemann-Liouville operator as

$$J^{\nu}_{a} f (x) := \frac{1}{\Gamma (\nu)} \odot \int_{a}^{x} (x - t)^{\nu-1} \odot f (t) \, dt, \; x \in [a, b] , \tag{11.23}$$

$$J^{0}_{a} f := f.$$

Also, we define the Fuzzy Fractional right Riemann-Liouville operator as

$$I^{\nu}_{b-} f (x) := \frac{1}{\Gamma (\nu)} \odot \int_{x}^{b} (t - x)^{\nu-1} \odot f (t) \, dt, \; x \in [a, b] , \tag{11.24}$$

$$I^{0}_{b-} f := f.$$

We mention

Definition 11.36 ([11]) Let $f : [a, b] \to \mathbb{R}_{\mathcal{F}}$ is called fuzzy absolutely continuous iff $\forall \, \epsilon > 0$, $\exists \, \delta > 0$ for every finite, pairwise disjoint, family

$$(c_k, d_k)^{n}_{k=1} \subseteq (a, b) \;\; \text{with} \;\; \sum_{k=1}^{n} (d_k - c_k) < \delta$$

we get

$$\sum_{k=1}^{n} D (f (d_k) , f (c_k)) < \epsilon. \tag{11.25}$$

We denote the related space of functions by $AC_{\mathcal{F}}([a,b])$.

If $f \in AC_{\mathcal{F}}([a,b])$, then $f \in C_{\mathcal{F}}([a,b])$.

It holds

Proposition 11.37 ([11]) $f \in AC_{\mathcal{F}}([a,b]) \Leftrightarrow f_{\pm}^{(r)} \in AEC([a,b])$, $\forall\, r \in [0,1]$ *(absolutely equicontinuous).*

We need

Definition 11.38 ([11]) We define the Fuzzy Fractional left Caputo derivative, $x \in [a,b]$.

Let $f \in C_{\mathcal{F}}^{n}([a,b])$, $n = \lceil \nu \rceil$, $\nu > 0$ ($\lceil \cdot \rceil$ denotes the ceiling). We define

$$D_{*a}^{\nu \mathcal{F}} f(x) := \frac{1}{\Gamma(n-\nu)} \odot \int_a^x (x-t)^{n-\nu-1} \odot f^{(n)}(t)\, dt \tag{11.26}$$

$$= \left\{ \left(\frac{1}{\Gamma(n-\nu)} \int_a^x (x-t)^{n-\nu-1} \left(f^{(n)}\right)_-^{(r)}(t)\, dt, \right. \right.$$

$$\left. \left. \frac{1}{\Gamma(n-\nu)} \int_a^x (x-t)^{n-\nu-1} \left(f^{(n)}\right)_+^{(r)}(t)\, dt \right) | 0 \le r \le 1 \right\} =$$

$$= \left\{ \left(\frac{1}{\Gamma(n-\nu)} \int_a^x (x-t)^{n-\nu-1} \left(f_-^{(r)}\right)^{(n)}(t)\, dt, \right. \right.$$

$$\left. \left. \frac{1}{\Gamma(n-\nu)} \int_a^x (x-t)^{n-\nu-1} \left(f_+^{(r)}\right)^{(n)}(t)\, dt \right) | 0 \le r \le 1 \right\}. \tag{11.27}$$

So, we get

$$\left[D_{*a}^{\nu \mathcal{F}} f(x)\right]^r = \left[\left(\frac{1}{\Gamma(n-\nu)} \int_a^x (x-t)^{n-\nu-1} \left(f_-^{(r)}\right)^{(n)}(t)\, dt, \right. \right.$$

$$\left. \left. \frac{1}{\Gamma(n-\nu)} \int_a^x (x-t)^{n-\nu-1} \left(f_+^{(r)}\right)^{(n)}(t)\, dt \right) \right], \quad 0 \le r \le 1. \tag{11.28}$$

That is

$$\left(D_{*a}^{\nu \mathcal{F}} f(x)\right)_{\pm}^{(r)} = \frac{1}{\Gamma(n-\nu)} \int_a^x (x-t)^{n-\nu-1} \left(f_{\pm}^{(r)}\right)^{(n)}(t)\, dt = \left(D_{*a}^{\nu}\left(f_{\pm}^{(r)}\right)\right)(x),$$

see [7, 30].

I.e. we get that

$$\left(D_{*a}^{\nu \mathcal{F}} f(x)\right)_{\pm}^{(r)} = \left(D_{*a}^{\nu}\left(f_{\pm}^{(r)}\right)\right)(x), \tag{11.29}$$

$\forall\, x \in [a, b]$, in short

$$\left(D_{*a}^{\nu\mathcal{F}} f\right)_{\pm}^{(r)} = D_{*a}^{\nu}\left(f_{\pm}^{(r)}\right), \ \forall\, r \in [0, 1]. \tag{11.30}$$

We need

Lemma 11.39 ([11]) *$D_{*a}^{\nu\mathcal{F}} f(x)$ is fuzzy continuous in $x \in [a, b]$.*

We need

Definition 11.40 ([11]) We define the Fuzzy Fractional right Caputo derivative, $x \in [a, b]$.

Let $f \in C_{\mathcal{F}}^{n}([a, b])$, $n = \lceil \nu \rceil$, $\nu > 0$. We define

$$D_{b-}^{\nu\mathcal{F}} f(x) := \frac{(-1)^n}{\Gamma(n-\nu)} \odot \int_x^b (t-x)^{n-\nu-1} \odot f^{(n)}(t)\, dt$$

$$= \left\{\left(\frac{(-1)^n}{\Gamma(n-\nu)} \int_x^b (t-x)^{n-\nu-1} \left(f^{(n)}\right)_{-}^{(r)}(t)\, dt,\right.\right.$$

$$\left.\left.\frac{(-1)^n}{\Gamma(n-\nu)} \int_x^b (t-x)^{n-\nu-1} \left(f^{(n)}\right)_{+}^{(r)}(t)\, dt\right) \middle| 0 \le r \le 1\right\} \tag{11.31}$$

$$= \left\{\left(\frac{(-1)^n}{\Gamma(n-\nu)} \int_x^b (t-x)^{n-\nu-1} \left(f_{-}^{(r)}\right)^{(n)}(t)\, dt,\right.\right.$$

$$\left.\left.\frac{(-1)^n}{\Gamma(n-\nu)} \int_x^b (t-x)^{n-\nu-1} \left(f_{+}^{(r)}\right)^{(n)}(t)\, dt\right) \middle| 0 \le r \le 1\right\}.$$

We get

$$\left[D_{b-}^{\nu\mathcal{F}} f(x)\right]^r = \left[\left(\frac{(-1)^n}{\Gamma(n-\nu)} \int_x^b (t-x)^{n-\nu-1} \left(f_{-}^{(r)}\right)^{(n)}(t)\, dt,\right.\right.$$

$$\left.\left.\frac{(-1)^n}{\Gamma(n-\nu)} \int_x^b (t-x)^{n-\nu-1} \left(f_{+}^{(r)}\right)^{(n)}(t)\, dt\right)\right], \quad 0 \le r \le 1.$$

That is

$$\left(D_{b-}^{\nu\mathcal{F}} f(x)\right)_{\pm}^{(r)} = \frac{(-1)^n}{\Gamma(n-\nu)} \int_x^b (t-x)^{n-\nu-1} \left(f_{\pm}^{(r)}\right)^{(n)}(t)\, dt = \left(D_{b-}^{\nu}\left(f_{\pm}^{(r)}\right)\right)(x),$$

see [6].

I.e. we get that

$$\left(D_{b-}^{\nu\mathcal{F}} f(x)\right)_{\pm}^{(r)} = \left(D_{b-}^{\nu}\left(f_{\pm}^{(r)}\right)\right)(x), \tag{11.32}$$

$\forall\, x \in [a, b]$, in short

$$\left(D_{b-}^{\nu\mathcal{F}} f\right)_{\pm}^{(r)} = D_{b-}^{\nu}\left(f_{\pm}^{(r)}\right), \ \forall\, r \in [0, 1]. \tag{11.33}$$

Clearly,

$$D_{b-}^{\nu}\left(f_{-}^{(r)}\right) \leq D_{b-}^{\nu}\left(f_{+}^{(r)}\right), \ \forall\, r \in [0, 1].$$

We need

Lemma 11.41 ([11]) $D_{b-}^{\nu\mathcal{F}} f(x)$ *is fuzzy continuous in* $x \in [a, b]$.

11.3 Real Neural Network Approximation

11.3.1 About the Arctangent Activation Function Neural Networks

We consider the

$$\arctan x = \int_0^x \frac{dz}{1+z^2}, \ x \in \mathbb{R}. \tag{11.34}$$

We will be using

$$h(x) := \frac{2}{\pi} \arctan\left(\frac{\pi}{2} x\right) = \frac{2}{\pi} \int_0^{\frac{\pi x}{2}} \frac{dz}{1+z^2}, \ x \in \mathbb{R}, \tag{11.35}$$

which is a sigmoid type function and it is strictly increasing. We have that

$$h(0) = 0, \ h(-x) = -h(x), h(+\infty) = 1, \ h(-\infty) = -1,$$

and

$$h'(x) = \frac{4}{4+\pi^2 x^2} > 0, \text{ all } x \in \mathbb{R}. \tag{11.36}$$

We consider the activation function

$$\psi_1(x) := \frac{1}{4}\left(h(x+1) - h(x-1)\right), \ x \in \mathbb{R}, \tag{11.37}$$

and we notice that

$$\psi_1(-x) = \psi_1(x), \tag{11.38}$$

it is an even function.

Since $x + 1 > x - 1$, then $h(x + 1) > h(x - 1)$, and $\psi_1(x) > 0$, all $x \in \mathbb{R}$.

We see that

$$\psi_1(0) = \frac{1}{\pi} \arctan \frac{\pi}{2} \cong 0.319. \tag{11.39}$$

Let $x > 0$, we have that

$$\psi_1'(x) = \frac{1}{4}\left(h'(x+1) - h'(x-1)\right) =$$

$$\frac{-4\pi^2 x}{\left(4 + \pi^2 (x+1)^2\right)\left(4 + \pi^2 (x-1)^2\right)} < 0. \tag{11.40}$$

That is

$$\psi_1'(x) < 0, \text{ for } x > 0. \tag{11.41}$$

That is ψ_1 is strictly decreasing on $[0, \infty)$ and clearly is strictly increasing on $(-\infty, 0]$, and $\psi_1'(0) = 0$.

Observe that

$$\begin{gathered} \lim_{x \to +\infty} \psi_1(x) = \tfrac{1}{4}\left(h(+\infty) - h(+\infty)\right) = 0, \\ \text{and} \\ \lim_{x \to -\infty} \psi_1(x) = \tfrac{1}{4}\left(h(-\infty) - h(-\infty)\right) = 0. \end{gathered} \tag{11.42}$$

That is the x-axis is the horizontal asymptote on ψ_1.

All in all, ψ_1 is a bell symmetric function with maximum $\psi_1(0) \cong 18.31$.

We need

Theorem 11.42 ([20], p. 286) *We have that*

$$\sum_{i=-\infty}^{\infty} \psi_1(x - i) = 1, \quad \forall\, x \in \mathbb{R}. \tag{11.43}$$

Theorem 11.43 ([20], p. 287) *It holds*

$$\int_{-\infty}^{\infty} \psi_1(x)\, dx = 1. \tag{11.44}$$

So that $\psi_1(x)$ is a density function on $\mathbb{R}$.

We mention

Theorem 11.44 ([20], p. 288) *Let* $0 < \alpha < 1$, *and* $n \in \mathbb{N}$ *with* $n^{1-\alpha} > 2$. *It holds*

$$\sum_{\begin{cases} k=-\infty \\ : |nx-k| \geq n^{1-\alpha} \end{cases}}^{\infty} \psi_1 (nx-k) < \frac{2}{\pi^2 \left(n^{1-\alpha}-2\right)} =: c_1 (\alpha, n) . \tag{11.45}$$

Denote by $\lfloor \cdot \rfloor$ the integral part of the number and by $\lceil \cdot \rceil$ the ceiling of the number. We further mention

Theorem 11.45 ([20]) *Let* $x \in [a,b] \subset \mathbb{R}$ *and* $n \in \mathbb{N}$ *so that* $\lceil na \rceil \leq \lfloor nb \rfloor$. *It holds*

$$\frac{1}{\sum_{k=\lceil na \rceil}^{\lfloor nb \rfloor} \psi_1 (nx-k)} < \frac{1}{\psi_1 (1)} \cong 4.9737 =: \alpha_1, \ \forall \, x \in [a,b] . \tag{11.46}$$

We make

Remark 11.46 ([20]) (i) We have that

$$\lim_{n \to \infty} \sum_{k=\lceil na \rceil}^{\lfloor nb \rfloor} \psi_1 (nx-k) \neq 1, \tag{11.47}$$

for at least some $x \in [a,b]$.

(ii) For large enough n we always obtain $\lceil na \rceil \leq \lfloor nb \rfloor$. Also $a \leq \frac{k}{n} \leq b$, iff $\lceil na \rceil \leq k \leq \lfloor nb \rfloor$. In general it holds (by (11.43)) that

$$\sum_{k=\lceil na \rceil}^{\lfloor nb \rfloor} \psi_1 (nx-k) \leq 1. \tag{11.48}$$

We give

Definition 11.47 ([20]) Let $f \in C([a,b])$ and $n \in \mathbb{N} : \lceil na \rceil \leq \lfloor nb \rfloor$. We define the real positive linear neural network operator

$${}_1A_n (f,x) := \frac{\sum_{k=\lceil na \rceil}^{\lfloor nb \rfloor} f\left(\frac{k}{n}\right) \psi_1 (nx-k)}{\sum_{k=\lceil na \rceil}^{\lfloor nb \rfloor} \psi_1 (nx-k)}, \ x \in [a,b] . \tag{11.49}$$

Clearly here ${}_1A_n (f,x) \in C([a,b])$. In [20] we studied the pointwise and uniform convergence of ${}_1A_n (f,x)$ to $f(x)$ with rates.

We mention

Theorem 11.48 ([20]) *Let* $f \in C([a,b]), \ 0 < \alpha < 1, \ n \in \mathbb{N} : n^{1-\alpha} > 2, \ x \in [a,b]$. *Then*

(i)

$$\left|{}_1A_n(f,x)-f(x)\right|\leq 4.9737\left[\omega_1\left(f,\frac{1}{n^{\alpha}}\right)+\frac{4\|f\|_{\infty}}{\pi^2\left(n^{1-\alpha}-2\right)}\right]=:\rho_1(f), \tag{11.50}$$

and

(ii)

$$\left\|{}_1A_n(f)-f\right\|_{\infty}\leq\rho_1(f). \tag{11.51}$$

We notice that $\lim_{n\to\infty}{}_1A_n(f)=f$, *pointwise and uniformly.*

We mention

Theorem 11.49 ([20]) *Let* $f\in C^N([a,b])$, $n,N\in\mathbb{N}$, $0<\alpha<1$, $x\in[a,b]$ *and* $n^{1-\alpha}>2$. *Then*

(i)

$$\left|{}_1A_n(f,x)-f(x)\right|\leq 4.9737\left\{\sum_{j=1}^{N}\frac{\left|f^{(j)}(x)\right|}{j!}\left[\frac{1}{n^{\alpha j}}+\frac{2(b-a)^j}{\pi^2\left(n^{1-\alpha}-2\right)}\right]+\right. \tag{11.52}$$

$$\left.\left[\omega_1\left(f^{(N)},\frac{1}{n^{\alpha}}\right)\frac{1}{n^{\alpha N}N!}+\frac{4\left\|f^{(N)}\right\|_{\infty}(b-a)^N}{N!\pi^2\left(n^{1-\alpha}-2\right)}\right]\right\}=:\gamma_{11}(f,x),$$

(ii) assume further $f^{(j)}(x_0)=0$, $j=1,\ldots,N$, *for some* $x_0\in[a,b]$, *it holds*

$$\left|{}_1A_n(f,x_0)-f(x_0)\right|\leq 4.9737\cdot$$

$$\left\{\omega_1\left(f^{(N)},\frac{1}{n^{\alpha}}\right)\frac{1}{n^{\alpha N}N!}+\frac{4\left\|f^{(N)}\right\|_{\infty}(b-a)^N}{N!\pi^2\left(n^{1-\alpha}-2\right)}\right\}=:\gamma_{12}(f), \tag{11.53}$$

and

(iii)

$$\left\|{}_1A_n(f)-f\right\|_{\infty}\leq 4.9737\left\{\sum_{j=1}^{N}\frac{\left\|f^{(j)}\right\|_{\infty}}{j!}\left[\frac{1}{n^{\alpha j}}+\frac{2(b-a)^j}{\pi^2\left(n^{1-\alpha}-2\right)}\right]+\right.$$

$$\left.\left[\omega_1\left(f^{(N)},\frac{1}{n^{\alpha}}\right)\frac{1}{n^{\alpha N}N!}+\frac{4\left\|f^{(N)}\right\|_{\infty}(b-a)^N}{N!\pi^2\left(n^{1-\alpha}-2\right)}\right]\right\}=:\gamma_{13}(f). \tag{11.54}$$

Again we obtain $\lim_{n\to\infty}{}_1A_n(f)=f$, *pointwise and uniformly.*

We also mention the real valued fractional approximation result by neural networks.

Theorem 11.50 ([20]) *Let* $\alpha > 0$, $N = \lceil \alpha \rceil$, $\alpha \notin \mathbb{N}$, $f \in C^N([a,b])$, $0 < \beta < 1$, $x \in [a,b]$, $n \in \mathbb{N} : n^{1-\beta} > 2$. *Then*

(i)

$$\left| {}_1A_n(f,x) - \sum_{j=1}^{N-1} \frac{f^{(j)}(x)}{j!} \, {}_1A_n\left((\cdot - x)^j\right)(x) - f(x) \right| \leq$$

$$\frac{(4.9737)}{\Gamma(\alpha+1)} \left\{ \frac{\left(\omega_1\left(D_{x-}^{\alpha} f, \frac{1}{n^{\beta}}\right)_{[a,x]} + \omega_1\left(D_{*x}^{\alpha} f, \frac{1}{n^{\beta}}\right)_{[x,b]} \right)}{n^{\alpha\beta}} + \right.$$

$$\left. \frac{2}{\pi^2\left(n^{1-\beta}-2\right)} \left(\left\| D_{x-}^{\alpha} f \right\|_{\infty,[a,x]} (x-a)^{\alpha} + \left\| D_{*x}^{\alpha} f \right\|_{\infty,[x,b]} (b-x)^{\alpha} \right) \right\} =: \delta_{11}(f,x), \tag{11.55}$$

(ii) if $f^{(j)}(x) = 0$, *for* $j = 1, \ldots, N-1$, *we have*

$$\left| {}_1A_n(f,x) - f(x) \right| \leq \frac{(4.9737)}{\Gamma(\alpha+1)}$$

$$\left\{ \frac{\left(\omega_1\left(D_{x-}^{\alpha} f, \frac{1}{n^{\beta}}\right)_{[a,x]} + \omega_1\left(D_{*x}^{\alpha} f, \frac{1}{n^{\beta}}\right)_{[x,b]} \right)}{n^{\alpha\beta}} + \right.$$

$$\left. \frac{2}{\pi^2\left(n^{1-\beta}-2\right)} \left(\left\| D_{x-}^{\alpha} f \right\|_{\infty,[a,x]} (x-a)^{\alpha} + \left\| D_{*x}^{\alpha} f \right\|_{\infty,[x,b]} (b-x)^{\alpha} \right) \right\} =: \delta_{12}(f,x), \tag{11.56}$$

(iii)

$$\left| {}_1A_n(f,x) - f(x) \right| \leq (4.9737) \cdot$$

$$\left\{ \sum_{j=1}^{N-1} \frac{\left| f^{(j)}(x) \right|}{j!} \left\{ \frac{1}{n^{\beta j}} + (b-a)^j \frac{2}{\pi^2\left(n^{1-\beta}-2\right)} \right\} + \right.$$

$$\frac{1}{\Gamma(\alpha+1)} \left\{ \frac{\left(\omega_1\left(D_{x-}^{\alpha} f, \frac{1}{n^{\beta}}\right)_{[a,x]} + \omega_1\left(D_{*x}^{\alpha} f, \frac{1}{n^{\beta}}\right)_{[x,b]} \right)}{n^{\alpha\beta}} + \right.$$

$$\left.\left.\frac{2}{\pi^2\left(n^{1-\beta}-2\right)}\left(\left\|D_{x-}^{\alpha}f\right\|_{\infty,[a,x]}(x-a)^{\alpha}+\left\|D_{*x}^{\alpha}f\right\|_{\infty,[x,b]}(b-x)^{\alpha}\right)\right\}\right\} =: \delta_{13}(f,x), \tag{11.57}$$

$\forall\, x \in [a,b]$,

and

(iv)

$$\|{}_1A_n f - f\|_{\infty} \leq (4.9737)\cdot$$

$$\left\{\sum_{j=1}^{N-1}\frac{\|f^{(j)}\|_{\infty}}{j!}\left\{\frac{1}{n^{\beta j}}+(b-a)^j\frac{2}{\pi^2\left(n^{1-\beta}-2\right)}\right\}+\right.$$

$$\frac{1}{\Gamma(\alpha+1)}\left\{\frac{\left(\sup_{x\in[a,b]}\omega_1\left(D_{x-}^{\alpha}f,\frac{1}{n^{\beta}}\right)_{[a,x]}+\sup_{x\in[a,b]}\omega_1\left(D_{*x}^{\alpha}f,\frac{1}{n^{\beta}}\right)_{[x,b]}\right)}{n^{\alpha\beta}}+\right.$$

$$\left.\left.\frac{2}{\pi^2\left(n^{1-\beta}-2\right)}(b-a)^{\alpha}\left(\sup_{x\in[a,b]}\left\|D_{x-}^{\alpha}f\right\|_{\infty,[a,x]}+\sup_{x\in[a,b]}\left\|D_{*x}^{\alpha}f\right\|_{\infty,[x,b]}\right)\right\}\right\} =: \delta_{14}(f). \tag{11.58}$$

Above, when $N=1$ the sum $\sum_{j=1}^{N-1}\cdot=0$.

As we see here we obtain the real valued fractionally type pointwise and uniform convergence with rates of ${}_1A_n \to I$ the unit operator, as $n \to \infty$.

11.3.2 About the Algebraic Activation Function Neural Networks

Here see also [21].

We consider the generator algebraic function

$$\varphi(x) = \frac{x}{\sqrt[2m]{1+x^{2m}}},\quad m \in \mathbb{N},\, x \in \mathbb{R}, \tag{11.59}$$

which is a sigmoidal type of function and is a strictly increasing function.

We see that $\varphi(-x) = -\varphi(x)$ with $\varphi(0)=0$. We get that

$$\varphi'(x) = \frac{1}{\left(1+x^{2m}\right)^{\frac{2m+1}{2m}}} > 0,\ \forall\, x \in \mathbb{R}, \tag{11.60}$$

proving φ as strictly increasing over $\mathbb{R}$, $\varphi'(x)=\varphi'(-x)$. We easily find that $\lim_{x\to+\infty}\varphi(x)=1$, $\varphi(+\infty)=1$, and $\lim_{x\to-\infty}\varphi(x)=-1$, $\varphi(-\infty)=-1$.

We consider the activation function

$$\psi_2(x) = \frac{1}{4}\left[\varphi(x+1) - \varphi(x-1)\right]. \tag{11.61}$$

Clearly it is $\psi_2(x) = \psi_2(-x)$, $\forall\, x \in \mathbb{R}$, so that ψ_2 is an even function and symmetric with respect to the y-axis. Clearly $\psi_2(x) > 0$, $\forall\, x \in \mathbb{R}$.

Also it is

$$\psi_2(0) = \frac{1}{2\sqrt[2m]{2}}. \tag{11.62}$$

By [21], we have that $\psi_2'(x) < 0$ for $x > 0$. That is ψ_2 is strictly decreasing over $(0, +\infty)$.

Clearly, ψ_2 is strictly increasing over $(-\infty, 0)$ and $\psi_2'(0) = 0$.

Furthermore we obtain that

$$\lim_{x \to +\infty} \psi_2(x) = \frac{1}{4}\left[\varphi(+\infty) - \varphi(+\infty)\right] = 0, \tag{11.63}$$

and

$$\lim_{x \to -\infty} \psi_2(x) = \frac{1}{4}\left[\varphi(-\infty) - \varphi(-\infty)\right] = 0. \tag{11.64}$$

That is the x-axis is the horizontal asymptote of ψ_2.

Conclusion, ψ_2 is a bell shape symmetric function with maximum

$$\psi_2(0) = \frac{1}{2\sqrt[2m]{2}}, \quad m \in \mathbb{N}. \tag{11.65}$$

We need

Theorem 11.51 ([21]) *We have that*

$$\sum_{i=-\infty}^{\infty} \psi_2(x - i) = 1, \quad \forall\, x \in \mathbb{R}. \tag{11.66}$$

Theorem 11.52 ([21]) *It holds*

$$\int_{-\infty}^{\infty} \psi_2(x)\, dx = 1. \tag{11.67}$$

Theorem 11.53 ([21]) *Let* $0 < \alpha < 1$, *and* $n \in \mathbb{N}$ *with* $n^{1-\alpha} > 2$. *It holds*

$$\sum_{\substack{k=-\infty \\ : |nx-k| \geq n^{1-\alpha}}}^{\infty} \psi_2(nx - k) < \frac{1}{4m\left(n^{1-\alpha} - 2\right)^{2m}} =: c_2(\alpha, n), \quad m \in \mathbb{N}. \tag{11.68}$$

Theorem 11.54 ([21]) *Let* $[a, b] \subset \mathbb{R}$ *and* $n \in \mathbb{N}$ *so that* $\lceil na \rceil \leq \lfloor nb \rfloor$. *It holds*

$$\frac{1}{\sum_{k=\lceil na \rceil}^{\lfloor nb \rfloor} \psi_2 (nx - k)} < 2 \left(\sqrt[2m]{1 + 4^m} \right) =: \alpha_2, \tag{11.69}$$

$\forall\, x \in [a, b]$, $m \in \mathbb{N}$.

We make

Remark 11.55 ([21]) (1) We have that

$$\lim_{n \to \infty} \sum_{k=\lceil na \rceil}^{\lfloor nb \rfloor} \psi_2 (nx - k) \neq 1, \quad \text{for at least some } x \in [a, b]. \tag{11.70}$$

(2) In general it holds that

$$\sum_{k=\lceil na \rceil}^{\lfloor nb \rfloor} \psi_2 (nx - k) \leq 1. \tag{11.71}$$

We mention

Definition 11.56 ([21]) Let $f \in C([a, b])$ and $n \in \mathbb{N} : \lceil na \rceil \leq \lfloor nb \rfloor$. We introduce and define the real positive valued linear neural network operator

$${}_2A_n (f, x) := \frac{\sum_{k=\lceil na \rceil}^{\lfloor nb \rfloor} f\left(\frac{k}{n}\right) \psi_2 (nx - k)}{\sum_{k=\lceil na \rceil}^{\lfloor nb \rfloor} \psi_2 (nx - k)}, \quad x \in [a, b]. \tag{11.72}$$

Clearly here ${}_2A_n (f, x) \in C([a, b])$.

We mention here about the pointwise and uniform convergence of ${}_2A_n (f, x)$ to $f(x)$ with rates.

Theorem 11.57 ([21]) *Let* $f \in C([a, b])$, $0 < \alpha < 1$, $n \in \mathbb{N} : n^{1-\alpha} > 2$, $x \in [a, b]$, $m \in \mathbb{N}$. *Then*

(i)

$$|{}_2A_n (f, x) - f(x)| \leq \left(\sqrt[2m]{1 + 4^m} \right) \left[2\omega_1 \left(f, \frac{1}{n^\alpha} \right) + \frac{\|f\|_\infty}{m \left(n^{1-\alpha} - 2\right)^{2m}} \right] =: \rho_2 (f), \tag{11.73}$$

and

(ii)

$$\|{}_2A_n (f) - f\|_\infty \leq \rho_2 (f). \tag{11.74}$$

We get that $\lim_{n\to\infty} {}_2A_n(f) = f$, *pointwise and uniformly.*

In the next we mention the high order neural network real valued approximation result by using the smoothness of f.

Theorem 11.58 ([21]) *Let* $f \in C^N([a,b])$, $n, N, m \in \mathbb{N}$, $0 < \alpha < 1$, $x \in [a,b]$ *and* $n^{1-\alpha} > 2$. *Then*

(i)

$$|{}_2A_n(f,x) - f(x)| \leq \left(\sqrt[2m]{1+4^m}\right)\left\{\sum_{j=1}^{N} \frac{\left\|f^{(j)}(x)\right\|}{j!}\left[\frac{2}{n^{\alpha j}} + \frac{(b-a)^j}{2m\left(n^{1-\alpha}-2\right)^{2m}}\right] + \right. \tag{11.75}$$

$$\left.\left[\omega_1\left(f^{(N)}, \frac{1}{n^\alpha}\right)\frac{2}{n^{\alpha N}N!} + \frac{\left\|f^{(N)}\right\|_\infty (b-a)^N}{N!m\left(n^{1-\alpha}-2\right)^{2m}}\right]\right\} =: \gamma_{21}(f,x),$$

(ii) assume further $f^{(j)}(x_0) = 0$, $j = 1, ..., N$, *for some* $x_0 \in [a,b]$, *it holds*

$$|{}_2A_n(f,x_0) - f(x_0)| \leq \left(\sqrt[2m]{1+4^m}\right)\cdot \tag{11.76}$$

$$\left[\omega_1\left(f^{(N)}, \frac{1}{n^\alpha}\right)\frac{2}{n^{\alpha N}N!} + \frac{\left\|f^{(N)}\right\|_\infty (b-a)^N}{N!m\left(n^{1-\alpha}-2\right)^{2m}}\right] =: \gamma_{22}(f),$$

and

(iii)

$$\|{}_2A_n(f) - f\|_\infty \leq \left(\sqrt[2m]{1+4^m}\right)\left\{\sum_{j=1}^{N} \frac{\left\|f^{(j)}\right\|_\infty}{j!}\left[\frac{2}{n^{\alpha j}} + \frac{(b-a)^j}{2m\left(n^{1-\alpha}-2\right)^{2m}}\right] + \right.$$

$$\left.\left[\omega_1\left(f^{(N)}, \frac{1}{n^\alpha}\right)\frac{2}{n^{\alpha N}N!} + \frac{\left\|f^{(N)}\right\|_\infty (b-a)^N}{N!m\left(n^{1-\alpha}-2\right)^{2m}}\right]\right\} =: \gamma_{23}(f). \tag{11.77}$$

We derive that $\lim_{n\to\infty} {}_2A_n(f) = f$, *pointwise and uniformly.*

The corresponding fractional approximation result follows:

Theorem 11.59 ([21]) *Let* $\alpha > 0$, $N = \lceil \alpha \rceil$, $\alpha \notin \mathbb{N}$, $f \in C^N([a,b])$, $0 < \beta < 1$, $m \in \mathbb{N}$, $x \in [a,b]$, $n \in \mathbb{N} : n^{1-\beta} > 2$. *Then*

(i)

$$\left|{}_2A_n(f,x) - \sum_{j=1}^{N-1} \frac{f^{(j)}(x)}{j!}\, {}_2A_n\left((\cdot - x)^j\right)(x) - f(x)\right| \leq$$

$$\frac{2\left(\sqrt[2m]{1+4^m}\right)}{\Gamma(\alpha+1)}\left\{\frac{\left(\omega_1\left(D_{x-}^{\alpha}f,\frac{1}{n^{\beta}}\right)_{[a,x]}+\omega_1\left(D_{*x}^{\alpha}f,\frac{1}{n^{\beta}}\right)_{[x,b]}\right)}{n^{\alpha\beta}}+\right.$$

$$\left.\frac{1}{4m\left(n^{1-\beta}-2\right)^{2m}}\left(\left\|D_{x-}^{\alpha}f\right\|_{\infty,[a,x]}(x-a)^{\alpha}+\left\|D_{*x}^{\alpha}f\right\|_{\infty,[x,b]}(b-x)^{\alpha}\right)\right\}=:\delta_{21}(f,x), \tag{11.78}$$

(ii) if $f^{(j)}(x)=0$, *for* $j=1,\dots,N-1$, *we have*

$$\left|{}_2A_n(f,x)-f(x)\right|\leq\frac{2\left(\sqrt[2m]{1+4^m}\right)}{\Gamma(\alpha+1)}$$

$$\left\{\frac{\left(\omega_1\left(D_{x-}^{\alpha}f,\frac{1}{n^{\beta}}\right)_{[a,x]}+\omega_1\left(D_{*x}^{\alpha}f,\frac{1}{n^{\beta}}\right)_{[x,b]}\right)}{n^{\alpha\beta}}+\right.$$

$$\left.\frac{1}{4m\left(n^{1-\beta}-2\right)^{2m}}\left(\left\|D_{x-}^{\alpha}f\right\|_{\infty,[a,x]}(x-a)^{\alpha}+\left\|D_{*x}^{\alpha}f\right\|_{\infty,[x,b]}(b-x)^{\alpha}\right)\right\}=:\delta_{22}(f,x), \tag{11.79}$$

(iii)

$$\left|{}_2A_n(f,x)-f(x)\right|\leq 2\left(\sqrt[2m]{1+4^m}\right)\cdot$$

$$\left\{\sum_{j=1}^{N-1}\frac{\left\|f^{(j)}(x)\right\|}{j!}\left\{\frac{1}{n^{\beta j}}+\frac{(b-a)^j}{4m\left(n^{1-\beta}-2\right)^{2m}}\right\}+\right.$$

$$\frac{1}{\Gamma(\alpha+1)}\left\{\frac{\left(\omega_1\left(D_{x-}^{\alpha}f,\frac{1}{n^{\beta}}\right)_{[a,x]}+\omega_1\left(D_{*x}^{\alpha}f,\frac{1}{n^{\beta}}\right)_{[x,b]}\right)}{n^{\alpha\beta}}+\right.$$

$$\left.\left.\frac{1}{4m\left(n^{1-\beta}-2\right)^{2m}}\left(\left\|D_{x-}^{\alpha}f\right\|_{\infty,[a,x]}(x-a)^{\alpha}+\left\|D_{*x}^{\alpha}f\right\|_{\infty,[x,b]}(b-x)^{\alpha}\right)\right\}\right\}=:\delta_{23}(f,x), \tag{11.80}$$

$\forall\, x\in[a,b]$,

and

(iv)

$$\left\|{}_2A_nf-f\right\|_{\infty}\leq 2\left(\sqrt[2m]{1+4^m}\right)\cdot$$

$$\left\{\sum_{j=1}^{N-1} \frac{\left\|f^{(j)}\right\|_{\infty}}{j!} \left\{\frac{1}{n^{\beta j}} + \frac{(b-a)^j}{4m\left(n^{1-\beta}-2\right)^{2m}}\right\} + \right.$$

$$\frac{1}{\Gamma(\alpha+1)} \left\{ \frac{\left(\sup_{x\in[a,b]} \omega_1 \left(D_{x-}^{\alpha} f, \frac{1}{n^{\beta}}\right)_{[a,x]} + \sup_{x\in[a,b]} \omega_1 \left(D_{*x}^{\alpha} f, \frac{1}{n^{\beta}}\right)_{[x,b]} \right)}{n^{\alpha\beta}} + \right.$$

$$\left. \left. \frac{(b-a)^{\alpha}}{4m\left(n^{1-\beta}-2\right)^{2m}} \left(\sup_{x\in[a,b]} \left\| D_{x-}^{\alpha} f \right\|_{\infty,[a,x]} + \sup_{x\in[a,b]} \left\| D_{*x}^{\alpha} f \right\|_{\infty,[x,b]} \right) \right\} \right\} =: \delta_{24}(f). \tag{11.81}$$

Above, when $N = 1$ the sum $\sum_{j=1}^{N-1} \cdot = 0$.

As we see here we obtain the real valued fractionally type pointwise and uniform convergence with rates of ${}_2A_n \to I$ the unit operator, as $n \to \infty$.

11.3.3 About the Gudermannian Activation Function Neural Networks

See also [22, 42].

Here we consider $gd(x)$ the Gudermannian function [22], which is a sigmoid function, as a generator function:

$$\sigma(x) = 2\arctan\left(\tanh\left(\frac{x}{2}\right)\right) = \int_0^x \frac{dt}{\cosh t} =: gd(x), x \in \mathbb{R}. \tag{11.82}$$

Let the normalized generator sigmoid function

$$f(x) := \frac{4}{\pi}\sigma(x) = \frac{4}{\pi}\int_0^x \frac{dt}{\cosh t} = \frac{8}{\pi}\int_0^x \frac{1}{e^t + e^{-t}} dt, \ x \in \mathbb{R}. \tag{11.83}$$

Here

$$f'(x) = \frac{4}{\pi \cosh x} > 0, \ \forall\, x \in \mathbb{R},$$

hence f is strictly increasing on $\mathbb{R}$.

Notice that $\tanh(-x) = -\tanh x$ and $\arctan(-x) = -\arctan x$, $x \in \mathbb{R}$.

So, here the neural network activation function will be:

$$\psi_3(x) = \frac{1}{4}\left[f(x+1) - f(x-1)\right], x \in \mathbb{R}. \tag{11.84}$$

By [22], we get that

$$\psi_3(x) = \psi_3(-x), \ \forall x \in \mathbb{R}, \tag{11.85}$$

i.e. it is even and symmetric with respect to the y-axis. Here we have $f(+\infty) = 1$, $f(-\infty) = -1$ and $f(0) = 0$. Clearly it is

$$f(-x) = -f(x), \ \forall x \in \mathbb{R}, \tag{11.86}$$

an odd function, symmetric with respect to the origin. Since $x + 1 > x - 1$, and $f(x + 1) > f(x - 1)$, we obtain $\psi_3(x) > 0$, $\forall x \in \mathbb{R}$.

By [22], we have that

$$\psi_3(0) = \frac{2}{\pi} gd(1) \cong 0.551. \tag{11.87}$$

By [22] ψ_3 is strictly decreasing on $(0, +\infty)$, and strictly increasing on $(-\infty, 0)$, and $\psi_3'(0) = 0$.

Also we have that

$$\lim_{x \to +\infty} \psi_3(x) = \lim_{x \to -\infty} \psi_3(x) = 0, \tag{11.88}$$

that is the x-axis is the horizontal asymptote for ψ_3.

Conclusion, ψ_3 is a bell shaped symmetric function with maximum $\psi_3(0) \cong 0.551$.

We mention

Theorem 11.60 ([22]) *It holds that*

$$\sum_{i=-\infty}^{\infty} \psi_3(x - i) = 1, \forall x \in \mathbb{R}. \tag{11.89}$$

Theorem 11.61 ([22]) *We have that*

$$\int_{-\infty}^{\infty} \psi_3(x) \, dx = 1. \tag{11.90}$$

So $\psi_3(x)$ is a density function.

Theorem 11.62 ([22]) *Let $0 < \alpha < 1$, and $n \in \mathbb{N}$ with $n^{1-\alpha} > 2$. It holds*

$$\sum_{\substack{k = -\infty \\ : |nx - k| \geq n^{1-\alpha}}}^{\infty} \psi_3(nx - k) < \frac{4}{\pi e^{(n^{1-\alpha} - 2)}} = \frac{4e^2}{\pi e^{n^{1-\alpha}}} =: c_3(\alpha, n). \tag{11.91}$$

Theorem 11.63 ([22]) *Let $[a, b] \subset \mathbb{R}$ and $n \in \mathbb{N}$ so that $\lceil na \rceil \leq \lfloor nb \rfloor$. It holds*

$$\frac{1}{\sum_{k=\lceil na \rceil}^{\lfloor nb \rfloor} \psi_3 (nx - k)} < 2.412 =: \alpha_3, \tag{11.92}$$

$\forall\, x \in [a, b]$.

We make

Remark 11.64 ([22]) (1) We have that

$$\lim_{n \to \infty} \sum_{k=\lceil na \rceil}^{\lfloor nb \rfloor} \psi_3 (nx - k) \neq 1, \quad \text{for at least some } x \in [a, b]. \tag{11.93}$$

(2) In general it holds

$$\sum_{k=\lceil na \rceil}^{\lfloor nb \rfloor} \psi_3 (nx - k) \leq 1. \tag{11.94}$$

Definition 11.65 ([22]) Let $f \in C([a, b])$ and $n \in \mathbb{N} : \lceil na \rceil \leq \lfloor nb \rfloor$. We define the real positive valued linear neural network operator

$${}_3A_n (f, x) := \frac{\sum_{k=\lceil na \rceil}^{\lfloor nb \rfloor} f\left(\frac{k}{n}\right) \psi_3 (nx - k)}{\sum_{k=\lceil na \rceil}^{\lfloor nb \rfloor} \psi_3 (nx - k)}, \quad x \in [a, b]. \tag{11.95}$$

Clearly here ${}_3A_n (f, x) \in C([a, b])$.

We mention here about the pointwise and uniform convergence of ${}_3A_n (f, x)$ to $f(x)$ with rates.

Theorem 11.66 ([22]) *Let $f \in C([a, b])$, $0 < \alpha < 1$, $n \in \mathbb{N} : n^{1-\alpha} > 2$, $x \in [a, b]$. Then*

(i)

$$\left|{}_3A_n (f, x) - f(x)\right| \leq 2.412 \left[\omega_1 \left(f, \frac{1}{n^\alpha}\right) + \frac{8 \|f\|_\infty}{\pi e^{(n^{1-\alpha}-2)}}\right] =: \rho_3 (f), \tag{11.96}$$

and

(ii)

$$\left\|{}_3A_n (f) - f\right\|_\infty \leq \rho_3 (f). \tag{11.97}$$

We get that $\lim_{n \to \infty} {}_3A_n (f) = f$*, pointwise and uniformly.*

In the next we mention the high order neural network real valued approximation result by using the smoothness of f.

Theorem 11.67 ([22]) *Let $f \in C^N([a,b])$, $n, N \in \mathbb{N}$, $0 < \alpha < 1$, $x \in [a,b]$ and $n^{1-\alpha} > 2$. Then*

(i)

$$|{}_3A_n(f,x) - f(x)| \leq 2.412 \left\{ \sum_{j=1}^{N} \frac{\left|f^{(j)}(x)\right|}{j!} \left[\frac{1}{n^{\alpha j}} + \frac{4(b-a)^j}{\pi e^{(n^{1-\alpha}-2)}} \right] + \right. \quad (11.98)$$

$$\left. \left[\omega_1\left(f^{(N)}, \frac{1}{n^{\alpha}}\right) \frac{1}{n^{\alpha N} N!} + \frac{8\left\|f^{(N)}\right\|_{\infty}(b-a)^N}{N! \pi e^{(n^{1-\alpha}-2)}} \right] \right\} =: \gamma_{31}(f,x),$$

(ii) assume further $f^{(j)}(x_0) = 0$, $j = 1, ..., N$, for some $x_0 \in [a,b]$, it holds

$$|{}_3A_n(f,x_0) - f(x_0)| \leq 2.412 \cdot$$

$$\left\{ \omega_1\left(f^{(N)}, \frac{1}{n^{\alpha}}\right) \frac{1}{n^{\alpha N} N!} + \frac{8\left\|f^{(N)}\right\|_{\infty}(b-a)^N}{N! \pi e^{(n^{1-\alpha}-2)}} \right\} =: \gamma_{32}(f), \quad (11.99)$$

and

(iii)

$$\|{}_3A_n(f) - f\|_{\infty} \leq 2.412 \left\{ \sum_{j=1}^{N} \frac{\left\|f^{(j)}\right\|_{\infty}}{j!} \left[\frac{1}{n^{\alpha j}} + \frac{4(b-a)^j}{\pi e^{(n^{1-\alpha}-2)}} \right] + \right.$$

$$\left. \left[\omega_1\left(f^{(N)}, \frac{1}{n^{\alpha}}\right) \frac{1}{n^{\alpha N} N!} + \frac{8\left\|f^{(N)}\right\|_{\infty}(b-a)^N}{N! \pi e^{(n^{1-\alpha}-2)}} \right] \right\} =: \gamma_{33}(f). \quad (11.100)$$

Again we obtain $\lim_{n \to \infty} {}_3A_n(f) = f$, pointwise and uniformly.

We also mention the following real valued fractional approximation result by neural networks.

Theorem 11.68 ([22]) *Let $\alpha > 0$, $N = \lceil \alpha \rceil$, $\alpha \notin \mathbb{N}$, $f \in C^N([a,b])$, $0 < \beta < 1$, $x \in [a,b]$, $n \in \mathbb{N} : n^{1-\beta} > 2$. Then*

(i)

$$\left| {}_3A_n(f,x) - \sum_{j=1}^{N-1} \frac{f^{(j)}(x)}{j!} \, {}_3A_n\left((\cdot - x)^j\right)(x) - f(x) \right| \leq$$

$$\frac{2.412}{\Gamma(\alpha+1)} \left\{ \frac{\left(\omega_1\left(D_{x-}^{\alpha} f, \frac{1}{n^{\beta}}\right)_{[a,x]} + \omega_1\left(D_{*x}^{\alpha} f, \frac{1}{n^{\beta}}\right)_{[x,b]}\right)}{n^{\alpha\beta}} + \right.$$

$$\frac{4}{\pi e^{\left(n^{1-\alpha}-2\right)}}\left(\left\|D_{x-}^{\alpha}f\right\|_{\infty,[a,x]}(x-a)^{\alpha}+\left\|D_{*x}^{\alpha}f\right\|_{\infty,[x,b]}(b-x)^{\alpha}\right)\Bigg\} =: \delta_{31}(f,x), \tag{11.101}$$

(ii) if $f^{(j)}(x)=0$, for $j=1,...,N-1$, we have

$$|{}_3A_n(f,x)-f(x)| \leq \frac{2.412}{\Gamma(\alpha+1)}$$

$$\left\{\frac{\left(\omega_1\left(D_{x-}^{\alpha}f,\frac{1}{n^{\beta}}\right)_{[a,x]}+\omega_1\left(D_{*x}^{\alpha}f,\frac{1}{n^{\beta}}\right)_{[x,b]}\right)}{n^{\alpha\beta}}+\right.$$

$$\left.\frac{4}{\pi^2\left(n^{1-\beta}-2\right)}\left(\left\|D_{x-}^{\alpha}f\right\|_{\infty,[a,x]}(x-a)^{\alpha}+\left\|D_{*x}^{\alpha}f\right\|_{\infty,[x,b]}(b-x)^{\alpha}\right)\right\} =: \delta_{32}(f,x), \tag{11.102}$$

(iii)

$$|{}_3A_n(f,x)-f(x)| \leq 2.412 \cdot$$

$$\left\{\sum_{j=1}^{N-1}\frac{\left|f^{(j)}(x)\right|}{j!}\left\{\frac{1}{n^{\beta j}}+(b-a)^{j}\frac{4}{\pi e^{\left(n^{1-\alpha}-2\right)}}\right\}+\right.$$

$$\frac{1}{\Gamma(\alpha+1)}\left\{\frac{\left(\omega_1\left(D_{x-}^{\alpha}f,\frac{1}{n^{\beta}}\right)_{[a,x]}+\omega_1\left(D_{*x}^{\alpha}f,\frac{1}{n^{\beta}}\right)_{[x,b]}\right)}{n^{\alpha\beta}}+\right.$$

$$\left.\left.\frac{4}{\pi e^{\left(n^{1-\alpha}-2\right)}}\left(\left\|D_{x-}^{\alpha}f\right\|_{\infty,[a,x]}(x-a)^{\alpha}+\left\|D_{*x}^{\alpha}f\right\|_{\infty,[x,b]}(b-x)^{\alpha}\right)\right\}\right\} =: \delta_{33}(f,x), \tag{11.103}$$

$\forall\, x \in [a,b]$,

and

(iv)

$$\|{}_3A_n f - f\|_{\infty} \leq 2.412 \cdot$$

$$\left\{\sum_{j=1}^{N-1}\frac{\left\|f^{(j)}\right\|_{\infty}}{j!}\left\{\frac{1}{n^{\beta j}}+(b-a)^{j}\frac{4}{\pi e^{\left(n^{1-\alpha}-2\right)}}\right\}+\right.$$

$$\frac{1}{\Gamma(\alpha+1)}\left\{\frac{\left(\sup_{x\in[a,b]}\omega_1\left(D_{x-}^{\alpha}f,\frac{1}{n^{\beta}}\right)_{[a,x]}+\sup_{x\in[a,b]}\omega_1\left(D_{*x}^{\alpha}f,\frac{1}{n^{\beta}}\right)_{[x,b]}\right)}{n^{\alpha\beta}}+\right.$$

$$\left.\frac{4}{\pi e^{\left(n^{1-\alpha}-2\right)}}(b-a)^{\alpha}\left(\sup_{x\in[a,b]}\left\|D_{x-}^{\alpha}f\right\|_{\infty,[a,x]}+\sup_{x\in[a,b]}\left\|D_{*x}^{\alpha}f\right\|_{\infty,[x,b]}\right)\right\}\right\}=:\delta_{34}(f). \tag{11.104}$$

Above, when $N=1$ the sum $\sum_{j=1}^{N-1}\cdot=0$.

As we see here we obtain real valued fractionally type pointwise and uniform convergence with rates of ${}_3A_n \to I$ the unit operator, as $n\to\infty$.

11.3.4 About the Generalized Symmetrical Activation Function Neural Networks

Here we consider the generalized symmetrical sigmoid function [23, 31]

$$f_1(x)=\frac{x}{\left(1+|x|^{\mu}\right)^{\frac{1}{\mu}}},\ \mu>0,\ x\in\mathbb{R}. \tag{11.105}$$

This has applications in immunology and protection from disease together with probability theory. It is also called a symmetrical protection curve.

The parameter μ is a shape parameter controlling how fast the curve approaches the asymptotes for a given slope at the inflection point. When $\mu=1$ f_1 is the absolute sigmoid function, and when $\mu=2$, f_1 is the square root sigmoid function. When $\mu=1.5$ the function approximates the arctangent function, when $\mu=2.9$ it approximates the logistic function, and when $\mu=3.4$ it approximates the error function. Parameter μ is estimated in the likelihood maximization [31]. For more see [31].

Next we study the particular generator sigmoid function

$$f_2(x)=\frac{x}{\left(1+|x|^{\lambda}\right)^{\frac{1}{\lambda}}},\ \lambda \text{ is an odd number, } x\in\mathbb{R}. \tag{11.106}$$

We have that $f_2(0)=0$, and

$$f_2(-x)=-f_2(x), \tag{11.107}$$

so f_2 is symmetric with respect to zero.

When $x\geq 0$, we get that [23]

$$f_2'(x) = \frac{1}{\left(1 + x^{\lambda}\right)^{\frac{\lambda+1}{\lambda}}} > 0, \quad (11.108)$$

that is f_2 is strictly increasing on $[0, +\infty)$ and f_2 is strictly increasing on $(-\infty, 0]$. Hence f_2 is strictly increasing on $\mathbb{R}$.

We also have $f_2(+\infty) = f_2(-\infty) = 1$.

Let us consider the activation function [23]:

$$\psi_4(x) = \frac{1}{4}[f_2(x+1) - f_2(x-1)] =$$

$$\frac{1}{4}\left[\frac{(x+1)}{\left(1 + |x+1|^{\lambda}\right)^{\frac{1}{\lambda}}} - \frac{(x-1)}{\left(1 + |x-1|^{\lambda}\right)^{\frac{1}{\lambda}}}\right]. \quad (11.109)$$

Clearly it holds [23]

$$\psi_4(x) = \psi_4(-x), \ \forall\, x \in \mathbb{R}. \quad (11.110)$$

and

$$\psi_4(0) = \frac{1}{2\sqrt[\lambda]{2}}, \quad (11.111)$$

and $\psi_4(x) > 0$, $\forall\, x \in \mathbb{R}$.

Following [23], we have that ψ_4 is strictly decreasing over $[0, +\infty)$, and ψ_4 is strictly increasing on $(-\infty, 0]$, by ψ_4-symmetry with respect to y-axis, and $\psi_4'(0) = 0$.

Clearly it is

$$\lim_{x \to +\infty} \psi_4(x) = \lim_{x \to -\infty} \psi_4(x) = 0, \quad (11.112)$$

therefore the x-axis is the horizontal asymptote of $\psi_4(x)$.

The value

$$\psi_4(0) = \frac{1}{2\sqrt[\lambda]{2}}, \ \lambda \text{ is an odd number}, \quad (11.113)$$

is the maximum of ψ_4, which is a bell shaped function.

We need

Theorem 11.69 ([23]) *It holds*

$$\sum_{i=-\infty}^{\infty} \psi_4(x - i) = 1, \ \forall\, x \in \mathbb{R}. \quad (11.114)$$

Theorem 11.70 ([23]) *We have that*

$$\int_{-\infty}^{\infty} \psi_4(x)\,dx = 1. \tag{11.115}$$

So that $\psi_4(x)$ is a density function on $\mathbb{R}$.
We need

Theorem 11.71 ([23]) *Let* $0 < \alpha < 1$, *and* $n \in \mathbb{N}$ *with* $n^{1-\alpha} > 2$. *It holds*

$$\sum_{\substack{j=-\infty \\ : |nx-j| \geq n^{1-\alpha}}}^{\infty} \psi_4(nx - j) < \frac{1}{2\lambda\left(n^{1-\alpha}-2\right)^{\lambda}} =: c_4(\alpha, n), \tag{11.116}$$

where $\lambda \in \mathbb{N}$ *is an odd number.*

Theorem 11.72 ([23]) *Let* $[a, b] \subset \mathbb{R}$ *and* $n \in \mathbb{N}$ *so that* $\lceil na \rceil \leq \lfloor nb \rfloor$. *Then*

$$\frac{1}{\sum_{k=\lceil na \rceil}^{\lfloor nb \rfloor} \psi_4(|nx - k|)} < 2\sqrt[\lambda]{1 + 2^{\lambda}} =: \alpha_4, \tag{11.117}$$

where λ *is an odd number,* $\forall\, x \in [a, b]$.

We make

Remark 11.73 ([23]) (1) We have that

$$\lim_{n\to\infty} \sum_{k=\lceil na \rceil}^{\lfloor nb \rfloor} \psi_4(nx - k) \neq 1, \quad \text{for at least some } x \in [a, b]. \tag{11.118}$$

(2) In general it holds

$$\sum_{k=\lceil na \rceil}^{\lfloor nb \rfloor} \psi_4(nx - k) \leq 1. \tag{11.119}$$

We mention

Definition 11.74 ([23]) Let $f \in C([a, b])$ and $n \in \mathbb{N} : \lceil na \rceil \leq \lfloor nb \rfloor$. We define the real positive valued linear neural network operator

$${}_4A_n(f, x) := \frac{\sum_{k=\lceil na \rceil}^{\lfloor nb \rfloor} f\left(\frac{k}{n}\right) \psi_4(nx - k)}{\sum_{k=\lceil na \rceil}^{\lfloor nb \rfloor} \psi_4(nx - k)}, \quad x \in [a, b]. \tag{11.120}$$

Clearly here ${}_4A_n(f,x) \in C([a,b])$.

We mention results about the pointwise and uniform convergence of ${}_4A_n(f,x)$ to $f(x)$ with rates.

Theorem 11.75 ([23]) *Let $f \in C([a,b])$, $0<\alpha<1$, $n \in \mathbb{N}: n^{1-\alpha}>2$, $x \in [a,b]$, $\lambda \in \mathbb{N}$ is odd. Then*

(i)

$$\left|{}_4A_n(f,x) - f(x)\right| \leq 2\sqrt[\lambda]{1+2^\lambda}\left[\omega_1\left(f,\frac{1}{n^\alpha}\right) + \frac{\|f\|_\infty}{\lambda\left(n^{1-\alpha}-2\right)^\lambda}\right] =: \rho_4(f), \tag{11.121}$$

and

(ii)

$$\left\|{}_4A_n(f) - f\right\|_\infty \leq \rho_4(f). \tag{11.122}$$

We get that $\lim_{n\to\infty} {}_4A_n(f) = f$, pointwise and uniformly.

In the next we mention about the high order neural network real valued approximation result by using the smoothness of f.

Theorem 11.76 ([23]) *Let $f \in C^N([a,b])$, $n, N \in \mathbb{N}$, λ is odd, $0<\alpha<1$, $x \in [a,b]$ and $n^{1-\alpha}>2$. Then*

(i)

$$\left|{}_4A_n(f,x) - f(x)\right| \leq \sqrt[\lambda]{1+2^\lambda}\left\{\sum_{j=1}^{N}\frac{\left|f^{(j)}(x)\right|}{j!}\left[\frac{2}{n^{\alpha j}} + \frac{(b-a)^j}{\lambda\left(n^{1-\alpha}-2\right)^\lambda}\right] + \right. \tag{11.123}$$

$$\left.\left[\omega_1\left(f^{(N)},\frac{1}{n^\alpha}\right)\frac{2}{n^{\alpha N}N!} + \frac{2\left\|f^{(N)}\right\|_\infty (b-a)^N}{N!\lambda\left(n^{1-\alpha}-2\right)^\lambda}\right]\right\} =: \gamma_{41}(f,x),$$

(ii) assume further $f^{(j)}(x_0)=0$, $j=1,\ldots,N$, for some $x_0 \in [a,b]$, it holds

$$\left|{}_4A_n(f,x_0) - f(x_0)\right| \leq \sqrt[\lambda]{1+2^\lambda}\cdot$$

$$\left[\omega_1\left(f^{(N)},\frac{1}{n^\alpha}\right)\frac{2}{n^{\alpha N}N!} + \frac{2\left\|f^{(N)}\right\|_\infty (b-a)^N}{N!\lambda\left(n^{1-\alpha}-2\right)^\lambda}\right] =: \gamma_{42}(f), \tag{11.124}$$

and

(iii)

$$\|{}_4A_n(f) - f\|_\infty \leq \sqrt[\lambda]{1+2^\lambda}\left\{\sum_{j=1}^{N}\frac{\|f^{(j)}\|_\infty}{j!}\left[\frac{2}{n^{\alpha j}}+\frac{(b-a)^j}{\lambda\left(n^{1-\alpha}-2\right)^\lambda}\right]+\right.$$

$$\left.\left[\omega_1\left(f^{(N)},\frac{1}{n^\alpha}\right)\frac{2}{n^{\alpha N}N!}+\frac{2\|f^{(N)}\|_\infty(b-a)^N}{N!\lambda\left(n^{1-\alpha}-2\right)^\lambda}\right]\right\} =: \gamma_{43}(f). \quad (11.125)$$

We derive that $\lim_{n\to\infty} {}_4A_n(f) = f$*, pointwise and uniformly.*

Next we mention the corresponding real valued fractional approximation result by neural networks.

Theorem 11.77 ([23]) *Let* $\alpha > 0$, $N = \lceil\alpha\rceil$, $\alpha \notin \mathbb{N}$, $f \in C^N([a,b])$, $0 < \beta < 1$, λ *is odd,* $x \in [a,b]$, $n \in \mathbb{N} : n^{1-\beta} > 2$*. Then*

(i)

$$\left|{}_4A_n(f,x) - \sum_{j=1}^{N-1}\frac{f^{(j)}(x)}{j!}\,{}_4A_n\left((\cdot - x)^j\right)(x) - f(x)\right| \leq$$

$$\frac{2\sqrt[\lambda]{1+2^\lambda}}{\Gamma(\alpha+1)}\left\{\frac{\left(\omega_1\left(D_{x-}^\alpha f,\frac{1}{n^\beta}\right)_{[a,x]}+\omega_1\left(D_{*x}^\alpha f,\frac{1}{n^\beta}\right)_{[x,b]}\right)}{n^{\alpha\beta}}+\right.$$

$$\left.\frac{1}{2\lambda\left(n^{1-\beta}-2\right)^\lambda}\left(\left\|D_{x-}^\alpha f\right\|_{\infty,[a,x]}(x-a)^\alpha+\left\|D_{*x}^\alpha f\right\|_{\infty,[x,b]}(b-x)^\alpha\right)\right\} =: \delta_{41}(f,x), \quad (11.126)$$

(ii) if $f^{(j)}(x) = 0$*, for* $j = 1, \ldots, N-1$*, we have*

$$|{}_4A_n(f,x) - f(x)| \leq \frac{2\sqrt[\lambda]{1+2^\lambda}}{\Gamma(\alpha+1)}$$

$$\left\{\frac{\left(\omega_1\left(D_{x-}^\alpha f,\frac{1}{n^\beta}\right)_{[a,x]}+\omega_1\left(D_{*x}^\alpha f,\frac{1}{n^\beta}\right)_{[x,b]}\right)}{n^{\alpha\beta}}+\right.$$

$$\left.\frac{1}{2\lambda\left(n^{1-\beta}-2\right)^\lambda}\left(\left\|D_{x-}^\alpha f\right\|_{\infty,[a,x]}(x-a)^\alpha+\left\|D_{*x}^\alpha f\right\|_{\infty,[x,b]}(b-x)^\alpha\right)\right\} =: \delta_{42}(f,x), \quad (11.127)$$

(iii)

$$|{}_4A_n(f,x) - f(x)| \leq 2\sqrt[\lambda]{1+2^\lambda}\cdot$$

$$\left\{\sum_{j=1}^{N-1} \frac{\left\|f^{(j)}(x)\right\|}{j!}\left\{\frac{1}{n^{\beta j}}+\frac{(b-a)^{j}}{2\lambda\left(n^{1-\beta}-2\right)^{\lambda}}\right\}+\right.$$

$$\frac{1}{\Gamma(\alpha+1)}\left\{\frac{\left(\omega_{1}\left(D_{x-}^{\alpha} f, \frac{1}{n^{\beta}}\right)_{[a, x]}+\omega_{1}\left(D_{* x}^{\alpha} f, \frac{1}{n^{\beta}}\right)_{[x, b]}\right)}{n^{\alpha \beta}}+\right.$$

$$\left.\left.\frac{1}{2\lambda\left(n^{1-\beta}-2\right)^{\lambda}}\left(\left\|D_{x-}^{\alpha} f\right\|_{\infty,[a, x]}(x-a)^{\alpha}+\left\|D_{* x}^{\alpha} f\right\|_{\infty,[x, b]}(b-x)^{\alpha}\right)\right\}\right\}=: \delta_{43}(f, x), \tag{11.128}$$

$\forall\, x \in [a, b]$,

and

(iv)

$$\|{}_4A_n f - f\|_\infty \leq 2\sqrt[\lambda]{1+2^\lambda}.$$

$$\left\{\sum_{j=1}^{N-1} \frac{\left\|f^{(j)}\right\|_{\infty}}{j!}\left\{\frac{1}{n^{\beta j}}+\frac{(b-a)^{j}}{2\lambda\left(n^{1-\beta}-2\right)^{\lambda}}\right\}+\right.$$

$$\frac{1}{\Gamma(\alpha+1)}\left\{\frac{\left(\sup_{x \in[a, b]} \omega_{1}\left(D_{x-}^{\alpha} f, \frac{1}{n^{\beta}}\right)_{[a, x]}+\sup_{x \in[a, b]} \omega_{1}\left(D_{* x}^{\alpha} f, \frac{1}{n^{\beta}}\right)_{[x, b]}\right)}{n^{\alpha \beta}}+\right.$$

$$\left.\left.\frac{(b-a)^{\alpha}}{2\lambda\left(n^{1-\beta}-2\right)^{\lambda}}\left(\sup_{x \in[a, b]}\left\|D_{x-}^{\alpha} f\right\|_{\infty,[a, x]}+\sup_{x \in[a, b]}\left\|D_{* x}^{\alpha} f\right\|_{\infty,[x, b]}\right)\right\}\right\}=: \delta_{44}(f). \tag{11.129}$$

Above, when $N = 1$ the sum $\sum_{j=1}^{N-1} \cdot = 0$.

As we see here we obtain real valued fractionally type pointwise and uniform convergence with rates of ${}_4A_n \rightarrow I$ the unit operator, as $n \rightarrow \infty$.

For further related results see [24–26].

11.4 Main Results: Approximation by Fuzzy Quasi-interpolation Neural Network Operators with Respect to Arctangent, Algebraic, Gudermannian and Generalized Symmetrical Activation Functions

Let $f \in C_{\mathcal{F}}([a,b])$ (fuzzy continuous functions on $[a,b] \subset \mathbb{R}$), $n \in \mathbb{N}$. We define the following Fuzzy Quasi-Interpolation Neural Network operators

$$ {}_jA_n^{\mathcal{F}}(f,x) = \sum_{k=\lceil na \rceil}^{\lfloor nb \rfloor *} f\left(\frac{k}{n}\right) \odot \frac{\psi_j(nx-k)}{\sum_{k=\lceil na \rceil}^{\lfloor nb \rfloor} \psi_j(nx-k)}, \tag{11.130} $$

$\forall\, x \in [a,b]$, see also (11.49), (11.72), (11.95) and (11.120); $j = 1,2,3,4$.

The fuzzy sums in (11.130) are finite.

Let $r \in [0,1]$, we observe that

$$ {}_j\left[A_n^{\mathcal{F}}(f,x)\right]^r = \sum_{k=\lceil na \rceil}^{\lfloor nb \rfloor} \left[f\left(\frac{k}{n}\right)\right]^r \left(\frac{\psi_j(nx-k)}{\sum_{k=\lceil na \rceil}^{\lfloor nb \rfloor} \psi_j(nx-k)}\right) = $$

$$ \sum_{k=\lceil na \rceil}^{\lfloor nb \rfloor} \left[f_-^{(r)}\left(\frac{k}{n}\right), f_+^{(r)}\left(\frac{k}{n}\right)\right] \left(\frac{\psi_j(nx-k)}{\sum_{k=\lceil na \rceil}^{\lfloor nb \rfloor} \psi_j(nx-k)}\right) = $$

$$ \left[\sum_{k=\lceil na \rceil}^{\lfloor nb \rfloor} f_-^{(r)}\left(\frac{k}{n}\right) \left(\frac{\psi_j(nx-k)}{\sum_{k=\lceil na \rceil}^{\lfloor nb \rfloor} \psi_j(nx-k)}\right), \sum_{k=\lceil na \rceil}^{\lfloor nb \rfloor} f_+^{(r)}\left(\frac{k}{n}\right) \left(\frac{\psi_j(nx-k)}{\sum_{k=\lceil na \rceil}^{\lfloor nb \rfloor} \psi_j(nx-k)}\right)\right] \tag{11.131} $$

$$ = \left[{}_jA_n\left(f_-^{(r)}, x\right), {}_jA_n\left(f_+^{(r)}, x\right)\right]. \tag{11.132} $$

We have proved that

$$ \left({}_jA_n^{\mathcal{F}}(f,x)\right)_{\pm}^{(r)} = {}_jA_n\left(f_{\pm}^{(r)}, x\right), \quad j = 1,2,3,4, \tag{11.133} $$

respectively, $\forall\, r \in [0,1]$, $\forall\, x \in [a,b]$.

Therefore we get

$$D\left({}_jA_n^{\mathcal{F}}(f,x), f(x)\right) =$$

$$\sup_{r\in[0,1]} \max\left\{\left|{}_jA_n\left(f_-^{(r)},x\right) - f_-^{(r)}(x)\right|, \left|{}_jA_n\left(f_+^{(r)},x\right) - f_+^{(r)}(x)\right|\right\}, \quad (11.134)$$

$\forall\, x \in [a,b]$; $j = 1,2,3,4$.

We present our first fuzzy neural network approximation result.

Theorem 11.78 *Let $f \in C_{\mathcal{F}}([a,b])$, $0 < \alpha < 1$, $x \in [a,b]$, $n \in \mathbb{N}$ with $n^{1-\alpha} > 2$; $j = 1,2,3,4$. Then*

(1)

$$D\left({}_jA_n^{\mathcal{F}}(f,x), f(x)\right) \le \alpha_j \left[\omega_1^{(\mathcal{F})}\left(f, \frac{1}{n^\alpha}\right) + 2c_j(\alpha,n)\, D^*\left(f,\widetilde{o}\right)\right] =: {}_jT_{1n}, \quad (11.135)$$

and

(2)

$$D^*\left({}_jA_n^{\mathcal{F}}(f), f\right) \le {}_jT_{1n}. \quad (11.136)$$

We notice that $\lim_{n\to\infty}\left({}_jA_n^{\mathcal{F}}(f)\right)(x) \overset{D}{\to} f(x)$, $\lim_{n\to\infty} {}_jA_n^{\mathcal{F}}(f) \overset{D^*}{\to} f$, pointwise and uniformly.

Above α_j are as in (11.46), (11.69), (11.92), (11.117), and $c_j(\alpha,n)$ are as in (11.45), (11.68), (11.91), (11.116); $j = 1,2,3,4$, respectively.

Proof We have that $f_\pm^{(r)} \in C([a,b])$, $\forall\, r \in [0,1]$. Hence by (11.50), (11.73), (11.96) and (11.121), we obtain ($j = 1,2,3,4$)

$$\left|{}_jA_n\left(f_\pm^{(r)},x\right) - f_\pm^{(r)}(x)\right| \le \alpha_j\left[\omega_1\left(f_\pm^{(r)}, \frac{1}{n^\alpha}\right) + 2c_j(\alpha,n)\left\|f_\pm^{(r)}\right\|_\infty\right] \quad (11.137)$$

(by Proposition 11.7 and $\left\|f_\pm^{(r)}\right\|_\infty \le D^*\left(f,\widetilde{o}\right)$)

$$\le \alpha_j\left[\omega_1^{(\mathcal{F})}\left(f, \frac{1}{n^\alpha}\right) + 2c_j(\alpha,n)\, D^*\left(f,\widetilde{o}\right)\right]. \quad (11.138)$$

Taking into account (11.134) the theorem is proved. ∎

We also give

Theorem 11.79 *Let $f \in C_{\mathcal{F}}^N([a,b])$, $N \in \mathbb{N}$, $0 < \alpha < 1$, $x \in [a,b]$, $n \in \mathbb{N}$ with $n^{1-\alpha} > 2$; $j = 1,2,3,4$. Then*

(1)

$$D\left({}_jA_n^{\mathcal{F}}(f,x), f(x)\right) \le \alpha_j$$

$$\left\{\sum_{j_*=1}^{N} \frac{D\left(f^{(j_*)}(x), \widetilde{o}\right)}{j_*!}\left[\frac{1}{n^{\alpha j_*}} + c_j(\alpha, n)(b-a)^{j_*}\right] + \right.$$

$$\left.\left[\omega_1^{(\mathcal{F})}\left(f^{(N)}, \frac{1}{n^\alpha}\right)\frac{1}{n^{\alpha N} N!} + 2c_j(\alpha, n) D^*\left(f^{(N)}, \widetilde{o}\right)\frac{(b-a)^N}{N!}\right]\right\}, \quad (11.139)$$

(2) assume further that $f^{(j_*)}(x_0) = \widetilde{o}$, $j_* = 1, ..., N$, *for some* $x_0 \in [a, b]$, *it holds*

$$D\left({}_jA_n^{\mathcal{F}}(f, x_0), f(x_0)\right) \leq$$

$$\left[\omega_1^{(\mathcal{F})}\left(f^{(N)}, \frac{1}{n^\alpha}\right)\frac{1}{n^{\alpha N} N!} + 2c_j(\alpha, n) D^*\left(f^{(N)}, \widetilde{o}\right)\frac{(b-a)^N}{N!}\right], \quad (11.140)$$

notice here the extremely high rate of convergence $n^{-(N+1)\alpha}$,
(3)

$$D^*\left({}_jA_n^{\mathcal{F}}(f), f\right) \leq \alpha_j$$

$$\left\{\sum_{j_*=1}^{N} \frac{D^*\left(f^{(j_*)}, \widetilde{o}\right)}{j_*!}\left[\frac{1}{n^{\alpha j_*}} + c_j(\alpha, n)(b-a)^{j_*}\right] + \right.$$

$$\left.\left[\omega_1^{(\mathcal{F})}\left(f^{(N)}, \frac{1}{n^\alpha}\right)\frac{1}{n^{\alpha N} N!} + 2c_j(\alpha, n) D^*\left(f^{(N)}, \widetilde{o}\right)\frac{(b-a)^N}{N!}\right]\right\}. \quad (11.141)$$

Proof Since $f \in C_{\mathcal{F}}^N([a, b])$, $N \geq 1$, we have that $f_\pm^{(r)} \in C^N([a, b])$, $\forall\, r \in [0, 1]$; $j = 1, 2, 3, 4$. Using (11.52), (11.75), (11.98), (11.123), we get

$$\left|{}_jA_n\left(f_\pm^{(r)}, x\right) - f_\pm^{(r)}(x)\right| \leq \alpha_j \quad (11.74)$$

$$\left\{\sum_{j_*=1}^{N} \frac{\left|\left(f_\pm^{(r)}\right)^{(j_*)}(x)\right|}{j_*!}\left[\frac{1}{n^{\alpha j_*}} + c_j(\alpha, n)(b-a)^{j_*}\right] + \right.$$

$$\left.\left[\omega_1\left(\left(f_\pm^{(r)}\right)^{(N)}, \frac{1}{n^\alpha}\right)\frac{1}{n^{\alpha N} N!} + 2c_j(\alpha, n)\left\|\left(f_\pm^{(r)}\right)^{(N)}\right\|_\infty \frac{(b-a)^N}{N!}\right]\right\} \quad (11.142)$$

(by Remark 11.12)

$$= \alpha_j \left\{\sum_{j_*=1}^{N} \frac{\left|\left(f^{(j_*)}\right)_\pm^{(r)}(x)\right|}{j_*!}\left[\frac{1}{n^{\alpha j_*}} + c_j(\alpha, n)(b-a)^{j_*}\right] + \right.$$

$$\left[\omega_1\left(\left(f^{(N)}\right)_{\pm}^{(r)}, \frac{1}{n^{\alpha}}\right)\frac{1}{n^{\alpha N}N!} + 2c_j(\alpha, n)\left\|\left(f^{(N)}\right)_{\pm}^{(r)}\right\|_{\infty}\frac{(b-a)^N}{N!}\right]\right\} \leq$$

$$\alpha_j\left\{\sum_{j_*=1}^{N}\frac{D\left(f^{(j_*)}(x), \widetilde{o}\right)}{j_*!}\left[\frac{1}{n^{\alpha j_*}} + c_j(\alpha, n)(b-a)^{j_*}\right] +\right.$$

$$\left.\left[\omega_1^{(\mathcal{F})}\left(f^{(N)}, \frac{1}{n^{\alpha}}\right)\frac{1}{n^{\alpha N}N!} + 2c_j(\alpha, n)D^*\left(f^{(N)}, \widetilde{o}\right)\frac{(b-a)^N}{N!}\right]\right\}, \tag{11.143}$$

by Proposition 11.7, $\left\|\left(f^{(N)}\right)_{\pm}^{(r)}\right\|_{\infty} \leq D^*\left(f^{(N)}, \widetilde{o}\right)$ and apply (11.134).

The theorem is proved. ■

Next we present

Theorem 11.80 *Let* $\alpha > 0$, $N = \lceil \alpha \rceil$, $\alpha \notin \mathbb{N}$, $f \in C_{\mathcal{F}}^{N}([a, b])$, $0 < \beta < 1$, $x \in [a, b]$, $n \in \mathbb{N}$, $n^{1-\beta} > 2$; $j = 1, 2, 3, 4$. *Then*

(i)

$$D\left({}_jA_n^{\mathcal{F}}(f, x), f(x)\right) \leq$$

$$\alpha_j\left\{\sum_{j_*=1}^{N-1}\frac{D\left(f^{(j_*)}(x), \widetilde{o}\right)}{j_*!}\left[\frac{1}{n^{\beta j_*}} + c_j(\beta, n)(b-a)^{j_*}\right] +\right.$$

$$\frac{1}{\Gamma(\alpha+1)}\left\{\frac{\left[\omega_1^{(\mathcal{F})}\left(\left(D_{x-}^{\alpha\mathcal{F}}f\right), \frac{1}{n^{\beta}}\right)_{[a,x]} + \omega_1^{(\mathcal{F})}\left(\left(D_{*x}^{\alpha\mathcal{F}}f\right), \frac{1}{n^{\beta}}\right)_{[x,b]}\right]}{n^{\alpha\beta}} + \right. \tag{11.144}$$

$$\left.\left.c_j(\beta, n)\left[D^*\left(\left(D_{x-}^{\alpha\mathcal{F}}f\right), \widetilde{o}\right)_{[a,x]}(x-a)^{\alpha} + D^*\left(\left(D_{*x}^{\alpha\mathcal{F}}f\right), \widetilde{o}\right)_{[x,b]}(b-x)^{\alpha}\right]\right\}\right\},$$

(ii) if $f^{(j)}(x_0) = 0$, $j = 1, ..., N-1$, *for some* $x_0 \in [a, b]$, *we have*

$$D\left({}_jA_n^{\mathcal{F}}(f, x_0), f(x_0)\right) \leq$$

$$\frac{\alpha_j}{\Gamma(\alpha+1)}\left\{\frac{\left[\omega_1^{(\mathcal{F})}\left(\left(D_{x_0-}^{\alpha\mathcal{F}}f\right), \frac{1}{n^{\beta}}\right)_{[a,x_0]} + \omega_1^{(\mathcal{F})}\left(\left(D_{*x_0}^{\alpha\mathcal{F}}f\right), \frac{1}{n^{\beta}}\right)_{[x_0,b]}\right]}{n^{\alpha\beta}} + \right. \tag{11.145}$$

$$\left.c_j(\beta, n)\left[D^*\left(\left(D_{x_0-}^{\alpha\mathcal{F}}f\right), \widetilde{o}\right)_{[a,x_0]}(x_0-a)^{\alpha} + D^*\left(\left(D_{*x_0}^{\alpha\mathcal{F}}f\right), \widetilde{o}\right)_{[x_0,b]}(b-x_0)^{\alpha}\right]\right\},$$

when $\alpha > 1$ notice here the extremely high rate of convergence at $n^{-(\alpha+1)\beta}$,
and
(iii)

$$D^{*}\left({}_{j}A_{n}^{\mathcal{F}}(f), f\right) \leq$$

$$\alpha_{j}\left\{\sum_{j_{*}=1}^{N-1} \frac{D^{*}\left(f^{(j_{*})}, \widetilde{o}\right)}{j_{*}!}\left[\frac{1}{n^{\beta j_{*}}}+c_{j}(\beta, n)(b-a)^{j_{*}}\right]+\right.$$

$$\frac{1}{\Gamma(\alpha+1)}\left\{\frac{\left[\sup_{x\in[a,b]} \omega_{1}^{(\mathcal{F})}\left(\left(D_{x-}^{\alpha\mathcal{F}} f\right), \frac{1}{n^{\beta}}\right)_{[a,x]}+\sup_{x\in[a,b]} \omega_{1}^{(\mathcal{F})}\left(\left(D_{*x}^{\alpha\mathcal{F}} f\right), \frac{1}{n^{\beta}}\right)_{[x,b]}\right]}{n^{\alpha\beta}}+\right.$$

$$\left.\left.c_{j}(\beta, n)(b-a)^{\alpha}\left[\sup_{x\in[a,b]} D^{*}\left(\left(D_{x-}^{\alpha\mathcal{F}} f\right), \widetilde{o}\right)_{[a,x]}+\sup_{x\in[a,b]} D^{*}\left(\left(D_{*x}^{\alpha\mathcal{F}} f\right), \widetilde{o}\right)_{[x,b]}\right]\right\}\right\}, \tag{11.146}$$

above, when $N = 1$ the sum $\sum_{j=1}^{N-1} \cdot = 0$.

As we see here we obtain fractionally the fuzzy pointwise and uniform convergence with rates of ${}_{j}A_{n}^{\mathcal{F}} \to I$ the unit operator, as $n \to \infty$.

Proof Here $f_{\pm}^{(r)} \in C^{N}([a, b])$, $\forall\, r \in [0, 1]$, and $D_{x-}^{\alpha\mathcal{F}} f$, $D_{*x}^{\alpha\mathcal{F}} f$ are fuzzy continuous over $[a, b]$, $\forall\, x \in [a, b]$, so that $\left(D_{x-}^{\alpha\mathcal{F}} f\right)_{\pm}^{(r)}, \left(D_{*x}^{\alpha\mathcal{F}} f\right)_{\pm}^{(r)} \in C([a, b])$, $\forall\, r \in [0, 1]$, $\forall\, x \in [a, b]$.

We observe by (11.57), (11.80), (11.103), (11.128), $\forall\, x \in [a, b]$, that (respectively in $\pm$)

$$\left|{}_{j}A_{n}\left(f_{\pm}^{(r)}, x\right) - f_{\pm}^{(r)}(x)\right| \leq \alpha_{j} \cdot$$

$$\left\{\sum_{j_{*}=1}^{N-1} \frac{\left|\left(f_{\pm}^{(r)}\right)^{(j_{*})}(x)\right|}{j_{*}!}\left\{\frac{1}{n^{\beta j_{*}}}+c_{j}(\beta, n)(b-a)^{j_{*}}\right\}+\right. \tag{11.147}$$

$$\frac{1}{\Gamma(\alpha+1)}\left\{\frac{\left(\omega_{1}\left(D_{x-}^{\alpha}\left(f_{\pm}^{(r)}\right), \frac{1}{n^{\beta}}\right)_{[a,x]}+\omega_{1}\left(D_{*x}^{\alpha}\left(f_{\pm}^{(r)}\right), \frac{1}{n^{\beta}}\right)_{[x,b]}\right)}{n^{\alpha\beta}}+\right.$$

$$\left.\left.c_{j}(\beta, n)\left(\left\|D_{x-}^{\alpha}\left(f_{\pm}^{(r)}\right)\right\|_{\infty,[a,x]}(x-a)^{\alpha}+\left\|D_{*x}^{\alpha}\left(f_{\pm}^{(r)}\right)\right\|_{\infty,[x,b]}(b-x)^{\alpha}\right)\right\}\right\} =$$

(by Remark 11.12, (11.30), (11.33))

$$\alpha_j \left\{ \sum_{j_*=1}^{N-1} \frac{\left| \left(f^{(j_*)}(x) \right)_{\pm}^{(r)} \right|}{j_*!} \left\{ \frac{1}{n^{\beta j_*}} + c_j(\beta, n)(b-a)^{j_*} \right\} + \right.$$

$$\frac{1}{\Gamma(\alpha+1)} \left\{ \frac{\left(\omega_1 \left(\left(D_{x-}^{\alpha \mathcal{F}} f \right)_{\pm}^{(r)}, \frac{1}{n^{\beta}} \right)_{[a,x]} + \omega_1 \left(\left(D_{*x}^{\alpha \mathcal{F}} f \right)_{\pm}^{(r)}, \frac{1}{n^{\beta}} \right)_{[x,b]} \right)}{n^{\alpha \beta}} + \right. \quad (11.148)$$

$$\left. \left. c_j(\beta, n) \left(\left\| \left(D_{x-}^{\alpha \mathcal{F}} f \right)_{\pm}^{(r)} \right\|_{\infty,[a,x]} (x-a)^{\alpha} + \left\| \left(D_{*x}^{\alpha \mathcal{F}} f \right)_{\pm}^{(r)} \right\|_{\infty,[x,b]} (b-x)^{\alpha} \right) \right\} \right\} \leq$$

$$\alpha_j \left\{ \sum_{j_*=1}^{N-1} \frac{D\left(f^{(j_*)}(x), \widetilde{o} \right)}{j_*!} \left\{ \frac{1}{n^{\beta j_*}} + c_j(\beta, n)(b-a)^{j_*} \right\} + \right.$$

$$\frac{1}{\Gamma(\alpha+1)} \left\{ \frac{\left[\omega_1^{(\mathcal{F})} \left(\left(D_{x-}^{\alpha \mathcal{F}} f \right), \frac{1}{n^{\beta}} \right)_{[a,x]} + \omega_1^{(\mathcal{F})} \left(\left(D_{*x}^{\alpha \mathcal{F}} f \right), \frac{1}{n^{\beta}} \right)_{[x,b]} \right]}{n^{\alpha \beta}} + \right. \quad (11.149)$$

$$\left. \left. c_j(\beta, n) \left[D^* \left(\left(D_{x-}^{\alpha \mathcal{F}} f \right), \widetilde{o} \right)_{[a,x]} (x-a)^{\alpha} + D^* \left(\left(D_{*x}^{\alpha \mathcal{F}} f \right), \widetilde{o} \right)_{[x,b]} (b-x)^{\alpha} \right] \right\} \right\},$$

along with (11.134) proving all inequalities of theorem.

Here we notice that

$$\left(D_{x-}^{\alpha \mathcal{F}} f \right)_{\pm}^{(r)}(t) = \left(D_{x-}^{\alpha} \left(f_{\pm}^{(r)} \right) \right)(t)$$

$$= \frac{(-1)^N}{\Gamma(N-\alpha)} \int_t^x (s-t)^{N-\alpha-1} \left(f_{\pm}^{(r)} \right)^{(N)}(s)\, ds,$$

where $a \leq t \leq x$.

Hence

$$\left| \left(D_{x-}^{\alpha \mathcal{F}} f \right)_{\pm}^{(r)}(t) \right| \leq \frac{1}{\Gamma(N-\alpha)} \int_t^x (s-t)^{N-\alpha-1} \left| \left(f_{\pm}^{(r)} \right)^{(N)}(s) \right| ds$$

$$\leq \frac{\left\| \left(f^{(N)} \right)_{\pm}^{(r)} \right\|_{\infty}}{\Gamma(N-\alpha+1)} (b-a)^{N-\alpha} \leq \frac{D^* \left(f^{(N)}, \widetilde{o} \right)}{\Gamma(N-\alpha+1)} (b-a)^{N-\alpha}.$$

So we have

$$\left| \left(D_{x-}^{\alpha \mathcal{F}} f \right)_{\pm}^{(r)}(t) \right| \leq \frac{D^* \left(f^{(N)}, \widetilde{o} \right)}{\Gamma(N-\alpha+1)} (b-a)^{N-\alpha},$$

all $a \leq t \leq x$.

And it holds

$$\left\| \left(D_{x-}^{\alpha \mathcal{F}} f \right)_{\pm}^{(r)} \right\|_{\infty,[a,x]} \leq \frac{D^* \left(f^{(N)}, \widetilde{o} \right)}{\Gamma (N - \alpha + 1)} (b - a)^{N-\alpha}, \tag{11.150}$$

that is

$$D^* \left(\left(D_{x-}^{\alpha \mathcal{F}} f \right), \widetilde{o} \right)_{[a,x]} \leq \frac{D^* \left(f^{(N)}, \widetilde{o} \right)}{\Gamma (N - \alpha + 1)} (b - a)^{N-\alpha},$$

and

$$\sup_{x \in [a,b]} D^* \left(\left(D_{x-}^{\alpha \mathcal{F}} f \right), \widetilde{o} \right)_{[a,x]} \leq \frac{D^* \left(f^{(N)}, \widetilde{o} \right)}{\Gamma (N - \alpha + 1)} (b - a)^{N-\alpha} < \infty. \tag{11.151}$$

Similarly we have

$$\left(D_{*x}^{\alpha \mathcal{F}} f \right)_{\pm}^{(r)} (t) = \left(D_{*x}^{\alpha} \left(f_{\pm}^{(r)} \right) \right) (t)$$

$$= \frac{1}{\Gamma (N - \alpha)} \int_x^t (t - s)^{N-\alpha-1} \left(f_{\pm}^{(r)} \right)^{(N)} (s) \, ds,$$

where $x \leq t \leq b$.

Hence

$$\left| \left(D_{*x}^{\alpha \mathcal{F}} f \right)_{\pm}^{(r)} (t) \right| \leq \frac{1}{\Gamma (N - \alpha)} \int_x^t (t - s)^{N-\alpha-1} \left| \left(f^{(N)} \right)_{\pm}^{(r)} (s) \right| ds \leq$$

$$\frac{\left\| \left(f^{(N)} \right)_{\pm}^{(r)} \right\|_{\infty}}{\Gamma (N - \alpha + 1)} (b - a)^{N-\alpha} \leq \frac{D^* \left(f^{(N)}, \widetilde{o} \right)}{\Gamma (N - \alpha + 1)} (b - a)^{N-\alpha},$$

$x \leq t \leq b$.

So we have

$$\left\| \left(D_{*x}^{\alpha \mathcal{F}} f \right)_{\pm}^{(r)} \right\|_{\infty,[x,b]} \leq \frac{D^* \left(f^{(N)}, \widetilde{o} \right)}{\Gamma (N - \alpha + 1)} (b - a)^{N-\alpha}, \tag{11.152}$$

that is

$$D^* \left(\left(D_{*x}^{\alpha \mathcal{F}} f \right), \widetilde{o} \right)_{[x,b]} \leq \frac{D^* \left(f^{(N)}, \widetilde{o} \right)}{\Gamma (N - \alpha + 1)} (b - a)^{N-\alpha},$$

and

$$\sup_{x \in [a,b]} D^* \left(\left(D_{*x}^{\alpha \mathcal{F}} f \right), \widetilde{o} \right)_{[x,b]} \leq \frac{D^* \left(f^{(N)}, \widetilde{o} \right)}{\Gamma (N - \alpha + 1)} (b - a)^{N-\alpha} < +\infty. \tag{11.153}$$

Furthermore we notice

$$\omega_1^{(\mathcal{F})}\left(\left(D_{x-}^{\alpha\mathcal{F}}f\right),\frac{1}{n^\beta}\right)_{[a,x]}=\sup_{\substack{s,t\in[a,x]\\|s-t|\leq\frac{1}{n^\beta}}}D\left(\left(D_{x-}^{\alpha\mathcal{F}}f\right)(s),\left(D_{x-}^{\alpha\mathcal{F}}f\right)(t)\right)\leq$$

$$\sup_{\substack{s,t\in[a,x]\\|s-t|\leq\frac{1}{n^\beta}}}\left\{D\left(\left(D_{x-}^{\alpha\mathcal{F}}f\right)(s),\widetilde{o}\right)+D\left(\left(D_{x-}^{\alpha\mathcal{F}}f\right)(t),\widetilde{o}\right)\right\}\leq 2D^*\left(\left(D_{x-}^{\alpha\mathcal{F}}f\right),\widetilde{o}\right)_{[a,x]}$$

$$\leq\frac{2D^*\left(f^{(N)},\widetilde{o}\right)}{\Gamma(N-\alpha+1)}(b-a)^{N-\alpha}.$$

Therefore it holds

$$\sup_{x\in[a,b]}\omega_1^{(\mathcal{F})}\left(\left(D_{x-}^{\alpha\mathcal{F}}f\right),\frac{1}{n^\beta}\right)_{[a,x]}\leq\frac{2D^*\left(f^{(N)},\widetilde{o}\right)}{\Gamma(N-\alpha+1)}(b-a)^{N-\alpha}<+\infty. \quad (11.154)$$

Similarly we observe

$$\omega_1^{(\mathcal{F})}\left(\left(D_{*x}^{\alpha\mathcal{F}}f\right),\frac{1}{n^\beta}\right)_{[x,b]}=\sup_{\substack{s,t\in[x,b]\\|s-t|\leq\frac{1}{n^\beta}}}D\left(\left(D_{*x}^{\alpha\mathcal{F}}f\right)(s),\left(D_{*x}^{\alpha\mathcal{F}}f\right)(t)\right)\leq$$

$$2D^*\left(\left(D_{*x}^{\alpha\mathcal{F}}f\right),\widetilde{o}\right)_{[x,b]}\leq\frac{2D^*\left(f^{(N)},\widetilde{o}\right)}{\Gamma(N-\alpha+1)}(b-a)^{N-\alpha}.$$

Consequently it holds

$$\sup_{x\in[a,b]}\omega_1^{(\mathcal{F})}\left(\left(D_{*x}^{\alpha\mathcal{F}}f\right),\frac{1}{n^\beta}\right)_{[x,b]}\leq\frac{2D^*\left(f^{(N)},\widetilde{o}\right)}{\Gamma(N-\alpha+1)}(b-a)^{N-\alpha}<+\infty. \quad (11.155)$$

So everything in the statements of the theorem makes sense.

The proof of the theorem is now completed. ■

Corollary 11.81 (to Theorem 11.80, $N=1$ case) *Let* $0<\alpha,\beta<1$, $f\in C_{\mathcal{F}}^1([a,b])$, $n\in\mathbb{N}$, $n^{1-\beta}>2$; $j=1,2,3,4$. *Then*

$$D^*\left({}_jA_n^{\mathcal{F}}(f),f\right)\leq$$

$$\frac{\alpha_j}{\Gamma(\alpha+1)}\left\{\frac{\left[\sup_{x\in[a,b]}\omega_1^{(\mathcal{F})}\left(\left(D_{x-}^{\alpha\mathcal{F}}f\right),\frac{1}{n^{\beta}}\right)_{[a,x]}+\sup_{x\in[a,b]}\omega_1^{(\mathcal{F})}\left(\left(D_{*x}^{\alpha\mathcal{F}}f\right),\frac{1}{n^{\beta}}\right)_{[x,b]}\right]}{n^{\alpha\beta}}+\right.$$

$$\left.c_j(\beta,n)(b-a)^{\alpha}\left[\sup_{x\in[a,b]}D^{*}\left(\left(D_{x-}^{\alpha\mathcal{F}}f\right),\widetilde{o}\right)_{[a,x]}+\sup_{x\in[a,b]}D^{*}\left(\left(D_{*x}^{\alpha\mathcal{F}}f\right),\widetilde{o}\right)_{[x,b]}\right]\right\}. \tag{11.156}$$

Proof By (11.146). ■

Finally we specialize to $\alpha = \frac{1}{2}$.

Corollary 11.82 (to Theorem 11.80) *Let* $0 < \beta < 1$, $f \in C_{\mathcal{F}}^{1}([a,b])$, $n \in \mathbb{N}$, $n^{1-\beta} > 2$; $j = 1, 2, 3, 4$. *Then*

$$D^{*}\left({}_{j}A_{n}^{\mathcal{F}}(f), f\right) \leq$$

$$\frac{2\alpha_j}{\sqrt{\pi}}\left\{\frac{\left[\sup_{x\in[a,b]}\omega_1^{(\mathcal{F})}\left(\left(D_{x-}^{\frac{1}{2}\mathcal{F}}f\right),\frac{1}{n^{\beta}}\right)_{[a,x]}+\sup_{x\in[a,b]}\omega_1^{(\mathcal{F})}\left(\left(D_{*x}^{\frac{1}{2}\mathcal{F}}f\right),\frac{1}{n^{\beta}}\right)_{[x,b]}\right]}{n^{\frac{\beta}{2}}}+\right.$$

$$\left.c_j(\beta,n)\sqrt{b-a}\left[\sup_{x\in[a,b]}D^{*}\left(\left(D_{x-}^{\frac{1}{2}\mathcal{F}}f\right),\widetilde{o}\right)_{[a,x]}+\sup_{x\in[a,b]}D^{*}\left(\left(D_{*x}^{\frac{1}{2}\mathcal{F}}f\right),\widetilde{o}\right)_{[x,b]}\right]\right\}. \tag{11.157}$$

Proof By (11.156). ■

Conclusion 11.83 *We have extended to the fuzzy setting all the main approximation theorems of Sect. 11.3.*

References

1. G.A. Anastassiou, Rate of convergence of some neural network operators to the unit-univariate case. J. Math. Anal. Appl. **212**, 237–262 (1997)
2. G.A. Anastassiou, *Quantitative Approximation* (Chapmann and Hall/CRC, Boca Raton, New York, 2001)
3. G.A. Anastassiou, Fuzzy approximation by fuzzy convolution type operators. Comput. Math. **48**, 1369–1386 (2004)
4. G.A. Anastassiou, Higher order Fuzzy Korovkin Theory via inequalities. Commun. Appl. Anal. **10**(2), 359–392 (2006)
5. G.A. Anastassiou, Fuzzy Korovkin theorems and inequalities. J. Fuzzy Math. **15**(1), 169–205 (2007)

6. G.A. Anastassiou, On right fractional calculus. Chaos, Solitons Fractals **42**, 365–376 (2009)
7. G.A. Anastassiou, *Fractional Differentiation Inequalities* (Springer, New York, 2009)
8. G.A. Anastassiou, Fractional Korovkin theory. Chaos, Solitons Fractals **42**(4), 2080–2094 (2009)
9. G.A. Anastassiou, *Fuzzy Mathematics: Approximation Theory* (Springer, Heildelberg, New York, 2010)
10. G.A. Anastassiou, *Intelligent Systems: Approximation by Artificial Neural Networks, Intelligent Systems Reference Library*, vol. 19 (Springer, Heidelberg, 2011)
11. G.A. Anastassiou, Fuzzy fractional Calculus and Ostrowski inequality. J. Fuzzy Math. **19**(3), 577–590 (2011)
12. G.A. Anastassiou, Fractional representation formulae and right fractional inequalities. Math. Comput. Model. **54**(11–12), 3098–3115 (2011)
13. G.A. Anastassiou, Univariate hyperbolic tangent neural network approximation. Math. Comput. Model. **53**, 1111–1132 (2011)
14. G.A. Anastassiou, Multivariate hyperbolic tangent neural network approximation. Comput. Math. **61**, 809–821 (2011)
15. G.A. Anastassiou, Multivariate sigmoidal neural network approximation. Neural Netw. **24**, 378–386 (2011)
16. G.A. Anastassiou, Univariate sigmoidal neural network approximation. J. Comput. Anal. Appl. **14**(4), 659–690 (2012)
17. G.A. Anastassiou, Fractional neural network approximation. Comput. Math. Appl. **64**(6), 1655–1676 (2012)
18. G.A. Anastassiou, Fuzzy fractional neural network approximation by fuzzy quasi-interpolation operators. J. Appl. Nonlinear Dyn. **2**(3), 235–259 (2013)
19. G.A. Anastassiou, *Intelligent Systems II: Complete Approximation by Neural Network Operators* (Springer, Heidelberg, New York, 2016)
20. G.A. Anastassiou, *Intelligent Computations: Abstract Fractional Calculus, Inequalities, Approximations* (Springer, Heidelberg, New York, 2018)
21. G.A. Anastassiou, Algebraic function based Banach space valued ordinary and fractional neural network approximations. New Trends Math. Sci. 10 special issues (1), 100–125 (2022)
22. G.A. Anastassiou, Gudermannian function activated Banach space valued ordinary and fractional neural network approximation. Adv. Nonlinear Var. Inequal. **25**(2), 27–64 (2022)
23. G.A. Anastassiou, *Generalized symmetrical sigmoid function activated Banach space valued ordinary and fractional neural network approximation*. Analele Universităţii Oradea, Fasc. Matematica, accepted for publication (2022)
24. G.A. Anastassiou, Abstract multivariate Gudermannian function activated neural network approximations. Panam. Math. J. accepted (2022)
25. G.A. Anastassiou, *General multivariate arctangent function activated neural network approximations*, submitted (2022)
26. G.A. Anastassiou, *Generalized symmetrical sigmoid function activated neural network multivariate approximation*, submitted (2022)
27. G.A. Anastassiou, *Fuzzy Fractional more sigmoid function activated neural network approximations revisited*, submitted (2022)
28. G.A. Anastassiou, R. Kozma, *Abstract multivariate algebraic function activated neural network approximations*, submitted (2022)
29. Z. Chen, F. Cao, The approximation operators with sigmoidal functions. Comput. Math. Appl. **58**, 758–765 (2009)
30. K. Diethelm, *The Analysis of Fractional Differential Equations*, Lecture Notes in Mathematics, vol. 2004. (Springer, Berlin, Heidelberg, 2010)
31. A.J. Dunning, J. Kensler, L. Goudeville, F. Bailleux, Some extensions in continuous methods for immunological correlates of protection. BMC Med. Res. Methodol. **15**(107) (2015). https://doi.org/10.1186/s12874-015-0096-9
32. A.M.A. El-Sayed, M. Gaber, On the finite Caputo and finite Riesz derivatives. Electron. J. Theor. Phys. **3**(12), 81–95 (2006)

33. G.S. Frederico, D.F.M. Torres, Fractional optimal control in the sense of Caputo and the fractional Noether's theorem. Int. Math. Forum **3**(10), 479–493 (2008)
34. S. Gal, Approximation theory in fuzzy setting, Chapter 13, in *Handbook of Analytic-Computational Methods in Applied Mathematics*. ed. by G. Anastassiou (Chapman & Hall/CRC, Boca Raton, New York, 2000), pp.617–666
35. R. Goetschel Jr., W. Voxman, Elementary fuzzy calculus. Fuzzy Sets Syst. **18**, 31–43 (1986)
36. S. Haykin, *Neural Networks: A Comprehensive Foundation*, 2nd edn. (Prentice Hall, New York, 1998)
37. O. Kaleva, Fuzzy differential equations. Fuzzy Sets Syst. **24**, 301–317 (1987)
38. Y.K. Kim, B.M. Ghil, Integrals of fuzzy-number-valued functions. Fuzzy Sets Syst. **86**, 213–222 (1997)
39. W. McCulloch, W. Pitts, A logical calculus of the ideas immanent in nervous activity. Bull. Math. Biophys. **7**, 115–133 (1943)
40. T.M. Mitchell, *Machine Learning* (WCB-McGraw-Hill, New York, 1997)
41. S.G. Samko, A.A. Kilbas, O.I. Marichev, *Fractional Integrals and Derivatives, Theory and Applications* (Gordon and Breach, Amsterdam, 1993) [English translation from the Russian, Integrals and Derivatives of Fractional Order and Some of Their Applications (Nauka i Tekhnika, Minsk, 1987)]
42. E.W. Weisstein, *Gudermannian*, MathWorld
43. Wu. Congxin, Gong Zengtai, On Henstock integrals of interval-valued functions and fuzzy valued functions. Fuzzy Sets Syst. **115**(3), 377–391 (2000)
44. C. Wu, Z. Gong, On Henstock integral of fuzzy-number-valued functions (I). Fuzzy Sets Syst. **3**(120), 523–532 (2001)
45. C. Wu, M. Ma, On embedding problem of fuzzy numer spaces: Part 1. Fuzzy Sets Syst. **44**, 33–38 (1991)

Chapter 12
Multivariate Fuzzy Approximation by Neural Network Operators Induced by Several Sigmoid Functions Revisited

Here are researched in terms of multivariate fuzzy complete approximation to the multivariate unit sequences of multivariate fuzzy arctangent-algebraic-Gudermannian-generalized symmetrical activation functions based neural network operators. These operators are multivariate fuzzy analogs of earlier studied multivariate Banach space valued ones. The derived results generalize earlier Banach space valued ones into the fuzzy level. Here the high order multivariate fuzzy pointwise and uniform convergences with rates to the multivariate fuzzy unit operator are given through multivariate fuzzy Jackson type inequalities involving the multivariate fuzzy moduli of continuity of the mth order ($m \geq 0$) H-fuzzy partial derivatives, of the involved multivariate fuzzy number valued function. The treated operators are of averaged, quasi-interpolation, Kantorovich and quadrature types at the multivariate fuzzy setting. It follows [29].

12.1 Introduction

The author in [1–3], see Chaps. 2–5, was the first to derive neural network approximations to continuous functions with rates by very specifically defined neural network operators of Cardaliaguet-Euvrard and "Squashing" types [31], by employing the modulus of continuity of the engaged function or its high order derivative, and deriving very tight Jackson type inequalities. He treats there both the univariate and multivariate cases. The defining these operators "bell-shaped" and "squashing" function are assumed to be of compact support.

The author motivated by [32], continued his studies on neural networks approximation by introducing and using the proper quasi-interpolation operators of sigmoidal and hyperbolic tangent type which resulted into [12–20], by treating both the univariate and multivariate cases.

© The Author(s), under exclusive license to Springer Nature Switzerland AG 2023
G. A. Anastassiou, *Banach Space Valued Neural Network*,
Studies in Computational Intelligence 1062,
https://doi.org/10.1007/978-3-031-16400-2_12

Continuation of the author's works ([4–11, 22] and especially of [21], Chap. 21) is this chapter, where multivariate fuzzy neural network approximation based on the arctangent-algebraic-Gudermannian-generalized symmetrical activation functions, which among others, results into higher rates of approximation. We involve the fuzzy partial derivatives of the multivariate fuzzy function under approximation or itself, and we establish tight multivariate fuzzy Jackson type inequalities. An extensive background is given on fuzzy multivariate analysis and real neural network approximation, all needed to present our results.

Our fuzzy multivariate feed-forward neural networks (FFNNs) are with one hidden layer. For neural networks in general you may read [36, 39, 40].

12.2 Fuzzy Real Analysis Background

See also [21], Chap. 21, pp. 466–473.

We need the following background

Definition 12.1 (*see* [44]) Let $\mu : \mathbb{R} \to [0, 1]$ with the following properties

(i) is normal, i.e., $\exists\, x_0 \in \mathbb{R};\ \mu(x_0) = 1$.

(ii) $\mu(\lambda x + (1-\lambda) y) \geq \min\{\mu(x), \mu(y)\}, \forall\, x, y \in \mathbb{R}, \forall\, \lambda \in [0, 1]$ (μ is called a convex fuzzy subset).

(iii) μ is upper semicontinuous on $\mathbb{R}$, i.e. $\forall\, x_0 \in \mathbb{R}$ and $\forall\, \varepsilon > 0$, $\exists$ neighborhood $V(x_0) : \mu(x) \leq \mu(x_0) + \varepsilon, \forall\, x \in V(x_0)$.

(iv) The set $\overline{\sup p(\mu)}$ is compact in $\mathbb{R}$, (where $\sup p(\mu) := \{x \in \mathbb{R} : \mu(x) > 0\}$).

We call μ a fuzzy real number. Denote the set of all μ with $\mathbb{R}_{\mathcal{F}}$.

E.g. $\chi_{\{x_0\}} \in \mathbb{R}_{\mathcal{F}}$, for any $x_0 \in \mathbb{R}$, where $\chi_{\{x_0\}}$ is the characteristic function at x_0.

For $0 < r \leq 1$ and $\mu \in \mathbb{R}_{\mathcal{F}}$ define

$$[\mu]^r := \{x \in \mathbb{R} : \mu(x) \geq r\} \tag{12.1}$$

and

$$[\mu]^0 := \overline{\{x \in \mathbb{R} : \mu(x) \geq 0\}}.$$

Then it is well known that for each $r \in [0, 1]$, $[\mu]^r$ is a closed and bounded interval on $\mathbb{R}$ [35].

For $u, v \in \mathbb{R}_{\mathcal{F}}$ and $\lambda \in \mathbb{R}$, we define uniquely the sum $u \oplus v$ and the product $\lambda \odot u$ by

$$[u \oplus v]^r = [u]^r + [v]^r, \quad [\lambda \odot u]^r = \lambda [u]^r, \forall\, r \in [0, 1],$$

where $[u]^r + [v]^r$ means the usual addition of two intervals (as substes of $\mathbb{R}$) and $\lambda [u]^r$ means the usual product between a scalar and a subset of $\mathbb{R}$ (see, e.g. [44]).

Notice $1 \odot u = u$ and it holds

$$u \oplus v = v \oplus u, \lambda \odot u = u \odot \lambda.$$

If $0 \leq r_1 \leq r_2 \leq 1$ then

$$[u]^{r_2} \subseteq [u]^{r_1}.$$

Actually $[u]^r = \left[u_-^{(r)}, u_+^{(r)}\right]$, where $u_-^{(r)} \leq u_+^{(r)}$, $u_-^{(r)}, u_+^{(r)} \in \mathbb{R}$, $\forall\, r \in [0, 1]$.

For $\lambda > 0$ one has $\lambda u_\pm^{(r)} = (\lambda \odot u)_\pm^{(r)}$, respectively.

Define $D : \mathbb{R}_\mathcal{F} \times \mathbb{R}_\mathcal{F} \to \mathbb{R}_\mathcal{F}$ by

$$D(u, v) := \sup_{r \in [0,1]} \max \left\{ \left| u_-^{(r)} - v_-^{(r)} \right|, \left| u_+^{(r)} - v_+^{(r)} \right| \right\}, \tag{12.2}$$

where

$$[v]^r = \left[v_-^{(r)}, v_+^{(r)}\right];\ u, v \in \mathbb{R}_\mathcal{F}.$$

We have that D is a metric on $\mathbb{R}_\mathcal{F}$.

Then $(\mathbb{R}_\mathcal{F}, D)$ is a complete metric space, see [44, 45].

Let $f, g : W \subseteq \mathbb{R}^m \to \mathbb{R}_\mathcal{F}$. We define the distance

$$D^*(f, g) = \sup_{x \in W} D(f(x), g(x)).$$

Remark 12.2 We determine and use

$$D^*(f, \widetilde{o}) = \sup_{x \in W} D(f(x), \widetilde{o}) =$$

$$\sup_{x \in W} \sup_{r \in [0,1]} \max \left\{ \left| f_-^{(r)}(x) \right|, \left| f_+^{(r)}(x) \right| \right\}.$$

By the principle of iterated suprema we find that

$$D^*(f, \widetilde{o}) = \sup_{r \in [0,1]} \max \left\{ \left\| f_-^{(r)} \right\|_\infty, \left\| f_+^{(r)} \right\|_\infty \right\}, \tag{12.3}$$

under the assumption $D^*(f, \widetilde{o}) < \infty$, that is f is a fuzzy bounded function.

Above $\|\cdot\|_\infty$ is the supremum norm of the function over $W \subseteq \mathbb{R}^m$.

Here Σ^* stands for fuzzy summation and $\widetilde{0} := \chi_{\{0\}} \in \mathbb{R}_\mathcal{F}$ is the neutral element with respect to $\oplus$, i.e.,

$$u \oplus \widetilde{0} = \widetilde{0} \oplus u = u,\ \forall\, u \in \mathbb{R}_\mathcal{F}.$$

We need

Remark 12.3 ([5]). Here $r \in [0, 1]$, $x_i^{(r)}, y_i^{(r)} \in \mathbb{R}$, $i = 1, ..., m \in \mathbb{N}$. Suppose that

$$\sup_{r\in[0,1]} \max\left(x_i^{(r)}, y_i^{(r)}\right) \in \mathbb{R}, \text{ for } i = 1, ..., m.$$

Then one sees easily that

$$\sup_{r\in[0,1]} \max\left(\sum_{i=1}^{m} x_i^{(r)}, \sum_{i=1}^{m} y_i^{(r)}\right) \leq \sum_{i=1}^{m} \sup_{r\in[0,1]} \max\left(x_i^{(r)}, y_i^{(r)}\right). \tag{12.4}$$

Definition 12.4 Let $f \in C(W)$, $W \subseteq \mathbb{R}^m$, $m \in \mathbb{N}$, which is bounded or uniformly continuous, we define $(h > 0)$

$$\omega_1(f, h) := \sup_{x,y\in W,\ \|x-y\|_\infty \leq h} |f(x) - f(y)|, \tag{12.5}$$

where $x = (x_1, ..., x_m)$, $y = (y_1, ..., y_m)$.

Definition 12.5 Let $f : W \to \mathbb{R}_{\mathcal{F}}$, $W \subseteq \mathbb{R}^m$, we define the fuzzy modulus of continuity of f by

$$\omega_1^{(\mathcal{F})}(f, h) = \sup_{x,y\in W,\ \|x-y\|_\infty \leq h} D(f(x), f(y)),\ h > 0. \tag{12.6}$$

where $x = (x_1, ..., x_m)$, $y = (y_1, ..., y_m)$.

For $f : W \to \mathbb{R}_{\mathcal{F}}$, $W \subseteq \mathbb{R}^m$, we use

$$[f]^r = \left[f_-^{(r)}, f_+^{(r)}\right], \tag{12.7}$$

where $f_\pm^{(r)} : W \to \mathbb{R}$, $\forall\, r \in [0, 1]$.
We need

Proposition 12.6 ([5]) *Let $f : W \to \mathbb{R}_{\mathcal{F}}$. Assume that $\omega_1^{\mathcal{F}}(f, \delta)$, $\omega_1\left(f_-^{(r)}, \delta\right)$, $\omega_1\left(f_+^{(r)}, \delta\right)$ are finite for any $\delta > 0$, $r \in [0, 1]$.*
Then

$$\omega_1^{(\mathcal{F})}(f, \delta) = \sup_{r\in[0,1]} \max\left\{\omega_1\left(f_-^{(r)}, \delta\right), \omega_1\left(f_+^{(r)}, \delta\right)\right\}. \tag{12.8}$$

We denote by $C_{\mathcal{F}}^U(W)$ the space of fuzzy uniformly continuous functions from $W \to \mathbb{R}_{\mathcal{F}}$, also $C_{\mathcal{F}}(W)$ is the space of fuzzy continuous functions on $W \subseteq \mathbb{R}^m$, and $C_B(W, \mathbb{R}_{\mathcal{F}})$ is the fuzzy continuous and bounded functions.
We mention

Proposition 12.7 ([7]) *Let $f \in C_{\mathcal{F}}^U(W)$, where $W \subseteq \mathbb{R}^m$ is convex. Then $\omega_1^{(\mathcal{F})}(f, \delta) < \infty$, for any $\delta > 0$.*

Proposition 12.8 ([7]) *It holds*

$$\lim_{\delta \to 0} \omega_1^{(\mathcal{F})}(f, \delta) = \omega_1^{(\mathcal{F})}(f, 0) = 0, \tag{12.9}$$

iff $f \in C_{\mathcal{F}}^{U}(W)$, $W \subseteq \mathbb{R}^m$, *where* W *is convex and compact.*

Proposition 12.9 ([7]) *Let* $f \in C_{\mathcal{F}}(W)$, $W \subseteq \mathbb{R}^m$ *open or compact. Then* $f_{\pm}^{(r)}$ *are equicontinuous with respect to* $r \in [0, 1]$ *over* W, *respectively in* $\pm$.

Note 12.10 It is clear by Propositions 12.6 and 12.8, that if $f \in C_{\mathcal{F}}^{U}(W)$, then $f_{\pm}^{(r)} \in C_U(W)$ (uniformly continuous on W). Also if $f \in C_B(W, \mathbb{R}_{\mathcal{F}})$, it implies by (12.3) that $f_{\pm}^{(r)} \in C_B(W)$ (continuous and bounded functions on W).

We need

Definition 12.11 Let $x, y \in \mathbb{R}_{\mathcal{F}}$. If there exists $z \in \mathbb{R}_{\mathcal{F}} : x = y \oplus z$, then we call z the H-difference on x and y, denoted $x - y$.

Definition 12.12 ([44]) Let $T := [x_0, x_0 + \beta] \subset \mathbb{R}$, with $\beta > 0$. A function $f : T \to \mathbb{R}_{\mathcal{F}}$ is H-difference at $x \in T$ if there exists an $f'(x) \in \mathbb{R}_{\mathcal{F}}$ such that the limits (with respect to D)

$$\lim_{h \to 0+} \frac{f(x+h) - f(x)}{h}, \quad \lim_{h \to 0+} \frac{f(x) - f(x-h)}{h} \tag{12.10}$$

exist and are equal to $f'(x)$.

We call f' the H-derivative or fuzzy derivative of f at x.

Above is assumed that the H-differences $f(x+h) - f(x)$, $f(x) - f(x-h)$ exists in $\mathbb{R}_{\mathcal{F}}$ in a neighborhood of x.

Definition 12.13 We denote by $C_{\mathcal{F}}^{N^*}(W)$, $N^* \in \mathbb{N}$, the space of all N^*-times fuzzy continuously differentiable functions from W into $\mathbb{R}_{\mathcal{F}}$, $W \subseteq \mathbb{R}^m$ open or compact which is convex.

Here fuzzy partial derivatives are defined via Definition 12.12 in the obvious way as in the ordinary real case.

We mention

Theorem 12.14 ([37]) *Let* $f : [a, b] \subseteq \mathbb{R} \to \mathbb{R}_{\mathcal{F}}$ *be* H*-fuzzy differentiable. Let* $t \in [a, b]$, $0 \leq r \leq 1$. *Clearly*

$$[f(t)]^r = \left[f(t)_-^{(r)}, f(t)_+^{(r)} \right] \subseteq \mathbb{R}.$$

Then $(f(t))_{\pm}^{(r)}$ *are differentiable and*

$$\left[f'(t)\right]^r = \left[\left(f(t)_-^{(r)}\right)', \left(f(t)_+^{(r)}\right)'\right].$$

I.e.

$$\left(f'\right)_\pm^{(r)} = \left(f_\pm^{(r)}\right)', \quad \forall\, r \in [0,1]. \tag{12.11}$$

Remark 12.15 (*see also* [6]) Let $f \in C^{N^*}([a,b], \mathbb{R}_{\mathcal{F}})$, $N^* \geq 1$. Then by Theorem 12.14 we obtain $f_\pm^{(r)} \in C^{N^*}([a,b])$ and

$$\left[f^{(i)}(t)\right]^r = \left[\left(f(t)_-^{(r)}\right)^{(i)}, \left(f(t)_+^{(r)}\right)^{(i)}\right],$$

for $i = 0, 1, 2, \ldots, N^*$, and in particular we have

$$\left(f^{(i)}\right)_\pm^{(r)} = \left(f_\pm^{(r)}\right)^{(i)}, \tag{12.12}$$

for any $r \in [0,1]$.

Let $f \in C_{\mathcal{F}}^{N^*}(W)$, $W \subseteq \mathbb{R}^m$, open or compact, which is convex, denote $f_{\widetilde{\alpha}} := \frac{\partial^{\widetilde{\alpha}} f}{\partial x^{\widetilde{\alpha}}}$, where $\widetilde{\alpha} := (\widetilde{\alpha}_1, \ldots, \widetilde{\alpha}_m)$, $\widetilde{\alpha}_i \in \mathbb{Z}^+$, $i = 1, \ldots, m$ and

$$0 < |\widetilde{\alpha}| := \sum_{i=1}^m \widetilde{\alpha}_i \leq N^*, \quad N^* > 1.$$

Then by Theorem 12.14 we get that

$$\left(f_\pm^{(r)}\right)_{\widetilde{\alpha}} = \left(f_{\widetilde{\alpha}}\right)_\pm^{(r)}, \; \forall\, r \in [0,1], \tag{12.13}$$

and any $\widetilde{\alpha} : |\widetilde{\alpha}| \leq N^*$. Here $f_\pm^{(r)} \in C^{N^*}(W)$.

Notation 12.16 We denote

$$\left(\sum_{i=1}^2 D\left(\frac{\partial}{\partial x_i}, \widetilde{o}\right)\right)^2 f(\overrightarrow{x}) := \tag{12.14}$$

$$D\left(\frac{\partial^2 f(x_1,x_2)}{\partial x_1^2}, \widetilde{o}\right) + D\left(\frac{\partial^2 f(x_1,x_2)}{\partial x_2^2}, \widetilde{o}\right) + 2D\left(\frac{\partial^2 f(x_1,x_2)}{\partial x_1 \partial x_2}, \widetilde{o}\right).$$

In general we denote ($j = 1, \ldots, N^*$)

$$\left(\sum_{i=1}^m D\left(\frac{\partial}{\partial x_i}, \widetilde{o}\right)\right)^j f(\overrightarrow{x}) := \tag{12.15}$$

$$\sum_{(j_1,\ldots,j_m)\in\mathbb{Z}_+^m:\sum_{i=1}^m j_i=j} \frac{j!}{j_1!j_2!\ldots j_m!} D\left(\frac{\partial^j f(x_1,\ldots,x_m)}{\partial x_1^{j_1}\partial x_2^{j_2}\ldots\partial x_m^{j_m}},\widetilde{0}\right).$$

We mention also a particular case of the Fuzzy Henstock integral ($\delta(x) = \frac{\delta}{2}$), see [44].

Definition 12.17 ([34], p. 644) Let $f : [a, b] \to \mathbb{R}_{\mathcal{F}}$. We say that f is Fuzzy-Riemann integrable to $I \in \mathbb{R}_{\mathcal{F}}$ if for any $\varepsilon > 0$, there exists $\delta > 0$ such that for any division $P = \{[u, v]; \xi\}$ of $[a, b]$ with the norms $\Delta(P) < \delta$, we have

$$D\left(\sum_P^{*}(v-u)\odot f(\xi), I\right) < \varepsilon.$$

We write

$$I := (FR)\int_a^b f(x)\,dx. \tag{12.16}$$

We mention

Theorem 12.18 ([35]) *Let $f : [a, b] \to \mathbb{R}_{\mathcal{F}}$ be fuzzy continuous. Then*

$$(FR)\int_a^b f(x)\,dx$$

exists and belongs to $\mathbb{R}_{\mathcal{F}}$, furthermore it holds

$$\left[(FR)\int_a^b f(x)\,dx\right]^r = \left[\int_a^b (f)_-^{(r)}(x)\,dx, \int_a^b (f)_+^{(r)}(x)\,dx\right],$$

$\forall\, r \in [0, 1]$.

For the definition of general fuzzy integral we follow [38] next.

Definition 12.19 Let (Ω, Σ, μ) be a complete σ-finite measure space. We call $F : \Omega \to R_{\mathcal{F}}$ measurable iff $\forall$ closed $B \subseteq \mathbb{R}$ the function $F^{-1}(B) : \Omega \to [0, 1]$ defined by

$$F^{-1}(B)(w) := \sup_{x\in B} F(w)(x), \text{ all } w \in \Omega$$

is measurable, see [38].

Theorem 12.20 ([38]) *For $F : \Omega \to \mathbb{R}_{\mathcal{F}}$,*

$$F(w) = \left\{\left(F_-^{(r)}(w), F_+^{(r)}(w)\right) | 0 \le r \le 1\right\},$$

the following are equivalent
(1) F is measurable,
(2) $\forall\, r \in [0,1]$, $F_-^{(r)}$, $F_+^{(r)}$ are measurable.

Following [38], given that for each $r \in [0,1]$, $F_-^{(r)}$, $F_+^{(r)}$ are integrable we have that the parametrized representation

$$\left\{\left(\int_A F_-^{(r)} d\mu, \int_A F_+^{(r)} d\mu\right) | 0 \leq r \leq 1\right\} \tag{12.17}$$

is a fuzzy real number for each $A \in \Sigma$.
The last fact leads to

Definition 12.21 ([38]) A measurable function $F : \Omega \to \mathbb{R}_{\mathcal{F}}$,

$$F(w) = \left\{\left(F_-^{(r)}(w), F_+^{(r)}(w)\right) | 0 \leq r \leq 1\right\}$$

is integrable if for each $r \in [0,1]$, $F_\pm^{(r)}$ are integrable, or equivalently, if $F_\pm^{(0)}$ are integrable.

In this case, the fuzzy integral of F over $A \in \Sigma$ is defined by

$$\int_A F d\mu := \left\{\left(\int_A F_-^{(r)} d\mu, \int_A F_+^{(r)} d\mu\right) | 0 \leq r \leq 1\right\}. \tag{12.18}$$

By [38], F is integrable iff $w \to \|F(w)\|_{\mathcal{F}}$ is real-valued integrable.
Here denote

$$\|u\|_{\mathcal{F}} := D\left(u, \widetilde{0}\right), \ \forall\, u \in \mathbb{R}_{\mathcal{F}}.$$

We need also

Theorem 12.22 ([38]) *Let $F, G : \Omega \to \mathbb{R}_{\mathcal{F}}$ be integrable. Then*
(1) Let $a, b \in \mathbb{R}$, then $aF + bG$ is integrable and for each $A \in \Sigma$,

$$\int_A (aF + bG) d\mu = a \int_A F d\mu + b \int_A G d\mu;$$

(2) $D(F, G)$ is a real- valued integrable function and for each $A \in \Sigma$,

$$D\left(\int_A F d\mu, \int_A G d\mu\right) \leq \int_A D(F, G) d\mu.$$

In particular,

$$\left\|\int_A F d\mu\right\|_{\mathcal{F}} \leq \int_A \|F\|_{\mathcal{F}} d\mu.$$

Above μ could be the multivariate Lebesgue measure, which we use in this chapter, with all the basic properties valid here too. Notice by [38], Fubini's theorem is valid for fuzzy integral (12.18).

Basically here we have that

$$\left[\int_A F d\mu\right]^r = \left[\int_A F_-^{(r)} d\mu, \int_A F_+^{(r)} d\mu\right], \tag{12.19}$$

i.e.

$$\left(\int_A F d\mu\right)_\pm^{(r)} = \int_A F_\pm^{(r)} d\mu, \ \forall\, r \in [0,1]. \tag{12.20}$$

12.3 About Neural Networks Background

12.3.1 About the Arctangent Activation Function

We consider the

$$\arctan x = \int_0^x \frac{dz}{1+z^2}, \ x \in \mathbb{R}. \tag{12.21}$$

We will be using

$$h(x) := \frac{2}{\pi} \arctan\left(\frac{\pi}{2} x\right) = \frac{2}{\pi} \int_0^{\frac{\pi x}{2}} \frac{dz}{1+z^2}, \ x \in \mathbb{R}, \tag{12.22}$$

which is a sigmoid type function and it is strictly increasing. We have that

$$h(0) = 0, \ h(-x) = -h(x), \ h(+\infty) = 1, \ h(-\infty) = -1,$$

and

$$h'(x) = \frac{4}{4+\pi^2 x^2} > 0, \text{ all } x \in \mathbb{R}. \tag{12.23}$$

We consider the activation function

$$\psi_1(x) := \frac{1}{4}\left(h(x+1) - h(x-1)\right), \ x \in \mathbb{R}, \tag{12.24}$$

and we notice that

$$\psi_1(-x) = \psi_1(x), \tag{12.25}$$

it is an even function.

Since $x+1>x-1$, then $h(x+1)>h(x-1)$, and $\psi_1(x)>0$, all $x\in\mathbb{R}$.
We see that

$$\psi_1(0)=\frac{1}{\pi}\arctan\frac{\pi}{2}\cong 0.319. \tag{12.26}$$

Let $x>0$, we have that

$$\psi_1'(x)=\frac{1}{4}\left(h'(x+1)-h'(x-1)\right)=$$
$$\frac{-4\pi^2 x}{\left(4+\pi^2(x+1)^2\right)\left(4+\pi^2(x-1)^2\right)}<0. \tag{12.27}$$

That is

$$\psi_1'(x)<0, \text{ for } x>0. \tag{12.28}$$

That is ψ_1 is strictly decreasing on $[0,\infty)$ and clearly is strictly increasing on $(-\infty,0]$, and $\psi_1'(0)=0$.

Observe that

$$\begin{gathered}\lim_{x\to+\infty}\psi_1(x)=\tfrac{1}{4}\left(h(+\infty)-h(+\infty)\right)=0,\\ \text{and}\\ \lim_{x\to-\infty}\psi_1(x)=\tfrac{1}{4}\left(h(-\infty)-h(-\infty)\right)=0.\end{gathered} \tag{12.29}$$

That is the x-axis is the horizontal asymptote on ψ_1.

All in all, ψ_1 is a bell symmetric function with maximum $\psi_1(0)\cong 18.31$.

We need

Theorem 12.23 ([22], p. 286) *We have that*

$$\sum_{i=-\infty}^{\infty}\psi_1(x-i)=1,\ \ \forall\, x\in\mathbb{R}. \tag{12.30}$$

Theorem 12.24 ([22], p. 287) *It holds*

$$\int_{-\infty}^{\infty}\psi_1(x)\,dx=1. \tag{12.31}$$

So that $\psi_1(x)$ is a density function on $\mathbb{R}$.

We mention

Theorem 12.25 ([22], p. 288) *Let* $0<\alpha<1$, *and* $n\in\mathbb{N}$ *with* $n^{1-\alpha}>2$. *It holds*

$$\sum_{\begin{cases} k=-\infty \\ : |nx-k| \geq n^{1-\alpha} \end{cases}}^{\infty} \psi_1(nx-k) < \frac{2}{\pi^2 \left(n^{1-\alpha}-2\right)} =: c_1(\alpha, n). \tag{12.32}$$

Denote by $\lfloor \cdot \rfloor$ the integral part of the number and by $\lceil \cdot \rceil$ the ceiling of the number. We need

Theorem 12.26 ([22], p. 289) *Let $x \in [a, b] \subset \mathbb{R}$ and $n \in \mathbb{N}$ so that $\lceil na \rceil \leq \lfloor nb \rfloor$. It holds*

$$\frac{1}{\sum_{k=\lceil na \rceil}^{\lfloor nb \rfloor} \psi_1(nx-k)} < \frac{1}{\psi_1(1)} \cong 4.9737 =: \alpha_1, \ \forall \, x \in [a, b]. \tag{12.33}$$

Note 12.27 ([22], pp. 290–291)
(i) We have that

$$\lim_{n \to \infty} \sum_{k=\lceil na \rceil}^{\lfloor nb \rfloor} \psi_1(nx-k) \neq 1, \tag{12.34}$$

for at least some $x \in [a, b]$.

(ii) For large enough $n \in \mathbb{N}$ we always obtain $\lceil na \rceil \leq \lfloor nb \rfloor$. Also $a \leq \frac{k}{n} \leq b$, iff $\lceil na \rceil \leq k \leq \lfloor nb \rfloor$.

In general, by Theorem 12.23, it holds

$$\sum_{k=\lceil na \rceil}^{\lfloor nb \rfloor} \psi_1(nx-k) \leq 1. \tag{12.35}$$

We introduce (see [27])

$$Z_1(x_1, ..., x_N) := Z_1(x) := \prod_{i=1}^{N} \psi_1(x_i), \ \ x = (x_1, ..., x_N) \in \mathbb{R}^N, \ N \in \mathbb{N}. \tag{12.36}$$

Denote by $a = (a_1, ..., a_N)$ and $b = (b_1, ..., b_N)$.

It has the properties:

(i) $Z_1(x) > 0, \ \forall \, x \in \mathbb{R}^N$,

(ii)

$$\sum_{k=-\infty}^{\infty} Z_1(x-k) := \sum_{k_1=-\infty}^{\infty} \sum_{k_2=-\infty}^{\infty} \ldots \sum_{k_N=-\infty}^{\infty} Z_1(x_1-k_1, ..., x_N-k_N) = 1, \tag{12.37}$$

where $k := (k_1, ..., k_n) \in \mathbb{Z}^N, \forall \, x \in \mathbb{R}^N$,

hence

(iii)

$$\sum_{k=-\infty}^{\infty} Z_1 (nx - k) = 1, \tag{12.38}$$

$\forall\, x \in \mathbb{R}^N;\ n \in \mathbb{N}$,

and

(iv)

$$\int_{\mathbb{R}^N} Z_1 (x)\, dx = 1, \tag{12.39}$$

that is Z_1 is a multivariate density function.

(v) It is clear that

$$\sum_{\substack{k = -\infty \\ \left\| \frac{k}{n} - x \right\|_\infty > \frac{1}{n^\beta}}}^{\infty} Z_1 (nx - k) < \frac{2}{\pi^2 \left(n^{1-\beta} - 2\right)} = c_1 (\beta, n), \tag{12.40}$$

$0 < \beta < 1,\ n \in \mathbb{N} : n^{1-\beta} > 2,\ x \in \mathbb{R}^N$.

(vi) By Theorem 12.26 we get that

$$0 < \frac{1}{\sum_{k=\lceil na \rceil}^{\lfloor nb \rfloor} Z_1 (nx - k)} < \frac{1}{(\psi_1 (1))^N} \cong (4.9737)^N =: \gamma_1 (N), \tag{12.41}$$

$\forall\, x \in \left(\prod_{i=1}^N [a_i, b_i]\right),\ n \in \mathbb{N}$.

Furthermore it holds

$$\lim_{n\to\infty} \sum_{k=\lceil na \rceil}^{\lfloor nb \rfloor} Z_1 (nx - k) \neq 1, \tag{12.42}$$

for at least some $x \in \left(\prod_{i=1}^N [a_i, b_i]\right)$.

Above it is $\|x\|_\infty := \max\{|x_1|, ..., |x_N|\}$, $x \in \mathbb{R}^N$, also set $\infty := (\infty, ..., \infty)$, $-\infty = (-\infty, ... - \infty)$ upon the multivariate context.

12.3.2 About the Algebraic Activation Function

Here see also [23].

We consider the generator algebraic function

$$\varphi (x) = \frac{x}{\sqrt[2m]{1 + x^{2m}}},\quad m \in \mathbb{N},\ x \in \mathbb{R}, \tag{12.43}$$

which is a sigmoidal type of function and is a strictly increasing function.

We see that $\varphi(-x) = -\varphi(x)$ with $\varphi(0) = 0$. We get that

$$\varphi'(x) = \frac{1}{\left(1 + x^{2m}\right)^{\frac{2m+1}{2m}}} > 0, \ \forall x \in \mathbb{R}, \tag{12.44}$$

proving φ as strictly increasing over $\mathbb{R}$, $\varphi'(x) = \varphi'(-x)$. We easily find that $\lim_{x \to +\infty} \varphi(x) = 1$, $\varphi(+\infty) = 1$, and $\lim_{x \to -\infty} \varphi(x) = -1$, $\varphi(-\infty) = -1$.

We consider the activation function

$$\psi_2(x) = \frac{1}{4}\left[\varphi(x+1) - \varphi(x-1)\right]. \tag{12.45}$$

Clearly it is $\psi_2(x) = \psi_2(-x)$, $\forall x \in \mathbb{R}$, so that ψ_2 is an even function and symmetric with respect to the y-axis. Clealry $\psi_2(x) > 0$, $\forall x \in \mathbb{R}$.

Also it is

$$\psi_2(0) = \frac{1}{2\sqrt[2m]{2}}. \tag{12.46}$$

By [23], we have that $\psi_2'(x) < 0$ for $x > 0$. That is ψ_2 is strictly decreasing over $(0, +\infty)$.

Clearly, ψ_2 is strictly increasing over $(-\infty, 0)$ and $\psi_2'(0) = 0$.

Furthermore we obtain that

$$\lim_{x \to +\infty} \psi_2(x) = \frac{1}{4}\left[\varphi(+\infty) - \varphi(+\infty)\right] = 0, \tag{12.47}$$

and

$$\lim_{x \to -\infty} \psi_2(x) = \frac{1}{4}\left[\varphi(-\infty) - \varphi(-\infty)\right] = 0. \tag{12.48}$$

That is the x-axis is the horizontal asymptote of ψ_2.

Conclusion, ψ_2 is a bell shape symmetric function with maximum

$$\psi_2(0) = \frac{1}{2\sqrt[2m]{2}}, \ m \in \mathbb{N}. \tag{12.49}$$

We need

Theorem 12.28 ([23]) *We have that*

$$\sum_{i=-\infty}^{\infty} \psi_2(x - i) = 1, \ \forall x \in \mathbb{R}. \tag{12.50}$$

Theorem 12.29 ([23]) *It holds*

$$\int_{-\infty}^{\infty} \psi_2(x)\,dx = 1. \tag{12.51}$$

Theorem 12.30 ([23]) *Let* $0 < \alpha < 1$, *and* $n \in \mathbb{N}$ *with* $n^{1-\alpha} > 2$. *It holds*

$$\sum_{\substack{k=-\infty \\ : |nx-k| \geq n^{1-\alpha}}}^{\infty} \psi_2(nx-k) < \frac{1}{4m\left(n^{1-\alpha}-2\right)^{2m}} =: c_2(\alpha, n), \quad m \in \mathbb{N}. \tag{12.52}$$

We need

Theorem 12.31 ([23]) *Let* $[a, b] \subset \mathbb{R}$ *and* $n \in \mathbb{N}$ *so that* $\lceil na \rceil \leq \lfloor nb \rfloor$. *It holds*

$$\frac{1}{\sum_{k=\lceil na \rceil}^{\lfloor nb \rfloor} \psi_2(nx-k)} < 2\left(\sqrt[2m]{1+4^m}\right) =: \alpha_2, \tag{12.53}$$

$\forall\, x \in [a, b]$, $m \in \mathbb{N}$.

Note 12.32 (1) By [23] we have that

$$\lim_{n\to\infty} \sum_{k=\lceil na \rceil}^{\lfloor nb \rfloor} \psi_2(nx-k) \neq 1, \tag{12.54}$$

for at least some $x \in [a, b]$.

(2) Let $[a, b] \subset \mathbb{R}$. For large $n \in \mathbb{N}$ we always have $\lceil na \rceil \leq \lfloor nb \rfloor$. Also $a \leq \frac{k}{n} \leq b$, iff $\lceil na \rceil \leq k \leq \lfloor nb \rfloor$.

In general it holds that

$$\sum_{k=\lceil na \rceil}^{\lfloor nb \rfloor} \psi_2(nx-k) \leq 1. \tag{12.55}$$

We introduce (see also [30])

$$Z_2(x_1, ..., x_N) := Z_2(x) := \prod_{i=1}^{N} \psi_2(x_i), \quad x = (x_1, ..., x_N) \in \mathbb{R}^N, \ N \in \mathbb{N}. \tag{12.56}$$

It has the properties:

(i) $Z_2(x) > 0$, $\forall\, x \in \mathbb{R}^N$,

(ii)

$$\sum_{k=-\infty}^{\infty} Z_2(x-k) := \sum_{k_1=-\infty}^{\infty} \sum_{k_2=-\infty}^{\infty} \ldots \sum_{k_N=-\infty}^{\infty} Z_2(x_1-k_1, \ldots, x_N-k_N) = 1, \tag{12.57}$$

where $k := (k_1, \ldots, k_n) \in \mathbb{Z}^N$, $\forall\, x \in \mathbb{R}^N$,

hence

(iii)

$$\sum_{k=-\infty}^{\infty} Z_2(nx-k) = 1, \tag{12.58}$$

$\forall\, x \in \mathbb{R}^N$; $n \in \mathbb{N}$,

and

(iv)

$$\int_{\mathbb{R}^N} Z_2(x)\, dx = 1, \tag{12.59}$$

that is Z_2 is a multivariate density function.

(v) It is clear that

$$\sum_{\begin{cases} k = -\infty \\ \left\| \frac{k}{n} - x \right\|_\infty > \frac{1}{n^\beta} \end{cases}}^{\infty} Z_2(nx-k) < \frac{1}{4m\left(n^{1-\beta}-2\right)^{2m}} = c_2(\beta, n), \tag{12.60}$$

$0 < \beta < 1$, $n \in \mathbb{N} : n^{1-\beta} > 2$, $x \in \mathbb{R}^N$, $m \in \mathbb{N}$.

(vi) By Theorem 12.31 we get that

$$0 < \frac{1}{\sum_{k=\lceil na \rceil}^{\lfloor nb \rfloor} Z_2(nx-k)} < \frac{1}{(\psi_2(1))^N} \cong \left[2\left(\sqrt[2m]{1+4^m}\right)\right]^N := \gamma_2(N), \tag{12.61}$$

$\forall\, x \in \left(\prod_{i=1}^N [a_i, b_i]\right)$, $n \in \mathbb{N}$.

Furthermore it holds

$$\lim_{n\to\infty} \sum_{k=\lceil na \rceil}^{\lfloor nb \rfloor} Z_2(nx-k) \neq 1, \tag{12.62}$$

for at least some $x \in \left(\prod_{i=1}^N [a_i, b_i]\right)$.

12.3.3 About the Gudermannian Activation Function

See also [24, 42].

Here we consider $gd(x)$ the Gudermannian function [42], which is a sigmoid function, as a generator function:

$$\sigma(x) = 2\arctan\left(\tanh\left(\frac{x}{2}\right)\right) = \int_0^x \frac{dt}{\cosh t} =: gd(x), \, x \in \mathbb{R}. \tag{12.63}$$

Let the normalized generator sigmoid function

$$f(x) := \frac{4}{\pi}\sigma(x) = \frac{4}{\pi}\int_0^x \frac{dt}{\cosh t} = \frac{8}{\pi}\int_0^x \frac{1}{e^t + e^{-t}}dt, \; x \in \mathbb{R}. \tag{12.64}$$

Here

$$f'(x) = \frac{4}{\pi\cosh x} > 0, \; \forall\, x \in \mathbb{R},$$

hence f is strictly increasing on $\mathbb{R}$.

Notice that $\tanh(-x) = -\tanh x$ and $\arctan(-x) = -\arctan x$, $x \in \mathbb{R}$.

So, here the neural network activation function will be:

$$\psi_3(x) = \frac{1}{4}\left[f(x+1) - f(x-1)\right], x \in \mathbb{R}. \tag{12.65}$$

By [24], we get that

$$\psi_3(x) = \psi_3(-x), \; \forall\, x \in \mathbb{R}, \tag{12.66}$$

i.e. it is even and symmetric with respect to the y-axis. Here we have $f(+\infty) = 1$, $f(-\infty) = -1$ and $f(0) = 0$. Clearly it is

$$f(-x) = -f(x), \; \forall\, x \in \mathbb{R}, \tag{12.67}$$

an odd function, symmetric with respect to the origin. Since $x + 1 > x - 1$, and $f(x+1) > f(x-1)$, we obtain $\psi_3(x) > 0$, $\forall\, x \in \mathbb{R}$.

By [24], we have that

$$\psi_3(0) = \frac{2}{\pi}gd(1) \cong 0.551. \tag{12.68}$$

By [24] ψ_3 is strictly decreasing on $(0, +\infty)$, and strictly increasing on $(-\infty, 0)$, and $\psi_3'(0) = 0$.

Also we have that

$$\lim_{x\to+\infty}\psi_3(x) = \lim_{x\to-\infty}\psi_3(x) = 0, \tag{12.69}$$

that is the x-axis is the horizontal asymptote for ψ_3.

Conclusion, ψ_3 is a bell shaped symmetric function with maximum $\psi_3(0) \cong 0.551$.

We need

Theorem 12.33 ([24]) *It holds that*

$$\sum_{i=-\infty}^{\infty} \psi_3(x-i) = 1, \ \forall\, x \in \mathbb{R}. \tag{12.70}$$

Theorem 12.34 ([24]) *We have that*

$$\int_{-\infty}^{\infty} \psi_3(x)\, dx = 1. \tag{12.71}$$

So $\psi_3(x)$ is a density function.

Theorem 12.35 ([24]) *Let $0 < \alpha < 1$, and $n \in \mathbb{N}$ with $n^{1-\alpha} > 2$. It holds*

$$\sum_{\substack{k=-\infty \\ : |nx-k| \geq n^{1-\alpha}}}^{\infty} \psi_3(nx-k) < \frac{4}{\pi e^{\left(n^{1-\alpha}-2\right)}} = \frac{4e^2}{\pi e^{n^{1-\alpha}}} =: c_3(\alpha, n). \tag{12.72}$$

Theorem 12.36 ([24]) *Let $[a,b] \subset \mathbb{R}$ and $n \in \mathbb{N}$, so that $\lceil na \rceil \leq \lfloor nb \rfloor$. It holds*

$$\frac{1}{\sum_{k=\lceil na \rceil}^{\lfloor nb \rfloor} \psi_3(nx-k)} < \frac{\pi}{gd(2)} \cong 2.412 =: \alpha_3, \tag{12.73}$$

$\forall\, x \in [a,b]$.

We make

Remark 12.37 ([24])

(i) We have that

$$\lim_{n\to\infty} \sum_{k=\lceil na \rceil}^{\lfloor nb \rfloor} \psi_3(nx-k) \neq 1, \tag{12.74}$$

for at least some $x \in [a,b]$.

(ii) Let $[a,b] \subset \mathbb{R}$. For large n we always have $\lceil na \rceil \leq \lfloor nb \rfloor$. Also $a \leq \frac{k}{n} \leq b$, iff $\lceil na \rceil \leq k \leq \lfloor nb \rfloor$.

In general it holds

$$\sum_{k=\lceil na \rceil}^{\lfloor nb \rfloor} \psi_3(nx-k) \leq 1. \tag{12.75}$$

We introduce (see also [26])

$$Z_3(x_1, ..., x_N) := Z_3(x) := \prod_{i=1}^{N} \psi_3(x_i), \quad x = (x_1, ..., x_N) \in \mathbb{R}^N, \ N \in \mathbb{N}. \tag{12.76}$$

It has the properties:

(i) $Z_3(x) > 0, \ \forall\, x \in \mathbb{R}^N$,

(ii)

$$\sum_{k=-\infty}^{\infty} Z_3(x-k) := \sum_{k_1=-\infty}^{\infty} \sum_{k_2=-\infty}^{\infty} ... \sum_{k_N=-\infty}^{\infty} Z_3(x_1-k_1, ..., x_N-k_N) = 1, \tag{12.77}$$

where $k := (k_1, ..., k_n) \in \mathbb{Z}^N, \forall\, x \in \mathbb{R}^N$,

hence

(iii)

$$\sum_{k=-\infty}^{\infty} Z_3(nx-k) = 1, \tag{12.78}$$

$\forall\, x \in \mathbb{R}^N; \ n \in \mathbb{N}$,

and

(iv)

$$\int_{\mathbb{R}^N} Z_3(x)\, dx = 1, \tag{12.79}$$

that is Z_3 is a multivariate density function.

(v) It is also clear that

$$\sum_{\begin{cases} k = -\infty \\ \left\| \frac{k}{n} - x \right\|_{\infty} > \frac{1}{n^{\beta}} \end{cases}}^{\infty} Z_3(nx-k) < \frac{4e^2}{\pi e^{n^{1-\beta}}} = c_3(\beta, n), \tag{12.80}$$

$0 < \beta < 1, n \in \mathbb{N} : n^{1-\beta} > 2, x \in \mathbb{R}^N, m \in \mathbb{N}$.

(vi) By Theorem 12.36 we get that

$$0 < \frac{1}{\sum_{k=\lceil na \rceil}^{\lfloor nb \rfloor} Z_3(nx-k)} < \left(\frac{\pi}{gd(2)} \right)^N \cong (2.412)^N =: \gamma_3(N), \tag{12.81}$$

$\forall\, x \in \left(\prod_{i=1}^{N} [a_i, b_i] \right), \ n \in \mathbb{N}$.

Furthermore it holds

$$\lim_{n \to \infty} \sum_{k=\lceil na \rceil}^{\lfloor nb \rfloor} Z_3 (nx - k) \neq 1, \tag{12.82}$$

for at least some $x \in \left(\prod_{i=1}^{N} [a_i, b_i] \right)$.

12.3.4 *About the Generalized Symmetrical Activation Function*

Here we consider the generalized symmetrical sigmoid function [25, 33]

$$f_1 (x) = \frac{x}{(1 + |x|^{\mu})^{\frac{1}{\mu}}}, \; \mu > 0, x \in \mathbb{R}. \tag{12.83}$$

This has applications in immunology and protection from disease together with probability theory. It is also called a symmetrical protection curve.

The parameter μ is a shape parameter controling how fast the curve approaches the asymptotes for a given slope at the inflection point. When $\mu = 1$ f_1 is the absolute sigmoid function, and when $\mu = 2$, f_1 is the square root sigmoid function. When $\mu = 1.5$ the function approximates the arctangent function, when $\mu = 2.9$ it approximates the logistic function, and when $\mu = 3.4$ it approximates the error function. Parameter μ is estimated in the likelihood maximization [33]. For more see [33].

Next we study the particular generator sigmoid function

$$f_2 (x) = \frac{x}{\left(1 + |x|^{\lambda}\right)^{\frac{1}{\lambda}}}, \; \lambda \text{ is an odd number, } x \in \mathbb{R}. \tag{12.84}$$

We have that $f_2 (0) = 0$, and

$$f_2 (-x) = -f_2 (x), \tag{12.85}$$

so f_2 is symmetric with respect to zero.

When $x \geq 0$, we get that [25]

$$f_2' (x) = \frac{1}{\left(1 + x^{\lambda}\right)^{\frac{\lambda+1}{\lambda}}} > 0, \tag{12.86}$$

that is f_2 is strictly increasing on $[0, +\infty)$ and f_2 is strictly increasing on $(-\infty, 0]$. Hence f_2 is strictly increasing on $\mathbb{R}$.

We also have $f_2 (+\infty) = f_2 (-\infty) = 1$.

Let us consider the activation function [25]:

$$\psi_4(x) = \frac{1}{4}\left[f_2(x+1) - f_2(x-1)\right] =$$

$$\frac{1}{4}\left[\frac{(x+1)}{\left(1+|x+1|^{\lambda}\right)^{\frac{1}{\lambda}}} - \frac{(x-1)}{\left(1+|x-1|^{\lambda}\right)^{\frac{1}{\lambda}}}\right]. \tag{12.87}$$

Clearly it holds [25]

$$\psi_4(x) = \psi_4(-x), \ \forall \, x \in \mathbb{R}. \tag{12.88}$$

and

$$\psi_4(0) = \frac{1}{2\sqrt[\lambda]{2}}, \tag{12.89}$$

and $\psi_4(x) > 0$, $\forall \, x \in \mathbb{R}$.

Following [25], we have that ψ_4 is strictly decreasing over $[0, +\infty)$, and ψ_4 is strictly increasing on $(-\infty, 0]$, by ψ_4-symmetry with respect to y-axis, and $\psi_4'(0) = 0$.

Clearly it is

$$\lim_{x \to +\infty} \psi_4(x) = \lim_{x \to -\infty} \psi_4(x) = 0, \tag{12.90}$$

therefore the x-axis is the horizontal asymptote of $\psi_4(x)$.

The value

$$\psi_4(0) = \frac{1}{2\sqrt[\lambda]{2}}, \ \lambda \text{ is an odd number,} \tag{12.91}$$

is the maximum of ψ_4, which is a bell shaped function.

We need

Theorem 12.38 ([25]) *It holds*

$$\sum_{i=-\infty}^{\infty} \psi_4(x-i) = 1, \ \forall \, x \in \mathbb{R}. \tag{12.92}$$

Theorem 12.39 ([25]) *We have that*

$$\int_{-\infty}^{\infty} \psi_4(x) \, dx = 1. \tag{12.93}$$

So that $\psi_4(x)$ is a density function on $\mathbb{R}$.

We need

Theorem 12.40 ([25]) *Let* $0 < \alpha < 1$, *and* $n \in \mathbb{N}$ *with* $n^{1-\alpha} > 2$. *It holds*

$$\sum_{\substack{j=-\infty \\ : |nx-j| \geq n^{1-\alpha}}}^{\infty} \psi_4 (nx - j) < \frac{1}{2\lambda \left(n^{1-\alpha} - 2\right)^{\lambda}} =: c_4 (\alpha, n), \tag{12.94}$$

where $\lambda \in \mathbb{N}$ is an odd number.

We also need

Theorem 12.41 ([25]) *Let $[a, b] \subset \mathbb{R}$ and $n \in \mathbb{N}$ so that $\lceil na \rceil \leq \lfloor nb \rfloor$. Then*

$$\frac{1}{\sum_{k=\lceil na \rceil}^{\lfloor nb \rfloor} \psi_4 (|nx - k|)} < 2\sqrt[\lambda]{1 + 2^{\lambda}} =: \alpha_4, \tag{12.95}$$

where λ is an odd number, $\forall\, x \in [a, b]$.

We make

Remark 12.42 ([25]) (1) We have that

$$\lim_{n \to \infty} \sum_{k=\lceil na \rceil}^{\lfloor nb \rfloor} \psi_4 (nx - k) \neq 1, \text{ for at least some } x \in [a, b]. \tag{12.96}$$

(2) Let $[a, b] \subset \mathbb{R}$. For large enough n we always obtain $\lceil na \rceil \leq \lfloor nb \rfloor$. Also $a \leq \frac{k}{n} \leq b$, iff $\lceil na \rceil \leq k \leq \lfloor nb \rfloor$.

In general it holds that

$$\sum_{k=\lceil na \rceil}^{\lfloor nb \rfloor} \psi_4 (nx - k) \leq 1. \tag{12.97}$$

We introduce (see also [28])

$$Z_4 (x_1, ..., x_N) := Z_4 (x) := \prod_{i=1}^{N} \psi_4 (x_i), \quad x = (x_1, ..., x_N) \in \mathbb{R}^N, \ N \in \mathbb{N}. \tag{12.98}$$

It has the properties:

(i) $Z_4 (x) > 0, \ \forall\, x \in \mathbb{R}^N$,

(ii)

$$\sum_{k=-\infty}^{\infty} Z_4 (x - k) := \sum_{k_1=-\infty}^{\infty} \sum_{k_2=-\infty}^{\infty} \ldots \sum_{k_N=-\infty}^{\infty} Z_4 (x_1 - k_1, ..., x_N - k_N) = 1, \tag{12.99}$$

where $k := (k_1, ..., k_n) \in \mathbb{Z}^N, \forall\, x \in \mathbb{R}^N$,

hence
(iii)

$$\sum_{k=-\infty}^{\infty} Z_4(nx-k) = 1, \tag{12.100}$$

$\forall\, x \in \mathbb{R}^N;\ n \in \mathbb{N}$,
and
(iv)

$$\int_{\mathbb{R}^N} Z_4(x)\,dx = 1, \tag{12.101}$$

that is Z_4 is a multivariate density function.
(v) It is clear that

$$\sum_{\substack{k=-\infty \\ \left\| \frac{k}{n} - x \right\|_\infty > \frac{1}{n^\beta}}}^{\infty} Z_4(nx-k) < \frac{1}{2\lambda\left(n^{1-\beta}-2\right)^\lambda} = c_4(\beta, n), \tag{12.102}$$

$0<\beta<1,\ n \in \mathbb{N} : n^{1-\beta} > 2,\ x \in \mathbb{R}^N,\ \lambda$ is odd.
(vi) By Theorem 12.41 we get that

$$0 < \frac{1}{\sum_{k=\lceil na \rceil}^{\lfloor nb \rfloor} Z_4(nx-k)} < \left(2\sqrt[\lambda]{1+2^\lambda}\right)^N =: \gamma_4(N), \tag{12.103}$$

$\forall\, x \in \left(\prod_{i=1}^N [a_i, b_i]\right),\ n \in \mathbb{N},\ \lambda$ is odd.
Furthermore it holds

$$\lim_{n\to\infty} \sum_{k=\lceil na \rceil}^{\lfloor nb \rfloor} Z_4(nx-k) \neq 1, \tag{12.104}$$

for at least some $x \in \left(\prod_{i=1}^N [a_i, b_i]\right)$.

Let $f \in C\left(\prod_{i=1}^N [a_i, b_i]\right)$, and $n \in \mathbb{N}$ such that $\lceil na_i \rceil \le \lfloor nb_i \rfloor$, $i = 1, ..., N$.

We define the multivariate averaged positive linear neural network operators ($x := (x_1, ..., x_N) \in \left(\prod_{i=1}^N [a_i, b_i]\right)$); $j = 1, 2, 3, 4$:

$${}_jA_n(f, x_1, ..., x_N) := {}_jA_n(f, x) := \frac{\sum_{k=\lceil na \rceil}^{\lfloor nb \rfloor} f\left(\frac{k}{n}\right) Z_j(nx-k)}{\sum_{k=\lceil na \rceil}^{\lfloor nb \rfloor} Z_j(nx-k)} = \tag{12.105}$$

$$\frac{\sum_{k_1=\lceil na_1\rceil}^{\lfloor nb_1\rfloor}\sum_{k_2=\lceil na_2\rceil}^{\lfloor nb_2\rfloor}\cdots\sum_{k_N=\lceil na_N\rceil}^{\lfloor nb_N\rfloor} f\left(\frac{k_1}{n},...,\frac{k_N}{n}\right)\left(\prod_{i=1}^{N}\psi_j\left(nx_i-k_i\right)\right)}{\prod_{i=1}^{N}\left(\sum_{k_i=\lceil na_i\rceil}^{\lfloor nb_i\rfloor}\psi_j\left(nx_i-k_i\right)\right)}.$$

For large enough $n \in \mathbb{N}$ we always obtain $\lceil na_i\rceil \leq \lfloor nb_i\rfloor$, $i = 1, ..., N$. Also $a_i \leq \frac{k_i}{n} \leq b_i$, iff $\lceil na_i\rceil \leq k_i \leq \lfloor nb_i\rfloor$, $i = 1, ..., N$.

When $f \in C_B\left(\mathbb{R}^N\right)$ we define $(j = 1, 2, 3, 4)$

$$ {}_jB_n\left(f, x\right) := {}_jB_n\left(f, x_1, ..., x_N\right) := \sum_{k=-\infty}^{\infty} f\left(\frac{k}{n}\right) Z_j\left(nx-k\right) := \tag{12.106}$$

$$\sum_{k_1=-\infty}^{\infty}\sum_{k_2=-\infty}^{\infty}\cdots\sum_{k_N=-\infty}^{\infty} f\left(\frac{k_1}{n},\frac{k_2}{n},...,\frac{k_N}{n}\right)\left(\prod_{i=1}^{N}\psi_j\left(nx_i-k_i\right)\right),$$

$n \in \mathbb{N}$, $\forall\, x \in \mathbb{R}^N$, $N \in \mathbb{N}$, the multivariate quasi-interpolation neural network operators.

Also for $f \in C_B\left(\mathbb{R}^N\right)$ we define the multivariate Kantorovich type neural network operators $(j = 1, 2, 3, 4)$

$$ {}_jC_n\left(f, x\right) := {}_jC_n\left(f, x_1, ..., x_N\right) := \sum_{k=-\infty}^{\infty}\left(n^N\int_{\frac{k}{n}}^{\frac{k+1}{n}} f\left(t\right)dt\right) Z_j\left(nx-k\right) := \tag{12.107}$$

$$\sum_{k_1=-\infty}^{\infty}\sum_{k_2=-\infty}^{\infty}\cdots\sum_{k_N=-\infty}^{\infty}\left(n^N\int_{\frac{k_1}{n}}^{\frac{k_1+1}{n}}\int_{\frac{k_2}{n}}^{\frac{k_2+1}{n}}\cdots\int_{\frac{k_N}{n}}^{\frac{k_N+1}{n}} f\left(t_1,...,t_N\right)dt_1...dt_N\right)$$

$$\left(\prod_{i=1}^{N}\psi_j\left(nx_i-k_i\right)\right),$$

$n \in \mathbb{N}$, $\forall\, x \in \mathbb{R}^N$.

Again for $f \in C_B\left(\mathbb{R}^N\right)$, $N \in \mathbb{N}$, we define the multivariate neural network operators of quadrature type ${}_jD_n\left(f, x\right)$, $n \in \mathbb{N}$, as follows. Let $\theta = (\theta_1, ..., \theta_N) \in \mathbb{N}^N$, $\overline{r} = (r_1, ..., r_N) \in \mathbb{Z}_+^N$, $w_{\overline{r}} = w_{r_1,r_2,...r_N} \geq 0$, such that $\sum_{\overline{r}=0}^{\theta} w_{\overline{r}} = \sum_{r_1=0}^{\theta_1}\sum_{r_2=0}^{\theta_2}\cdots\sum_{r_N=0}^{\theta_N} w_{r_1,r_2,...r_N} = 1$; $k \in \mathbb{Z}^N$ and

$$\delta_{nk}\left(f\right) := \delta_{n,k_1,k_2,...,k_N}\left(f\right) := \sum_{\overline{r}=0}^{\theta} w_{\overline{r}} f\left(\frac{k}{n}+\frac{\overline{r}}{n\theta}\right) :=$$

$$\sum_{r_1=0}^{\theta_1}\sum_{r_2=0}^{\theta_2}\cdots\sum_{r_N=0}^{\theta_N} w_{r_1,r_2,\ldots r_N} f\left(\frac{k_1}{n}+\frac{r_1}{n\theta_1}, \frac{k_2}{n}+\frac{r_2}{n\theta_2}, \ldots, \frac{k_N}{n}+\frac{r_N}{n\theta_N}\right), \quad (12.108)$$

where $\frac{\overline{r}}{\theta} := \left(\frac{r_1}{\theta_1}, \frac{r_2}{\theta_2}, \ldots, \frac{r_N}{\theta_N}\right)$; $j = 1, 2, 3, 4$.

We put

$${}_jD_n(f, x) := {}_jD_n(f, x_1, \ldots, x_N) := \sum_{k=-\infty}^{\infty} \delta_{nk}(f) Z_j(nx-k) := \quad (12.109)$$

$$\sum_{k_1=-\infty}^{\infty}\sum_{k_2=-\infty}^{\infty}\cdots\sum_{k_N=-\infty}^{\infty} \delta_{n,k_1,k_2,\ldots,k_N}(f)\left(\prod_{i=1}^{N}\psi_j(nx_i-k_i)\right),$$

$\forall\, x \in \mathbb{R}^N$.

Let $f \in C^m\left(\prod_{i=1}^{N}[a_i, b_i]\right)$, $m, N \in \mathbb{N}$. Here f_α denotes a partial derivative of f, $\alpha := (\alpha_1, \ldots, \alpha_N)$, $\alpha_i \in \mathbb{Z}_+$, $i = 1, \ldots, N$, and $|\alpha| := \sum_{i=1}^{N}\alpha_i = l$, where $l = 0, 1, \ldots, m$. We write also $f_\alpha := \frac{\partial^\alpha f}{\partial x^\alpha}$ and we say it is of order l.

We denote

$$\omega_{1,m}^{\max}(f_\alpha, h) := \max_{\alpha:|\alpha|=m}\omega_1(f_\alpha, h).$$

Call also

$$\|f_\alpha\|_{\infty,m}^{\max} := \max_{\alpha:|\alpha|=m}\left\{\|f_\alpha\|_\infty\right\},$$

where $\|\cdot\|_\infty$ is the supremum norm.

In [26–28, 30], $j = 1, 2, 3, 4$, we studied the basic approximation properties of ${}_jA_n$, ${}_jB_n$, ${}_jC_n$, ${}_jD_n$ neural network operators and as well of their iterates for Banach space valued functions. That is, the quantitative pointwise and uniform convergence of these operators to the unit operator I.

We need

Theorem 12.43 (see [26–28, 30]) *Let* $f \in C\left(\prod_{i=1}^{N}[a_i, b_i]\right)$, $0 < \beta < 1$, $x \in \left(\prod_{i=1}^{N}[a_i, b_i]\right)$, $N, n \in \mathbb{N}$ *with* $n^{1-\beta} > 2$; $j = 1, 2, 3, 4$. *Then*

(1)

$$\left|{}_jA_n(f, x) - f(x)\right| \leq \gamma_j(N)\left[\omega_1\left(f, \frac{1}{n^\beta}\right) + 2c_j(\beta, n)\|f\|_\infty\right] =: \lambda_{j1}, \quad (12.110)$$

and

(2)

$$\left\|{}_jA_n(f) - f\right\|_\infty \leq \lambda_{j1}. \quad (12.111)$$

We notice that $\lim_{n\to\infty} {}_jA_n(f) = f$, *pointwise and uniformly.*

We need

Theorem 12.44 (see [26–28, 30]) *Let* $f \in C_B(\mathbb{R}^N)$, $0 < \beta < 1$, $x \in \mathbb{R}^N$, $N, n \in \mathbb{N}$ *with* $n^{1-\beta} > 2$; $j = 1, 2, 3, 4$. *Then*

(1)

$$\left|{}_jB_n(f,x) - f(x)\right| \leq \omega_1\left(f, \frac{1}{n^\beta}\right) + 2c_j(\beta, n)\|f\|_\infty =: \lambda_{j2}, \tag{12.112}$$

(2)

$$\left\|{}_jB_n(f) - f\right\|_\infty \leq \lambda_{j2}. \tag{12.113}$$

Given that $f \in \left(C_U(\mathbb{R}^N) \cap C_B(\mathbb{R}^N)\right)$, *we obtain* $\lim_{n\to\infty} {}_jB_n(f) = f$, *uniformly.*

We also need

Theorem 12.45 (see [26–28, 30]) *Let* $f \in C_B(\mathbb{R}^N)$, $0 < \beta < 1$, $x \in \mathbb{R}^N$, $N, n \in \mathbb{N}$ *with* $n^{1-\beta} > 2$; $j = 1, 2, 3, 4$. *Then*

(1)

$$\left|{}_jC_n(f,x) - f(x)\right| \leq \omega_1\left(f, \frac{1}{n} + \frac{1}{n^\beta}\right) + 2c_j(\beta, n)\|f\|_\infty =: \lambda_{j3}, \tag{12.114}$$

(2)

$$\left\|{}_jC_n(f) - f\right\|_\infty \leq \lambda_{j3}. \tag{12.115}$$

Given that $f \in \left(C_U(\mathbb{R}^N) \cap C_B(\mathbb{R}^N)\right)$, *we obtain* $\lim_{n\to\infty} {}_jC_n(f) = f$, *uniformly.*

We also need

Theorem 12.46 (see [26–28, 30]) *Let* $f \in C_B(\mathbb{R}^N)$, $0 < \beta < 1$, $x \in \mathbb{R}^N$, $N, n \in \mathbb{N}$ *with* $n^{1-\beta} > 2$; $j = 1, 2, 3, 4$. *Then*

(1)

$$\left|{}_jD_n(f,x) - f(x)\right| \leq \omega_1\left(f, \frac{1}{n} + \frac{1}{n^\beta}\right) + 2c_j(\beta, n)\|f\|_\infty = \lambda_{j3}, \tag{12.116}$$

(2)

$$\left\|{}_jD_n(f) - f\right\|_\infty \leq \lambda_{j3}. \tag{12.117}$$

Given that $f \in \left(C_U(\mathbb{R}^N) \cap C_B(\mathbb{R}^N)\right)$, *we obtain* $\lim_{n\to\infty} {}_jD_n(f) = f$, *uniformly.*

We finally mention (similar to [21], p. 481).

Theorem 12.47 *Let* $f \in C^m\left(\prod_{i=1}^N [a_i, b_i]\right)$, $0 < \beta < 1$, $n, m, N \in \mathbb{N}$, $n^{1-\beta} > 2$, $x \in \left(\prod_{i=1}^N [a_i, b_i]\right)$; $j = 1, 2, 3, 4$. *Then*

(i)

$$\left| {}_jA_n(f, x) - f(x) - \sum_{j_*=1}^{m} \left(\sum_{|\alpha|=j_*} \left(\frac{f_\alpha(x)}{\prod_{i=1}^N \alpha_i!} \right) {}_jA_n \left(\prod_{i=1}^N (\cdot - x_i)^{\alpha_i}, x \right) \right) \right| \leq \tag{12.118}$$

$$\gamma_j(N) \cdot \left\{ \frac{N^m}{m! n^{m\beta}} \omega_{1,m}^{\max}\left(f_\alpha, \frac{1}{n^\beta}\right) + \left(\frac{\|b-a\|_\infty^m \|f_\alpha\|_{\infty,m}^{\max} N^m}{m!} \right) 2c_j(\beta, n) \right\},$$

(ii)

$$\left| {}_jA_n(f, x) - f(x) \right| \leq \gamma_j(N) \cdot \tag{12.119}$$

$$\left\{ \sum_{j_*=1}^{m} \left(\sum_{|\alpha|=j_*} \left(\frac{|f_\alpha(x)|}{\prod_{i=1}^N \alpha_i!} \right) \left[\frac{1}{n^{\beta j_*}} + \left(\prod_{i=1}^N (b_i - a_i)^{\alpha_i} \right) c_j(\beta, n) \right] \right) \right.$$

$$\left. + \frac{N^m}{m! n^{m\beta}} \omega_{1,m}^{\max}\left(f_\alpha, \frac{1}{n^\beta}\right) + \left(\frac{\|b-a\|_\infty^m \|f_\alpha\|_{\infty,m}^{\max} N^m}{m!} \right) 2c_j(\beta, n) \right\},$$

(iii)

$$\left\| {}_jA_n(f) - f \right\|_\infty \leq \gamma_j(N) \cdot \tag{12.120}$$

$$\left\{ \sum_{j_*=1}^{m} \left(\sum_{|\alpha|=j_*} \left(\frac{\|f_\alpha\|_\infty}{\prod_{i=1}^N \alpha_i!} \right) \left[\frac{1}{n^{\beta j_*}} + \left(\prod_{i=1}^N (b_i - a_i)^{\alpha_i} \right) c_j(\beta, n) \right] \right) \right.$$

$$\left. + \frac{N^m}{m! n^{m\beta}} \omega_{1,m}^{\max}\left(f_\alpha, \frac{1}{n^\beta}\right) + \left(\frac{\|b-a\|_\infty^m \|f_\alpha\|_{\infty,m}^{\max} N^m}{m!} \right) 2c_j(\beta, n) \right\},$$

(iv) additionally assume $f_\alpha(x_0) = 0$, *for all* $\alpha : |\alpha| = 1, ..., m$; $x_0 \in \left(\prod_{i=1}^N [a_i, b_i]\right)$, *then*

$$\left| {}_jA_n(f, x_0) - f(x_0) \right| \leq \gamma_j(N) \left\{ \frac{N^m}{m! n^{m\beta}} \omega_{1,m}^{\max}\left(f_\alpha, \frac{1}{n^\beta}\right) + \right. \tag{12.121}$$

$$\left. \left(\frac{\|b-a\|_\infty^m \|f_\alpha\|_{\infty,m}^{\max} N^m}{m!} \right) 2c_j(\beta, n) \right\},$$

notice in the last the extremely high rate of convergence at $n^{-\beta(m+1)}$.

12.4 Main Results: Fuzzy Multivariate Neural Network Approximation Based on the Arctangent, Algebraic, Gudermannian and Generalized Symmetrical Activation Functions

We define the following Fuzzy multivariate Neural Network Operators ${}_jA_n^{\mathcal{F}}$, ${}_jB_n^{\mathcal{F}}$, ${}_jC_n^{\mathcal{F}}$, ${}_jD_n^{\mathcal{F}}$, $j = 1, 2, 3, 4$, based on the arctangent, algebraic, Gudermannian and generalized symmetrical activation functions. These are analogs of the real ${}_jA_n$, ${}_jB_n$, ${}_jC_n$, ${}_jD_n$, see (12.105), (12.106), (12.107) and (12.109), respectively.

Let $f \in C_{\mathcal{F}}\left(\prod_{i=1}^{N}[a_i, b_i]\right)$, $N \in \mathbb{N}$, we set

$$ {}_jA_n^{\mathcal{F}}(f, x_1, ..., x_N) := {}_jA_n^{\mathcal{F}}(f, x) := \frac{\sum_{k=\lceil na \rceil}^{\lfloor nb \rfloor *} f\left(\frac{k}{n}\right) \odot Z_j(nx - k)}{\sum_{k=\lceil na \rceil}^{\lfloor nb \rfloor} Z_j(nx - k)} $$

$$ = \frac{\sum_{k_1=\lceil na_1 \rceil}^{\lfloor nb_1 \rfloor *} \cdots \sum_{k_N=\lceil na_N \rceil}^{\lfloor nb_N \rfloor *} f\left(\frac{k_1}{n}, ..., \frac{k_N}{n}\right) \odot \left(\prod_{i=1}^{N} \psi_j(nx_i - k_i)\right)}{\prod_{i=1}^{N}\left(\sum_{k_i=\lceil na_i \rceil}^{\lfloor nb_i \rfloor} \psi_j(nx_i - k_i)\right)}, \qquad (12.122) $$

$x \in \prod_{i=1}^{N}[a_i, b_i]$, $n \in \mathbb{N}$.

Let $f \in C_B\left(\mathbb{R}^N, \mathbb{R}_{\mathcal{F}}\right)$, we put

$$ {}_jB_n^{\mathcal{F}}(f, x) := {}_jB_n^{\mathcal{F}}(f, x_1, ..., x_N) := \sum_{k=-\infty}^{\infty *} f\left(\frac{k}{n}\right) \odot Z_j(nx - k) $$

$$:= \sum_{k_1=-\infty}^{\infty *} \cdots \sum_{k_N=-\infty}^{\infty *} f\left(\frac{k_1}{n}, ..., \frac{k_N}{n}\right) \odot \left(\prod_{i=1}^{N} \psi_j(nx_i - k_i)\right), \qquad (12.123) $$

$x \in \mathbb{R}^N$, $n \in \mathbb{N}$.

Let $f \in C_B\left(\mathbb{R}^N, \mathbb{R}_{\mathcal{F}}\right)$, we define the multivariate fuzzy Kantorovich type neural network operator,

$$ {}_jC_n^{\mathcal{F}}(f, x) := {}_jC_n^{\mathcal{F}}(f, x_1, ..., x_N) := \sum_{k=-\infty}^{\infty *} \left(n^N \odot \int_{\frac{k}{n}}^{\frac{k+1}{n}} f(t)\, dt\right) \odot Z_j(nx - k) := $$

$$\sum_{k_1=-\infty}^{\infty *} \dots \sum_{k_N=-\infty}^{\infty *} \left(n^N \odot \int_{\frac{k_1}{n}}^{\frac{k_1+1}{n}} \dots \int_{\frac{k_N}{n}}^{\frac{k_N+1}{n}} f(t_1, ..., t_N) \, dt_1 ... dt_N \right)$$

$$\odot \left(\prod_{i=1}^{N} \psi_j (nx_i - k_i) \right), \tag{12.124}$$

$x \in \mathbb{R}^N$, $n \in \mathbb{N}$.

Let $f \in C_B(\mathbb{R}^N, \mathbb{R}_{\mathcal{F}})$, we define the multivariate fuzzy quadrature type neural network operator. Let here

$$\delta_{nk}^{\mathcal{F}}(f) := \delta_{n,k_1,...,k_N}^{\mathcal{F}}(f) := \sum_{\overline{r}=0}^{\theta *} w_{\overline{r}} \odot f\left(\frac{k}{n} + \frac{\overline{r}}{n\theta}\right) := \tag{12.125}$$

$$\sum_{r_1=0}^{\theta_1 *} \dots \sum_{r_N=0}^{\theta_N *} w_{r_1,...r_N} \odot f\left(\frac{k_1}{n} + \frac{r_1}{n\theta_1}, ..., \frac{k_N}{n} + \frac{r_N}{n\theta_N}\right).$$

We put

$${}_jD_n^{\mathcal{F}}(f, x) := {}_jD_n^{\mathcal{F}}(f, x_1, ..., x_N) := \sum_{k=-\infty}^{\infty *} \delta_{nk}^{\mathcal{F}}(f) \odot Z_j(nx - k) :=$$

$$\sum_{k_1=-\infty}^{\infty *} \dots \sum_{k_N=-\infty}^{\infty *} \delta_{n,k_1,...,k_N}^{\mathcal{F}}(f) \odot \left(\prod_{i=1}^{N} \psi_j (nx_i - k_i) \right), \tag{12.126}$$

$x \in \mathbb{R}^N$, $n \in \mathbb{N}$.

We can put together all ${}_jB_n^{\mathcal{F}}$, ${}_jC_n^{\mathcal{F}}$, ${}_jD_n^{\mathcal{F}}$ fuzzy operators as follows:

$${}_jL_n^{\mathcal{F}}(f, x) := \sum_{k=-\infty}^{\infty *} l_{nk}^{\mathcal{F}}(f) \odot Z_j(nx - k), \;\; j = 1, 2, 3, 4, \tag{12.127}$$

where

$$l_{nk}^{\mathcal{F}}(f) := \begin{cases} f\left(\frac{k}{n}\right), \text{ if } {}_jL_n^{\mathcal{F}} = {}_jB_n^{\mathcal{F}}, \\ n^N \odot \int_{\frac{k}{n}}^{\frac{k+1}{n}} f(t) \, dt, \text{ if } {}_jL_n^{\mathcal{F}} = {}_jC_n^{\mathcal{F}}, \\ \delta_{nk}^{\mathcal{F}}(f), \text{ if } {}_jL_n^{\mathcal{F}} = {}_jD_n^{\mathcal{F}}, \end{cases} \tag{12.128}$$

$x \in \mathbb{R}^N$, $n \in \mathbb{N}$.

Similarly, we can put together all ${}_jB_n$, ${}_jC_n$, ${}_jD_n$ real operators as

$$ {}_jL_n(f,x) := \sum_{k=-\infty}^{\infty} l_{nk}(f)\, Z_j(nx-k)\,, \ j=1,2,3,4, \tag{12.129} $$

where

$$ l_{nk}(f) := \begin{cases} f\left(\frac{k}{n}\right), \text{if } {}_jL_n = {}_jB_n, \\ n^N \int_{\frac{k}{n}}^{\frac{k+1}{n}} f(t)\,dt, \text{if } {}_jL_n = {}_jC_n, \\ \delta_{nk}(f), \text{if } {}_jL_n = {}_jD_n, \end{cases} \tag{12.130} $$

$x \in \mathbb{R}^N$, $n \in \mathbb{N}$.

Let $r \in [0,1]$, we observe that ($j = 1,2,3,4$):

$$ \left[{}_jA_n^{\mathcal{F}}(f,x)\right]^r = \sum_{k=\lceil na \rceil}^{\lfloor nb \rfloor} \left[f\left(\frac{k}{n}\right)\right]^r \left(\frac{Z_j(nx-k)}{\sum_{k=\lceil na \rceil}^{\lfloor nb \rfloor} Z_j(nx-k)} \right) = $$

$$ \sum_{k=\lceil na \rceil}^{\lfloor nb \rfloor} \left[f_-^{(r)}\left(\frac{k}{n}\right), f_+^{(r)}\left(\frac{k}{n}\right)\right] \left(\frac{Z_j(nx-k)}{\sum_{k=\lceil na \rceil}^{\lfloor nb \rfloor} Z_j(nx-k)} \right) = \tag{12.131} $$

$$ \left[\sum_{k=\lceil na \rceil}^{\lfloor nb \rfloor} f_-^{(r)}\left(\frac{k}{n}\right) \left(\frac{Z_j(nx-k)}{\sum_{k=\lceil na \rceil}^{\lfloor nb \rfloor} Z_j(nx-k)} \right), \sum_{k=\lceil na \rceil}^{\lfloor nb \rfloor} f_+^{(r)}\left(\frac{k}{n}\right) \left(\frac{Z_j(nx-k)}{\sum_{k=\lceil na \rceil}^{\lfloor nb \rfloor} Z_j(nx-k)} \right) \right] $$

$$ = \left[{}_jA_n\left(f_-^{(r)}, x\right), \ {}_jA_n\left(f_+^{(r)}, x\right) \right]. \tag{12.132} $$

We have proved that

$$ \left({}_jA_n^{\mathcal{F}}(f,x)\right)_{\pm}^{(r)} = {}_jA_n\left(f_{\pm}^{(r)}, x\right), \ j=1,2,3,4, \tag{12.133} $$

respectively, $\forall\, r \in [0,1]$, $\forall\, x \in \left(\prod_{i=1}^{N} [a_i, b_i] \right)$.

Similarly, as in [21], pp. 485–489, a lengthy proof (see Remark 21.31 and proof of (21.76) there) it holds that

$$ \left({}_jL_n^{\mathcal{F}}(f,x)\right)_{\pm}^{(r)} = {}_jL_n\left(f_{\pm}^{(r)}, x\right), \ j=1,2,3,4, \tag{12.134} $$

respectively, $\forall\, r \in [0,1]$, $\forall\, x \in \mathbb{R}^N$.

Based on (12.133) and (12.134) now one may write ($j = 1, 2, 3, 4$)

$$D\left({}_{j}A_{n}^{\mathcal{F}}(f, x), f(x)\right) = \tag{12.135}$$

$$\sup_{r\in[0,1]} \max\left\{\left|\left({}_{j}A_{n}\left(f_{-}^{(r)}, x\right)\right) - f_{-}^{(r)}(x)\right|, \left|{}_{j}A_{n}\left(f_{+}^{(r)}, x\right) - f_{+}^{(r)}(x)\right|\right\},$$

and

$$D\left({}_{j}L_{n}^{\mathcal{F}}(f, x), f(x)\right) =$$

$$\sup_{r\in[0,1]} \max\left\{\left|\left({}_{j}L_{n}\left(f_{-}^{(r)}, x\right)\right) - f_{-}^{(r)}(x)\right|, \left|{}_{j}L_{n}\left(f_{+}^{(r)}, x\right) - f_{+}^{(r)}(x)\right|\right\}. \tag{12.136}$$

We present

Theorem 12.48 *Let* $f \in C_{\mathcal{F}}\left(\prod_{i=1}^{N}[a_i, b_i]\right)$, $0 < \beta < 1$, $x \in \left(\prod_{i=1}^{N}[a_i, b_i]\right)$, $N, n \in \mathbb{N}$ *with* $n^{1-\beta} > 2$; $j = 1, 2, 3, 4$. *Then*

(1)

$$D\left({}_{j}A_{n}^{\mathcal{F}}(f, x), f(x)\right) \leq$$

$$\gamma_j(N)\left[\omega_1^{(\mathcal{F})}\left(f, \frac{1}{n^\beta}\right) + 2c_j(\beta, n)\, D^*\left(f, \widetilde{o}\right)\right] =: \rho_{j1}, \tag{12.137}$$

and

(2)

$$D^*\left({}_{j}A_{n}^{\mathcal{F}}(f), f\right) \leq \rho_{j1}. \tag{12.138}$$

We notice that ${}_{j}A_{n}^{\mathcal{F}}(f, x) \stackrel{D}{\rightarrow} f(x)$, *and* ${}_{j}A_{n}^{\mathcal{F}}(f) \stackrel{D^*}{\rightarrow} f$, *as* $n \rightarrow \infty$, *quantitatively with rates.*

Proof Since $f \in C_{\mathcal{F}}\left(\prod_{i=1}^{N}[a_i, b_i]\right)$ we have that $f_{\pm}^{(r)} \in C\left(\prod_{i=1}^{N}[a_i, b_i]\right)$, $\forall\, r \in [0, 1]$. Hence by (12.110) we obtain

$$\left|{}_{j}A_{n}\left(f_{\pm}^{(r)}, x\right) - f_{\pm}^{(r)}(x)\right| \leq \gamma_j(N)\left[\omega_1\left(f_{\pm}^{(r)}, \frac{1}{n^\beta}\right) + 2c_j(\beta, n)\left\|f_{\pm}^{(r)}\right\|_{\infty}\right]$$

$$\stackrel{\text{(by (12.8), (12.3))}}{\leq} \gamma_j(N)\left[\omega_1^{(\mathcal{F})}\left(f, \frac{1}{n^\beta}\right) + 2c_j(\beta, n)\, D^*\left(f, \widetilde{o}\right)\right]. \tag{12.139}$$

By (12.135) now we are proving the claim. ∎

We give

Theorem 12.49 *Let $f \in C_B\left(\mathbb{R}^N, \mathbb{R}_{\mathcal{F}}\right)$, $0 < \beta < 1$, $x \in \mathbb{R}^N$, $N, n \in \mathbb{N}$, with $n^{1-\beta} > 2$; $j = 1, 2, 3, 4$. Then*

(1)

$$D\left({}_jB_n^{\mathcal{F}}(f, x), f(x)\right) \leq \tag{12.140}$$

$$\omega_1^{(\mathcal{F})}\left(f, \frac{1}{n^\beta}\right) + 2c_j(\beta, n) D^*(f, \widetilde{o}) =: \rho_{j2},$$

and

(2)

$$D^*\left({}_jB_n^{\mathcal{F}}(f), f\right) \leq \rho_{j2}. \tag{12.141}$$

Proof Similar to Theorem 12.48. We use (12.112) and (12.136), along with (12.3) and (12.8). ■

We also present

Theorem 12.50 *All as in Theorem 12.49, $j = 1, 2, 3, 4$. Then*

(1)

$$D\left({}_jC_n^{\mathcal{F}}(f, x), f(x)\right) \leq$$

$$\omega_1^{(\mathcal{F})}\left(f, \frac{1}{n} + \frac{1}{n^\beta}\right) + 2c_j(\beta, n) D^*(f, \widetilde{o}) =: \rho_{j3}, \tag{12.142}$$

and

(2)

$$D^*\left({}_jC_n^{\mathcal{F}}(f), f\right) \leq \rho_{j3}. \tag{12.143}$$

Proof Similar to Theorem 12.48. We use (12.114) and (12.136), along with (12.3) and (12.8). ■

We also give

Theorem 12.51 *All as in Theorem 12.49, $j = 1, 2, 3, 4$. Then*

(1)

$$D\left({}_jD_n^{\mathcal{F}}(f, x), f(x)\right) \leq$$

$$\omega_1^{(\mathcal{F})}\left(f, \frac{1}{n} + \frac{1}{n^\beta}\right) + 2c_j(\beta, n) D^*(f, \widetilde{o}) = \rho_{j3}, \tag{12.144}$$

and

(2)

$$D^*\left({}_jD_n^{\mathcal{F}}(f), f\right) \leq \rho_{j3}. \tag{12.145}$$

Proof Similar to Theorem 12.48. We use (12.116) and (12.136), along with (12.3) and (12.8). ∎

Note 12.52 By Theorems 12.49, 12.50, 12.51 for $f \in \left(C_B\left(\mathbb{R}^N, \mathbb{R}_{\mathcal{F}}\right) \cap C_{\mathcal{F}}^U\left(\mathbb{R}^N\right)\right)$, we obtain $\lim_{n\to\infty} D\left({}_jL_n^{\mathcal{F}}(f,x), f(x)\right) = 0$, and $\lim_{n\to\infty} D^*\left({}_jL_n^{\mathcal{F}}(f), f\right) = 0$, $j = 1, 2, 3, 4$, quantitatively with rates, where ${}_jL_n^{\mathcal{F}}$ is as in (12.127) and (12.128).

Notation 12.53 Let $f \in C_{\mathcal{F}}^m\left(\prod_{i=1}^N [a_i, b_i]\right)$, $m, N \in \mathbb{N}$. Here f_α denotes a fuzzy partial derivative with all related notation similar to the real case, see also Remark 12.15 and Notation 12.16. We denote

$$\omega_{1,m}^{(\mathcal{F})\max}(f_\alpha, h) := \max_{\alpha:|\alpha|=m} \omega_1^{(\mathcal{F})}(f_\alpha, h), \quad h > 0. \tag{12.146}$$

Call also

$$D_m^{*\max}(f_\alpha, \widetilde{o}) := \max_{\alpha:|\alpha|=m} \left\{D^*(f_\alpha, \widetilde{o})\right\}. \tag{12.147}$$

We finally present

Theorem 12.54 *Let* $f \in C_{\mathcal{F}}^m\left(\prod_{i=1}^N [a_i, b_i]\right)$, $0 < \beta < 1$, $n, m, N \in \mathbb{N}$ *with* $n^{1-\beta} > 2$, *and* $x \in \left(\prod_{i=1}^N [a_i, b_i]\right)$, $j = 1, 2, 3, 4$. *Then*

(1)

$$D\left({}_jA_n^{\mathcal{F}}(f,x), f(x)\right) \leq \gamma_j(N) \cdot$$

$$\left\{\sum_{j_*=1}^m \left(\sum_{|\alpha|=j_*} \frac{D(f_\alpha(x), \widetilde{o})}{\prod_{i=1}^N \alpha_i!} \left[\frac{1}{n^{\beta j_*}} + \left(\prod_{i=1}^N (b_i - a_i)^{\alpha_i}\right) c_j(\beta, n)\right]\right)\right.$$

$$\left. + \frac{N^m}{m! n^{m\beta}} \omega_{1,m}^{(\mathcal{F})\max}\left(f_\alpha, \frac{1}{n^\beta}\right) + \left(\frac{\|b-a\|_\infty^m D_m^{*\max}(f_\alpha, \widetilde{o}) N^m}{m!}\right) 2c_j(\beta, n)\right\}, \tag{12.148}$$

(2)

$$D^*\left({}_jA_n^{\mathcal{F}}(f), f\right) \leq \gamma_j(N) \cdot$$

$$\left\{\sum_{j_*=1}^m \left(\sum_{|\alpha|=j_*} \frac{D^*(f_\alpha, \widetilde{o})}{\prod_{i=1}^N \alpha_i!} \left[\frac{1}{n^{\beta j_*}} + \left(\prod_{i=1}^N (b_i - a_i)^{\alpha_i}\right) c_j(\beta, n)\right]\right)\right.$$

$$+\frac{N^m}{m!n^{m\beta}}\omega_{1,m}^{(\mathcal{F})\max}\left(f_\alpha,\frac{1}{n^\beta}\right)+\left(\frac{\|b-a\|_\infty^m D_m^{*\max}\left(f_\alpha,\widetilde{o}\right)N^m}{m!}\right)2c_j\left(\beta,n\right)\Bigg\},\tag{12.149}$$

(3) additionally assume that $f_\alpha(x_0)=\widetilde{o}$, *for all* $\alpha:|\alpha|=1,...,m$; $x_0\in\left(\prod_{i=1}^N[a_i,b_i]\right)$, *then*

$$D\left({}_jA_n^{\mathcal{F}}(f,x_0),f(x_0)\right)\leq\gamma_j(N)\left\{\frac{N^m}{m!n^{m\beta}}\omega_{1,m}^{(\mathcal{F})\max}\left(f_\alpha,\frac{1}{n^\beta}\right)+\right.$$

$$\left.\left(\frac{\|b-a\|_\infty^m D_m^{*\max}\left(f_\alpha,\widetilde{o}\right)N^m}{m!}\right)2c_j\left(\beta,n\right)\right\},\tag{12.150}$$

notice in the last the extremely high rate of convergence at $n^{-\beta(m+1)}$.

Above we derive quantitatively with rates the high speed approximation of $D\left({}_jA_n^{\mathcal{F}}(f,x),f(x)\right)\to 0$, *as* $n\to\infty$.

Also we establish with rates that $D^*\left({}_jA_n^{\mathcal{F}}(f),f\right)\to 0$, *as* $n\to\infty$, *involving the fuzzy smoothness of* f.

Proof Here $f_\pm^{(r)}\in C^m\left(\prod_{i=1}^N[a_i,b_i]\right)$. We observe that

$$\left|{}_jA_n\left(f_\pm^{(r)},x\right)-f_\pm^{(r)}(x)\right|\overset{(12.119)}{\leq}\gamma_j(N)\cdot$$

$$\left\{\sum_{j_*=1}^m\left(\sum_{|\alpha|=j_*}\left(\frac{\left|\left(f_\pm^{(r)}\right)_\alpha(x)\right|}{\prod_{i=1}^N\alpha_i!}\right)\left[\frac{1}{n^{\beta j_*}}+\left(\prod_{i=1}^N(b_i-a_i)^{\alpha_i}\right)c_j(\beta,n)\right]\right)\right.$$

$$+\frac{N^m}{m!n^{m\beta}}\omega_{1,m}^{\max}\left(\left(f_\pm^{(r)}\right)_\alpha,\frac{1}{n^\beta}\right)+$$

$$\left.\left(\frac{\|b-a\|_\infty^m\left\|\left(f_\pm^{(r)}\right)_\alpha\right\|_{\infty,m}^{\max}N^m}{m!}\right)2c_j(\beta,n)\right\}\overset{(12.13)}{=}\tag{12.151}$$

$$\gamma_j(N)\left\{\sum_{j_*=1}^m\left(\sum_{|\alpha|=j_*}\left(\frac{\left|(f_\alpha)_\pm^{(r)}(x)\right|}{\prod_{i=1}^N\alpha_i!}\right)\left[\frac{1}{n^{\beta j_*}}+\left(\prod_{i=1}^N(b_i-a_i)^{\alpha_i}\right)c_j(\beta,n)\right]\right)\right.$$

$$+\frac{N^m}{m!n^{m\beta}}\omega_{1,m}^{\max}\left((f_\alpha)_{\pm}^{(r)}, \frac{1}{n^\beta}\right)+$$

$$\left(\frac{\|b-a\|_\infty^m \left\|(f_\alpha)_{\pm}^{(r)}\right\|_{\infty,m}^{\max} N^m}{m!}\right)2c_j(\beta,n)\Bigg\} \overset{\text{(by (12.3), (12.8))}}{\leq} \quad (12.152)$$

$$\gamma_j(N)\left\{\sum_{j_*=1}^{m}\left(\sum_{|\alpha|=j_*}\left(\frac{D(f_\alpha(x),\widetilde{o})}{\prod_{i=1}^{N}\alpha_i!}\right)\left[\frac{1}{n^{\beta j_*}}+\left(\prod_{i=1}^{N}(b_i-a_i)^{\alpha_i}\right)c_j(\beta,n)\right]\right)\right.$$

$$+\frac{N^m}{m!n^{m\beta}}\omega_{1,m}^{(\mathcal{F})\max}\left(f_\alpha, \frac{1}{n^\beta}\right)+$$

$$\left.\left(\frac{\|b-a\|_\infty^m D_m^{*\max}(f_\alpha,\widetilde{o})N^m}{m!}\right)2c_j(\beta,n)\right\} =: T_j, \quad (12.153)$$

respectively in $\pm$.

We have proved that

$$\left|{}_jA_n\left(f_{\pm}^{(r)},x\right)-f_{\pm}^{(r)}(x)\right| \leq T_j, \quad (12.154)$$

$\forall\, r \in [0,1]$, respectively in $\pm$.

Using (12.135) we obtain

$$D\left({}_jA_n^{\mathcal{F}}(f,x), f(x)\right) \leq T_j, \quad (12.155)$$

proving the theorem. ■

References

1. G.A. Anastassiou, Rate of convergence of some neural network operators to the unit-univariate case. J. Math. Anal. Appl. **212**, 237–262 (1997)
2. G.A. Anastassiou, Rate of convergence of some multivariate neural network operators to the unit. Comput. Math. **40**, 1–19 (2000)
3. G.A. Anastassiou, *Quantitative Approximations* (Chapman and Hall/CRC, Boca Raton, New York, 2001)
4. G.A. Anastassiou, Higher order fuzzy approximation by fuzzy wavelet type and neural network operators. Comput. Math. **48**, 1387–1401 (2004)
5. G.A. Anastassiou, Fuzzy approximation by fuzzy convolution type operators. Comput. Math. **48**, 1369–1386 (2004)

6. G.A. Anastassiou, Higher order fuzzy Korovkin Theory via inequalities. Commun. Appl. Anal. **10**(2), 359–392 (2006)
7. G.A. Anastassiou, Fuzzy Korovkin Theorems and inequalities. J. Fuzzy Math. **15**(1), 169–205 (2007)
8. G.A. Anastassiou, On right fractional calculus. Chaos Solitons Fractals **42**, 365–376 (2009)
9. G.A. Anastassiou, *Fractional Differentiation Inequalities* (Springer, New York, 2009)
10. G.A. Anastassiou, Fractional Korovkin theory. Chaos Solitons Fractals **42**(4), 2080–2094 (2009)
11. G.A. Anastassiou, *Fuzzy Mathematics: Approximation Theory* (Springer, Heildelberg, New York, 2010)
12. G.A. Anastassiou, *Intelligent Systems: Approximation by Artificial Neural Networks, Intelligent Systems Reference Library*, vol. 19 (Springer, Heidelberg, 2011)
13. G.A. Anastassiou, Higher order multivariate fuzzy approximation by multivariate fuzzy wavelet type and neural network operators. J. Fuzzy Math. **19**(3), 601–618 (2011)
14. G.A. Anastassiou, Univariate hyperbolic tangent neural network approximation. Math. Comput. Model. **53**, 1111–1132 (2011)
15. G.A. Anastassiou, Multivariate hyperbolic tangent neural network approximation. Comput. Math. **61**, 809–821 (2011)
16. G.A. Anastassiou, Multivariate sigmoidal neural network approximation. Neural Netw. **24**, 378–386 (2011)
17. G.A. Anastassiou, Univariate sigmoidal neural network approximation. J. Comput. Anal. Appl. **14**(4), 659–690 (2012)
18. G.A. Anastassiou, Approximation by neural network iterates, in *Advances in Applied Mathematics and Approximation Theory: Contributions from AMAT 2012*, ed. by G. Anastassiou, O. Duman (Springer, NY, 2013), pp. 1–20
19. G.A. Anastassiou, High degree multivariate fuzzy approximation by quasi-interpolation neural network operators. Discontin. Nonlinearity Complex. **2**(2), 125–146 (2013)
20. G.A. Anastassiou, Rate of convergence of some multivariate neural network operators to the unit, revisited. J. Comput. Anal. Appl. **15**(7), 1300–1309 (2013)
21. G.A. Anastassiou, *Intelligent Systems II: Complete Approximation by Neural Network Operators* (Springer, Heidelberg, New York, 2016)
22. G.A. Anastassiou, *Intelligent Computations: Abstract Fractional Calculus* (Inequalities, Approximations, Springer, Heidelberg, New York, 2018)
23. G.A. Anastassiou, Algebraic function based Banach space valued ordinary and fractional neural network approximations. New Trends Math. Sci. **10** special issues (1), 100–125 (2022)
24. G.A. Anastassiou, Gudermannian function activated Banach space valued ordinary and fractional neural network approximation. Adv. Nonlinear Var. Inequal. **25**(2), 27–64 (2022)
25. G.A. Anastassiou, *Generalized Symmetrical Sigmoid Function Activated Banach Space Valued Ordinary and Fractional Neural Network Approximation* (Fasc. Matematica, Accepted for Publication, Analele Universităţii Oradea, 2022)
26. G.A. Anastassiou, Abstract multivariate Gudermannian function activated neural network approximations. Panamerican Math. J. (2022)
27. G.A. Anastassiou, *General Multivariate Arctangent Function Activated Neural Network Approximations*. Submitted (2022)
28. G.A. Anastassiou, *Generalized Symmetrical Sigmoid Function Activated Neural Network Multivariate Approximation*. Submitted (2022)
29. G.A. Anastassiou, *Multivariate Fuzzy Approximation by Neural Network Operators Activated by Several Sigmoid Functions Revisited*. Submitted (2022)
30. G.A. Anastassiou, R. Kozma, *Abstract Multivariate Algebraic Function Activated Neural Network Approximations*. Submitted (2022)
31. P. Cardaliaguet, G. Euvrard, Approximation of a function and its derivative with a neural network. Neural Netw. **5**, 207–220 (1992)
32. Z. Chen, F. Cao, The approximation operators with sigmoidal functions. Comput. Math. Appl. **58**, 758–765 (2009)

33. A.J. Dunning, J. Kensler, L. Goudeville, F. Bailleux, Some extensions in continuous methods for immunological correlates of protection. BMC Med. Res. Methodol. **15**(107) (2015). https://doi.org/10.1186/s12874-015-0096-9
34. S. Gal, Approximation theory in fuzzy setting, Chapter 13, in *Handbook of Analytic-Computational Methods in Applied Mathematics*, ed. by G (Chapman & Hall/CRC, Boca Raton, New York, Anastassiou, 2000), pp. 617–666
35. R. Goetschel Jr., W. Voxman, Elementary fuzzy calculus. Fuzzy Sets Syst. **18**, 31–43 (1986)
36. S. Haykin, *Neural Networks: A Comprehensive Foundation*, 2nd edn. (Prentice Hall, New York, 1998)
37. O. Kaleva, Fuzzy differential equations. Fuzzy Sets Syst. **24**, 301–317 (1987)
38. Y.K. Kim, B.M. Ghil, Integrals of fuzzy-number-valued functions. Fuzzy Sets Syst. **86**, 213–222 (1997)
39. W. McCulloch, W. Pitts, A logical calculus of the ideas immanent in nervous activity. Bull. Math. Biophys. **7**, 115–133 (1943)
40. T.M. Mitchell, *Machine Learning* (WCB-McGraw-Hill, New York, 1997)
41. S.G. Samko, A.A. Kilbas, O.I. Marichev, *Fractional Integrals and Derivatives, Theory and Applications*, (Gordon and Breach, Amsterdam, 1993) [English translation from the Russian, Integrals and Derivatives of Fractional Order and Some of Their Applications (Nauka i Tekhnika, Minsk, 1987)]
42. E.W. Weisstein, *Gudermannian*, MathWorld
43. Wu. Congxin, Gong Zengtai, On Henstock integrals of interval-valued functions and fuzzy valued functions. Fuzzy Sets Syst. **115**(3), 377–391 (2000)
44. C. Wu, Z. Gong, On Henstock integral of fuzzy-number-valued functions (I). Fuzzy Sets Syst. **3**(120), 523–532 (2001)
45. C. Wu, M. Ma, On embedding problem of fuzzy number spaces: part 1. Fuzzy Sets Syst. **44**, 33–38 (1991)

Chapter 13
Multivariate Fuzzy-Random and Stochastic Various Activation Functions Activated Neural Network Approximations

In this chapter we study the degree of approximation of multivariate pointwise and uniform convergences in the q-mean to the Fuzzy-Random unit operator of multivariate Fuzzy-Random Quasi-Interpolation arctangent, algebraic, Gudermannian and generalized symmetric activation functions based neural network operators. These multivariate Fuzzy-Random operators arise in a natural way among multivariate Fuzzy-Random neural networks. The rates are given through multivariate Probabilistic-Jackson type inequalities involving the multivariate Fuzzy-Random modulus of continuity of the engaged multivariate Fuzzy-Random function. The plain stochastic extreme analog of this theory is also met in detail for the stochastic analogs of the operators: the stochastic full quasi-interpolation operators, the stochastic Kantorovich type operators and the stochastic quadrature type operators. It follows [27].

13.1 Fuzzy-Random Functions and Stochastic Processes Background

See also [18], Chap. 22, pp. 497–501.

We start with

Definition 13.1 (*see* [36]) Let $\mu : \mathbb{R} \to [0, 1]$ with the following properties:

(i) is normal, i.e., $\exists\, x_0 \in \mathbb{R} : \mu(x_0) = 1$.

(ii) $\mu(\lambda x + (1-\lambda) y) \geq \min\{\mu(x), \mu(y)\}, \forall\, x, y \in \mathbb{R}, \forall\, \lambda \in [0, 1]$ (μ is called a convex fuzzy subset).

(iii) μ is upper semicontinuous on $\mathbb{R}$, i.e., $\forall\, x_0 \in \mathbb{R}$ and $\forall\, \varepsilon > 0$, $\exists$ neighborhood $V(x_0) : \mu(x) \leq \mu(x_0) + \varepsilon, \forall\, x \in V(x_0)$.

© The Author(s), under exclusive license to Springer Nature Switzerland AG 2023

G. A. Anastassiou, *Banach Space Valued Neural Network*,
Studies in Computational Intelligence 1062,
https://doi.org/10.1007/978-3-031-16400-2_13

(iv) the set $\overline{\text{supp}(\mu)}$ is compact in $\mathbb{R}$ (where $\text{supp}(\mu) := \{x \in \mathbb{R}; \mu(x) > 0\}$).

We call μ a fuzzy real number. Denote the set of all μ with $\mathbb{R}_{\mathcal{F}}$.

E.g., $\chi_{\{x_0\}} \in \mathbb{R}_{\mathcal{F}}$, for any $x_0 \in \mathbb{R}$, where $\chi_{\{x_0\}}$ is the characteristic function at x_0.

For $0 < r \leq 1$ and $\mu \in \mathbb{R}_{\mathcal{F}}$ define $[\mu]^r := \{x \in \mathbb{R} : \mu(x) \geq r\}$ and $[\mu]^0 := \overline{\{x \in \mathbb{R} : \mu(x) > 0\}}$.

Then it is well known that for each $r \in [0, 1]$, $[\mu]^r$ is a closed and bounded interval of $\mathbb{R}$. For $u, v \in \mathbb{R}_{\mathcal{F}}$ and $\lambda \in \mathbb{R}$, we define uniquely the sum $u \oplus v$ and the product $\lambda \odot u$ by

$$[u \oplus v]^r = [u]^r + [v]^r, \quad [\lambda \odot u]^r = \lambda [u]^r, \quad \forall\, r \in [0, 1],$$

where $[u]^r + [v]^r$ means the usual addition of two intervals (as subsets of $\mathbb{R}$) and $\lambda [u]^r$ means the usual product between a scalar and a subset of $\mathbb{R}$ (see, e.g., [36]). Notice $1 \odot u = u$ and it holds $u \oplus v = v \oplus u$, $\lambda \odot u = u \odot \lambda$. If $0 \leq r_1 \leq r_2 \leq 1$ then $[u]^{r_2} \subseteq [u]^{r_1}$. Actually $[u]^r = \left[u_-^{(r)}, u_+^{(r)}\right]$, where $u_-^{(r)} < u_+^{(r)}$, $u_-^{(r)}, u_+^{(r)} \in \mathbb{R}$, $\forall$ $r \in [0, 1]$.

Define

$$D : \mathbb{R}_{\mathcal{F}} \times \mathbb{R}_{\mathcal{F}} \to \mathbb{R}_+ \cup \{0\}$$

by

$$D(u, v) := \sup_{r \in [0,1]} \max \left\{ \left| u_-^{(r)} - v_-^{(r)} \right|, \left| u_+^{(r)} - v_+^{(r)} \right| \right\},$$

where $[v]^r = \left[v_-^{(r)}, v_+^{(r)}\right]$; $u, v \in \mathbb{R}_{\mathcal{F}}$. We have that D is a metric on $\mathbb{R}_{\mathcal{F}}$. Then $(\mathbb{R}_{\mathcal{F}}, D)$ is a complete metric space, see [36], with the properties

$$\begin{gathered} D(u \oplus w, v \oplus w) = D(u, v), \quad \forall\, u, v, w \in \mathbb{R}_{\mathcal{F}}, \\ D(k \odot u, k \odot v) = |k|\, D(u, v), \quad \forall\, u, v \in \mathbb{R}_{\mathcal{F}}, \forall\, k \in \mathbb{R}, \\ D(u \oplus v, w \oplus e) \leq D(u, w) + D(v, e), \quad \forall\, u, v, w, e \in \mathbb{R}_{\mathcal{F}}. \end{gathered} \tag{13.1}$$

Let (M, d) metric space and $f, g : M \to \mathbb{R}_{\mathcal{F}}$ be fuzzy real number valued functions. The distance between f, g is defined by

$$D^*(f, g) := \sup_{x \in M} D(f(x), g(x)).$$

On $\mathbb{R}_{\mathcal{F}}$ we define a partial order by "$\leq$": $u, v \in \mathbb{R}_{\mathcal{F}}$, $u \leq v$ iff $u_-^{(r)} \leq v_-^{(r)}$ and $u_+^{(r)} \leq v_+^{(r)}$, $\forall\, r \in [0, 1]$.

$\overset{*}{\sum}$ denotes the fuzzy summation, $\widetilde{o} := \chi_{\{0\}} \in \mathbb{R}_{\mathcal{F}}$ the neutral element with respect to $\oplus$. For more see also [37, 38].

We need

Definition 13.2 (*see also* [31], *Definition* 13.16, p. 654) Let $(X, \mathcal{B}, P)$ be a probability space. A fuzzy-random variable is a $\mathcal{B}$-measurable mapping $g : X \rightarrow \mathbb{R}_{\mathcal{F}}$ (i.e., for any open set $U \subseteq \mathbb{R}_{\mathcal{F}}$, in the topology of $\mathbb{R}_{\mathcal{F}}$ generated by the metric D, we have

$$g^{-1}(U) = \{s \in X; g(s) \in U\} \in \mathcal{B}). \tag{13.2}$$

The set of all fuzzy-random variables is denoted by $\mathcal{L}_{\mathcal{F}}(X, \mathcal{B}, P)$. Let $g_n, g \in \mathcal{L}_{\mathcal{F}}(X, \mathcal{B}, P)$, $n \in \mathbb{N}$ and $0 < q < +\infty$. We say $g_n(s) \overset{\text{“}q\text{-mean”}}{\underset{n\to+\infty}{\rightarrow}} g(s)$ if

$$\lim_{n\to+\infty} \int_X D(g_n(s), g(s))^q P(ds) = 0. \tag{13.3}$$

Remark 13.3 (*see* [31], p. 654) If $f, g \in \mathcal{L}_{\mathcal{F}}(X, \mathcal{B}, P)$, let us denote $F : X \rightarrow \mathbb{R}_+ \cup \{0\}$ by $F(s) = D(f(s), g(s))$, $s \in X$. Here, F is $\mathcal{B}$-measurable, because $F = G \circ H$, where $G(u, v) = D(u, v)$ is continuous on $\mathbb{R}_{\mathcal{F}} \times \mathbb{R}_{\mathcal{F}}$, and $H : X \rightarrow \mathbb{R}_{\mathcal{F}} \times \mathbb{R}_{\mathcal{F}}$, $H(s) = (f(s), g(s))$, $s \in X$, is $\mathcal{B}$-measurable. This shows that the above convergence in q-mean makes sense.

Definition 13.4 (*see* [31], p. 654, *Definition* 13.17) Let $(T, \mathcal{T})$ be a topological space. A mapping $f : T \rightarrow \mathcal{L}_{\mathcal{F}}(X, \mathcal{B}, P)$ will be called fuzzy-random function (or fuzzy-stochastic process) on T. We denote $f(t)(s) = f(t, s)$, $t \in T$, $s \in X$.

Remark 13.5 (*see* [31], p. 655) Any usual fuzzy real function $f : T \rightarrow \mathbb{R}_{\mathcal{F}}$ can be identified with the degenerate fuzzy-random function $f(t, s) = f(t)$, $\forall\, t \in T$, $s \in X$.

Remark 13.6 (*see* [31], p. 655) Fuzzy-random functions that coincide with probability one for each $t \in T$ will be consider equivalent.

Remark 13.7 (*see* [31], p. 655) Let $f, g : T \rightarrow \mathcal{L}_{\mathcal{F}}(X, \mathcal{B}, P)$. Then $f \oplus g$ and $k \odot f$ are defined pointwise, i.e.,

$$(f \oplus g)(t, s) = f(t, s) \oplus g(t, s),$$
$$(k \odot f)(t, s) = k \odot f(t, s), \ t \in T, s \in X, \ k \in \mathbb{R}.$$

Definition 13.8 (*see also Definition* 13.18, pp. 655–656, [31]) For a fuzzy-random function $f : W \subseteq \mathbb{R}^N \rightarrow \mathcal{L}_{\mathcal{F}}(X, \mathcal{B}, P)$, $N \in \mathbb{N}$, we define the (first) fuzzy-random modulus of continuity

$$\Omega_1^{(\mathcal{F})}(f, \delta)_{L^q} =$$

$$\sup\left\{\left(\int_X D^q(f(x, s), f(y, s)) P(ds)\right)^{\frac{1}{q}} : x, y \in W, \ \|x - y\|_\infty \le \delta\right\},$$

$0 < \delta$, $1 \le q < \infty$.

Definition 13.9 ([16]) Here $1 \leq q < +\infty$. Let $f : W \subseteq \mathbb{R}^N \rightarrow \mathcal{L}_{\mathcal{F}}(X, \mathcal{B}, P)$, $N \in \mathbb{N}$, be a fuzzy random function. We call f a (q-mean) uniformly continuous fuzzy random function over W, iff $\forall\, \varepsilon > 0\, \exists\, \delta > 0$:whenever $\|x - y\|_\infty \leq \delta$, $x, y \in W$, implies that

$$\int_X (D(f(x,s), f(y,s)))^q P(ds) \leq \varepsilon.$$

We denote it as $f \in C_{FR}^{U_q}(W)$.

Proposition 13.10 ([16]) *Let $f \in C_{FR}^{U_q}(W)$, where $W \subseteq \mathbb{R}^N$ is convex.*
Then $\Omega_1^{(\mathcal{F})}(f, \delta)_{L^q} < \infty$, any $\delta > 0$.

Proposition 13.11 ([16]) *Let $f, g : W \subseteq \mathbb{R}^N \rightarrow \mathcal{L}_{\mathcal{F}}(X, \mathcal{B}, P)$, $N \in \mathbb{N}$, be fuzzy random functions. It holds*
(i) $\Omega_1^{(\mathcal{F})}(f, \delta)_{L^q}$ is nonnegative and nondecreasing in $\delta > 0$.
(ii) $\lim_{\delta \downarrow 0} \Omega_1^{(\mathcal{F})}(f, \delta)_{L^q} = \Omega_1^{(\mathcal{F})}(f, 0)_{L^q} = 0$, iff $f \in C_{FR}^{U_q}(W)$.

We mention

Definition 13.12 (*see also* [6]) Let $f(t, s)$ be a random function (stochastic process) from $W \times (X, \mathcal{B}, P)$, $W \subseteq \mathbb{R}^N$, into $\mathbb{R}$, where $(X, \mathcal{B}, P)$ is a probability space. We define the q-mean multivariate first modulus of continuity of f by

$$\Omega_1(f, \delta)_{L^q} :=$$

$$\sup \left\{ \left(\int_X |f(x,s) - f(y,s)|^q P(ds) \right)^{\frac{1}{q}} : x, y \in W,\ \|x - y\|_\infty \leq \delta \right\}, \quad (13.4)$$

$\delta > 0$, $1 \leq q < \infty$.

The concept of f being (q-mean) uniformly continuous random function is defined the same way as in Definition 13.9, just replace D by $|\cdot|$, etc. We denote it as $f \in C_{\mathbb{R}}^{U_q}(W)$.

Similar properties as in Propositions 13.10 and 13.11 are valid for $\Omega_1(f, \delta)_{L^q}$. Also we have

Proposition 13.13 ([3]) *Let $Y(t, \omega)$ be a real valued stochastic process such that Y is continuous in $t \in [a, b]$. Then Y is jointly measurable in (t, ω).*

According to [29], p. 94 we have the following

Definition 13.14 Let $(Y, \mathcal{T})$ be a topological space, with its σ-algebra of Borel sets $\mathcal{B} := \mathcal{B}(Y, \mathcal{T})$ generated by $\mathcal{T}$. If $(X, \mathcal{S})$ is a measurable space, a function $f : X \rightarrow Y$ is called measurable iff $f^{-1}(B) \in \mathcal{S}$ for all $B \in \mathcal{B}$.

By Theorem 4.1.6 of [29], p. 89 f as above is measurable iff

$$f^{-1}(C) \in \mathcal{S} \text{ for all } C \in \mathcal{T}.$$

We mention

Theorem 13.15 (see [29], p. 95) *Let $(X, \mathcal{S})$ be a measurable space and (Y, d) be a metric space. Let f_n be measurable functions from X into Y such that for all $x \in X$, $f_n(x) \to f(x)$ in Y. Then f is measurable. I.e., $\lim_{n\to\infty} f_n = f$ is measurable.*

We need also

Proposition 13.16 ([16]) *Let f, g be fuzzy random variables from $\mathcal{S}$ into $\mathbb{R}_{\mathcal{F}}$. Then*
(i) Let $c \in \mathbb{R}$, then $c \odot f$ is a fuzzy random variable.
(ii) $f \oplus g$ is a fuzzy random variable.

Proposition 13.17 *Let $Y\left(\overrightarrow{t}, \omega\right)$ be a real valued multivariate random function (stochastic process) such that Y is continuous in $\overrightarrow{t} \in \prod_{i=1}^{N}[a_i, b_i]$. Then Y is jointly measurable in $\left(\overrightarrow{t}, \omega\right)$ and $\int_{\prod_{i=1}^{N}[a_i,b_i]} Y\left(\overrightarrow{t}, \omega\right) d\overrightarrow{t}$ is a real valued random variable.*

Proof Similar to Proposition 18.14, p. 353 of [7]. ∎

13.2 About Neural Networks Background

13.2.1 About the Arctangent Activation Function

We consider the

$$\arctan x = \int_0^x \frac{dz}{1+z^2}, \ x \in \mathbb{R}. \tag{13.5}$$

We will be using

$$h(x) := \frac{2}{\pi}\arctan\left(\frac{\pi}{2}x\right) = \frac{2}{\pi}\int_0^{\frac{\pi x}{2}} \frac{dz}{1+z^2}, \ x \in \mathbb{R}, \tag{13.6}$$

which is a sigmoid type function and it is strictly increasing. We have that

$$h(0) = 0, \ h(-x) = -h(x), \ h(+\infty) = 1, \ h(-\infty) = -1,$$

and

$$h'(x) = \frac{4}{4+\pi^2 x^2} > 0, \text{ all } x \in \mathbb{R}. \tag{13.7}$$

We consider the activation function

$$\psi_1(x) := \frac{1}{4}(h(x+1) - h(x-1)), \ x \in \mathbb{R}, \tag{13.8}$$

and we notice that

$$\psi_1(-x) = \psi_1(x), \tag{13.9}$$

it is an even function.

Since $x + 1 > x - 1$, then $h(x+1) > h(x-1)$, and $\psi_1(x) > 0$, all $x \in \mathbb{R}$.

We see that

$$\psi_1(0) = \frac{1}{\pi} \arctan \frac{\pi}{2} \cong 0.319. \tag{13.10}$$

Let $x > 0$, we have that

$$\psi_1'(x) = \frac{1}{4}\left(h'(x+1) - h'(x-1)\right) =$$

$$\frac{-4\pi^2 x}{\left(4 + \pi^2 (x+1)^2\right)\left(4 + \pi^2 (x-1)^2\right)} < 0. \tag{13.11}$$

That is

$$\psi_1'(x) < 0, \text{ for } x > 0. \tag{13.12}$$

That is ψ_1 is strictly decreasing on $[0, \infty)$ and clearly is strictly increasing on $(-\infty, 0]$, and $\psi_1'(0) = 0$.

Observe that

$$\begin{aligned} &\lim_{x \to +\infty} \psi_1(x) = \tfrac{1}{4}(h(+\infty) - h(+\infty)) = 0, \\ &\text{and} \\ &\lim_{x \to -\infty} \psi_1(x) = \tfrac{1}{4}(h(-\infty) - h(-\infty)) = 0. \end{aligned} \tag{13.13}$$

That is the x-axis is the horizontal asymptote on ψ_1.

All in all, ψ_1 is a bell symmetric function with maximum $\psi_1(0) \cong 0.319$.

We need

Theorem 13.18 ([19], p. 286) *We have that*

$$\sum_{i=-\infty}^{\infty} \psi_1(x - i) = 1, \ \forall x \in \mathbb{R}. \tag{13.14}$$

Theorem 13.19 ([19], p. 287) *It holds*

$$\int_{-\infty}^{\infty} \psi_1(x)\,dx = 1. \tag{13.15}$$

So that $\psi_1(x)$ is a density function on $\mathbb{R}$.
We mention

Theorem 13.20 ([19], p. 288) *Let* $0 < \alpha < 1$, *and* $n \in \mathbb{N}$ *with* $n^{1-\alpha} > 2$. *It holds*

$$\sum_{\begin{cases} k=-\infty \\ : |nx-k| \geq n^{1-\alpha} \end{cases}}^{\infty} \psi_1(nx-k) < \frac{2}{\pi^2\left(n^{1-\alpha}-2\right)} =: c_1(\alpha, n). \tag{13.16}$$

Denote by $\lfloor\cdot\rfloor$ the integral part of the number and by $\lceil\cdot\rceil$ the ceiling of the number.
We need

Theorem 13.21 ([19], p. 289) *Let* $x \in [a,b] \subset \mathbb{R}$ *and* $n \in \mathbb{N}$ *so that* $\lceil na\rceil \leq \lfloor nb\rfloor$. *It holds*

$$\frac{1}{\sum_{k=\lceil na\rceil}^{\lfloor nb\rfloor} \psi_1(nx-k)} < \frac{1}{\psi_1(1)} \cong \mathbf{4.9737} =: \alpha_1, \ \forall\, \mathbf{x} \in [a,b]. \tag{13.17}$$

Note 13.22 ([19], pp. 290–291)
(i) We have that

$$\lim_{n\to\infty} \sum_{k=\lceil na\rceil}^{\lfloor nb\rfloor} \psi_1(nx-k) \neq 1, \tag{13.18}$$

for at least some $x \in [a,b]$.
(ii) For large enough $n \in \mathbb{N}$ we always obtain $\lceil na\rceil \leq \lfloor nb\rfloor$. Also $a \leq \frac{k}{n} \leq b$, iff $\lceil na\rceil \leq k \leq \lfloor nb\rfloor$.
In general, by Theorem 13.18, it holds

$$\sum_{k=\lceil na\rceil}^{\lfloor nb\rfloor} \psi_1(nx-k) \leq 1. \tag{13.19}$$

We introduce (see [24])

$$Z_1(x_1,...,x_N) := Z_1(x) := \prod_{i=1}^{N} \psi_1(x_i), \quad x = (x_1,...,x_N) \in \mathbb{R}^N, \ N \in \mathbb{N}. \tag{13.20}$$

Denote by $a = (a_1,...,a_N)$ and $b = (b_1,...,b_N)$.
It has the properties:

(i) $Z_1(x) > 0, \ \forall x \in \mathbb{R}^N$,
(ii)

$$\sum_{k=-\infty}^{\infty} Z_1(x-k) := \sum_{k_1=-\infty}^{\infty} \sum_{k_2=-\infty}^{\infty} \ldots \sum_{k_N=-\infty}^{\infty} Z_1(x_1-k_1, \ldots, x_N-k_N) = 1, \tag{13.21}$$

where $k := (k_1, \ldots, k_n) \in \mathbb{Z}^N, \forall x \in \mathbb{R}^N$,
hence
(iii)

$$\sum_{k=-\infty}^{\infty} Z_1(nx-k) = 1, \tag{13.22}$$

$\forall x \in \mathbb{R}^N; n \in \mathbb{N}$,
and
(iv)

$$\int_{\mathbb{R}^N} Z_1(x)\, dx = 1, \tag{13.23}$$

that is Z_1 is a multivariate density function.
(v) It is clear that

$$\sum_{\substack{k=-\infty \\ \left\| \frac{k}{n} - x \right\|_\infty > \frac{1}{n^\beta}}}^{\infty} Z_1(nx-k) < \frac{2}{\pi^2 \left(n^{1-\beta} - 2\right)} = c_1(\beta, n), \tag{13.24}$$

$0 < \beta < 1, n \in \mathbb{N} : n^{1-\beta} > 2, x \in \mathbb{R}^N$.
(vi) By Theorem 13.21 we get that

$$0 < \frac{1}{\sum_{k=\lceil na \rceil}^{\lfloor nb \rfloor} Z_1(nx-k)} < \frac{1}{(\psi_1(1))^N} \cong (4.9737)^N =: \gamma_1(N), \tag{13.25}$$

$\forall x \in \left(\prod_{i=1}^N [a_i, b_i]\right), \ n \in \mathbb{N}$.
Furthermore it holds

$$\lim_{n\to\infty} \sum_{k=\lceil na \rceil}^{\lfloor nb \rfloor} Z_1(nx-k) \neq 1, \tag{13.26}$$

for at least some $x \in \left(\prod_{i=1}^N [a_i, b_i]\right)$.

Above it is $\|x\|_\infty := \max\{|x_1|, \ldots, |x_N|\}$, $x \in \mathbb{R}^N$, also set $\infty := (\infty, \ldots, \infty)$, $-\infty = (-\infty, \ldots -\infty)$ upon the multivariate context.

13.2.2 About the Algebraic Activation Function

Here see also [20].

We consider the generator algebraic function

$$\varphi(x) = \frac{x}{\sqrt[2m]{1 + x^{2m}}}, \quad m \in \mathbb{N}, x \in \mathbb{R}, \tag{13.27}$$

which is a sigmoidal type of function and is a strictly increasing function.

We see that $\varphi(-x) = -\varphi(x)$ with $\varphi(0) = 0$. We get that

$$\varphi'(x) = \frac{1}{\left(1 + x^{2m}\right)^{\frac{2m+1}{2m}}} > 0, \ \forall\, x \in \mathbb{R}, \tag{13.28}$$

proving φ as strictly increasing over $\mathbb{R}$, $\varphi'(x) = \varphi'(-x)$. We easily find that $\lim_{x \to +\infty} \varphi(x) = 1$, $\varphi(+\infty) = 1$, and $\lim_{x \to -\infty} \varphi(x) = -1$, $\varphi(-\infty) = -1$.

We consider the activation function

$$\psi_2(x) = \frac{1}{4}\left[\varphi(x+1) - \varphi(x-1)\right]. \tag{13.29}$$

Clearly it is $\psi_2(x) = \psi_2(-x)$, $\forall\, x \in \mathbb{R}$, so that ψ_2 is an even function and symmetric with respect to the y-axis. Clealry $\psi_2(x) > 0$, $\forall\, x \in \mathbb{R}$.

Also it is

$$\psi_2(0) = \frac{1}{2\sqrt[2m]{2}}. \tag{13.30}$$

By [20], we have that $\psi_2'(x) < 0$ for $x > 0$. That is ψ_2 is strictly decreasing over $(0, +\infty)$.

Clearly, ψ_2 is strictly increasing over $(-\infty, 0)$ and $\psi_2'(0) = 0$.

Furthermore we obtain that

$$\lim_{x \to +\infty} \psi_2(x) = \frac{1}{4}\left[\varphi(+\infty) - \varphi(+\infty)\right] = 0, \tag{13.31}$$

and

$$\lim_{x \to -\infty} \psi_2(x) = \frac{1}{4}\left[\varphi(-\infty) - \varphi(-\infty)\right] = 0. \tag{13.32}$$

That is the x-axis is the horizontal asymptote of ψ_2.

Conclusion, ψ_2 is a bell shape symmetric function with maximum

$$\psi_2(0) = \frac{1}{2\sqrt[2m]{2}}, \quad m \in \mathbb{N}. \tag{13.33}$$

We need

Theorem 13.23 ([20]) *We have that*

$$\sum_{i=-\infty}^{\infty} \psi_2 (x - i) = 1, \ \forall x \in \mathbb{R}. \tag{13.34}$$

Theorem 13.24 ([20]) *It holds*

$$\int_{-\infty}^{\infty} \psi_2 (x) \, dx = 1. \tag{13.35}$$

Theorem 13.25 ([20]) *Let* $0 < \alpha < 1$, *and* $n \in \mathbb{N}$ *with* $n^{1-\alpha} > 2$. *It holds*

$$\sum_{\begin{cases} k = -\infty \\ : |nx - k| \geq n^{1-\alpha} \end{cases}}^{\infty} \psi_2 (nx - k) < \frac{1}{4m \left(n^{1-\alpha} - 2\right)^{2m}} =: c_2 (\alpha, n), \ m \in \mathbb{N}. \tag{13.36}$$

We need

Theorem 13.26 ([20]) *Let* $[a, b] \subset \mathbb{R}$ *and* $n \in \mathbb{N}$ *so that* $\lceil na \rceil \leq \lfloor nb \rfloor$. *It holds*

$$\frac{1}{\sum_{k=\lceil na \rceil}^{\lfloor nb \rfloor} \psi_2 (nx - k)} < 2 \left(\sqrt[2m]{1 + 4^m} \right) =: \alpha_2, \tag{13.37}$$

$\forall \, x \in [a, b]$, $m \in \mathbb{N}$.

Note 13.27 (1) By [20] we have that

$$\lim_{n \to \infty} \sum_{k=\lceil na \rceil}^{\lfloor nb \rfloor} \psi_2 (nx - k) \neq 1, \tag{13.38}$$

for at least some $x \in [a, b]$.

(2) Let $[a, b] \subset \mathbb{R}$. For large $n \in \mathbb{N}$ we always have $\lceil na \rceil \leq \lfloor nb \rfloor$. Also $a \leq \frac{k}{n} \leq b$, iff $\lceil na \rceil \leq k \leq \lfloor nb \rfloor$.

In general it holds that

$$\sum_{k=\lceil na \rceil}^{\lfloor nb \rfloor} \psi_2 (nx - k) \leq 1. \tag{13.39}$$

We introduce (see also [25])

$$Z_2(x_1, ..., x_N) := Z_2(x) := \prod_{i=1}^{N} \psi_2(x_i), \quad x = (x_1, ..., x_N) \in \mathbb{R}^N, \ N \in \mathbb{N}. \tag{13.40}$$

It has the properties:

(i) $Z_2(x) > 0, \ \forall x \in \mathbb{R}^N$,

(ii)

$$\sum_{k=-\infty}^{\infty} Z_2(x-k) := \sum_{k_1=-\infty}^{\infty} \sum_{k_2=-\infty}^{\infty} ... \sum_{k_N=-\infty}^{\infty} Z_2(x_1 - k_1, ..., x_N - k_N) = 1, \tag{13.41}$$

where $k := (k_1, ..., k_n) \in \mathbb{Z}^N$, $\forall x \in \mathbb{R}^N$,

hence

(iii)

$$\sum_{k=-\infty}^{\infty} Z_2(nx-k) = 1, \tag{13.42}$$

$\forall x \in \mathbb{R}^N$; $n \in \mathbb{N}$,

and

(iv)

$$\int_{\mathbb{R}^N} Z_2(x)\, dx = 1, \tag{13.43}$$

that is Z_2 is a multivariate density function.

(v) It is clear that

$$\sum_{\begin{cases} k = -\infty \\ \left\| \frac{k}{n} - x \right\|_\infty > \frac{1}{n^\beta} \end{cases}}^{\infty} Z_2(nx-k) < \frac{1}{4m\left(n^{1-\beta} - 2\right)^{2m}} = c_2(\beta, n), \tag{13.44}$$

$0 < \beta < 1$, $n \in \mathbb{N} : n^{1-\beta} > 2$, $x \in \mathbb{R}^N$, $m \in \mathbb{N}$.

(vi) By Theorem 13.26 we get that

$$0 < \frac{1}{\sum_{k=\lceil na \rceil}^{\lfloor nb \rfloor} Z_2(nx-k)} < \frac{1}{(\psi_2(1))^N} \cong \left[2\left(\sqrt[2m]{1+4^m}\right)\right]^N := \gamma_2(N), \tag{13.45}$$

$\forall x \in \left(\prod_{i=1}^{N} [a_i, b_i]\right)$, $n \in \mathbb{N}$.

Furthermore it holds

$$\lim_{n\to\infty} \sum_{k=\lceil na \rceil}^{\lfloor nb \rfloor} Z_2 (nx - k) \neq 1, \tag{13.46}$$

for at least some $x \in \left(\prod_{i=1}^{N} [a_i, b_i]\right)$.

13.2.3 About the Gudermannian Activation Function

See also [21, 35].

Here we consider $gd(x)$ the Gudermannian function [35], which is a sigmoid function, as a generator function:

$$\sigma(x) = 2\arctan\left(\tanh\left(\frac{x}{2}\right)\right) = \int_0^x \frac{dt}{\cosh t} =: gd(x), \ x \in \mathbb{R}. \tag{13.47}$$

Let the normalized generator sigmoid function

$$f(x) := \frac{4}{\pi}\sigma(x) = \frac{4}{\pi}\int_0^x \frac{dt}{\cosh t} = \frac{8}{\pi}\int_0^x \frac{1}{e^t + e^{-t}} dt, \ x \in \mathbb{R}. \tag{13.48}$$

Here

$$f'(x) = \frac{4}{\pi \cosh x} > 0, \ \forall \, x \in \mathbb{R},$$

hence f is strictly increasing on $\mathbb{R}$.

Notice that $\tanh(-x) = -\tanh x$ and $\arctan(-x) = -\arctan x$, $x \in \mathbb{R}$.

So, here the neural network activation function will be:

$$\psi_3(x) = \frac{1}{4}\left[f(x+1) - f(x-1)\right], \ x \in \mathbb{R}. \tag{13.49}$$

By [21], we get that

$$\psi_3(x) = \psi_3(-x), \ \forall \, x \in \mathbb{R}, \tag{13.50}$$

i.e. it is even and symmetric with respect to the y-axis. Here we have $f(+\infty) = 1$, $f(-\infty) = -1$ and $f(0) = 0$. Clearly it is

$$f(-x) = -f(x), \ \forall \, x \in \mathbb{R}, \tag{13.51}$$

an odd function, symmetric with respect to the origin. Since $x + 1 > x - 1$, and $f(x+1) > f(x-1)$, we obtain $\psi_3(x) > 0$, $\forall \, x \in \mathbb{R}$.

By [21], we have that

$$\psi_3(0) = \frac{2}{\pi} gd(1) \cong 0.551. \tag{13.52}$$

By [21] ψ_3 is strictly decreasing on $(0, +\infty)$, and strictly increasing on $(-\infty, 0)$, and $\psi_3'(0) = 0$.

Also we have that

$$\lim_{x \to +\infty} \psi_3(x) = \lim_{x \to -\infty} \psi_3(x) = 0, \tag{13.53}$$

that is the x-axis is the horizontal asymptote for ψ_3.

Conclusion, ψ_3 is a bell shaped symmetric function with maximum $\psi_3(0) \cong 0.551$.

We need

Theorem 13.28 ([21]) *It holds that*

$$\sum_{i=-\infty}^{\infty} \psi_3(x - i) = 1, \ \forall\, x \in \mathbb{R}. \tag{13.54}$$

Theorem 13.29 ([21]) *We have that*

$$\int_{-\infty}^{\infty} \psi_3(x)\, dx = 1. \tag{13.55}$$

So $\psi_3(x)$ is a density function.

Theorem 13.30 ([21]) *Let $0 < \alpha < 1$, and $n \in \mathbb{N}$ with $n^{1-\alpha} > 2$. It holds*

$$\sum_{\substack{k=-\infty \\ : |nx-k| \geq n^{1-\alpha}}}^{\infty} \psi_3(nx - k) < \frac{4}{\pi e^{\left(n^{1-\alpha}-2\right)}} = \frac{4e^2}{\pi e^{n^{1-\alpha}}} =: c_3(\alpha, n). \tag{13.56}$$

Theorem 13.31 ([21]) *Let $[a, b] \subset \mathbb{R}$ and $n \in \mathbb{N}$, so that $\lceil na \rceil \leq \lfloor nb \rfloor$. It holds*

$$\frac{1}{\sum_{k=\lceil na \rceil}^{\lfloor nb \rfloor} \psi_3(nx - k)} < \frac{\pi}{gd(2)} \cong 2.412 =: \alpha_3, \tag{13.57}$$

$\forall\, x \in [a, b]$.

We make

Remark 13.32 ([21])
(i) We have that

$$\lim_{n\to\infty} \sum_{k=\lceil na\rceil}^{\lfloor nb\rfloor} \psi_3(nx-k) \neq 1, \tag{13.58}$$

for at least some $x \in [a,b]$.

(ii) Let $[a,b] \subset \mathbb{R}$. For large n we always have $\lceil na\rceil \leq \lfloor nb\rfloor$. Also $a \leq \frac{k}{n} \leq b$, iff $\lceil na\rceil \leq k \leq \lfloor nb\rfloor$.

In general it holds

$$\sum_{k=\lceil na\rceil}^{\lfloor nb\rfloor} \psi_3(nx-k) \leq 1. \tag{13.59}$$

We introduce (see also [23])

$$Z_3(x_1,\ldots,x_N) := Z_3(x) := \prod_{i=1}^{N} \psi_3(x_i), \quad x = (x_1,\ldots,x_N) \in \mathbb{R}^N, \ N \in \mathbb{N}. \tag{13.60}$$

It has the properties:
(i) $Z_3(x) > 0$, $\forall\, x \in \mathbb{R}^N$,
(ii)

$$\sum_{k=-\infty}^{\infty} Z_3(x-k) := \sum_{k_1=-\infty}^{\infty} \sum_{k_2=-\infty}^{\infty} \ldots \sum_{k_N=-\infty}^{\infty} Z_3(x_1-k_1,\ldots,x_N-k_N) = 1, \tag{13.61}$$

where $k := (k_1,\ldots,k_n) \in \mathbb{Z}^N$, $\forall\, x \in \mathbb{R}^N$,
hence
(iii)

$$\sum_{k=-\infty}^{\infty} Z_3(nx-k) = 1, \tag{13.62}$$

$\forall\, x \in \mathbb{R}^N$; $n \in \mathbb{N}$,
and
(iv)

$$\int_{\mathbb{R}^N} Z_3(x)\,dx = 1, \tag{13.63}$$

that is Z_3 is a multivariate density function.

(v) It is also clear that

$$\sum_{\begin{cases} k=-\infty \\ \left\| \frac{k}{n}-x \right\|_{\infty} > \frac{1}{n^{\beta}} \end{cases}}^{\infty} Z_3(nx-k) < \frac{4e^2}{\pi e^{n^{1-\beta}}} = c_3(\beta, n), \tag{13.64}$$

$0 < \beta < 1, n \in \mathbb{N} : n^{1-\beta} > 2, x \in \mathbb{R}^N, m \in \mathbb{N}$.

(vi) By Theorem 13.31 we get that

$$0 < \frac{1}{\sum_{k=\lceil na \rceil}^{\lfloor nb \rfloor} Z_3(nx-k)} < \left(\frac{\pi}{gd(2)} \right)^N \cong (2.412)^N =: \gamma_3(N), \tag{13.65}$$

$\forall\, x \in \left(\prod_{i=1}^{N} [a_i, b_i] \right)$, $n \in \mathbb{N}$.

Furthermore it holds

$$\lim_{n \to \infty} \sum_{k=\lceil na \rceil}^{\lfloor nb \rfloor} Z_3(nx-k) \neq 1, \tag{13.66}$$

for at least some $x \in \left(\prod_{i=1}^{N} [a_i, b_i] \right)$.

13.2.4 About the Generalized Symmetrical Activation Function

Here we consider the generalized symmetrical sigmoid function ([22, 30])

$$f_1(x) = \frac{x}{\left(1 + |x|^{\mu}\right)^{\frac{1}{\mu}}}, \ \mu > 0, x \in \mathbb{R}. \tag{13.67}$$

This has applications in immunology and protection from disease together with probability theory. It is also called a symmetrical protection curve.

The parameter μ is a shape parameter controling how fast the curve approaches the asymptotes for a given slope at the inflection point. When $\mu = 1$ f_1 is the absolute sigmoid function, and when $\mu = 2$, f_1 is the square root sigmoid function. When $\mu = 1.5$ the function approximates the arctangent function, when $\mu = 2.9$ it approximates the logistic function, and when $\mu = 3.4$ it approximates the error function. Parameter μ is estimated in the likelihood maximization ([30]). For more see [30].

Next we study the particular generator sigmoid function

$$f_2(x) = \frac{x}{\left(1 + |x|^{\lambda}\right)^{\frac{1}{\lambda}}}, \ \lambda \text{ is an odd number}, x \in \mathbb{R}. \tag{13.68}$$

We have that $f_2(0) = 0$, and

$$f_2(-x) = -f_2(x), \tag{13.69}$$

so f_2 is symmetric with respect to zero.

When $x \geq 0$, we get that ([22])

$$f_2'(x) = \frac{1}{\left(1 + x^\lambda\right)^{\frac{\lambda+1}{\lambda}}} > 0, \tag{13.70}$$

that is f_2 is strictly increasing on $[0, +\infty)$ and f_2 is strictly increasing on $(-\infty, 0]$. Hence f_2 is strictly increasing on $\mathbb{R}$.

We also have $f_2(+\infty) = f_2(-\infty) = 1$.

Let us consider the activation function ([22]):

$$\psi_4(x) = \frac{1}{4}\left[f_2(x+1) - f_2(x-1)\right] =$$

$$\frac{1}{4}\left[\frac{(x+1)}{\left(1 + |x+1|^\lambda\right)^{\frac{1}{\lambda}}} - \frac{(x-1)}{\left(1 + |x-1|^\lambda\right)^{\frac{1}{\lambda}}}\right]. \tag{13.71}$$

Clearly it holds ([22])

$$\psi_4(x) = \psi_4(-x), \ \forall\, x \in \mathbb{R}. \tag{13.72}$$

and

$$\psi_4(0) = \frac{1}{2\sqrt[\lambda]{2}}, \tag{13.73}$$

and $\psi_4(x) > 0$, $\forall\, x \in \mathbb{R}$.

Following [22], we have that ψ_4 is strictly decreasing over $[0, +\infty)$, and ψ_4 is strictly increasing on $(-\infty, 0]$, by ψ_4-symmetry with respect to y-axis, and $\psi_4'(0) = 0$.

Clearly it is

$$\lim_{x \to +\infty} \psi_4(x) = \lim_{x \to -\infty} \psi_4(x) = 0, \tag{13.74}$$

therefore the x-axis is the horizontal asymptote of $\psi_4(x)$.

The value

$$\psi_4(0) = \frac{1}{2\sqrt[\lambda]{2}}, \ \lambda \text{ is an odd number}, \tag{13.75}$$

is the maximum of ψ_4, which is a bell shaped function.

We need

Theorem 13.33 ([22]) *It holds*

$$\sum_{i=-\infty}^{\infty} \psi_4 (x - i) = 1, \ \forall\, x \in \mathbb{R}. \tag{13.76}$$

Theorem 13.34 ([22]) *We have that*

$$\int_{-\infty}^{\infty} \psi_4 (x)\, dx = 1. \tag{13.77}$$

So that $\psi_4 (x)$ is a density function on $\mathbb{R}$.
We need

Theorem 13.35 ([22]) *Let* $0 < \alpha < 1$, *and* $n \in \mathbb{N}$ *with* $n^{1-\alpha} > 2$. *It holds*

$$\sum_{\substack{j=-\infty \\ : |nx - j| \geq n^{1-\alpha}}}^{\infty} \psi_4 (nx - j) < \frac{1}{2\lambda \left(n^{1-\alpha} - 2\right)^{\lambda}} =: c_4 (\alpha, n), \tag{13.78}$$

where $\lambda \in \mathbb{N}$ *is an odd number.*

We also need

Theorem 13.36 ([22]) *Let* $[a, b] \subset \mathbb{R}$ *and* $n \in \mathbb{N}$ *so that* $\lceil na \rceil \leq \lfloor nb \rfloor$. *Then*

$$\frac{1}{\sum_{k=\lceil na \rceil}^{\lfloor nb \rfloor} \psi_4 (|nx - k|)} < 2\sqrt[\lambda]{1 + 2^{\lambda}} =: \alpha_4, \tag{13.79}$$

where λ *is an odd number,* $\forall\, x \in [a, b]$.

We make

Remark 13.37 ([22]) (1) We have that

$$\lim_{n \to \infty} \sum_{k=\lceil na \rceil}^{\lfloor nb \rfloor} \psi_4 (nx - k) \neq 1, \quad \text{for at least some } x \in [a, b]. \tag{13.80}$$

(2) Let $[a, b] \subset \mathbb{R}$. For large enough n we always obtain $\lceil na \rceil \leq \lfloor nb \rfloor$. Also $a \leq \frac{k}{n} \leq b$, iff $\lceil na \rceil \leq k \leq \lfloor nb \rfloor$.

In general it holds that

$$\sum_{k=\lceil na \rceil}^{\lfloor nb \rfloor} \psi_4 (nx - k) \leq 1. \tag{13.81}$$

We introduce (see also [26])

$$Z_4(x_1, ..., x_N) := Z_4(x) := \prod_{i=1}^{N} \psi_4(x_i), \quad x = (x_1, ..., x_N) \in \mathbb{R}^N, \ N \in \mathbb{N}. \tag{13.82}$$

It has the properties:

(i) $Z_4(x) > 0, \ \forall\, x \in \mathbb{R}^N$,

(ii)

$$\sum_{k=-\infty}^{\infty} Z_4(x-k) := \sum_{k_1=-\infty}^{\infty} \sum_{k_2=-\infty}^{\infty} ... \sum_{k_N=-\infty}^{\infty} Z_4(x_1-k_1, ..., x_N-k_N) = 1, \tag{13.83}$$

where $k := (k_1, ..., k_n) \in \mathbb{Z}^N, \forall\, x \in \mathbb{R}^N$,

hence

(iii)

$$\sum_{k=-\infty}^{\infty} Z_4(nx-k) = 1, \tag{13.84}$$

$\forall\, x \in \mathbb{R}^N$; $n \in \mathbb{N}$,

and

(iv)

$$\int_{\mathbb{R}^N} Z_4(x)\, dx = 1, \tag{13.85}$$

that is Z_4 is a multivariate density function.

(v) It is clear that

$$\sum_{\substack{k=-\infty \\ \left\| \frac{k}{n} - x \right\|_{\infty} > \frac{1}{n^{\beta}}}}^{\infty} Z_4(nx-k) < \frac{1}{2\lambda\left(n^{1-\beta}-2\right)^{\lambda}} = c_4(\beta, n), \tag{13.86}$$

$0 < \beta < 1, n \in \mathbb{N} : n^{1-\beta} > 2, x \in \mathbb{R}^N$, λ is odd.

(vi) By Theorem 13.36 we get that

$$0 < \frac{1}{\sum_{k=\lceil na \rceil}^{\lfloor nb \rfloor} Z_4(nx-k)} < \left(2\sqrt[\lambda]{1+2^{\lambda}}\right)^N =: \gamma_4(N), \tag{13.87}$$

$\forall\, x \in \left(\prod_{i=1}^{N} [a_i, b_i]\right)$, $n \in \mathbb{N}$, λ is odd.

Furthermore it holds

$$\lim_{n\to\infty} \sum_{k=\lceil na \rceil}^{\lfloor nb \rfloor} Z_4(nx-k) \neq 1, \tag{13.88}$$

for at least some $x \in \left(\prod_{i=1}^{N}[a_i, b_i]\right)$.

Set

$$\lceil na \rceil := (\lceil na_1 \rceil, ..., \lceil na_N \rceil),$$

$$\lfloor nb \rfloor := (\lfloor nb_1 \rfloor, ..., \lfloor nb_N \rfloor),$$

where $a := (a_1, ..., a_N)$, $b := (b_1, ..., b_N)$, $k := (k_1, ..., k_N)$.

Let $f \in C\left(\prod_{i=1}^{N}[a_i, b_i]\right)$, and $n \in \mathbb{N}$ such that $\lceil na_i \rceil \leq \lfloor nb_i \rfloor$, $i = 1, ..., N$.

We define the multivariate averaged positive linear quasi-interpolation neural network operators ($x := (x_1, ..., x_N) \in \left(\prod_{i=1}^{N}[a_i, b_i]\right)$); $j = 1, 2, 3, 4$:

$${}_jA_n(f, x_1, ..., x_N) := {}_jA_n(f, x) := \frac{\sum_{k=\lceil na \rceil}^{\lfloor nb \rfloor} f\left(\frac{k}{n}\right) Z_j(nx-k)}{\sum_{k=\lceil na \rceil}^{\lfloor nb \rfloor} Z_j(nx-k)} = \tag{13.89}$$

$$\frac{\sum_{k_1=\lceil na_1 \rceil}^{\lfloor nb_1 \rfloor} \sum_{k_2=\lceil na_2 \rceil}^{\lfloor nb_2 \rfloor} \cdots \sum_{k_N=\lceil na_N \rceil}^{\lfloor nb_N \rfloor} f\left(\frac{k_1}{n}, ..., \frac{k_N}{n}\right)\left(\prod_{i=1}^{N} \psi_j(nx_i - k_i)\right)}{\prod_{i=1}^{N}\left(\sum_{k_i=\lceil na_i \rceil}^{\lfloor nb_i \rfloor} \psi_j(nx_i - k_i)\right)}.$$

For large enough $n \in \mathbb{N}$ we always obtain $\lceil na_i \rceil \leq \lfloor nb_i \rfloor$, $i = 1, ..., N$. Also $a_i \leq \frac{k_i}{n} \leq b_i$, iff $\lceil na_i \rceil \leq k_i \leq \lfloor nb_i \rfloor$, $i = 1, ..., N$.

When $f \in C_B(\mathbb{R}^N)$ we define ($j = 1, 2, 3, 4$)

$${}_jB_n(f, x) := {}_jB_n(f, x_1, ..., x_N) := \sum_{k=-\infty}^{\infty} f\left(\frac{k}{n}\right) Z_j(nx-k) := \tag{13.90}$$

$$\sum_{k_1=-\infty}^{\infty} \sum_{k_2=-\infty}^{\infty} \cdots \sum_{k_N=-\infty}^{\infty} f\left(\frac{k_1}{n}, \frac{k_2}{n}, ..., \frac{k_N}{n}\right)\left(\prod_{i=1}^{N} \psi_j(nx_i - k_i)\right),$$

$n \in \mathbb{N}$, $\forall\, x \in \mathbb{R}^N$, $N \in \mathbb{N}$, the multivariate full quasi-interpolation neural network operators.

Also for $f \in C_B(\mathbb{R}^N)$ we define the multivariate Kantorovich type neural network operators ($j = 1, 2, 3, 4$)

$${}_jC_n(f, x) := {}_jC_n(f, x_1, ..., x_N) := \sum_{k=-\infty}^{\infty}\left(n^N \int_{\frac{k}{n}}^{\frac{k+1}{n}} f(t)\, dt\right) Z_j(nx-k) := \tag{13.91}$$

$$\sum_{k_1=-\infty}^{\infty}\sum_{k_2=-\infty}^{\infty}\ldots\sum_{k_N=-\infty}^{\infty}\left(n^N\int_{\frac{k_1}{n}}^{\frac{k_1+1}{n}}\int_{\frac{k_2}{n}}^{\frac{k_2+1}{n}}\ldots\int_{\frac{k_N}{n}}^{\frac{k_N+1}{n}}f(t_1,\ldots,t_N)\,dt_1\ldots dt_N\right)$$

$$\cdot\left(\prod_{i=1}^{N}\psi_j(nx_i-k_i)\right),$$

$n\in\mathbb{N}$, $\forall\, x\in\mathbb{R}^N$.

Again for $f\in C_B(\mathbb{R}^N)$, $N\in\mathbb{N}$, we define the multivariate neural network operators of quadrature type ${}_jD_n(f,x)$, $n\in\mathbb{N}$, as follows. Let $\theta=(\theta_1,\ldots,\theta_N)\in\mathbb{N}^N$, $\overline{r}=(r_1,\ldots,r_N)\in\mathbb{Z}_+^N$, $w_{\overline{r}}=w_{r_1,r_2,\ldots r_N}\geq 0$, such that $\sum_{\overline{r}=0}^{\theta}w_{\overline{r}}=\sum_{r_1=0}^{\theta_1}\sum_{r_2=0}^{\theta_2}\ldots\sum_{r_N=0}^{\theta_N}w_{r_1,r_2,\ldots r_N}=1$; $k\in\mathbb{Z}^N$ and

$$\delta_{nk}(f):=\delta_{n,k_1,k_2,\ldots,k_N}(f):=\sum_{\overline{r}=0}^{\theta}w_{\overline{r}}f\left(\frac{k}{n}+\frac{\overline{r}}{n\theta}\right):=$$

$$\sum_{r_1=0}^{\theta_1}\sum_{r_2=0}^{\theta_2}\ldots\sum_{r_N=0}^{\theta_N}w_{r_1,r_2,\ldots r_N}f\left(\frac{k_1}{n}+\frac{r_1}{n\theta_1},\frac{k_2}{n}+\frac{r_2}{n\theta_2},\ldots,\frac{k_N}{n}+\frac{r_N}{n\theta_N}\right),\quad(13.92)$$

where $\frac{\overline{r}}{\theta}:=\left(\frac{r_1}{\theta_1},\frac{r_2}{\theta_2},\ldots,\frac{r_N}{\theta_N}\right)$; $j=1,2,3,4$.

We put

$${}_jD_n(f,x):={}_jD_n(f,x_1,\ldots,x_N):=\sum_{k=-\infty}^{\infty}\delta_{nk}(f)\,Z_j(nx-k):=\quad(13.93)$$

$$\sum_{k_1=-\infty}^{\infty}\sum_{k_2=-\infty}^{\infty}\ldots\sum_{k_N=-\infty}^{\infty}\delta_{n,k_1,k_2,\ldots,k_N}(f)\left(\prod_{i=1}^{N}\psi_j(nx_i-k_i)\right),$$

$\forall\, x\in\mathbb{R}^N$.

For the next we need, for $f\in C\left(\prod_{i=1}^{N}[a_i,b_i]\right)$ the first multivariate modulus of continuity

$$\omega_1(f,h):=\sup_{\substack{x,y\in\prod_{i=1}^{N}[a_i,b_i]\\ \|x-y\|_\infty\leq h}}|f(x)-f(y)|,\ h>0.\quad(13.94)$$

It holds that

$$\lim_{h\to 0}\omega_1(f,h)=0. \tag{13.95}$$

Similarly it is defined for $f\in C_B\left(\mathbb{R}^N\right)$ (continuous and bounded functions on $\mathbb{R}^N$) the $\omega_1(f,h)$, and it has the property (13.95), given that $f\in C_U\left(\mathbb{R}^N\right)$ (uniformly continuous functions on $\mathbb{R}^N$).

We mention

Theorem 13.38 (see [23–26]) *Let* $f\in C\left(\prod_{i=1}^N[a_i,b_i]\right)$, $0<\beta<1$, $x\in\left(\prod_{i=1}^N[a_i,b_i]\right)$, $N,n\in\mathbb{N}$ *with* $n^{1-\beta}>2$; $j=1,2,3,4$. *Then*

(1)

$$\left|{}_jA_n(f,x)-f(x)\right|\le\gamma_j(N)\left[\omega_1\left(f,\frac{1}{n^\beta}\right)+2c_j(\beta,n)\|f\|_\infty\right]=:\lambda_{j1}, \tag{13.96}$$

and

(2)

$$\left\|{}_jA_n(f)-f\right\|_\infty\le\lambda_{j1}. \tag{13.97}$$

We notice that $\lim_{n\to\infty}{}_jA_n(f)=f$, *pointwise and uniformly.*

In this chapter we extend Theorem 13.38 to the fuzzy-random level.

We mention

Theorem 13.39 (see [23–26]) *Let* $f\in C_B\left(\mathbb{R}^N\right)$, $0<\beta<1$, $x\in\mathbb{R}^N$, $N,n\in\mathbb{N}$ *with* $n^{1-\beta}>2$; $j=1,2,3,4$. *Then*

(1)

$$\left|{}_jB_n(f,x)-f(x)\right|\le\omega_1\left(f,\frac{1}{n^\beta}\right)+2c_j(\beta,n)\|f\|_\infty=:\lambda_{j2}, \tag{13.98}$$

(2)

$$\left\|{}_jB_n(f)-f\right\|_\infty\le\lambda_{j2}. \tag{13.99}$$

Given that $f\in\left(C_U\left(\mathbb{R}^N\right)\cap C_B\left(\mathbb{R}^N\right)\right)$, *we obtain* $\lim_{n\to\infty}{}_jB_n(f)=f$, *uniformly.*

We also need

Theorem 13.40 (see [23–26]) *Let* $f\in C_B\left(\mathbb{R}^N\right)$, $0<\beta<1$, $x\in\mathbb{R}^N$, $N,n\in\mathbb{N}$ *with* $n^{1-\beta}>2$; $j=1,2,3,4$. *Then*

(1)

$$\left|{}_jC_n(f,x)-f(x)\right|\le\omega_1\left(f,\frac{1}{n}+\frac{1}{n^\beta}\right)+2c_j(\beta,n)\|f\|_\infty=:\lambda_{j3}, \tag{13.100}$$

(2)

$$\left\| {}_jC_n(f) - f \right\|_\infty \le \lambda_{j3}. \tag{13.101}$$

Given that $f \in \left(C_U\left(\mathbb{R}^N\right) \cap C_B\left(\mathbb{R}^N\right)\right)$, *we obtain* $\lim_{n\to\infty} {}_jC_n(f) = f$, *uniformly.*

We also need

Theorem 13.41 (see [23–26]) *Let* $f \in C_B\left(\mathbb{R}^N\right)$, $0 < \beta < 1$, $x \in \mathbb{R}^N$, $N, n \in \mathbb{N}$ *with* $n^{1-\beta} > 2$; $j = 1, 2, 3, 4$. *Then*

(1)

$$\left| {}_jD_n(f, x) - f(x) \right| \le \omega_1\left(f, \frac{1}{n} + \frac{1}{n^\beta}\right) + 2c_j(\beta, n)\|f\|_\infty = \lambda_{j3}, \tag{13.102}$$

(2)

$$\left\| {}_jD_n(f) - f \right\|_\infty \le \lambda_{j3}. \tag{13.103}$$

Given that $f \in \left(C_U\left(\mathbb{R}^N\right) \cap C_B\left(\mathbb{R}^N\right)\right)$, *we obtain* $\lim_{n\to\infty} {}_jD_n(f) = f$, *uniformly.*

In this chapter we extend Theorems 13.39, 13.40 and 13.41 to the random level.

We are also motivated by [1–16] and continuing [17]. For general knowledge on neural networks we recommend [32–34].

13.3 Main Results

(I) q**-mean Approximation by Fuzzy-Random arctangent, algebraic, Gudermannian and generalized symmetric activation functions based Quasi-Interpolation Neural Network Operators**

All terms and assumptions here as in Sections 13.1 and 13.2.

Let $f \in C_{\mathcal{FR}}^{U_q}\left(\prod_{i=1}^{N}[a_i, b_i]\right)$, $1 \le q < +\infty$, $n, N \in \mathbb{N}$, $0 < \beta < 1$, $\overrightarrow{x} \in \left(\prod_{i=1}^{N}[a_i, b_i]\right)$, $(X, \mathcal{B}, P)$ probability space, $s \in X$; $j = 1, 2, 3, 4$.

We define the following multivariate fuzzy random arctangent, algebraic, Gudermannian and generalized symmetric activation functions based quasi-interpolation linear neural network operators

$$\left({}_jA_n^{\mathcal{FR}}(f)\right)\left(\overrightarrow{x}, s\right) := \sum_{\overrightarrow{k}=\lceil na \rceil}^{\lfloor nb \rfloor *} f\left(\frac{\overrightarrow{k}}{n}, s\right) \odot \frac{Z_j\left(n\overrightarrow{x} - \overrightarrow{k}\right)}{\sum_{\overrightarrow{k}=\lceil na \rceil}^{\lfloor nb \rfloor} Z_j\left(n\overrightarrow{x} - \overrightarrow{k}\right)}, \tag{13.104}$$

(see also (13.89).

We present

Theorem 13.42 *Let* $f \in C_{\mathcal{FR}}^{U_q}\left(\prod_{i=1}^{N}[a_i,b_i]\right)$, $0<\beta<1$, $\overrightarrow{x} \in \left(\prod_{i=1}^{N}[a_i,b_i]\right)$, $n, N \in \mathbb{N}$, *with* $n^{1-\beta} > 2$, $1 \le q < +\infty$. *Assume that* $\int_X \left(D^*\left(f(\cdot,s),\widetilde{o}\right)\right)^q P(ds) < \infty$; $j = 1,2,3,4$. *Then*

(1)

$$\left(\int_X D^q\left(\left({}_jA_n^{\mathcal{FR}}(f)\right)(\overrightarrow{x},s), f(\overrightarrow{x},s)\right) P(ds)\right)^{\frac{1}{q}} \le \tag{13.105}$$

$$\gamma_j(N)\left\{\Omega_1\left(f,\frac{1}{n^\beta}\right)_{L^q} + 2c_j(\beta,n)\left(\int_X \left(D^*\left(f(\cdot,s),\widetilde{o}\right)\right)^q P(ds)\right)^{\frac{1}{q}}\right\} =: \lambda_{j1}^{(\mathcal{FR})},$$

(2)

$$\left\|\left(\int_X D^q\left(\left({}_jA_n^{\mathcal{FR}}(f)\right)(\overrightarrow{x},s), f(\overrightarrow{x},s)\right) P(ds)\right)^{\frac{1}{q}}\right\|_{\infty,\left(\prod_{i=1}^{N}[a_i,b_i]\right)} \le \lambda_{j1}^{(\mathcal{FR})}, \tag{13.106}$$

where $\gamma_j(N)$ *as in (13.25), (13.45), (13.65) and (13.87) and* $c_j(\beta,n)$ *as in (13.24), (13.44), (13.64) and (13.86).*

Proof We notice that

$$D\left(f\left(\frac{\overrightarrow{k}}{n},s\right), f(\overrightarrow{x},s)\right) \le D\left(f\left(\frac{\overrightarrow{k}}{n},s\right),\widetilde{o}\right) + D\left(f(\overrightarrow{x},s),\widetilde{o}\right) \tag{13.107}$$

$$\le 2D^*\left(f(\cdot,s),\widetilde{o}\right).$$

Hence

$$D^q\left(f\left(\frac{\overrightarrow{k}}{n},s\right), f(\overrightarrow{x},s)\right) \le 2^q D^{*q}\left(f(\cdot,s),\widetilde{o}\right), \tag{13.108}$$

and

$$\left(\int_X D^q\left(f\left(\frac{\overrightarrow{k}}{n},s\right), f(\overrightarrow{x},s)\right) P(ds)\right)^{\frac{1}{q}} \le 2\left(\int_X \left(D^*\left(f(\cdot,s),\widetilde{o}\right)\right)^q P(ds)\right)^{\frac{1}{q}}. \tag{13.109}$$

We observe that

$$D\left(\left({}_jA_n^{\mathcal{FR}}(f)\right)(\overrightarrow{x},s), f(\overrightarrow{x},s)\right) = \tag{13.110}$$

$$D\left(\sum_{\vec{k}=\lceil na\rceil}^{\lfloor nb\rfloor *} f\left(\frac{\vec{k}}{n}, s\right) \odot \frac{Z_j(nx-k)}{\sum\limits_{\vec{k}=\lceil na\rceil}^{\lfloor nb\rfloor} Z_j(nx-k)}, f(\vec{x}, s) \odot 1\right) =$$

$$D\left(\sum_{\vec{k}=\lceil na\rceil}^{\lfloor nb\rfloor *} f\left(\frac{\vec{k}}{n}, s\right) \odot \frac{Z_j(nx-k)}{\sum\limits_{\vec{k}=\lceil na\rceil}^{\lfloor nb\rfloor} Z_j(nx-k)}, f(\vec{x}, s) \odot \frac{\sum\limits_{\vec{k}=\lceil na\rceil}^{\lfloor nb\rfloor} Z_j(nx-k)}{\sum\limits_{\vec{k}=\lceil na\rceil}^{\lfloor nb\rfloor} Z_j(nx-k)}\right) = \quad (13.111)$$

$$D\left(\sum_{\vec{k}=\lceil na\rceil}^{\lfloor nb\rfloor *} f\left(\frac{\vec{k}}{n}, s\right) \odot \frac{Z_j(nx-k)}{\sum\limits_{\vec{k}=\lceil na\rceil}^{\lfloor nb\rfloor} Z_j(nx-k)}, \sum_{\vec{k}=\lceil na\rceil}^{\lfloor nb\rfloor *} f(\vec{x}, s) \odot \frac{Z_j(nx-k)}{\sum\limits_{\vec{k}=\lceil na\rceil}^{\lfloor nb\rfloor} Z_j(nx-k)}\right)$$

$$\leq \sum_{\vec{k}=\lceil na\rceil}^{\lfloor nb\rfloor} \left(\frac{Z_j(nx-k)}{\sum\limits_{\vec{k}=\lceil na\rceil}^{\lfloor nb\rfloor} Z_j(nx-k)}\right) D\left(f\left(\frac{\vec{k}}{n}, s\right), f(\vec{x}, s)\right). \quad (13.112)$$

So that

$$D\left(\left({}_jA_n^{\mathcal{FR}}(f)\right)(\vec{x}, s), f(\vec{x}, s)\right) \leq$$

$$\sum_{\vec{k}=\lceil na\rceil}^{\lfloor nb\rfloor} \left(\frac{Z_j(nx-k)}{\sum\limits_{\vec{k}=\lceil na\rceil}^{\lfloor nb\rfloor} Z_j(nx-k)}\right) D\left(f\left(\frac{\vec{k}}{n}, s\right), f(\vec{x}, s)\right) = \quad (13.113)$$

$$\sum_{\substack{\vec{k}=\lceil na\rceil \\ \left\|\frac{\vec{k}}{n}-\vec{x}\right\|_\infty \leq \frac{1}{n^\beta}}}^{\lfloor nb\rfloor} \left(\frac{Z_j(nx-k)}{\sum\limits_{\vec{k}=\lceil na\rceil}^{\lfloor nb\rfloor} Z_j(nx-k)}\right) D\left(f\left(\frac{\vec{k}}{n}, s\right), f(\vec{x}, s)\right) +$$

$$\sum_{\substack{\vec{k}=\lceil na\rceil \\ \left\|\frac{\vec{k}}{n}-\vec{x}\right\|_\infty > \frac{1}{n^\beta}}}^{\lfloor nb\rfloor} \left(\frac{Z_j(nx-k)}{\sum\limits_{\vec{k}=\lceil na\rceil}^{\lfloor nb\rfloor} Z_j(nx-k)}\right) D\left(f\left(\frac{\vec{k}}{n}, s\right), f(\vec{x}, s)\right).$$

Hence it holds

$$\left(\int_X D^q\left(\left({}_jA_n^{\mathcal{FR}}(f)\right)(\overrightarrow{x},s), f(\overrightarrow{x},s)\right)P(ds)\right)^{\frac{1}{q}} \leq \tag{13.114}$$

$$\sum_{\substack{\overrightarrow{k}=\lceil na\rceil \\ \left\|\frac{\overrightarrow{k}}{n}-\overrightarrow{x}\right\|_\infty \leq \frac{1}{n^\beta}}}^{\lfloor nb\rfloor} \left(\frac{Z_j(nx-k)}{\sum_{\overrightarrow{k}=\lceil na\rceil}^{\lfloor nb\rfloor} Z_j(nx-k)}\right)\left(\int_X D^q\left(f\left(\frac{\overrightarrow{k}}{n},s\right), f(\overrightarrow{x},s)\right)P(ds)\right)^{\frac{1}{q}} +$$

$$\sum_{\substack{\overrightarrow{k}=\lceil na\rceil \\ \left\|\frac{\overrightarrow{k}}{n}-\overrightarrow{x}\right\|_\infty > \frac{1}{n^\beta}}}^{\lfloor nb\rfloor} \left(\frac{Z_j(nx-k)}{\sum_{\overrightarrow{k}=\lceil na\rceil}^{\lfloor nb\rfloor} Z_j(nx-k)}\right)\left(\int_X D^q\left(f\left(\frac{\overrightarrow{k}}{n},s\right), f(\overrightarrow{x},s)\right)P(ds)\right)^{\frac{1}{q}} \leq$$

$$\left(\frac{1}{\sum_{\overrightarrow{k}=\lceil na\rceil}^{\lfloor nb\rfloor} Z_j(nx-k)}\right)\cdot\left\{\Omega_1^{(\mathcal{F})}\left(f,\frac{1}{n^\beta}\right)_{L^q} + \right. \tag{13.115}$$

$$\left. 2\left(\int_X \left(D^*\left(f(\cdot,s),\widetilde{o}\right)\right)^q P(ds)\right)^{\frac{1}{q}}\left(\sum_{\substack{\overrightarrow{k}=\lceil na\rceil \\ \left\|\frac{\overrightarrow{k}}{n}-\overrightarrow{x}\right\|_\infty > \frac{1}{n^\beta}}}^{\lfloor nb\rfloor} Z_j(nx-k)\right)\right\}$$

(by (13.24), (13.25); (13.44), (13.45); (13.64), (13.65); (13.86), (13.87))

$$\leq \gamma_j(N)\left\{\Omega_1^{(\mathcal{F})}\left(f,\frac{1}{n^\beta}\right)_{L^q} + 2c_j(\beta,n)\left(\int_X \left(D^*\left(f(\cdot,s),\widetilde{o}\right)\right)^q P(ds)\right)^{\frac{1}{q}}\right\}. \tag{13.116}$$

We have proved claim. ■

Conclusion 13.43 *By Theorem 13.42 we obtain the pointwise and uniform convergences with rates in the q-mean and D-metric of the operator* ${}_jA_n^{\mathcal{FR}}$ *to the unit operator for* $f \in C_{\mathcal{FR}}^{U_q}\left(\prod_{i=1}^N [a_i,b_i]\right)$, $j = 1, 2, 3, 4$.

(II) 1-mean Approximation by Stochastic arctangent, algebraic, Gudermannian and generalized symmetric activation functions based full Quasi-Interpolation Neural Network Operators

Let $g \in C_{\mathcal{R}}^{U_1}\left(\mathbb{R}^N\right)$, $0 < \beta < 1$, $\overrightarrow{x} \in \mathbb{R}^N$, $n, N \in \mathbb{N}$, with $\|g\|_{\infty,\mathbb{R}^N,X} < \infty$, $(X, \mathcal{B}, P)$ probability space, $s \in X$.

We define

$$ {}_jB_n^{(\mathcal{R})}(g)\left(\overrightarrow{x}, s\right) := \sum_{\overrightarrow{k}=-\infty}^{\infty} g\left(\frac{\overrightarrow{k}}{n}, s\right) Z_j\left(n\overrightarrow{x} - \overrightarrow{k}\right), \quad j = 1, 2, 3, 4, \tag{13.117} $$

(see also (13.90)).

We give

Theorem 13.44 *Let $g \in C_{\mathcal{R}}^{U_1}\left(\mathbb{R}^N\right)$, $0 < \beta < 1$, $\overrightarrow{x} \in \mathbb{R}^N$, $n, N \in \mathbb{N}$, with $n^{1-\beta} > 2$, $\|g\|_{\infty,\mathbb{R}^N,X} < \infty$; $j = 1, 2, 3, 4$. Then*

(1)

$$ \int_X \left|\left({}_jB_n^{(\mathcal{R})}(g)\right)\left(\overrightarrow{x}, s\right) - g\left(\overrightarrow{x}, s\right)\right| P(ds) \leq \tag{13.118} $$

$$ \left\{\Omega_1\left(g, \frac{1}{n^\beta}\right)_{L^1} + 2c_j(\beta, n)\|g\|_{\infty,\mathbb{R}^N,X}\right\} =: \mu_{j1}^{(\mathcal{R})}, $$

(2)

$$ \left\|\int_X \left|\left({}_jB_n^{(\mathcal{R})}(g)\right)\left(\overrightarrow{x}, s\right) - g\left(\overrightarrow{x}, s\right)\right| P(ds)\right\|_{\infty,\mathbb{R}^N} \leq \mu_{j1}^{(\mathcal{R})}. \tag{13.119} $$

Proof Since $\|g\|_{\infty,\mathbb{R}^N,X} < \infty$, then

$$ \left|g\left(\frac{\overrightarrow{k}}{n}, s\right) - g\left(\overrightarrow{x}, s\right)\right| \leq 2\|g\|_{\infty,\mathbb{R}^N,X} < \infty. \tag{13.120} $$

Hence

$$ \int_X \left|g\left(\frac{\overrightarrow{k}}{n}, s\right) - g\left(\overrightarrow{x}, s\right)\right| P(ds) \leq 2\|g\|_{\infty,\mathbb{R}^N,X} < \infty. \tag{13.121} $$

We observe that

$$ \left({}_jB_n^{(\mathcal{R})}(g)\right)\left(\overrightarrow{x}, s\right) - g\left(\overrightarrow{x}, s\right) = $$

$$ \sum_{\overrightarrow{k}=-\infty}^{\infty} g\left(\frac{\overrightarrow{k}}{n}, s\right) Z_j(nx - k) - g\left(\overrightarrow{x}, s\right) \sum_{\overrightarrow{k}=-\infty}^{\infty} Z_j(nx - k) = \tag{13.122} $$

$$\left(\sum_{\overrightarrow{k}=-\infty}^{\infty} g\left(\frac{\overrightarrow{k}}{n}, s\right) - g\left(\overrightarrow{x}, s\right) \right) Z_j\left(nx - k\right).$$

However it holds

$$\sum_{\overrightarrow{k}=-\infty}^{\infty} \left| g\left(\frac{\overrightarrow{k}}{n}, s\right) - g\left(\overrightarrow{x}, s\right) \right| Z_j\left(nx - k\right) \leq 2\left\|g\right\|_{\infty, \mathbb{R}^N, X} < \infty. \tag{13.123}$$

Hence

$$\left|\left({}_jB_n^{(\mathcal{R})}(g)\right)\left(\overrightarrow{x}, s\right) - g\left(\overrightarrow{x}, s\right)\right| \leq$$

$$\sum_{\overrightarrow{k}=-\infty}^{\infty} \left| g\left(\frac{\overrightarrow{k}}{n}, s\right) - g\left(\overrightarrow{x}, s\right) \right| Z_j\left(nx - k\right) = \tag{13.124}$$

$$\sum_{\substack{\overrightarrow{k}=-\infty \\ \left\|\frac{\overrightarrow{k}}{n} - \overrightarrow{x}\right\|_{\infty} \leq \frac{1}{n^{\beta}}}}^{\infty} \left| g\left(\frac{\overrightarrow{k}}{n}, s\right) - g\left(\overrightarrow{x}, s\right) \right| Z_j\left(nx - k\right) +$$

$$\sum_{\substack{\overrightarrow{k}=-\infty \\ \left\|\frac{\overrightarrow{k}}{n} - \overrightarrow{x}\right\|_{\infty} > \frac{1}{n^{\beta}}}}^{\infty} \left| g\left(\frac{\overrightarrow{k}}{n}, s\right) - g\left(\overrightarrow{x}, s\right) \right| Z_j\left(nx - k\right).$$

Furthermore it holds

$$\left(\int_X \left|\left({}_jB_n^{(\mathcal{R})}(g)\right)\left(\overrightarrow{x}, s\right) - g\left(\overrightarrow{x}, s\right)\right| P\left(ds\right) \right) \leq$$

$$\sum_{\substack{\overrightarrow{k}=-\infty \\ \left\|\frac{\overrightarrow{k}}{n} - \overrightarrow{x}\right\|_{\infty} \leq \frac{1}{n^{\beta}}}}^{\infty} \left(\int_X \left| g\left(\frac{\overrightarrow{k}}{n}, s\right) - g\left(\overrightarrow{x}, s\right) \right| P\left(ds\right) \right) Z_j\left(nx - k\right) + \tag{13.125}$$

$$\sum_{\substack{\overrightarrow{k}=-\infty \\ \left\|\frac{\overrightarrow{k}}{n} - \overrightarrow{x}\right\|_{\infty} > \frac{1}{n^{\beta}}}}^{\infty} \left(\int_X \left| g\left(\frac{\overrightarrow{k}}{n}, s\right) - g\left(\overrightarrow{x}, s\right) \right| P\left(ds\right) \right) Z_j\left(nx - k\right) \leq$$

$$\Omega_1\left(g,\frac{1}{n^{\beta}}\right)_{L^1}+2\,\|g\|_{\infty,\mathbb{R}^N,X}\sum_{\substack{\overrightarrow{k}=-\infty\\ \left\|\frac{\overrightarrow{k}}{n}-\overrightarrow{x}\right\|_{\infty}>\frac{1}{n^{\beta}}}}^{\infty}Z_j\left(nx-k\right)\leq$$

$$\Omega_1\left(g,\frac{1}{n^{\beta}}\right)_{L^1}+2c_j\left(\beta,n\right)\|g\|_{\infty,\mathbb{R}^N,X},$$

proving the claim. ■

Conclusion 13.45 *By Theorem 13.44 we obtain pointwise and uniform convergences with rates in the* 1*-mean of random operators* ${}_jB_n^{(\mathcal{R})}$ *to the unit operator for* $g\in C_{\mathcal{R}}^{U_1}\left(\mathbb{R}^N\right)$, $j=1,2,3,4$.

(III) 1-mean Approximation by Stochastic arctangent, algebraic, Gudermannian and generalized symmetric activation functions based multivariate Kantorovich type neural network operator

Let $g\in C_{\mathcal{R}}^{U_1}\left(\mathbb{R}^N\right)$, $0<\beta<1$, $\overrightarrow{x}\in\mathbb{R}^N$, $n,N\in\mathbb{N}$, with $\|g\|_{\infty,\mathbb{R}^N,X}<\infty$, $(X,\mathcal{B},P)$ probability space, $s\in X$.

We define ($j=1,2,3,4$):

$${}_jC_n^{(\mathcal{R})}\left(g\right)\left(\overrightarrow{x},s\right):=\sum_{\overrightarrow{k}=-\infty}^{\infty}\left(n^N\int_{\frac{\overrightarrow{k}}{n}}^{\frac{\overrightarrow{k}+1}{n}}g\left(\overrightarrow{t},s\right)d\overrightarrow{t}\right)Z_j\left(n\overrightarrow{x}-\overrightarrow{k}\right),\tag{13.126}$$

(see also (13.91).

We present

Theorem 13.46 *Let* $g\in C_{\mathcal{R}}^{U_1}\left(\mathbb{R}^N\right)$, $0<\beta<1$, $\overrightarrow{x}\in\mathbb{R}^N$, $n,N\in\mathbb{N}$, *with* $n^{1-\beta}>2$; $j=1,2,3,4$, $\|g\|_{\infty,\mathbb{R}^N,X}<\infty$. *Then*

(1)

$$\int_X\left|\left({}_jC_n^{(\mathcal{R})}\left(g\right)\right)\left(\overrightarrow{x},s\right)-g\left(\overrightarrow{x},s\right)\right|P\left(ds\right)\leq$$

$$\left[\Omega_1\left(g,\frac{1}{n}+\frac{1}{n^{\beta}}\right)_{L^1}+2c_j\left(\beta,n\right)\|g\|_{\infty,\mathbb{R}^N,X}\right]=:\gamma_{j1}^{(\mathcal{R})},\tag{13.127}$$

(2)

$$\left\|\int_X\left|\left({}_jC_n^{(\mathcal{R})}\left(g\right)\right)\left(\overrightarrow{x},s\right)-g\left(\overrightarrow{x},s\right)\right|P\left(ds\right)\right\|_{\infty,\mathbb{R}^N}\leq\gamma_{j1}^{(\mathcal{R})}.\tag{13.128}$$

Proof Since $\|g\|_{\infty,\mathbb{R}^N,X}<\infty$, then

$$\left| n^N \int_{\frac{\vec{k}}{n}}^{\frac{\vec{k}+1}{n}} g\left(\vec{t}, s\right) d\vec{t} - g\left(\vec{x}, s\right) \right| = \left| n^N \int_{\frac{\vec{k}}{n}}^{\frac{\vec{k}+1}{n}} \left(g\left(\vec{t}, s\right) - g\left(\vec{x}, s\right)\right) d\vec{t} \right| \leq$$

$$n^N \int_{\frac{\vec{k}}{n}}^{\frac{\vec{k}+1}{n}} \left| g\left(\vec{t}, s\right) - g\left(\vec{x}, s\right) \right| d\vec{t} \leq 2 \|g\|_{\infty, \mathbb{R}^N, X} < \infty. \tag{13.129}$$

Hence

$$\int_X \left| n^N \int_{\frac{\vec{k}}{n}}^{\frac{\vec{k}+1}{n}} g\left(\vec{t}, s\right) d\vec{t} - g\left(\vec{x}, s\right) \right| P(ds) \leq 2 \|g\|_{\infty, \mathbb{R}^N, X} < \infty. \tag{13.130}$$

We observe that

$$\left({}_jC_n^{(\mathcal{R})}(g) \right) \left(\vec{x}, s\right) - g\left(\vec{x}, s\right) =$$

$$\sum_{\vec{k}=-\infty}^{\infty} \left(n^N \int_{\frac{\vec{k}}{n}}^{\frac{\vec{k}+1}{n}} g\left(\vec{t}, s\right) d\vec{t} \right) Z_j\left(n\vec{x} - \vec{k}\right) - g\left(\vec{x}, s\right) =$$

$$\sum_{\vec{k}=-\infty}^{\infty} \left(n^N \int_{\frac{\vec{k}}{n}}^{\frac{\vec{k}+1}{n}} g\left(\vec{t}, s\right) d\vec{t} \right) Z_j\left(n\vec{x} - \vec{k}\right) - g\left(\vec{x}, s\right) \sum_{\vec{k}=-\infty}^{\infty} Z_j\left(n\vec{x} - \vec{k}\right) = \tag{13.131}$$

$$\sum_{\vec{k}=-\infty}^{\infty} \left[\left(n^N \int_{\frac{\vec{k}}{n}}^{\frac{\vec{k}+1}{n}} g\left(\vec{t}, s\right) d\vec{t} \right) - g\left(\vec{x}, s\right) \right] Z_j\left(n\vec{x} - \vec{k}\right) =$$

$$\sum_{\vec{k}=-\infty}^{\infty} \left[n^N \int_{\frac{\vec{k}}{n}}^{\frac{\vec{k}+1}{n}} \left(g\left(\vec{t}, s\right) - g\left(\vec{x}, s\right)\right) d\vec{t} \right] Z_j\left(n\vec{x} - \vec{k}\right).$$

However it holds

$$\sum_{\vec{k}=-\infty}^{\infty} \left[n^N \int_{\frac{\vec{k}}{n}}^{\frac{\vec{k}+1}{n}} \left| g\left(\vec{t}, s\right) - g\left(\vec{x}, s\right) \right| d\vec{t} \right] Z_j\left(n\vec{x} - \vec{k}\right) \leq 2 \|g\|_{\infty, \mathbb{R}^N, X} < \infty. \tag{13.132}$$

Hence

$$\left| \left({}_jC_n^{(\mathcal{R})}(g) \right) \left(\vec{x}, s\right) - g\left(\vec{x}, s\right) \right| \leq$$

$$\sum_{\vec{k}=-\infty}^{\infty} \left[n^N \int_{\frac{\vec{k}}{n}}^{\frac{\vec{k}+1}{n}} \left| g\left(\vec{t}, s\right) - g\left(\vec{x}, s\right) \right| d\vec{t} \right] Z_j\left(n\vec{x} - \vec{k}\right) = \tag{13.133}$$

$$\sum_{\substack{\overrightarrow{k}=-\infty \\ \left\|\frac{\overrightarrow{k}}{n}-\overrightarrow{x}\right\|_{\infty}\leq\frac{1}{n^{\beta}}}}^{\infty} \left[n^{N}\int_{\frac{\overrightarrow{k}}{n}}^{\frac{\overrightarrow{k}+1}{n}} \left|g\left(\overrightarrow{t},s\right)-g\left(\overrightarrow{x},s\right)\right| d\overrightarrow{t}\right] Z_{j}\left(n\overrightarrow{x}-\overrightarrow{k}\right)+ \tag{13.134}$$

$$\sum_{\substack{\overrightarrow{k}=-\infty \\ \left\|\frac{\overrightarrow{k}}{n}-\overrightarrow{x}\right\|_{\infty}>\frac{1}{n^{\beta}}}}^{\infty} \left[n^{N}\int_{\frac{\overrightarrow{k}}{n}}^{\frac{\overrightarrow{k}+1}{n}} \left|g\left(\overrightarrow{t},s\right)-g\left(\overrightarrow{x},s\right)\right| d\overrightarrow{t}\right] Z_{j}\left(n\overrightarrow{x}-\overrightarrow{k}\right)=$$

$$\sum_{\substack{\overrightarrow{k}=-\infty \\ \left\|\frac{\overrightarrow{k}}{n}-\overrightarrow{x}\right\|_{\infty}\leq\frac{1}{n^{\beta}}}}^{\infty} \left[n^{N}\int_{0}^{\frac{1}{n}} \left|g\left(\overrightarrow{t}+\frac{\overrightarrow{k}}{n},s\right)-g\left(\overrightarrow{x},s\right)\right| d\overrightarrow{t}\right] Z_{j}\left(n\overrightarrow{x}-\overrightarrow{k}\right)+$$

$$\sum_{\substack{\overrightarrow{k}=-\infty \\ \left\|\frac{\overrightarrow{k}}{n}-\overrightarrow{x}\right\|_{\infty}>\frac{1}{n^{\beta}}}}^{\infty} \left[n^{N}\int_{0}^{\frac{1}{n}} \left|g\left(\overrightarrow{t}+\frac{\overrightarrow{k}}{n},s\right)-g\left(\overrightarrow{x},s\right)\right| d\overrightarrow{t}\right] Z_{j}\left(n\overrightarrow{x}-\overrightarrow{k}\right). \tag{13.135}$$

Furthermore it holds

$$\left(\int_{X}\left|\left({}_{j}C_{n}^{(\mathcal{R})}(g)\right)\left(\overrightarrow{x},s\right)-g\left(\overrightarrow{x},s\right)\right|P(ds)\right)\underset{\text{(by Fubini's theorem)}}{\leq}$$

$$\sum_{\substack{\overrightarrow{k}=-\infty \\ \left\|\frac{\overrightarrow{k}}{n}-\overrightarrow{x}\right\|_{\infty}\leq\frac{1}{n^{\beta}}}}^{\infty} \left[n^{N}\int_{0}^{\frac{1}{n}} \left(\int_{X}\left|g\left(\overrightarrow{t}+\frac{\overrightarrow{k}}{n},s\right)-g\left(\overrightarrow{x},s\right)\right| P(ds)\right) d\overrightarrow{t}\right] Z_{j}\left(n\overrightarrow{x}-\overrightarrow{k}\right)+$$

$$\sum_{\substack{\overrightarrow{k}=-\infty \\ \left\|\frac{\overrightarrow{k}}{n}-\overrightarrow{x}\right\|_{\infty}>\frac{1}{n^{\beta}}}}^{\infty} \left[n^{N}\int_{0}^{\frac{1}{n}} \left(\int_{X}\left|g\left(\overrightarrow{t}+\frac{\overrightarrow{k}}{n},s\right)-g\left(\overrightarrow{x},s\right)\right| P(ds)\right) d\overrightarrow{t}\right] Z_{j}\left(n\overrightarrow{x}-\overrightarrow{k}\right)\leq \tag{13.136}$$

$$\Omega_{1}\left(g,\frac{1}{n}+\frac{1}{n^{\beta}}\right)_{L^{1}}+2\left\|g\right\|_{\infty,\mathbb{R}^{N},X}\sum_{\substack{\overrightarrow{k}=-\infty \\ \left\|\frac{\overrightarrow{k}}{n}-\overrightarrow{x}\right\|_{\infty}>\frac{1}{n^{\beta}}}}^{\infty} Z_{j}\left(n\overrightarrow{x}-\overrightarrow{k}\right)\leq$$

$$\Omega_{1}\left(g,\frac{1}{n}+\frac{1}{n^{\beta}}\right)_{L^{1}}+2c_{j}\left(\beta,n\right)\left\|g\right\|_{\infty,\mathbb{R}^{N},X}, \tag{13.137}$$

proving the claim. ■

Conclusion 13.47 *By Theorem 13.46 we obtain pointwise and uniform convergences with rates in the* 1*-mean of random operators* ${}_jC_n^{(\mathcal{R})}$ *to the unit operator for* $g \in C_{\mathcal{R}}^{U_1}(\mathbb{R}^N)$, $j = 1, 2, 3, 4$.

(IV) 1-mean Approximation by Stochastic arctangent, algebraic, Gudermannian and generalized symmetric activation functions based multivariate quadrature type neural network operator

Let $g \in C_{\mathcal{R}}^{U_1}(\mathbb{R}^N)$, $0 < \beta < 1$, $\overrightarrow{x} \in \mathbb{R}^N$, $n, N \in \mathbb{N}$, with $\|g\|_{\infty,\mathbb{R}^N,X} < \infty$, $(X, \mathcal{B}, P)$ probability space, $s \in X$, $j = 1, 2, 3, 4$.

We define

$$ {}_jD_n^{(\mathcal{R})}(g)\left(\overrightarrow{x}, s\right) := \sum_{\overrightarrow{k}=-\infty}^{\infty} \left(\delta_{n\overrightarrow{k}}(g)\right)(s)\, Z_j\left(n\overrightarrow{x} - \overrightarrow{k}\right), \tag{13.138} $$

where

$$ \left(\delta_{n\overrightarrow{k}}(g)\right)(s) := \sum_{\overrightarrow{r}=0}^{\overrightarrow{\theta}} w_{\overrightarrow{r}}\, g\left(\frac{\overrightarrow{k}}{n} + \frac{\overrightarrow{r}}{n\overrightarrow{\theta}}, s\right), \tag{13.139} $$

(see also (13.92) and (13.93)).

We finally give

Theorem 13.48 *Let* $g \in C_{\mathcal{R}}^{U_1}(\mathbb{R}^N)$, $0 < \beta < 1$, $\overrightarrow{x} \in \mathbb{R}^N$, $n, N \in \mathbb{N}$, *with* $n^{1-\beta} > 2$; $j = 1, 2, 3, 4$, $\|g\|_{\infty,\mathbb{R}^N,X} < \infty$. *Then*

(1)

$$ \int_X \left|\left({}_jD_n^{(\mathcal{R})}(g)\right)\left(\overrightarrow{x}, s\right) - g\left(\overrightarrow{x}, s\right)\right| P(ds) \leq $$

$$ \left\{\Omega_1\left(g, \frac{1}{n} + \frac{1}{n^{\beta}}\right)_{L^1} + 2c_j(\beta, n)\|g\|_{\infty,\mathbb{R}^N,X}\right\} =: \gamma_{j1}^{(\mathcal{R})}, \tag{13.140} $$

(2)

$$ \left\|\int_X \left|\left({}_jD_n^{(\mathcal{R})}(g)\right)\left(\overrightarrow{x}, s\right) - g\left(\overrightarrow{x}, s\right)\right| P(ds)\right\|_{\infty,\mathbb{R}^N} \leq \gamma_{j1}^{(\mathcal{R})}. \tag{13.141} $$

Proof Notice that

$$ \left|\left(\delta_{n\overrightarrow{k}}(g)\right)(s) - g\left(\overrightarrow{x}, s\right)\right| = $$

$$ \left|\sum_{\overrightarrow{r}=0}^{\overrightarrow{\theta}} w_{\overrightarrow{r}}\left(g\left(\frac{\overrightarrow{k}}{n} + \frac{\overrightarrow{r}}{n\overrightarrow{\theta}}, s\right) - g\left(\overrightarrow{x}, s\right)\right)\right| \leq $$

$$\sum_{\vec{r}=0}^{\vec{\theta}} w_{\vec{r}} \left| g\left(\frac{\vec{k}}{n} + \frac{\vec{r}}{n\vec{\theta}}, s\right) - g\left(\vec{x}, s\right) \right| \leq 2 \|g\|_{\infty, \mathbb{R}^N, X} < \infty. \tag{13.142}$$

Hence

$$\int_X \left| \left(\delta_{n\vec{k}}(g) \right)(s) - g\left(\vec{x}, s\right) \right| P(ds) \leq 2 \|g\|_{\infty, \mathbb{R}^N, X} < \infty. \tag{13.143}$$

We observe that

$$\left({}_jD_n^{(\mathcal{R})}(g) \right)\left(\vec{x}, s\right) - g\left(\vec{x}, s\right) =$$

$$\sum_{\vec{k}=-\infty}^{\infty} \left(\delta_{n\vec{k}}(g) \right)(s)\, Z_j\left(n\vec{x} - \vec{k}\right) - g\left(\vec{x}, s\right) =$$

$$\sum_{\vec{k}=-\infty}^{\infty} \left(\left(\delta_{n\vec{k}}(g) \right)(s) - g\left(\vec{x}, s\right) \right) Z_j\left(n\vec{x} - \vec{k}\right). \tag{13.144}$$

Thus

$$\left| {}_jD_n^{(\mathcal{R})}(g)\left(\vec{x}, s\right) - g\left(\vec{x}, s\right) \right| \leq$$

$$\sum_{\vec{k}=-\infty}^{\infty} \left| \left(\delta_{n\vec{k}}(g) \right)(s) - g\left(\vec{x}, s\right) \right| Z_j\left(n\vec{x} - \vec{k}\right) \leq 2 \|g\|_{\infty, \mathbb{R}^N, X} < \infty. \tag{13.145}$$

Hence it holds

$$\left| \left({}_jD_n^{(\mathcal{R})}(g) \right)\left(\vec{x}, s\right) - g\left(\vec{x}, s\right) \right| \leq$$

$$\sum_{\vec{k}=-\infty}^{\infty} \left| \left(\delta_{n\vec{k}}(g) \right)(s) - g\left(\vec{x}, s\right) \right| Z_j\left(n\vec{x} - \vec{k}\right) =$$

$$\sum_{\substack{\vec{k}=-\infty \\ \left\| \frac{\vec{k}}{n} - \vec{x} \right\|_\infty \leq \frac{1}{n^\beta}}}^{\infty} \left| \left(\delta_{n\vec{k}}(g) \right)(s) - g\left(\vec{x}, s\right) \right| Z_j\left(n\vec{x} - \vec{k}\right) +$$

$$\sum_{\substack{\vec{k}=-\infty \\ \left\| \frac{\vec{k}}{n} - \vec{x} \right\|_\infty > \frac{1}{n^\beta}}}^{\infty} \left| \left(\delta_{n\vec{k}}(g) \right)(s) - g\left(\vec{x}, s\right) \right| Z_j\left(n\vec{x} - \vec{k}\right). \tag{13.146}$$

Furthermore we derive

$$\left(\int_X \left|\left({}_jD_n^{(\mathcal{R})}(g)\right)(\overrightarrow{x},s) - g(\overrightarrow{x},s)\right| P(ds)\right) \leq$$

$$\sum_{\substack{\overrightarrow{k}=-\infty \\ \left\|\frac{\overrightarrow{k}}{n}-\overrightarrow{x}\right\|_\infty \leq \frac{1}{n^\beta}}}^{\infty} \sum_{\overrightarrow{r}=0}^{\overrightarrow{\theta}} w_{\overrightarrow{r}} \left(\int_X \left| g\left(\frac{\overrightarrow{k}}{n} + \frac{\overrightarrow{r}}{n\overrightarrow{\theta}}, s\right) - g(\overrightarrow{x},s)\right| P(ds)\right) Z_j\left(n\overrightarrow{x} - \overrightarrow{k}\right) \tag{13.147}$$

$$+\left(\sum_{\substack{\overrightarrow{k}=-\infty \\ \left\|\frac{\overrightarrow{k}}{n}-\overrightarrow{x}\right\|_\infty > \frac{1}{n^\beta}}}^{\infty} Z_j\left(n\overrightarrow{x} - \overrightarrow{k}\right)\right) 2\|g\|_{\infty,\mathbb{R}^N,X} \leq$$

$$\Omega_1\left(g, \frac{1}{n} + \frac{1}{n^\beta}\right)_{L^1} + 2c_j(\beta,n)\|g\|_{\infty,\mathbb{R}^N,X}, \tag{13.148}$$

proving the claim. ■

Conclusion 13.49 *From Theorem 13.48 we obtain pointwise and uniform convergences with rates in the* 1*-mean of random operators* ${}_jD_n^{(\mathcal{R})}$ *to the unit operator for* $g \in C_{\mathcal{R}}^{U_1}(\mathbb{R}^N)$, $j = 1, 2, 3, 4$.

References

1. G.A. Anastassiou, Rate of convergence of Fuzzy neural network operators, univariate case. J. Fuzzy Math. **10**(3), 755–780 (2002)
2. G.A. Anastassiou, Higher order fuzzy approximation by fuzzy wavelet type and neural network operators. Comput. Math. **48**, 1387–1401 (2004)
3. G.A. Anastassiou, Univariate fuzzy-random neural network approximation operators, in *Computers and Mathematics with Applications, Special issue/Proceedings*, ed. by G. Anastassiou of special session "Computational Methods in Analysis", AMS meeting in Orlando, Florida, November 2002, vol. 48, pp. 1263–1283 (2004)
4. G.A. Anastassiou, Higher order Fuzzy Korovkin Theory via inequalities. Commun. Appl. Anal. **10**(2), 359–392
5. G.A. Anastassiou, Fuzzy Korovkin Theorems and inequalities. J. Fuzzy Math. **15**(1), 169–205 (2007)
6. G.A. Anastassiou, Multivariate Stochastic Korovkin Theory given quantitatively. Math. Comput. Model. **48**, 558–580 (2008)
7. G.A. Anastassiou, *Fuzzy Mathematics: Approximation Theory* (Springer, Heidelberg, New York, 2010)
8. G.A. Anastassiou, *Inteligent Systems: Approximation by Artificial Neural Networks* (Springer, Heidelberg, 2011)

9. G.A. Anastassiou, Univariate hyperbolic tangent neural network approximation. Math. Comput. Model. **53**, 1111–1132 (2011)
10. G.A. Anastassiou, Multivariate hyperbolic tangent neural network approximation. Comput. Math. **61**, 809–821 (2011)
11. G.A. Anastassiou, Multivariate sigmoidal neural network approximation. Neural Netw. **24**, 378–386 (2011)
12. G.A. Anastassiou, Higher order multivariate fuzzy approximation by multivariate fuzzy wavelet type and neural network operators. J. Fuzzy Math. **19**(3), 601–618 (2011)
13. G.A. Anastassiou, Univariate sigmoidal neural network approximation. J. Comput. Anal. Appl. **14**(4), 659–690 (2012)
14. G.A. Anastassiou, Rate of convergence of some multivariate neural network operators to the unit, revisited. J. Comput. Anal. Appl. **15**(7), 1300–1309 (2013)
15. G.A. Anastassiou, Higher order multivariate fuzzy approximation by basic neural network operators. CUBO **16**(3), 21–35 (2014)
16. G.A. Anastassiou, Multivariate Fuzzy-Random Quasi-interpolation neural network approximation operators. J. Fuzzy Math. **22**(1), 167–184 (2014)
17. G. Anastassiou, Multivariate Fuzzy-Random error function based Neural Network Approximation. J. Fuzzy Math. **23**(4), 917–935 (2015)
18. G.A. Anastassiou, *Intelligent Systems II: Complete Approximation by Neural Network Operators* (Springer, Heidelberg, New York, 2016)
19. G.A. Anastassiou, *Intelligent Computations: Abstract Fractional Calculus* (Inequalities, Approximations, Springer, Heidelberg, New York, 2018)
20. G.A. Anastassiou, Algebraic function based Banach space valued ordinary and fractional neural network approximations. New Trends Math. Sci. **10** special issues (1), 100–125 (2022)
21. G.A. Anastassiou, Gudermannian function activated Banach space valued ordinary and fractional neural network approximation. Adv. Nonlinear Var. Inequal. **25**(2), 27–64 (2022)
22. G.A. Anastassiou, *Generalized Symmetrical Sigmoid Function Activated Banach Space Valued Ordinary and Fractional Neural Network Approximation* (Fasc. Matematica, Accepted for Publication, Analele Universităţii Oradea, 2022)
23. G.A. Anastassiou, Abstract multivariate Gudermannian function activated neural network approximations. Panamerican Math. J. (2022)
24. G.A. Anastassiou, *General Multivariate Arctangent Function Activated Neural Network Approximations*. Submitted (2022)
25. G.A. Anastassiou, R. Kozma, *Abstract Multivariate Agebraic Function Activated Neural Network Approximations*. Submitted (2022)
26. G.A. Anastassiou, *Generalized Symmetrical Sigmoid Function Activated Neural Network Multivariate Approximation*. Submitted (2022)
27. G.A. Anastassiou, *Multivariate Fuzzy-Random and Stochastic Arctangent, Algebraic, Gudermannian and Generalized Symmetric Activation Functions Induced Neural Network Approximations*. Submitted (2022)
28. Z. Chen, F. Cao, The approximation operators with sigmoidal functions. Comput. Math. Appl. **58**, 758–765 (2009)
29. R.M. Dudley, *Real Analysis and Probability* (Wadsworth & Brooks/Cole Mathematics Series, Pacific Grove, CA, 1989)
30. A.J. Dunning, J. Kensler, L. Goudeville, F. Bailleux, Some extensions in continuous methods for immunological correlates of protection. BMC Med. Res. Methodol. **15**(107) (2015). https://doi.org/10.1186/s12874-015-0096-9
31. S. Gal, Approximation theory in fuzzy setting, Chapter 13 in *Handbook of Analytic-Computational Methods in Applied Mathematics*, ed. by G. Anastassiou (Chapman & Hall/CRC, Boca Raton, New York, 2000), pp. 617–666
32. S. Haykin, *Neural Networks: A Comprehensive Foundation*, 2nd edn. (Prentice Hall, New York, 1998)
33. T.M. Mitchell, *Machine Learning* (WCB-McGraw-Hill, New York, 1997)

34. W. McCulloch, W. Pitts, A logical calculus of the ideas immanent in nervous activity. Bull. Math. Biophys. **7**, 115–133 (1943)
35. E.W. Weisstein, *Gudermannian*, MathWorld
36. Wu. Congxin, Gong Zengtai, On Henstock integral of interval-valued functions and fuzzy valued functions. Fuzzy Sets Syst. **115**(3), 377–391 (2000)
37. C. Wu, Z. Gong, On Henstock integral of fuzzy-number-valued functions (I). Fuzzy Sets Syst. **3**(120), 523–532 (2001)
38. C. Wu, M. Ma, On embedding problem of fuzzy number space: part 1. Fuzzy Sets Syst. **44**, 33–38 (1991)

Conclusion

This monograph is the continuation of author's earlier two monographs. The generalization and modernization here is that the neural networks are Banach space valued, induced by a great variety of activation functions deriving from the arctangent, algebraic, Gudermannian and generalized symmetric sigmoid functions.

Thus, in this monograph all presented is original work by the author given at a very general level to cover a maximum number of different kinds of Neural Networks: giving ordinary, fractional, fuzzy and stochastic approximations. Namely it is presented here univariate, fractional and multivariate approximations. Iterated-sequential approximations are also covered. The new and important feature here is that the functions under approximation are Banach space valued.

© The Editor(s) (if applicable) and The Author(s), under exclusive license to Springer Nature Switzerland AG 2023

G. A. Anastassiou, *Banach Space Valued Neural Network*,
Studies in Computational Intelligence 1062,
https://doi.org/10.1007/978-3-031-16400-2

The manufacturer's authorised representative in the EU is Springer Nature Customer Service Centre GmbH, Europaplatz 3, 69115 Heidelberg, Germany. If you have any concerns regarding our products, please contact ProductSafety@springernature.com

Printed and bound by CPI Group (UK) Ltd, Croydon, CR0 4YY

28/09/2026

02224061-0003